INSTRUCTOR'S RESOURCE MANUAL
for
MENDENHALL, BEAVER, AND BEAVER's

A COURSE IN BUSINESS STATISTICS

Fourth Edition

Barbara M. Beaver
University of California, Riverside

Duxbury Press
An Imprint of Wadsworth Publishing Company
I(T)P® An International Thomson Publishing Company

Belmont • Albany • Bonn • Boston • Cincinnati • Detroit • London • Madrid • Melbourne
Mexico City • New York • Paris • San Francisco • Singapore • Tokyo • Toronto • Washington

COPYRIGHT © 1996 by Wadsworth Publishing Company
A Division of International Thomson Publishing Inc.
I(T)P The ITP logo is a registered trademark under license.
Duxbury Press and the leaf logo are trademarks used under license.

Printed in the United States of America
1 2 3 4 5 6 7 8 9 10

For more information, contact Duxbury Press at Wadsworth Publishing Company.

Wadsworth Publishing Company
10 Davis Drive
Belmont, California 94002, USA

International Thomson Publishing
Europe
Berkshire House 168-173
High Holborn
London, WC1V 7AA, England

Thomas Nelson Australia
102 Dodds Street
South Melbourne 3205
Victoria, Australia

Nelson Canada
1120 Birchmount Road
Scarborough, Ontario
Canada M1K 5G4

International Thomson Editores
Campos Eliseos 385, Piso 7
Col. Polanco
11560 México D.F. México

International Thomson Publishing
GmbH
Königswinterer Strasse 418
53227 Bonn, Germany

International Thomson Publishing Asia
221 Henderson Road
#05-10 Henderson Building
Singapore 0315

International Thomson Publishing Japan
Hirakawacho Kyowa Building, 3F
2-2-1 Hirakawacho
Chiyoda-ku, Tokyo 102, Japan

All rights reserved. Instructors of classes adopting *A Course in Business Statistics,* by Mendenhall, Beaver, and Beaver, as a required text may reproduce materials for classroom use. Otherwise, no part of this work covered by the copyright hereon may be reproduced or used in any form or by any means—graphic, electronic, or mechanical, including photocopying, recording, taping, or information storage and retrieval systems—without the written permission of the publisher.

ISBN 0-534-26510-3

Contents

1	What Is Statistics?	1
2	Describing Sets of Data	3
3	Probability and Discrete Probability Distributions	38
4	Useful Discrete Probability Distributions	59
5	The Normal and Other Continuous Probability Distributions	78
6	Sampling Distributions	105
7	Estimation of Means and Proportions	118
8	Tests of Hypotheses for Means and Proportions	144
9	The Analysis of Variance	192
10	Quality Control	217
11	Linear Regression and Correlation	230
12	Multiple Regression Analysis	259
13	Time Series and Index Numbers	275
14	Sampling Methods	310
15	The Chi-Square Goodness-of-Fit Test	324
16	Nonparametric Statistics	343
Appendix III	Additional Cases Studies	368
Transparency Masters		378

CHAPTER 1
What Is Statistics?

1.1

a. The population of interest is the population of measurements consisting of the number of a particular item that will be demanded by customers in a particular month, measured for all months, in the past and in the future.

b. The manager can possibly obtain monthly demand data for past months, but it is impossible to obtain the future data.

c. Although it is not possible to itemize the entire population, the manager may be able to use information about past demand for a single item to draw conclusions about possible demand for that item in the future.

1.2

a. The population of interest is the population of measurements consisting of the appraisals of the land by all experienced appraisers who might be asked to appraise the land.

b. Although the population might be quite large, it is possible to get appraisals from all existing appraisers, and to enumerate the population.

c. Yes, the populations are different. Buyers might tend to underestimate the appraisal, sellers might overestimate, and the public at large might be inexperienced at appraisals (hence they would either underestimate or overestimate). Experienced appraisers should give more unbiased and accurate appraisals.

d. More than one appraisal should be used in order to get a more accurate estimate of the "value of the land" as measured by the population of appraisals.

1.3

a. The population consists of measurements on the variable of interest (top speed, range before recharging, time to accelerate from 0 to 60) for all *Tropicas* that have been manufactured to date, or that will be manufactured in the future.

b. The figures given in the article were based on a sample taken from the population, since the population does not exist in fact, but is instead hypothetical (not all *Tropicas* have yet been manufactured).

1.4

a. The population of interest consists of a series of measurements (yes or no) in answer to the question "Would you be interested in buying the *Tropica*?" taken on all residents (of driving age) in the state of California.

b. The sample could be obtained using a list of registered drivers, perhaps supplied by the Department of Motor Vehicles. Once the drivers have been selected, their answers (yes or no) would be obtained by telephone or mail interviews.

c. No. The residents of Los Angeles County, who reside in an urban area, would not necessarily be representative of all drivers in the state. Los Angeles County residents, many of whom are commuters, will be more interested in speed and range before recharging than would rural drivers in some of the eastern or central counties in California.

1.5

a. The population of interest consists of a series of measurements (damaged or not damaged) taken for each pine tree in Yosemite National Park at a particular point in time (either 1985 or 1990).

b. The percentages are probably estimated.

c. One possible method is to divide the park into a number of grids and to sample trees randomly within each grid. If the trees are numbered by the park rangers using some method of categorization, this numbering system might be employed.

1.6

a. The percentage return of mutual stock funds is a function of the economic conditions during the past year. The characterization of this population will vary, depending on the year under consideration by the student. The population does exist and can be enumerated for a given time period.

b. The percentage return of mutual stock funds over the next twelve months is conceptual and cannot be enumerated.

1.7

a. The population consists of shopper opinions (prefer or do not prefer background music) for all shoppers who might patronize the supermarket, now or in the future.

b. It would not be possible to enumerate the entire population, since the identities of the future shoppers would not be known to the sampler.

c. No, but it will serve as an estimate of the population percentage.

1.8

a. The population consists of the valuation criteria of stocks of 8000 public companies as listed in *Barron's* magazine in November, 1990. The sample of valuation criteria for thirty-four stocks is drawn based on various valuation criteria under which a stock is considered "cheap."

b. The sample is not random, but is specifically chosen to include the stocks which are believed to be "cheap."

Chapter 2
Describing Sets of Data

2.1

a. "Amount of time" is a *quantitative variable* since a numerical quantity (1 hour, 1.5 hours, etc.) is measured.

b. "Number of new employees" is a *quantitative variable* since a numerical quantity (1, 2, etc.) is measured.

c. "Rating of a CEO" is a *qualitative variable* since a quality (excellent, good, fair, poor) is measured.

d. "State of residence" is a *qualitative variable* since a quality (CA, MT, AL, etc.) is measured.

2.2

a. "Appraised value" is a *discrete* variable, since it can take on only a countable number of values ($.01, .02,..., 1.00,$ etc.)

b. "Time" is a *continuous* variable, taking on any values associated with an interval on the real line.

c. "Number of miles", being integer-valued, is a *discrete* variable.

d. "Number of accounts" is integer-valued and hence *discrete*.

e. "Time" is a *continuous* variable.

2.3

a. The pie chart is constructed by partitioning the circle into four parts, according to the total contributed by each part. Since the total number of people is fifty, the total number in category A represents $11/50 = .22$ or 22% of the total. Thus, this category will be represented by a sector angle of $.22(360) = 79.2°$. The other sector angles are shown below.

Category	Frequency	Fraction of Total	Sector Angle
A	11	.22	79.2°
B	14	.28	100.8°
C	20	.40	144.0°
D	5	.10	36.0°

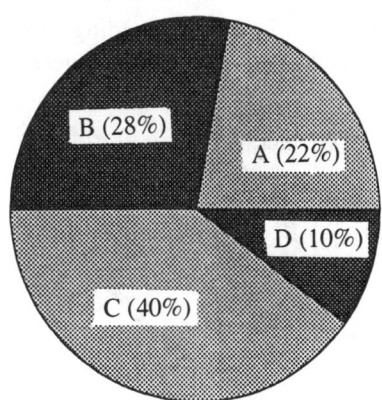

Figure 2.1

b. The bar graph represents each category as a bar with height equal to the frequency of occurrence of that category.

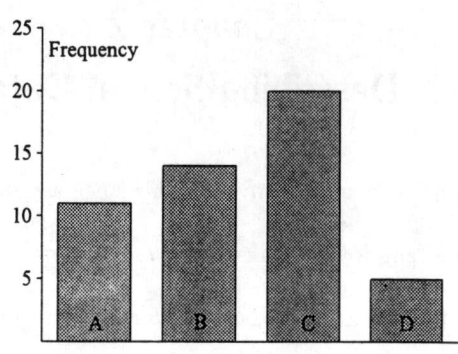

Figure 2.2

c. Yes, the shape will change depending on the order of presentation. The order is unimportant.

2.4

The line graph charts "day" on the horizontal axis and "price" on the vertical axis. The line graph shown below indicates that the stock price decreases on all but one successive day.

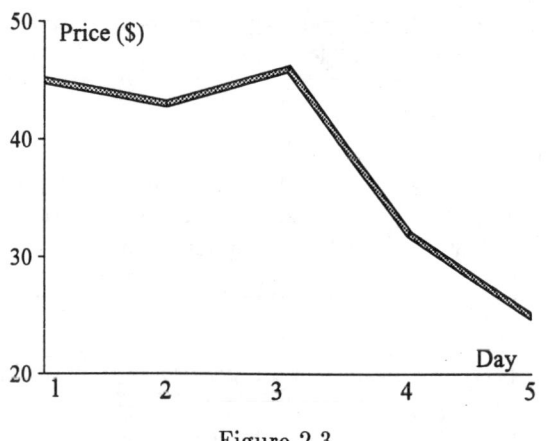

Figure 2.3

2.5

a. The line graph charts the line corresponding to the consumer price index for total services over the three years of interest. The vertical axis measures the size of the index.

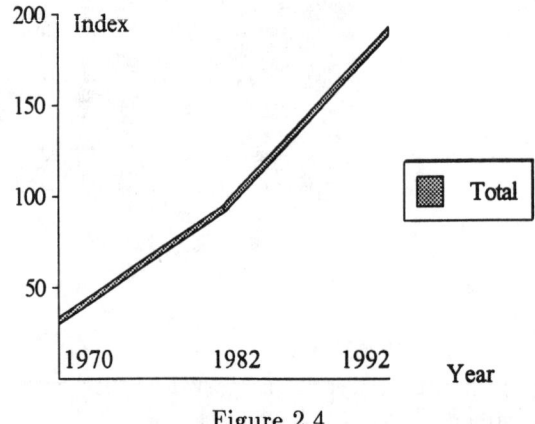

Figure 2.4

4

b. The figure below shows the two consumer price indexes (each shaded differently), each measured for the three years of interest. The vertical axis still measures the size of the index.

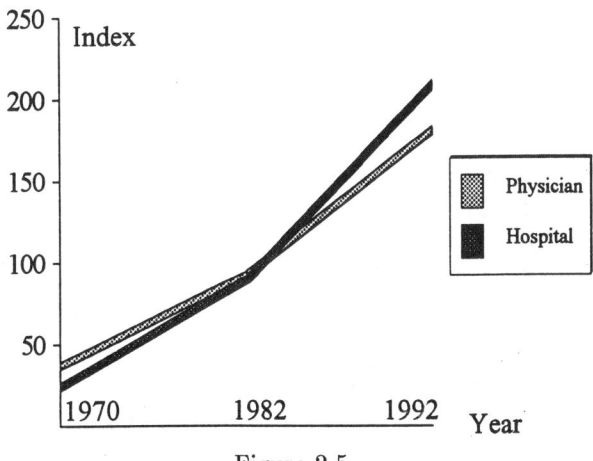

Figure 2.5

c. Using the line graphs in Figures 2.4 and 2.5, we see that the "Total Services" index and the "Physicians Fees" index behaved similarly over the period, both increasing dramatically, but at the same rate. The "Hospital Room Charges" index also increased dramatically over the period, but the rate of increase was greater from 1982 to 1992 than the rate of increase for the other two indexes.

2.6

This is similar to Exercise 2.3. The pie chart and bar graph are shown below.

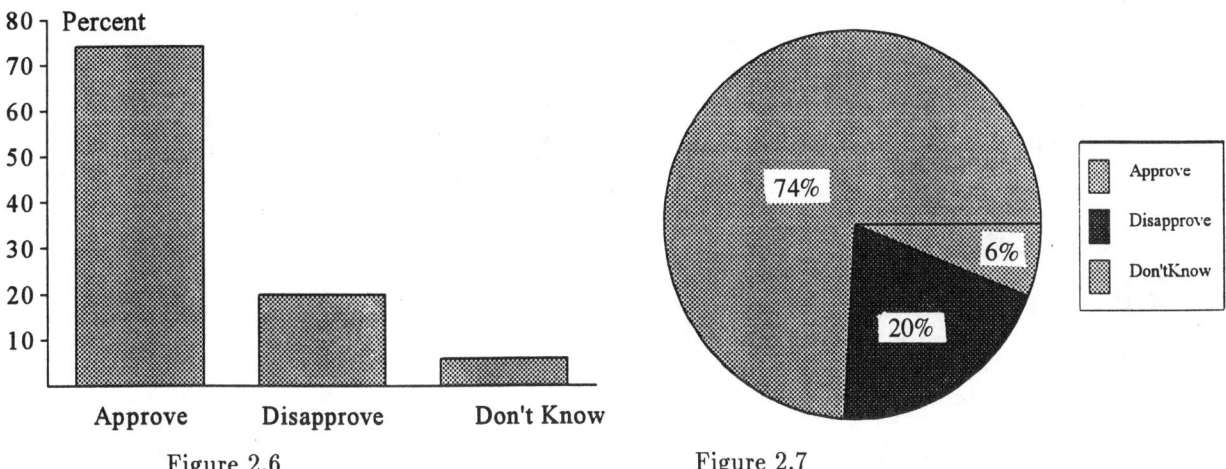

Figure 2.6   Figure 2.7

Since there are only three categories for this qualitative variable, either method of presentation is probably equally effective. The bar graph is easier to construct.

2.7

a. A relative frequency histogram is to be constructed for the data given. Since the interval boundaries are specified in the exercise, the next step is to construct a tally for the data.

Class i	Class Boundaries	Tally	f_i	Relative Frequency, f_i/n
1	0.5–1.5	1	1	1/30
2	1.5–2.5	111	3	3/30
3	2.5–3.5	1111	5	5/30
4	3.5–4.5	1111 11	7	7/30
5	4.5–5.5	1111 11	7	7/30
6	5.5–6.5	1111	4	4/30
7	6.5–7.5	11	2	2/30
8	7.5–8.5	1	1	1/30

The relative frequency histogram is now constructed with the horizontal axis representing classes and the vertical axis measuring relative frequency.

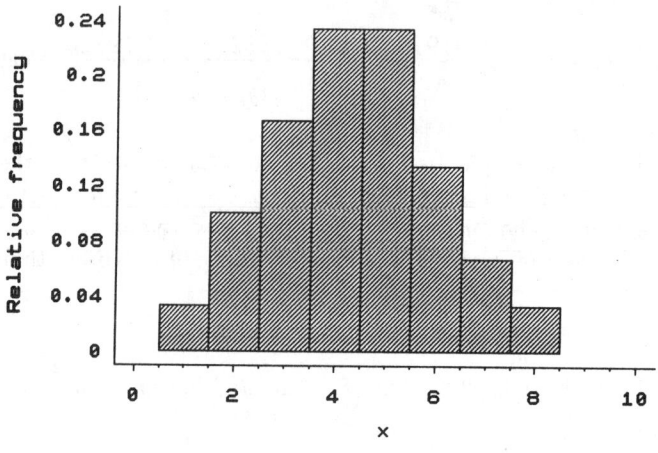

Figure 2.8

b. From the frequency table, $1 + 3 + 5 + 7 = 16$ measurements are less than 4.5. Hence, the fraction of measurements in this interval is 16/30.

c. Since fourteen measurements fall between 3.5 and 5.5, the fraction of measurements in this interval is 14/30.

2.8

a. For n = 50, use between eight and ten classes.

b.

Class i	Class Boundaries	Tally	f_i	Relative Frequency, f_i/n
1	1.55–2.05	11	2	.04
2	2.05–2.55	1111	5	.10
3	2.55–3.05	1111	5	.10
4	3.05–3.55	1111	5	.10
5	3.55–4.05	1111 1111 1111	14	.28
6	4.05–4.55	1111 1	6	.12
7	4.55–5.05	1111 1	6	.12
8	5.05–5.55	11	2	.04
9	5.55–6.05	111	3	.06
10	6.05–6.55	11	2	.04

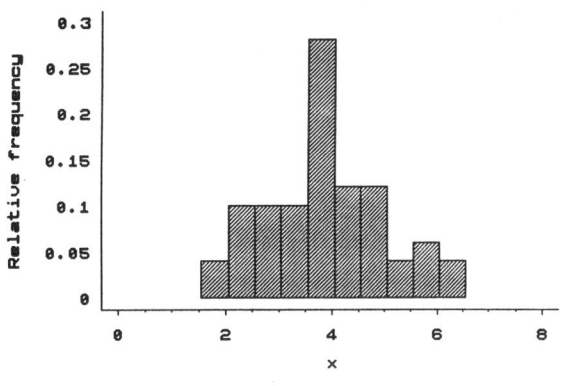

Figure 2.9

c. From b, the fraction less than 5.05 is that fraction lying in classes 1–7, or $(2 + 5 + \cdots + 6 + 6)/50 = 43/50 = .86$.

d. From b, the fraction larger than 3.55 lies in classes 5–10, or $(14 + 6 + \cdots + 3 + 2)/50 = 33/50 = .66$.

2.9

a. Since the variable of interest can only take the values 0, 1, or 2, the classes can be chosen as the integer values 0, 1, and 2. The table below shows the classes, their corresponding frequencies, and their relative frequencies. The relative frequency histogram is shown in Figure 2.10.

Value	Frequency	Relative Frequency
0	5	.25
1	9	.45
2	6	.30

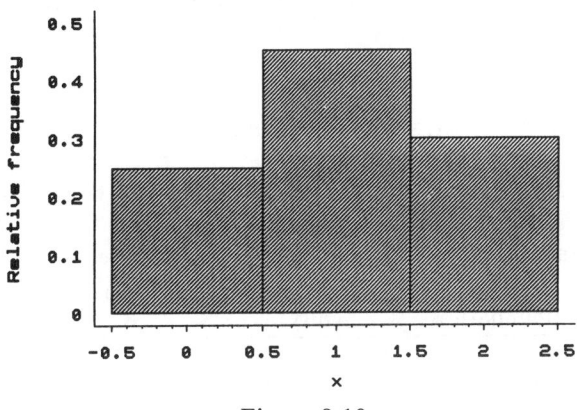

Figure 2.10

b. Using the table in part a, the proportion of measurements greater than 1 is the same as the proportion of "2"s, or .30.

c. The proportion of measurements less than 2 is the same as the proportion of "0"s and "1"s, or $.25 + .45 = .70$.

d. The probability of selecting a "2" in a random selection from these twenty measurements is $6/20 = .30$.

2.10

Each student will obtain a slightly different histogram, depending on the choice of class intervals. One example, generated by the EXECUSTAT computer software package, is shown below.

a.

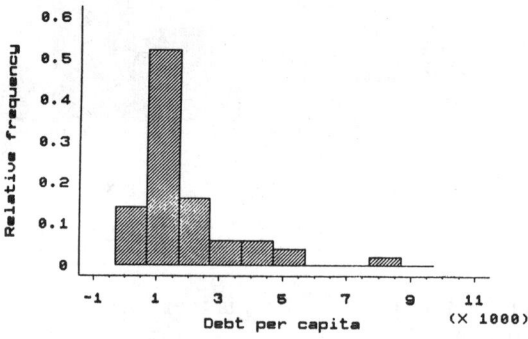

Figure 2.11

b.

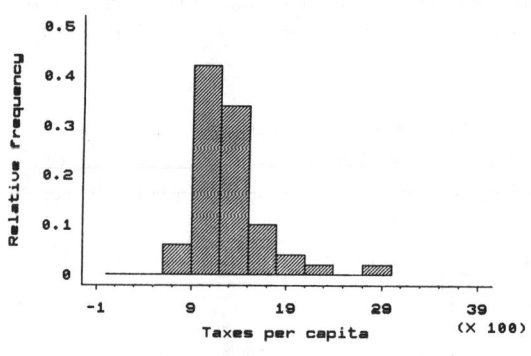

Figure 2.12

c. Notice the similarities in the shapes of the two histograms. Both are slightly skewed to the right.

2.11

The data ranges from .2 to 5.2, or 5.0 units. Since the number of class intervals should be between five and twenty, we choose to use eleven class intervals, with each class interval having length .50 ($5.0/11 = .45$, which, rounded to the nearest convenient fraction, is .50). We must now select interval boundaries such that no measurement can fall on a boundary point. The subintervals .05 to .55, .55 to 1.05, and so on, are convenient and a tally is constructed.

Class i	Class Boundaries	Tally	f_i	Relative Frequency, f_i/n
1	0.05–0.55	⊬⊬⊬⊬ ⊬⊬⊬⊬	10	.167
2	0.55–1.05	⊬⊬⊬⊬ ⊬⊬⊬⊬ ⊬⊬⊬⊬	15	.250
3	1.05–1.55	⊬⊬⊬⊬ ⊬⊬⊬⊬ ⊬⊬⊬⊬	15	.250
4	1.55–2.05	⊬⊬⊬⊬ ⊬⊬⊬⊬	10	.167
5	2.05–2.55	1111	4	.067
6	2.55–3.05	1	1	.017
7	3.05–3.55	11	2	.033
8	3.55–4.05	1	1	.017
9	4.05–4.55	1	1	.017
10	4.55–5.05		0	.000
11	5.05–5.55	1	1	.017

a. The relative frequency histogram is shown in Figure 2.13.

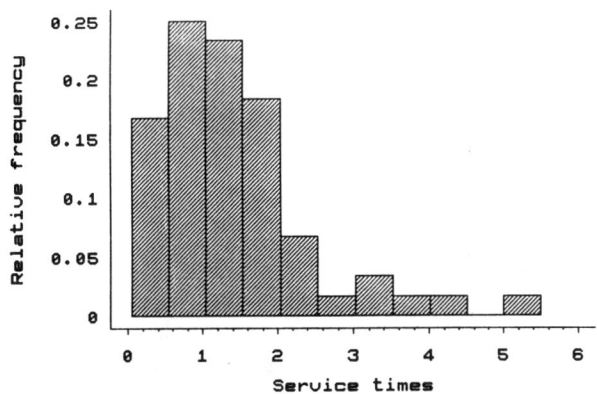

Figure 2.13

b. Looking at the original data, we see that twenty-five customers waited one minute or less. Therefore, the fraction of service times less than or equal to one is $25/60 = .4167$.

2.12

This is similar to Exercise 2.8. The range of the data is $32 - 0 = 32$. We choose to use eleven class intervals of length 3 ($32/11 = 2.9$, which, when rounded to the next largest integer, is 3). The subintervals -0.5 to 2.5, 2.5 to 5.5, 5.5 to 8.5, and so on, are convenient and the tally is shown below.

Class i	Class Boundaries	Tally	f_i	Relative Frequency, f_i/n
1	−0.5–2.5	⊮⊮ ⊮⊮ 1111	14	14/50
2	2.5–5.5	⊮⊮ ⊮⊮	10	10/50
3	5.5–8.5	⊮⊮ 1111	9	9/50
4	8.5–11.5	1111	4	4/50
5	11.5–14.5	1111	4	4/50
6	14.5–17.5	1	1	1/50
7	17.5–20.5	1111	4	4/50
8	20.5–23.5	1	1	1/50
9	23.5–26.5	11	2	2/50
10	26.5–29.5		0	0/50
11	29.5–32.5	1	1	1/50

The relative frequency histogram is shown in Figure 2.14.

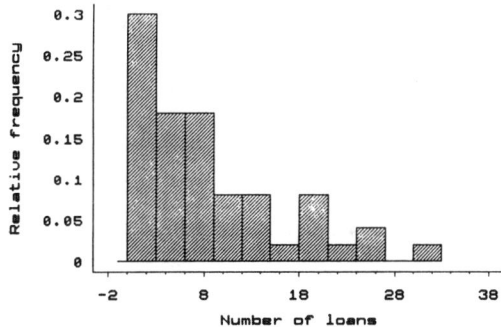

Figure 2.14

b. Looking at the data, we see that thirty-six commercial banks and/or lending institutions granted ten or fewer loans. Therefore, the fraction is 36/50 = .72.

2.13

Histograms will vary from student to student. A typical histogram is shown in the EXECUSTAT output, Figure 2.15.

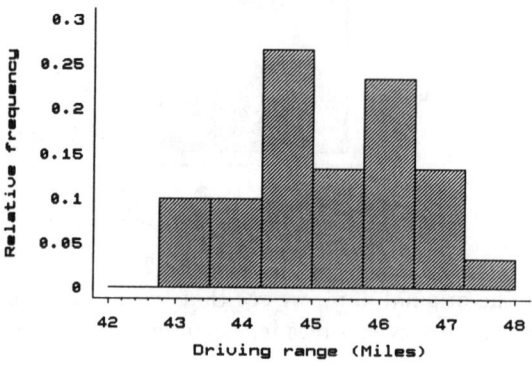

Figure 2.15

2.14

The most obvious choice of a stem is to use the ones digit. The portion of the observation to the right of the ones digit constitutes the leaf. Observations are classified by row according to stem and also within each stem according to relative magnitude. The stem and leaf display is shown below.

```
1 | 6 8
2 | 1 2 5 5 5 7 8 8 9 9
3 | 1 1 4 5 5 6 6 6 7 7 7 7 8 9 9 9     leaf digit = 0.1
4 | 0 0 0 1 2 2 3 4 5 6 7 8 9 9 9       1 2 represents 1.2
5 | 1 1 6 6 7
6 | 1 2
```

a. The stem and leaf display has a more peaked mound-shaped distribution than the distribution shown in Figure 2.9 because of the smaller number of groups.

b. From the stem and leaf display, the sixth-smallest observation is 2.5 (2 5).

2.15

The stems are split, with the leaf digits 0 to 4 belonging to the first part of the stem and the leaf digits 5 to 9 belonging to the second. The stem and leaf display shown below improves the presentation of the data, but only slightly.

```
1 | 6 8
2 | 1 2
2 | 5 5 5 7 8 8 9 9
3 | 1 1 4
3 | 5 5 6 6 6 7 7 7 7 8 9 9 9           leaf digit = 0.1
4 | 0 0 0 1 2 2 3 4
4 | 5 6 7 8 9 9 9                       1 2 represents 1.2
5 | 1 1
5 | 6 6 7
6 | 1 2
```

2.16

a. Use the first two digits as the stem and the ones digit as the leaf.

b.
```
64 | 6 8
65 | 0 2 2 3 3 4 5 8
66 | 2 6 7 9                leaf digit = 1.0
67 | 0 1 1 4 7 8 8 9 9      12 3 represents 123.0
68 | 2
```

c. The data are somewhat mound-shaped; there are no unusually high or low observations. From part b, the smallest observation is 646 and the largest is 682.

2.17

Stem and leaf displays will vary depending on the student's choice of stem and leaf units. A MINITAB printout shown in Figure 2.16 uses the thousands unit as the stem and the hundreds unit as the leaf, truncating the remainder of the number. Each stem is divided into two parts, creating seventeen stems. The resulting display is very similar in shape to the relative frequency histogram in Figure 2.11.

```
Stem-and-leaf of C1      N = 50
Leaf Unit = 100

    1    0 2
   13    0 566677789999
  (15)   1 001111122222344
   22    1 5566789
   15    2 1133
   11    2 577
    8    3 1
    7    3 69
    5    4 00
    3    4
    3    5 11
    1    5
    1    6
    1    6
    1    7
    1    7
    1    8 4
```

Figure 2.16

2.18

Use the ones digit as the stem, and the portion to the right of the ones digit as the leaf, dividing each stem into two parts.

```
0 | 2 2 3 3 4 4 4
0 | 5 5 6 6 6 6 7 7 7 8 8 8 8 9 9
1 | 0 0 1 1 1 1 1 1 1 2 2 2 3 3 4 4
1 | 6 6 7 7 8 8 8 8 9 9
2 | 1 2 3
2 | 5 8                    leaf digit = 0.1
3 | 1 1                    1 2 represents 1.2
3 | 6
4 |
4 | 5
5 | 2
```

2.19

Use the tens digit as the stem and the ones digit as the leaf, dividing each stem into two parts.

```
0 | 0 0 1 1 1 1 1 2 2 2 2 2 2 2 3 3 3 4 4 4 4 4
0 | 5 5 6 6 6 7 7 8 8 8 8 9 9 9
1 | 1 2 3 4 4
1 | 6 8 8 8 9
2 | 3 4
2 | 6                leaf digit = 0.1
3 | 2                1 2 represents 1.2
```

2.20

Refer to Exercise 2.19. The MINITAB program divides each stem into five parts; hence, the displays look quite different.

2.21

Stem and leaf displays will vary depending on the student's choice of stem and leaf units. A MINITAB printout shown in Figure 2.17 uses the portion to the left of the ones unit as the stem and the ones unit as the leaf, rounding off the remainder of the number.

```
Stem-and-leaf of C1        N = 50
Leaf Unit = 1.0

    3     0 999
  (28)    1 0000011122222233344444668889
   19     2 112226
   13     3 007788
    7     4
    7     5 07
    5     6 25
    3     7
    3     8
    3     9
    3    10 14
    1    11
    1    12
    1    13 3

MTB >
```

Figure 2.17

2.22

a. The scatterplot shown below plots the five measurements along the horizontal axis. Since there are two "1"s, the corresponding dots are placed one above the other. The approximate center of the data appears to be around 2.

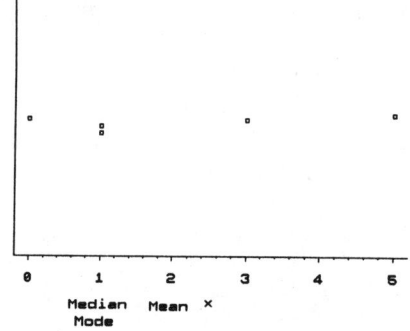

Figure 2.18

b. The mean is the sum of the measurements divided by the number of measurements, or

$$\overline{x} = \frac{\sum_{i=1}^{n} x_i}{n} = \frac{0 + 5 + 1 + 1 + 3}{5} = \frac{10}{5} = 2$$

To calculate the median, the observations are first ranked from smallest to largest: 0, 1, 1, 3, 5. Then since n = 5 is odd, the median is the (n + 1)/2 = 3rd measurement, or m = 1. The mode is the measurement occurring most frequently, or mode = 1.

c. The three measures in part b are located on the scatterplot. Since the median and mode are to the left of the mean, we conclude that the measurements are skewed to the right.

2.23

a. The mean is

$$\overline{x} = \frac{\sum_{i=1}^{n} x_i}{n} = \frac{3 + 1 + \cdots + 5}{8} = \frac{31}{8} = 3.875$$

b. To calculate the median, the observations are first ranked from smallest to largest:
1, 3, 3, 4, 4, 5, 5, 6
Since n = 8 is even, the median is the average of the n/2 = 4th and (n + 1)/2 = 5th measurements, or m = (4 + 4)/2 = 4.

c. Since the mean is slightly smaller than the median, we conclude that the measurements are skewed slightly to the left. The scatterplot shown below confirms this conclusion.

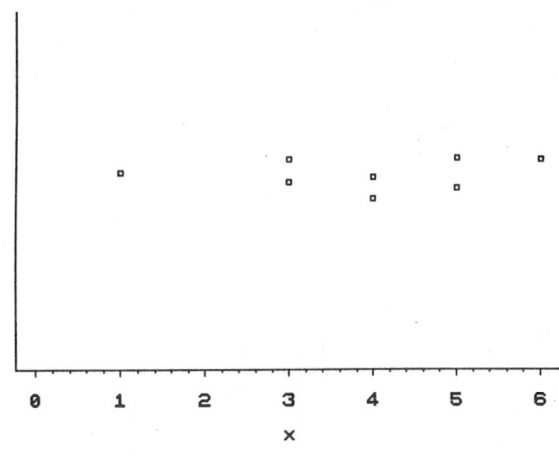

Figure 2.19

2.24

a. $\overline{x} = \frac{\sum_{i=1}^{n} x_i}{n} = \frac{58}{10} = 5.8$

b. The ranked observations are: 2, 3, 4, 5, 5, 6, 6, 8, 9, 10. Since n = 10 is even, the median is halfway between the 5th and 6th ordered observations, or m = (5 + 6)/2 = 5.5.

c. There are two measurements, 5 and 6, which both occur twice. Since this is the highest frequency of occurrence for the data set, we say that the set is *bimodal* with modes at 5 and 6.

2.25

a. $\overline{x} = \frac{\sum_{i=1}^{n} x_i}{n} = \frac{54.5}{7} = 7.786$

b. The ranked observations are: 7.4, 7.5, 7.7, 7.8, 7.9, 8.0, 8.2. Since n = 7 is odd, the median is the (n + 1)/2 = 4th ranked observation, or m = 7.8.

c. The consumer would more likely be interested in the company with the highest or maximum satisfaction rating.

2.26

a. There appear to be two companies with unusually high earnings per share ($1.28 and $1.64). Hence, the data may be skewed to the right.

b. $\bar{x} = \frac{\sum_{i=1}^{n} x_i}{n} = \frac{11.41}{20} = .5705$

To find the median, the observations are ranked from smallest to largest:

.10	.21	.29	.29	.29
.32	.33	.33	.43	.44
.54	.56	.56	.62	.72
.73	.84	.89	1.28	1.64

Since n = 20 is even, the median is the average of the $\frac{n}{2}$ = 10th and $\frac{n}{2} + 1$ = 11th ranked observations, or m = (.44 + .54)/2 = .49.

The mode is the observation that occurs most frequently, or mode = .29.

c. The relative frequency histogram generated using the EXECUSTAT software package is shown in Figure 2.20, with the mean, median, and mode located along the horizontal axis. Note that the mean is larger than the median, indicating that the data is skewed to the right.

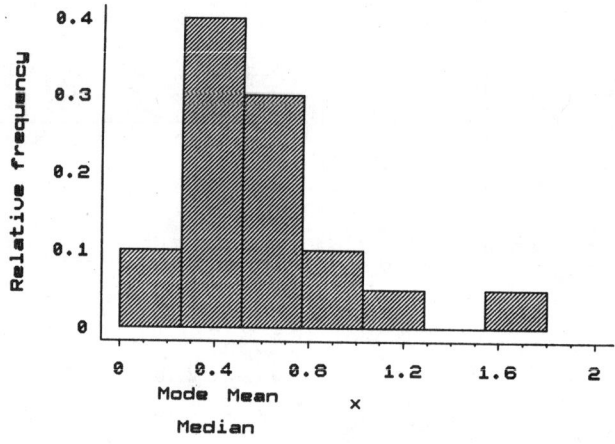

Figure 2.20

2.27

a. $\bar{x} = \frac{\sum_{i=1}^{n} x_i}{n} = \frac{787}{10} = 78.7$ **b.** $\bar{x} = \frac{\sum_{i=1}^{n} x_i}{n} = \frac{2731}{10} = 273.1$

c. The average overall value rating and the average estimated street price are not very useful, since a consumer would generally pick his preferred brand, using a combination of high value rating and/or low estimated street price. Once this choice is made, the consumer could then use the averages to compare his choice to the overall market.

2.28 The *Wall Street Journal* is probably referring to the average number of cubes used per glass measured for some population that they have chosen.

2.29

a. $\bar{x} = \dfrac{\sum_{i=1}^{n} x_i}{n} = \dfrac{92124}{50} = 1842.48$

b. The per capita U.S. aid for the fifty states must be ranked according to magnitude as shown below.

192	911	1160	1611	2657
453	913	1187	1615	2706
558	924	1210	1698	3083
558	998	1213	1762	3644
621	1015	1225	1920	3882
662	1079	1300	2114	4002
669	1092	1400	2135	4040
743	1097	1431	2266	5125
809	1106	1457	2331	5140
857	1138	1457	2540	8418

Since n = 50, the median is the average of the 25th and 26th ordered measurements or $(1225 + 1300)/2 = 1262.5$.

c. The mean is much larger than the median. Hence, there must be some unusually large measurements that are inflating the value of the mean, and we conclude that the distribution is skewed to the right.

2.30

a. $\bar{x} = \dfrac{\sum_{i=1}^{n} x_i}{n} = \dfrac{91}{10} = 9.1$

b. To find the median, the observations are ranked from smallest to largest:

5	6	8	9	9
10	10	10	11	13

Since n = 10 is even, the median is the average of the $\frac{n}{2} = $ 5th and $\frac{n}{2} + 1 = $ 6th ranked observations, or $m = (9 + 10)/2 = 9.5$.

c. Since the mean and the median are almost the same, the data is nearly symmetric. Therefore, either the mean or the median provides a good measure of the center of the distribution.

2.31

a. $\bar{x} = \dfrac{\sum_{i=1}^{n} x_i}{n} = \dfrac{17}{8} = 2.125$

b.

x_i	$(x_i - \bar{x})$	$(x_i - \bar{x})^2$
4	1.875	3.515625
1	−1.125	1.265625
3	.875	.765625
1	−1.125	1.265625
3	.875	.765625
1	−1.125	1.265625
2	−.125	.015625
2	−.125	.015625
	0.000	8.875000

$s^2 = \dfrac{\sum_{i=1}^{n} (x_i - \bar{x})^2}{n-1} = \dfrac{8.875}{7} = 1.267857$

$s = \sqrt{s^2} = \sqrt{1.267857} = 1.126$

c. Calculate $\sum_{i=1}^{n} x_i^2 = 4^2 + 1^2 + \cdots + 2^2 + 2^2 = 45$. Then

$$s^2 = \frac{\sum x_i^2 - \frac{(\sum x_i)^2}{n}}{n-1} = \frac{45 - \frac{(17)^2}{8}}{7} = \frac{8.875}{7} = 1.2679$$

$$s = \sqrt{1.2679} = 1.126$$

The results of parts a and b are identical.

2.32

a. $\overline{x} = \frac{\sum x_i}{n} = \frac{12}{5} = 2.4$

b. Calculate $\sum_{i=1}^{n} x_i^2 = 2^2 + 1^2 + \cdots + 5^2 = 40$. Then

$$s^2 = \frac{\sum x_i^2 - \frac{(\sum x_i)^2}{n}}{n-1} = \frac{40 - \frac{(12)^2}{5}}{4} = \frac{11.2}{4} = 2.8 \text{ and}$$

$$s = \sqrt{2.8} = 1.673$$

c. The range approximation for n = 5 is $s \approx \frac{R}{2.5} = \frac{5-1}{2.5} = 1.6$. Notice that the range approximation is quite accurate when adjusted for the small sample size.

2.33

a. $\overline{x} = \frac{\sum_{i=1}^{n} x_i}{n} = \frac{12}{6} = 2$

b. The following table will be helpful in calculating MAD and s:

x_i	$(x_i - \overline{x})$	$(x_i - \overline{x})^2$	$|x_i - \overline{x}|$
3	1	1	1
0	−2	4	2
2	0	0	0
2	0	0	0
1	−1	1	1
4	2	4	2
	0.0	10	6

Then $\text{MAD} = \frac{\sum_{i=1}^{n} |x_i - \overline{x}|}{n} = \frac{6}{6} = 1.0$

c. $s^2 = \frac{\sum_{i=1}^{n} (x_i - \overline{x})^2}{n-1} = \frac{10}{5} = 2$ or $s^2 = \frac{\sum x_i^2 - \frac{(\sum x_i)^2}{n}}{n-1} = \frac{34 - \frac{(12)^2}{6}}{5} = 2$

and $s = \sqrt{s^2} = \sqrt{2} = 1.414$.

2.34

a. $\overline{x} = \frac{\sum_{i=1}^{n} x_i}{n} = \frac{10}{5} = 2$

b.

x_i	$(x_i - \bar{x})$	$(x_i - \bar{x})^2$
2	0	0
4	2	4
0	−2	4
3	1	1
1	−1	1
	0.0	10

Then $s^2 = \dfrac{\sum_{i=1}^{n}(x_i - \bar{x})^2}{n-1} = \dfrac{10}{4} = 2.5$ or $s^2 = \dfrac{\sum x_i^2 - \dfrac{(\sum x_i)^2}{n}}{n-1} = \dfrac{30 - \dfrac{(10)^2}{5}}{4} = 2.5$

and $s = \sqrt{s^2} = \sqrt{2.5} = 1.581$.

c. $CV = \dfrac{s}{\bar{x}}(100) = \dfrac{1.581}{2}(100) = 79.1$

2.35

a. First calculate the intervals:
$\bar{x} \pm s = 36 \pm 3$ or 33 to 39,
$\bar{x} \pm 2s = 36 \pm 2(3)$ or 30 to 42,
$\bar{x} \pm 3s = 36 \pm 3(3)$ or 27 to 45.

According to the Empirical Rule, approximately 68% of the measurements will fall in the interval 33 to 39; approximately 95% of the measurements will fall between 30 and 42; almost all of the measurements will fall between 27 and 45.

b. If no prior information as to the shape of the distribution is available, we use Tchebysheff's Theorem. We would expect at least $(1 - 1/1^2) = 0$ of the measurements to fall in the interval 33 to 39; at least $(1 - 1/2^2) = 3/4$ of the measurements to fall in the interval 30 to 42; at least $(1 - 1/3^2) = 8/9$ of the measurements to fall in the interval 27 to 45.

2.36

Since we have no prior information as to the shape of the distribution of weekly wages, Tchebysheff's Theorem conservatively estimates that at least 8/9 of the measurements will fall within 3 standard deviations of the mean—that is, $\mu \pm 3\sigma = 364 \pm 78$ or $286 to $442.

2.37

a. The distribution of weekly wages is probably symmetric about the mean, with some higher and some lower wages.

b. According to the Empirical Rule, approximately 95% of the measurements will fall in the interval $\mu \pm 2\sigma = 364 \pm 52$ or 312 to 416. Hence, approximately 5% of the measurements will fall outside this interval. Because of the symmetry of a mound-shaped distribution, we can infer that half of 5%, or 2.5%, will be less than 312.

c. According to the Empirical Rule, almost all of the measurements will fall in the interval $\mu \pm 3\sigma = 364 \pm 78$ or $286 to 442. Hence, almost no contractors pay their workers less than $260. You could justly be accused of underpaying your workers.

2.38

a. First calculate the intervals:
$\bar{x} \pm s = 15.92 \pm .04$ or 15.88 to 15.96,
$\bar{x} \pm 2s = 15.92 \pm 2(.04)$ or 15.84 to 16.00,
$\bar{x} \pm 3s = 15.92 \pm 3(.04)$ or 15.80 to 16.04.

Using Tchebysheff's Theorem, we would expect at least $(1 - 1/1^2) = 0$ of the measurements to fall in the interval 15.88 to 15.96; at least $(1 - 1/2^2) = 3/4$ of the measurements to fall in the interval 15.84 to 16.00; at least $(1 - 1/3^2) = 8/9$ of the measurements to fall in the interval 15.80 to 16.04.

b. According to the Empirical Rule, approximately 68% of the measurements will fall in the interval 15.88 to 15.96; approximately 95% of the measurements will fall between 15.84 and 16.00; almost all of the measurements will fall between 15.80 and 16.04. Since mound-shaped distributions are so frequent, if we do have a sample size of 30 or greater, we expect the sample distribution to be mound-shaped. Therefore, in this exercise, we would expect the Empirical Rule to be suitable for describing the set of data.

c. If the inspector had used a sample size of 4 for this experiment, the distribution would not be mound-shaped. Any possible histogram we could construct would be nonmound-shaped. We can use at most 4 classes, each with frequency 1, and we will not obtain a histogram that is even close to mound-shaped. Therefore, the Empirical Rule would not be suitable for describing n = 4 measurements.

2.39

a. The range of the data is $R = 112 - 11 = 101$. Hence, the adjusted range approximation is
$$s \approx \frac{R}{3} = \frac{101}{3} = 33.67$$

b. Calculate $\sum x_i = 485$ and $\sum x_i^2 = 31{,}945$. Then
$$\bar{x} = \frac{485}{10} = 48.5$$

$$s^2 = \frac{31{,}945 - \frac{(485)^2}{10}}{9} = 935.83333 \quad \text{and} \quad s = \sqrt{935.83333} = 30.59$$

The approximated s found in part a is close to the actual value of s.

2.40

a. Calculate the range as $R = 15 - 1 = 14$. Using the range approximation, $s \approx R/4 = 14/4 = 3.5$.

b. Calculate $n = 25$, $\sum x_i = 155.5$, $\sum x_i^2 = 1260.75$. Then
$$\bar{x} = \frac{\sum x_i}{n} = \frac{155.5}{25} = 6.22$$

$$s^2 = \frac{\sum x_i^2 - \frac{(\sum x_i)^2}{n}}{n-1} = \frac{1260.75 - 967.21}{24} = 12.231 \quad \text{and}$$

$s = \sqrt{s^2} = 3.497$, which is very close to the approximation found in part a.

c. Calculate $\bar{x} \pm 2s = 6.22 \pm 6.994$ or -0.774 to 13.214. From the original data, twenty-four measurements or $(24/25)100 = 96\%$ of the measurements fall in this interval. This is close to the percentage given by the Empirical Rule.

2.41

a. We choose to use twelve classes of length 1.0. The tally and the relative frequency histogram follow.

Class i	Class Boundaries	Tally	f_i	Relative Frequency, f_i/n
1	1.5 − 2.5	1	1	1/70
2	2.5 − 3.5	1	1	1/70
3	3.5 − 4.5	111	3	3/70
4	4.5 − 5.5	1111	5	5/70
5	5.5 − 6.5	1111	5	5/70
6	6.5 − 7.5	1111 1111 11	12	12/70
7	7.5 − 8.5	1111 1111 1111 111	18	18/70
8	8.5 − 9.5	1111 1111 1111	15	15/70
9	9.5 −10.5	1111 1	6	6/70
10	10.5 −11.5	111	3	3/70
11	11.5 −12.5		0	0/70
12	12.5 −13.5	1	1	1/70

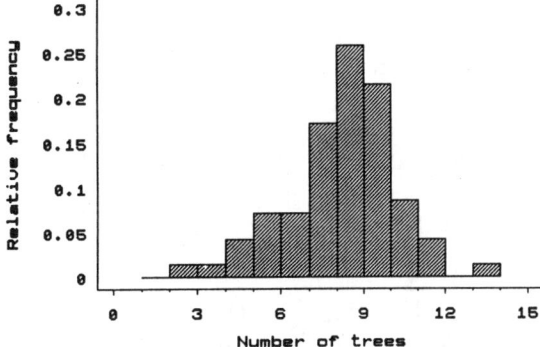

Figure 2.21

b. Calculate n = 70, $\sum x_i = 541$, $\sum x_i^2 = 4453$. Then

$$\overline{x} = \frac{\sum x_i}{n} = \frac{541}{70} = 7.729 \text{ is an estimate of } \mu.$$

c. The sample variance is

$$s^2 = \frac{\sum x_i^2 - \frac{\left(\sum x_i\right)^2}{n}}{n-1} = \frac{4453 - 4181.1571}{69} = 3.9398 \text{ and}$$

$$s = \sqrt{3.9398} = 1.985.$$

The three intervals, $\overline{x} \pm ks$ for k = 1,2,3 are calculated below. The table shows the actual percentage of measurements falling in a particular interval as well as the percentage predicted by Tchebysheff's Theorem and the Empirical Rule. Note that the Empirical Rule should be fairly accurate, as indicated by the mound-shape of the histogram in Figure 2.21.

k	$\overline{x} \pm ks$	Interval Boundaries	Fraction in Interval	Tchebysheff	Empirical Rule
1	7.729 ± 1.985	5.744 to 9.714	50/70 = .71	at least 0	≈.68
2	7.729 ± 3.970	3.759 to 11.699	67/70 = .96	at least .75	≈.95
3	7.729 ± 5.955	1.774 to 13.684	70/70 = 1.00	at least .89	≈1.00

2.42

a. Calculate $n = 7$, $\sum x_i = 54.5$, $\sum x_i^2 = 424.79$. Then

$$\bar{x} = \frac{\sum x_i}{n} = \frac{54.5}{7} = 7.79$$

$$s^2 = \frac{\sum x_i^2 - \frac{(\sum x_i)^2}{n}}{n-1} = \frac{424.79 - \frac{(54.5)^2}{7}}{6} = .078095 \text{ and}$$

$$s = \sqrt{s^2} = .279$$

b. $\text{MAD} = \frac{\sum |x_i - \bar{x}|}{n} = \frac{|7.5 - 7.79| + |7.9 - 7.79| + \cdots + |8.0 - 7.79|}{7} = \frac{1.51}{7} = .22$

c. Tchebysheff's Theorem is appropriate for any set of data, but the Empirical Rule is appropriate only when the data is mound-shaped. Since there are only $n = 7$ observations, the data cannot be mound-shaped, and the Empirical Rule is not appropriate.

2.43

a. Calculate the range as $R = 5.2 - .2 = 5.0$. Using the range approximation, $s \approx R/4 = 5/4 = 1.25$.

b. Calculate $n = 60$, $\sum x_i = 82$, $\sum x_i^2 = 171.38$. Then

$$\bar{x} = \frac{\sum x_i}{n} = \frac{82}{60} = 1.3667$$

$$s^2 = \frac{\sum x_i^2 - \frac{(\sum x_i)^2}{n}}{n-1} = \frac{171.38 - \frac{(82)^2}{60}}{59} = 1.00531 \text{ and } s = \sqrt{s^2} = 1.00265$$

c-d. The three intervals, $\bar{x} \pm ks$ for $k = 1, 2, 3$ are calculated below. The table shows the actual percentage of measurements falling in a particular interval as well as the percentage predicted by Tchebysheff's Theorem and the Empirical Rule.

k	$\bar{x} \pm ks$	Interval Boundaries	Fraction in Interval	Tchebysheff	Empirical Rule
1	1.367 ± 1.003	0.36 to 2.37	$53/60 = .88$	at least 0	$\approx .68$
2	1.367 ± 2.005	$-.64$ to 3.37	$57/60 = .95$	at least .75	$\approx .95$
3	1.367 ± 3.008	-1.64 to 4.37	$58/60 = .97$	at least .89	≈ 1.00

2.44

a. Calculate the range as $R = 132.8 - 9.2 = 123.6$. Using the range approximation, $s \approx R/4 = 123.6/4 = 30.9$.

b. Calculate $n = 50$, $\sum x_i = 1311.5$, $\sum x_i^2 = 68,093.15$. Then

$$\bar{x} = \frac{\sum x_i}{n} = \frac{1311.5}{50} = 26.23$$

$$s^2 = \frac{\sum x_i^2 - \frac{(\sum x_i)^2}{n}}{n-1} = \frac{68,093.15 - \frac{(1311.5)^2}{50}}{49} = 687.60214 \text{ and}$$

$$s = \sqrt{s^2} = 26.22$$

c-d. The three intervals, $\bar{x} \pm ks$ for k = 1,2,3 are calculated below. The table shows the actual percentage of measurements falling in a particular interval as well as the percentage predicted by Tchebysheff's Theorem and the Empirical Rule. The results are consistent with Tchebysheff's Theorem, but not the Empirical Rule.

k	$\bar{x} \pm ks$	Interval Boundaries	Fraction in Interval	Tchebysheff	Empirical Rule
1	26.23 ± 26.22	0.01 to 52.45	44/50 = .88	at least 0	$\approx .68$
2	26.23 ± 52.44	-26.21 to 78.67	47/50 = .94	at least .75	$\approx .95$
3	26.23 ± 78.66	-52.43 to 104.89	49/50 = .98	at least .89	≈ 1.00

2.45

a. Calculate n = 15, $\sum x_i = 21$, $\sum x_i^2 = 49$. Then

$$\bar{x} = \frac{\sum x_i}{n} = \frac{21}{15} = 1.4$$

$$s^2 = \frac{\sum x_i^2 - \frac{(\sum x_i)^2}{n}}{n-1} = \frac{49 - 29.4}{14} = 1.4$$

b. Using the frequency table and the grouped formulas, calculate

$$\sum x_i f_i = 0(4) + 1(5) + 2(2) + 3(4) = 21$$

$$\sum x_i^2 f_i = 0^2(4) + 1^2(5) + 2^2(2) + 3^2(4) = 49$$

Then, as in part a,

$$\bar{x} = \frac{\sum x_i f_i}{n} = \frac{21}{15} = 1.4$$

$$s^2 = \frac{\sum x_i^2 f_i - \frac{(\sum x_i f_i)^2}{n}}{n-1} = \frac{49 - 29.4}{14} = 1.4$$

2.46

a. The first variable (x) is the first number in the pair and is plotted on the horizontal axis, while the second variable (y) is the second number in the pair and is plotted on the vertical axis. The scatterplot is shown in Figure 2.22.

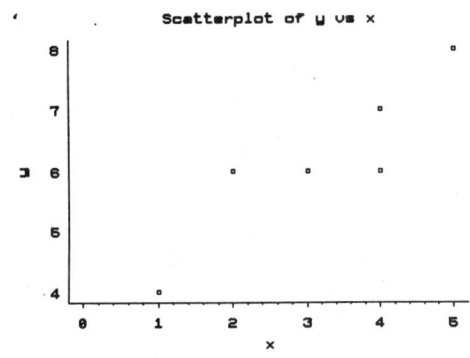

Figure 2.22

b. There appears to be a positive relationship between x and y; that is, as x increases, so does y.

2.47

a. This is similar to Exercise 2.46. The scatterplot is shown in Figure 2.23.

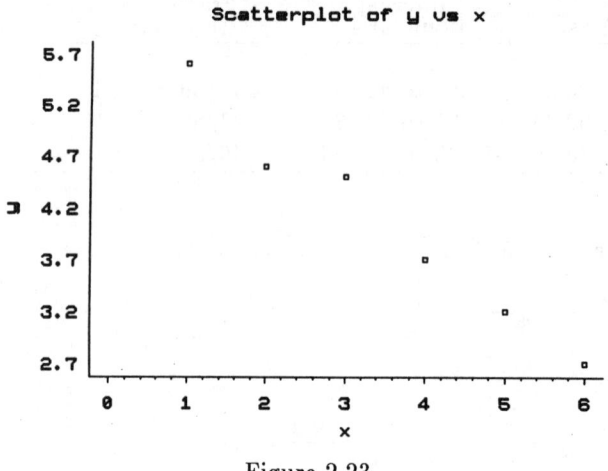

Figure 2.23

b. There appears to be a negative relationship between x and y; that is, as x increases, y decreases.

c. The following preliminary calculations are necessary:

$$n = 6 \qquad \sum_{i=1}^{n} x_i = 21 \qquad \sum_{i=1}^{n} x_i^2 = 91$$

$$\sum_{i=1}^{n} y_i = 24.3 \qquad \sum_{i=1}^{n} y_i^2 = 103.99 \qquad \sum_{i=1}^{n} x_i y_i = 75.3$$

Then $\quad s_x = \sqrt{\dfrac{\Sigma x_i^2 - \dfrac{(\Sigma x_i)^2}{n}}{n-1}} = 1.8708$

(which can also be found using the statistical function on your calculator). Similarly,

$$s_y = \sqrt{\dfrac{\Sigma y_i^2 - \dfrac{(\Sigma y_i)^2}{n}}{n-1}} = 1.0559$$

and

$$s_{xy} = \dfrac{\Sigma x_i y_i - \dfrac{(\Sigma x_i)(\Sigma y_i)}{n}}{n-1} = \dfrac{75.3 - \dfrac{(21)(24.3)}{6}}{5} = \dfrac{-9.75}{5} = -1.95$$

Finally,

$$r = \dfrac{s_{xy}}{s_x s_y} = \dfrac{-1.95}{(1.8708)(1.0559)} = -.987$$

Since r is negative and very close to -1, there is a strong negative relationship between x and y.

2.48

a-b. The scatterplot is shown in Figure 2.24. Notice that there is a negative relationship between x and y.

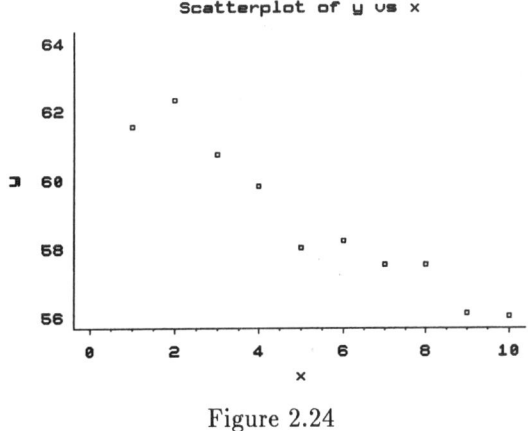

Figure 2.24

2.49

a-b. The scatterplots, generated by EXECUSTAT, are shown in Figures 2.25(a) and (b).

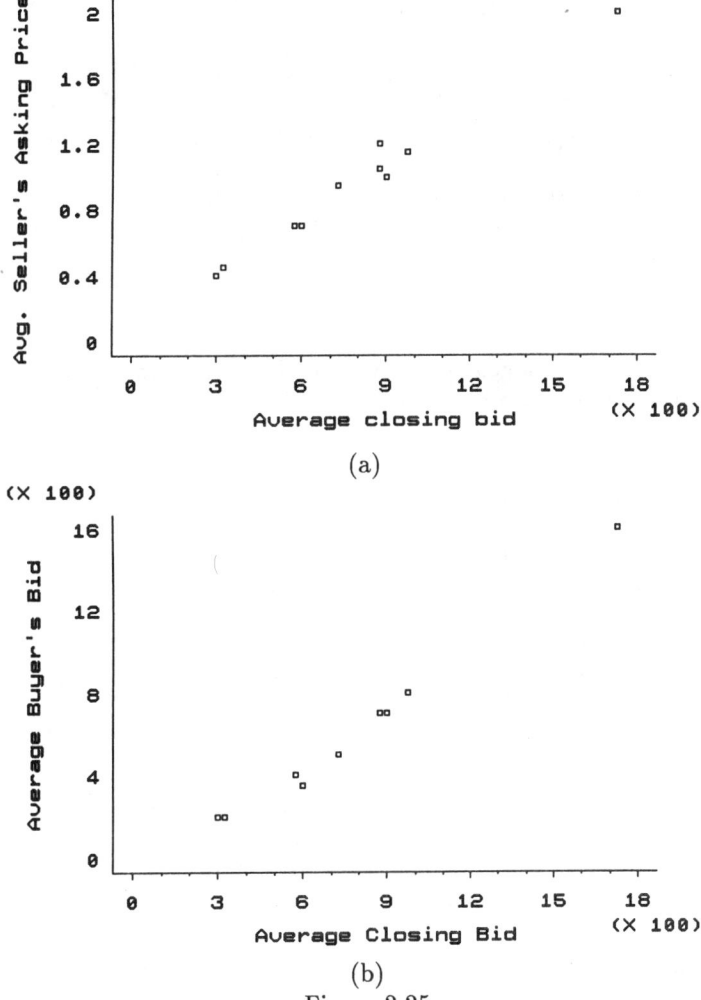

(a)

(b)

Figure 2.25

c. There is a strong positive relationship between both pairs of variables.

2.50

Notice that all three correlation coefficients are positive and very close to 1, indicating strong positive correlation between any of the three pairs of variables. Therefore, the behavior of any of the three variables will give the consumer a good idea of the behavior of either of the other two.

2.51

The stacked bar graph is shown in Figure 2.26, with the height of the bars representing the total taxes paid in Boston and Kansas City for a family of four with an annual income of $100,000 in 1992. The segments of the bars represent the different types of taxes paid. The proportion of the total taxes attributable to the four types of taxes differs for the two cities.

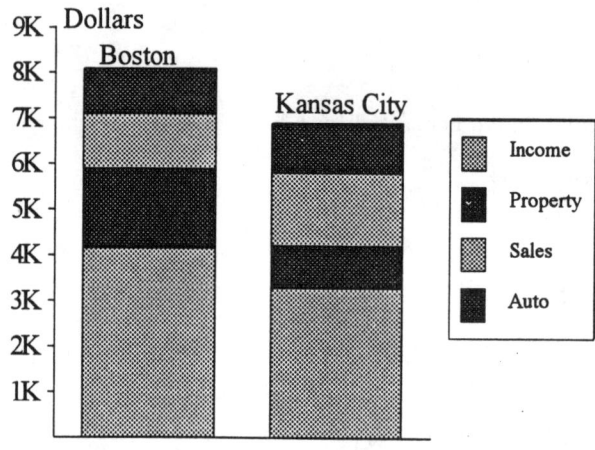

Figure 2.26

2.52

a-b. The scatterplot is generated by EXECUSTAT. Notice the lack of any type of trend or relationship between the two variables.

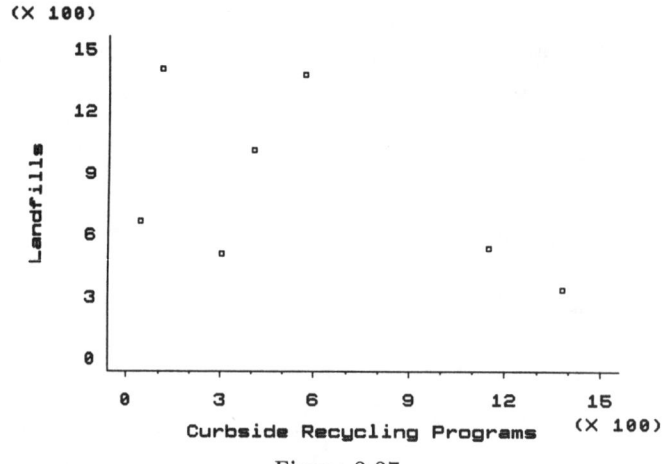

Figure 2.27

c. The difference may be due to the amount of open or unused area in each of the regions as well as to the amount of political interest in environmental concerns.

2.53

a. Calculate $n = 13$, $\sum x_i = 71$, $\sum x_i^2 = 439$. Then

$$\overline{x} = \frac{\sum x_i}{n} = \frac{71}{13} = 5.4615$$

$$s^2 = \frac{\sum x_i^2 - \frac{(\sum x_i)^2}{n}}{n-1} = \frac{439 - \frac{(71)^2}{13}}{12} = 4.269 \text{ and } s = \sqrt{s^2} = 2.066$$

b. The smallest and largest observations are 2 and 9, respectively. The corresponding z-scores are then

$$z = \frac{2 - \overline{x}}{s} = \frac{2 - 5.4615}{2.066} = -1.68 \quad \text{and}$$

$$z = \frac{9 - \overline{x}}{s} = \frac{9 - 5.4615}{2.066} = 1.71.$$

Neither of these z-scores exceeds 2 in absolute value; hence, neither of the observations is unusually large or small.

2.54

Calculate $n = 15$, $\sum x_i = 151$, $\sum x_i^2 = 1987$. Then

$$\overline{x} = \frac{\sum x_i}{n} = \frac{151}{15} = 10.0667$$

$$s^2 = \frac{\sum x_i^2 - \frac{(\sum x_i)^2}{n}}{n-1} = \frac{1987 - \frac{(151)^2}{15}}{14} = 33.35238 \text{ and } s = \sqrt{s^2} = 5.775$$

The largest observation is 19, with corresponding z-score,

$$z = \frac{19 - \overline{x}}{s} = \frac{19 - 10.0667}{5.775} = 1.55$$

This z-score does not exceed 2 in absolute value; hence, it is not unusually large.

2.55

Using the definition of a percentile, 90% of the students taking the test scored lower than you; only 10% scored higher.

2.56

a. Calculate $n = 50$, $\sum x_i = 92124$, $\sum x_i^2 = 280{,}879{,}356$. Then

$$\overline{x} = \frac{\sum x_i}{n} = \frac{92124}{50} = 1842.48$$

$$s^2 = \frac{\sum x_i^2 - \frac{(\sum x_i)^2}{n}}{n-1} = \frac{280{,}879{,}356 - \frac{(92124)^2}{50}}{49} = 2{,}268{,}218.949$$

$$s = \sqrt{s^2} = 1506.0607$$

b. Answers will vary from student to student, depending on the state in which they live.

2.57

Using the definition of percentiles given in the text, we can conclude that:
1. $20,000 is the 60th percentile
2. $30,000 is the 78th percentile
3. $40,000 is the 87th percentile

2.58

The z-score is calculated as

$$z = \frac{x - \bar{x}}{s} = \frac{430 - 410}{14} = 1.43$$

This is not an unusually high z-score, since it does not exceed $z = 2$.

2.59

The ordered data are: 12, 18, 22, 23, 24, 25, 25, 26, 26, 27, 28

For $n = 11$, the position of the median is $(n + 1)/2 = (11 + 1)/2 = 6$ and the depth of the median is $d(M) = 6$. The position of the hinges is then

$$\frac{d(M) + 1}{2} = \frac{6 + 1}{2} = 3.5$$

The lower hinge is halfway between the 3rd and 4th smallest observations or $(22 + 23)/2 = 22.5$, and the upper hinge is halfway between the 3rd and 4th largest observations or $(26 + 26)/2 = 26$. Then

$$\text{H-spread} = \text{upper hinge} - \text{lower hinge} = 26 - 22.5 = 3.5$$

The inner fences are: lower hinge $- 1.5(\text{H-spread}) = 22.5 - 5.25 = 17.25$
upper hinge $+ 1.5(\text{H-spread}) = 26 + 5.25 = 31.25$

The outer fences are: lower hinge $- 3(\text{H-spread}) = 22.5 - 10.5 = 12$
upper hinge $+ 3(\text{H-spread}) = 26 + 10.5 = 36.5$

The only observation falling outside the inner fences is $x = 12$, which lies on the boundary of the outer fence. It is a mild outlier.

2.60

The ordered data are: 2, 3, 4, 5, 6, 6, 6, 7, 8, 9, 9, 10, 22. For $n = 13$, the position of the median is $(n + 1)/2 = (13 + 1)/2 = 7$ and the depth of the median is $d(M) = 7$. The position of the hinges is then

$$\frac{d(M) + 1}{2} = \frac{7 + 1}{2} = 4$$

The lower hinge is the 4th smallest observation or 5, and the upper hinge is the 4th largest observation or 9. Then

$$\text{H-spread} = \text{upper hinge} - \text{lower hinge} = 9 - 5 = 4$$

The inner fences are: lower hinge $- 1.5(\text{H-spread}) = 5 - 6 = -1$
upper hinge $+ 1.5(\text{H-spread}) = 9 + 6 = 15$

The outer fences are: lower hinge $- 3(\text{H-spread}) = 5 - 12 = -7$
upper hinge $+ 3(\text{H-spread}) = 9 + 12 = 21$

The box plot is shown in Figure 2.28. The value $x = 22$ lies outside the outer fence and is an extreme outlier.

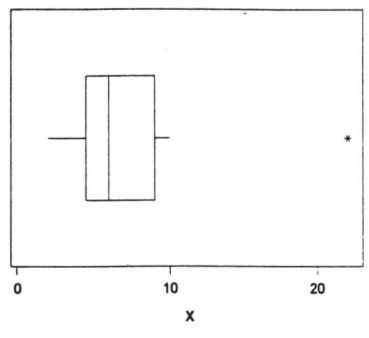

Figure 2.28

2.61

The ordered data are given in the text. For n = 50, the position of the median is $(n + 1)/2 = (50 + 1)/2 = 25.5$ and the depth of the median is $d(M) = 25$. The position of the hinges is then

$$d(H) = \frac{d(M) + 1}{2} = \frac{25 + 1}{2} = 13$$

The lower hinge is the 13th smallest observation or 11.8, and the upper hinge is the 13th largest observation or 29.9. Then

H-spread = upper hinge − lower hinge = 29.9 − 11.8 = 18.1

The inner fences are: lower hinge − 1.5(H-spread) = 11.8 − 27.15 = −15.35
upper hinge + 1.5(H-spread) = 29.9 + 27.15 = 57.05

The outer fences are: lower hinge − 3(H-spread) = 11.8 − 54.3 = −42.5
upper hinge + 3(H-spread) = 29.9 + 54.3 = 84.2

The box plot is shown in Figure 2.29. The values x = 100.8, 103.5, and 132.8 lie outside the outer fence and are extreme outliers; the values x = 62.2 and 65.1 lie between the inner and outer fences and are mild outliers.

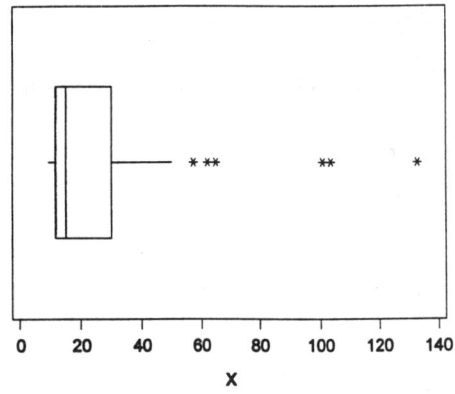

Figure 2.29

2.62

The ranked data are shown below:

6.5	10.1	14.8	21.1	29.0	42.1
6.5	10.1	15.8	21.6	30.3	53.3
7.6	10.5	15.8	22.0	30.6	69.0
8.9	12.0	17.1	25.3	33.9	79.0
8.9	12.1	19.3	25.4	34.1	
10.0	13.0	19.6	27.9	34.3	

a. For n = 34, the position of the median is $(n+1)/2 = (34+1)/2 = 17.5$ and the depth of the median is $d(M) = 17$. The position of the hinges is then

$$d(H) = \frac{d(M)+1}{2} = \frac{17+1}{2} = 9$$

The lower hinge is the 9th smallest observation or 10.5, and the upper hinge is the 9th largest observation or 30.3. Then

$$\text{H-spread} = \text{upper hinge} - \text{lower hinge} = 30.3 - 10.5 = 19.8$$

The inner fences are: lower hinge $- 1.5(\text{H-spread}) = 10.5 - 29.7 = -19.2$
upper hinge $+ 1.5(\text{H-spread}) = 30.3 + 29.7 = 60.0$

The outer fences are: lower hinge $- 3(\text{H-spread}) = 10.5 - 59.4 = -48.9$
upper hinge $+ 3(\text{H-spread}) = 30.3 + 59.4 = 89.7$

b. The values $x = 69.0$ and $x = 79.0$ are mild outliers.

2.63

The distribution of scores is slightly skewed to the right, with the score for the *Lexus* classified as an outlier.

2.64

a. "Ethnic origin" is a *qualitative variable* since a quality (ethnic origin) is measured.

b. "Score" is a *quantitative variable* since a numerical quantity (0-100) is measured.

c. "Type of establishment" is a *qualitative variable* since a category (Carl's Jr., McDonalds or Burger King) is measured.

d. "Mercury concentration" is a *quantitative variable* since a numerical quantity is measured.

2.65

a. The number of new clients acquired by a law firm in a month is a discrete variable since it can take only the values 0, 1, 2....

b. The shelf life of a particular drug is a continuous variable, since it can be any of the infinite number of positive real values.

c. The weight of a railway carload of wheat is a continuous variable.

d. The velocity of a pitched baseball is a continuous variable, as were the variables described in parts b and c.

e. Again, this is a discrete variable since it can take only the values 0, 1, 2, ….

2.66

To determine whether a distribution is likely to be skewed, look for the likelihood of observing extremely large or extremely small values of the variable of interest.

a. The distribution of nonsecured loan sizes might be skewed (a few extremely large loans are possible).

b. The distribution of secured loan sizes is not likely to contain unusually large or small values.

c. Not likely to be skewed.

d. If a package is dropped, it is likely that all the shells will be broken. Hence, a few large numbers of broken shells is possible. The distribution will be skewed.

2.67-2.68

The answers will vary from student to student. However, if the student uses between eight and ten classes, the relative frequency distribution should be fairly mound-shaped with mean close to 5 and standard deviation close to 1.58. The Empirical Rule and Tchebysheff's Theorem should both apply for this data.

2.69

a. The range of the data is $R = 4 - 0 = 4$. Hence,

$$s \approx \frac{R}{2.5} = \frac{4}{2.5} = 1.6$$

b. Calculate $n = 5$, $\sum x_i = 9$, $\sum x_i^2 = 25$. Then

$$s^2 = \frac{\sum x_i^2 - \frac{\left(\sum x_i\right)^2}{n}}{n-1} = \frac{25 - \frac{(9)^2}{5}}{4} = 2.2 \quad \text{and}$$

$$s = \sqrt{s^2} = 1.483$$

which is close to the approximation in part a.

2.70

a. The ordered measurements are: 0, 1, 2, 3, 5, 7. Then

$$\bar{x} = \frac{\sum x_i}{n} = \frac{18}{6} = 3$$

is the sample mean. The median is the average of the two middle observations, or

$$m = \frac{2+3}{2} = 2.5$$

There is no mode, since all observations occur only once.

b. Calculate the range as $R = 7 - 0 = 7$. Using the range approximation, $s \approx R/4 = 7/4 = 1.75$. Using the table given in Exercise 2.69, with $n = 6$ we would estimate $s \approx R/2.5 = 7/2.5 = 2.8$.

c. The following table will be helpful in calculating MAD and s:

x_i	$(x_i - \bar{x})$	$(x_i - \bar{x})^2$	$|x_i - \bar{x}|$
0	−3	9	3
1	−2	4	2
2	−1	1	1
3	0	0	0
5	2	4	2
7	4	16	4
	0	34	12

Then $\quad \text{MAD} = \dfrac{\sum_{i=1}^{n} |x_i - \bar{x}|}{n} = \dfrac{12}{6} = 2$

$s^2 = \dfrac{\sum_{i=1}^{n}(x_i - \bar{x})^2}{n-1} = \dfrac{34}{5} = 6.8$ and

$s = \sqrt{s^2} = \sqrt{6.8} = 2.60768$ (The adjusted range approximation, R/2.5, is fairly accurate)

d. $\quad \text{CV} = \dfrac{s}{\bar{x}}(100) = \dfrac{2.60768}{3}(100) = 86.92$

2.71

a. Calculate the range as R = 71 − 40 = 31. Using the adjusted range approximation from Exercise 2.69, $s \approx R/3 = 31/3 = 10.333$.

b. Calculate n = 10, $\sum x_i = 592$, $\sum x_i^2 = 36{,}104$. Then

$\bar{x} = \dfrac{\sum x_i}{n} = \dfrac{592}{10} = 59.2$

$s^2 = \dfrac{\sum x_i^2 - \dfrac{(\sum x_i)^2}{n}}{n-1} = \dfrac{36{,}104 - \dfrac{(592)^2}{10}}{9} = 107.5111$ and

$s = \sqrt{s^2} = 10.369$

The range approximation is fairly accurate.

2.72

Answers will vary depending on the student's choice of stem and leaf. As an example, a stem and leaf display was constructed using the MINITAB command STEM and is shown in Figure 2.30.

```
MTB > stem c1

Stem-and-leaf of C1      N = 45
Leaf Unit = 1.0

     9   0 146689999
   (15)  1 0133356666667889
    21   2 001122223459
     9   3 127
     6   4 11
     4   5 4
     3   6 2
     2   7 5
     1   8
     1   9 8
```

Figure 2.30

2.73

The ranked data are shown below:

1	10	16	21	29
4	11	16	21	31
6	13	17	22	37
6	13	18	22	41
8	13	18	22	41
9	15	19	22	54
9	16	20	23	62
9	16	20	24	75
9	16	21	25	98

For n = 45, the position of the median is $(n+1)/2 = (45+1)/2 = 23$ and the depth of the median is $d(M) = 23$. The position of the hinges is then

$$d(H) = \frac{d(M) + 1}{2} = \frac{23 + 1}{2} = 12$$

The lower hinge is the 12th smallest observation or 13, and the upper hinge is the 12th largest observation or 23. Then

$$\text{H-spread} = \text{upper hinge} - \text{lower hinge} = 23 - 13 = 10$$

The inner fences are: lower hinge $- 1.5(\text{H-spread}) = 13 - 15 = -2$
upper hinge $+ 1.5(\text{H-spread}) = 23 + 15 = 38$
The outer fences are: lower hinge $- 3(\text{H-spread}) = 13 - 30 = -17$
upper hinge $+ 3(\text{H-spread}) = 13 + 30 = 43$

The values x = 41 and 41 are mild outliers, while the values x = 54, 62, 75, and 98 are extreme outliers.

2.74

Using the MINITAB output, we find $\bar{x} = 22.89$ and $s = 18.66$. The three intervals, $\bar{x} \pm ks$ for k = 1,2,3 are calculated below. The table shows the actual fraction of measurements falling in a particular interval as well as the percentage predicted by Tchebysheff's Theorem and the Empirical Rule.

k	$\bar{x} \pm ks$	Interval Boundaries	Fraction in Interval	Tchebysheff	Empirical Rule
1	22.89 ± 18.66	4.23 to 41.55	39/45 = .87	at least 0	≈.68
2	22.89 ± 37.32	−14.43 to 60.21	42/45 = .93	at least .75	≈.95
3	22.89 ± 55.98	−33.09 to 78.87	44/45 = .98	at least .89	≈1.00

The data are consistent with Tchebysheff's Theorem, but the Empirical Rule is not too accurate. This is because the data are not mound-shaped, but skewed to the right.

2.75

Answers will vary from student to student, depending on the choice of class boundaries. As an example, we calculate the range to be R = 3.7 − 1.8 = 1.9. We could use ten classes of length .2, or we could use seven classes of length .3. Using the latter, we obtain the tabulation:

Class i	Class Boundaries	Tally	f_i	Relative Frequency, f_i/n
1	1.75–2.05	111	3	3/25
2	2.05–2.35	1111	4	4/25
3	2.35–2.65	1111 11	7	7/25
4	2.65–2.95	1111	4	4/25
5	2.95–3.25	111	3	3/25
6	3.25–3.55	111	3	3/25
7	3.55–3.85	1	1	1/25

Here, the lowest point of division is chosen just below the smallest observed measurement The relative frequency histogram is shown in Figure 2.31.

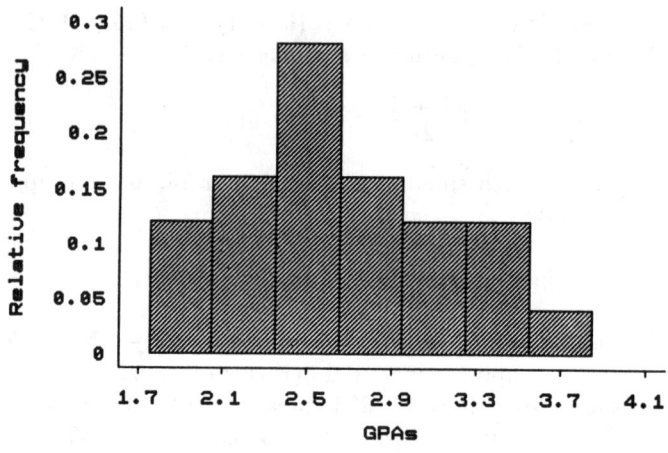

Figure 2.31

2.76

a. Refer to Exercise 2.75. Calculate n = 25, $\sum x_i = 66.3$, $\sum x_i^2 = 181.87$. Then

$$\bar{x} = \frac{\sum x_i}{n} = \frac{66.3}{25} = 2.652$$

$$s^2 = \frac{\sum x_i^2 - \frac{(\sum x_i)^2}{n}}{n-1} = \frac{181.87 - \frac{(66.3)^2}{25}}{24} = .252 \text{ and}$$

$$s = \sqrt{s^2} = .502$$

b-c. The three intervals, $\bar{x} \pm ks$ for $k = 1,2,3$ are calculated below. The table shows the actual fraction of measurements falling in a particular interval as well as the percentage predicted by Tchebysheff's Theorem and the Empirical Rule.

k	$\bar{x} \pm ks$	Interval Boundaries	Fraction in Interval	Tchebysheff	Empirical Rule
1	2.652 ± .502	2.150 to 3.154	17/25 = .68	at least 0	≈.68
2	2.652 ± 1.004	1.648 to 3.656	24/25 = .96	at least .75	≈.95
3	2.652 ± 1.506	1.146 to 4.158	25/25 = 1.00	at least .89	≈1.00

The data are consistent with Tchebysheff's Theorem and the Empirical Rule.

2.77

a. For the 13 gallon bags, calculate

$$\bar{x} = \frac{\sum x_i}{n} = \frac{91}{10} = 9.1$$

$$s^2 = \frac{\sum x_i^2 - \frac{(\sum x_i)^2}{n}}{n-1} = \frac{877 - \frac{(91)^2}{10}}{9} = 5.4333 \text{ and}$$

$$s = \sqrt{s^2} = 2.33$$

while for the 30 gallon bags,

$$\bar{x} = \frac{\sum x_i}{n} = \frac{194}{10} = 19.4$$

$$s^2 = \frac{\sum x_i^2 - \frac{(\sum x_i)^2}{n}}{n-1} = \frac{4148 - \frac{(194)^2}{10}}{9} = 42.7111 \text{ and}$$

$$s = \sqrt{s^2} = 6.535$$

b. For the 13 gallon bags, $CV = (s/\bar{x})100 = (2.33/9.1)100 = 25.6$ while for the 30 gallon bags, $CV = (s/\bar{x})100 = (6.535/19.4)100 = 33.7$.

c. From part b, unit prices of the 30 gallon bags are more variable.

2.78

a. The percentage of colleges that have between 15 and 35 business professors corresponds to the fraction of measurements expected to lie within two standard deviations of the mean. Tchebsheff's Theorem states that this fraction will be at least 3/4 or 75%.

b. If the population is normally distributed, the Empirical Rule is appropriate and the desired fraction is calculated. Referrring to the normal distribution shown in Figure 2.32, the fraction of area lying between 25 and 30 is .34, so that the fraction of colleges having more than 30 business professors is $.5 - .34 = .16$.

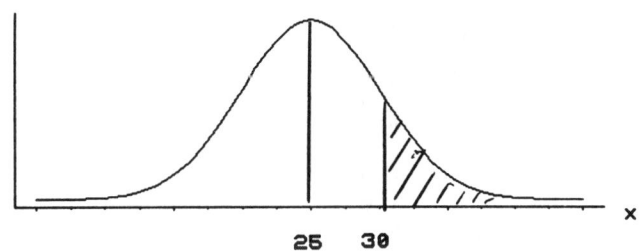

Figure 2.32

2.79

The range approximation gives

$$s \approx \frac{R}{4} = \frac{23-6}{4} = 4.25$$

which is more than twice as small as the experimenter's value. One would tend to doubt his computation.

2.80

The calculations are shown below.

k	$\bar{x} \pm ks$	Interval Boundaries	Empirical Rule
1	4.86 ± 1.581	3.279 to 6.441	≈.68
2	4.86 ± 3.162	1.698 to 8.022	≈.95
3	4.86 ± 4.743	0.117 to 9.603	≈1.00

2.81

a. It is known that duration times are approximately normal, with mean 75 and standard deviation 20. In order to determine the probability that a commercial lasts less than 35 seconds, we must determine the fraction of the curve that lies within the shaded area. Using the Empirical Rule, the fraction of the area between 35 and 75 is half of .95 or .475. Hence, the fraction below 35 would be .5 − .475 or .025.

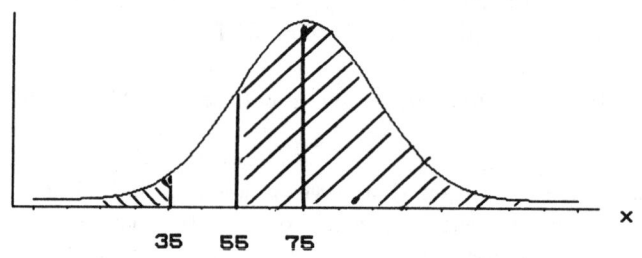

Figure 2.33

b. The fraction of the curve area that lies above the 55-second mark may again be determined by using the Empirical Rule. The fraction between 55 and 75 is .34 and the fraction above 75 is .5. Hence, the probability that a commercial lasts longer than 55 seconds is .5 + .34 = .84.

2.82

a. The relative frequency histogram for these data is shown in Figure 2.34.

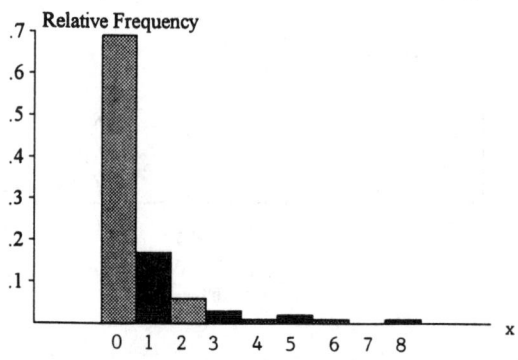

Figure 2.34

34

b. Refer to the formulas given in Exercise 2.45.

Calculate $n = 100$, $\sum x_i f_i = 66$, $\sum x_i^2 f_i = 234$. Then

$$\bar{x} = \frac{\sum x_i f_i}{n} = \frac{66}{100} = .66$$

$$s^2 = \frac{\sum x_i^2 f_i - \frac{\left(\sum x_i f_i\right)^2}{n}}{n-1} = \frac{234 - \frac{(66)^2}{100}}{99} = 1.9236$$

and

$$s = \sqrt{s^2} = 1.387$$

c. The fraction of the equipment counts falling within two and three standard deviations of the mean are found in the following table.

k	$\bar{x} \pm ks$	Interval Boundaries	Fraction
2	.66 ± 2.78	−2.12 to 3.44	95/100
3	.66 ± 4.17	−3.51 to 4.83	96/100

These results do agree with Tchebysheff's Theorem and the Empirical Rule.

2.83

a. The four variables measured are year (qualitative), sales (quantitative discrete), number of stores (quantitative discrete), and number of employees (quantitative discrete).

b. The three line graphs are shown below. Note the increase in all three variables over the ten-year period.

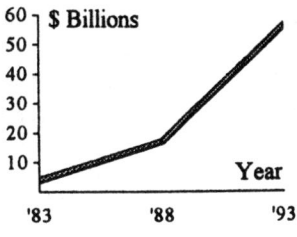
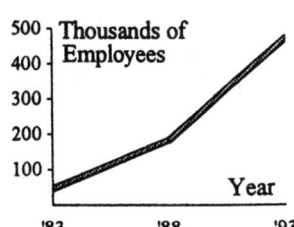
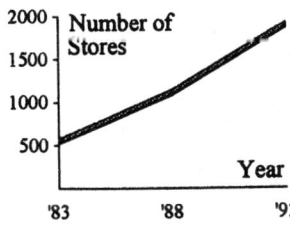

Figure 2.35

c-d. The scatterplots below show the positive relationship between all three pairs of variables, as one might guess.

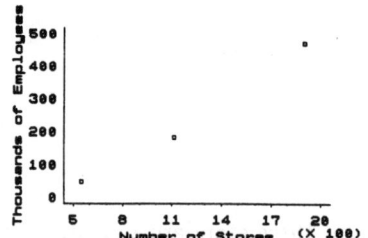 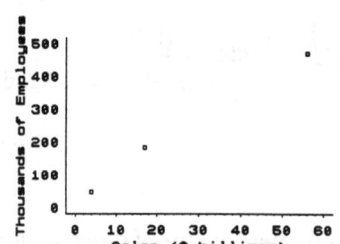 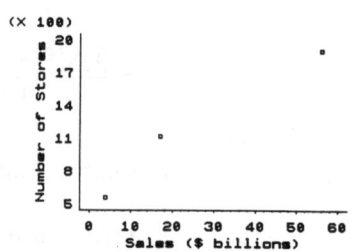

Figure 2.36

2.84

a. The interval $53 to $85 represents the mean plus or minus 2 standard deviations ($\bar{x} \pm 2s$). Using Tchebysheff's Theorem, we can say that at least 3/4 of the measurements will lie in this interval.

b. From part a, if at least 3/4 of the measurements are in the interval $53 to $85, then at most 1/4 are outside of that interval. Hence, the most that can be said about the number of measurements that exceed $85 is that it is at most 1/4.

2.85

a. If the prices of the game CDs are approximately mound-shaped, the Empirical Rule can be used. Refer to Figure 2.37. The proportion of prices from $61 to $85 is the proportion of the curve from one standard deviation below the mean to two standard deviations above the mean. Since the proportion between $53 and $85 is .95, according to the Empirical Rule, the proportion between $69 and $85 will be half of .95 or .475. Similarly, the fraction between $61 and $69 is half of .68 or .34. The proportion of measurements in the interval $61 to $85 is then

$$.475 + .34 = .815$$

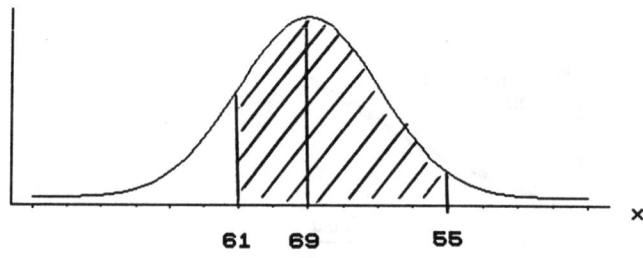

Figure 2.37

b. Refer to Figure 2.38. The proportion of the curve area between $61 and $69 is .34, while the proportion of the curve to the left of the mean ($69) is .5. Hence, the proportion of the curve to the left of $61 is $.5 - .34 = .16$.

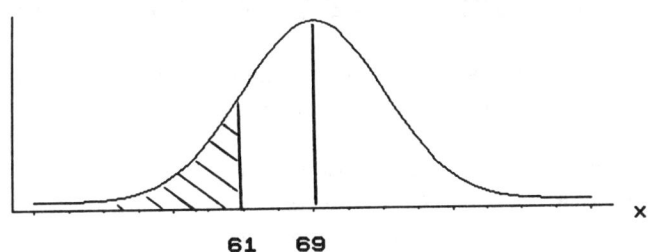

Figure 2.38

2.86

a. The variables are all quantitative, and there are three variables measured on each experimental unit (city). The data are multivariate.

b. Answers will vary from student to student.

2.87

a. A bar graph would probably be the most effective technique, possibly stacked with the percentage that are not expanding.

b-c. Calculate $\bar{x} = \frac{\sum x_i}{n} = \frac{190}{9} = 21.11$,

which does not equal the value given in the article. However, the listing in the exercise is based on groupings that do not all have equal sizes. The number of manufacturing companies may be, for example, much greater than the number of finance companies. Hence, the average percentage that we calculated above will not equal the overall percentage calculated for all 2072 companies as a group.

CHAPTER 3
Probability and Discrete Probability Distributions

3.1 This experiment involves tossing a single die and observing the outcome. The sample space for this experiment consists of the following simple events:

E_1: Observe a 1 E_4: Observe a 4
E_2: Observe a 2 E_5: Observe a 5
E_3: Observe a 3 E_6: Observe a 6

A through F are compound events and are composed in the following manner:

A: (E_2) D: contains no simple events
B: (E_1, E_3, E_5) E: (E_1, E_2, E_3, E_5)
C: (E_1, E_2, E_3) F: (E_2)

To find the probability of an event, we sum the probabilities assigned to the simple events in that event. Since the simple events E_i, $i = 1, 2, \ldots, 6$ are equally likely, $P(E_i) = 1/6$, so that

$P(A) = \frac{1}{6}$ $P(B) = \frac{3}{6} = \frac{1}{2}$ $P(C) = \frac{3}{6} = \frac{1}{2}$

$P(D) = 0$ $P(E) = \frac{4}{6} = \frac{2}{3}$ $P(F) = P(E_2) = \frac{1}{6}$

3.2 The assignment of probabilities is valid in this experiment, since each probability lies between 0 and 1 and since the sum of the four probabilities equals 1.

3.3 In order to have a valid assignment of probabilities, we must have

$$\sum_S P(E_i) = 1$$

Since $P(E_1) + P(E_2) + P(E_3) + P(E_4) = 4(.15) = .6$, we must have $P(E_5) = 1 - .6 = .4$.

3.4 It is given that $P(E_1) = P(E_2) = .15$ and $P(E_3) = .40$. Since $\sum_S P(E_i) = 1$, we know that

$P(E_4) + P(E_5) = 1 - .15 - .15 - .40 = .30$ (i)

Also, it is given that $P(E_4) = 2P(E_5)$.   (ii)

We have two equations in two unknowns that can be solved simultaneously for $P(E_4)$ and $P(E_5)$. Substituting equation (ii) into equation (i), we have

$2P(E_5) + P(E_5) = .3$

$3P(E_5) = .3$ so that $P(E_5) = .1$

Then, from (i), $P(E_4) + .1 = .3$ and $P(E_4) = .2$.

3.5

It is given that $P(E_1) = .45$ and that $3P(E_2) = .45$, so that $P(E_2) = .15$. Since $\sum_S P(E_i) = 1$, the remaining eight simple events must have probabilities whose sum is

$$P(E_3) + P(E_4) + \ldots + P(E_{10}) = 1 - .45 - .15 = .4$$

Since it is given that they are equiprobable,

$$P(E_i) = \frac{4}{8} = .05 \text{ for } i = 3, 4, \ldots, 10$$

3.6

a. There are thirty-eight simple events, each corresponding to a single outcome of the wheel's spin. The thirty-eight simple events are indicated below.

E_1: Observe a 1
E_2: Observe a 2
$\vdots$
E_{36}: Observe a 36
E_{37}: Observe a 0
E_{38}: Observe a 00

b. Since any pocket is just as likely as any other, $P(E_i) = 1/38$.

c. The event A contains two simple events, E_{37} and E_{38}. Then
$P(A) = P(E_{37}) + P(E_{38}) = 2/38 = 1/19$.

d. Define event B as the event that you win on a single spin. Since you have bet on the numbers 1 through 18, event B contains eighteen simple events, $E_1, E_2, \ldots, E_{18}$. Then
$P(B) = P(E_1) + P(E_2) + \ldots + P(E_{18}) = 18/38 = 9/19$.

3.7

a. The tree diagram is similar to Figure 3.1 in the text.

1st Drilling	2nd Drilling	Outcome	Probability
Hit	Hit	E_1 = HH	.01
Hit	Miss	E_2 = HM	?
Miss	Hit	E_3 = MH	.09
Miss	Miss	E_4 = MM	.81

b. It is required that $\sum_S P(E_i) = 1$. Hence, $P(E_2) = 1 - .01 - .09 - .81 = .09$.

c. The company will hit on at least one of the two drillings if it hits on the first, the second, or both. The associated simple events are E_1, E_2 and E_3 and

$$P[\text{hits on at least one}] = P(E_1) + P(E_2) + P(E_3) = .19$$

3.8

The two-way table given in the exercise shows the four possible simple events and their associated probabilities. For clarity, define the four simple events as follows:

E_1: adult is male and lives in the suburbs
E_2: adult is male and lives in the city
E_3: adult is female and lives in the suburbs
E_4: adult is female and lives in the city

a. P[customer resides in suburbs] = $P(E_1) + P(E_3)$ = .17 + .67 = .84
b. P[customer is female and lives in the city] = $P(E_4)$ = .12
c. P[customer is male] = $P(E_1) + P(E_2)$ = .17 + .04 = .21

3.9

Define the following events:

G: stock market gains 100 points or more
L: stock market loses 100 points or more
N: stock market changes less than 100 points

The experiment consists of pairs of forecasts, each of which will be one of the three events defined above. A tree diagram will assist in finding these simple events.

a. GG   LG   NG
GL   LL   NL
GN   LN   NN

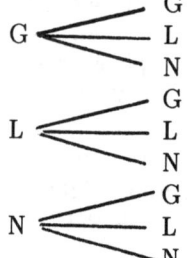

b. The simple events in A are GG, GN, NG, GL, and LG.

c. The simple events in B are GG, LL, and NN.

d. Since each analyst is as likely to select any one of the three choices as any other, $P(E_i)$ = 1/9 and $P(A)$ = 5/9.

e. $P(B)$ = 3/9 = 1/3.

3.10

a. Experiment: A taster tastes and ranks three varieties of tea, A, B, and C, according to preference.

b. Simple events in S are in triplet form:

E_1: (A, B, C) E_2: (A, C, B) E_3: (B, A, C)
E_4: (B, C, A) E_5: (C, B, A) E_6: (C, A, B)

Here the tea in the 1st position is the most desirable, the tea in the 2nd position is the next most desirable, and the tea in the 3rd is the least desirable.

c. Define the events   D: variety A is ranked 1st
F: variety A is ranked 3rd
Then $P(D) = P(E_1) + P(E_2)$ = 1/6 + 1/6 = 1/3.

The probability that A is least desirable is

$$P(F) = P(E_4) + P(E_5) = 1/6 + 1/6 = 1/3.$$

3.11

a. Experiment: Four union men, two from a minority group, are assigned to four one-man jobs, two of which are the most desirable and two of which are the least desirable.

b. Sample space: Let us assume that jobs 1 and 2 are the most desirable ones. Define M_1 and M_2 to be the minority workers and W_1 and W_2 to be the other two workers. A typical simple event is $(M_1M_2W_1W_2)$, which implies that minority workers 1 and 2 are assigned jobs 1 and 2, while the other workers are assigned jobs 3 and 4. There are twenty-four simple events.

E_1: $(M_1M_2W_1W_2)$ E_7: $(M_2M_1W_1W_2)$ E_{13}: $(W_1M_1M_2W_2)$ E_{19}: $(W_2M_1M_2W_1)$
E_2: $(M_1W_1M_2W_2)$ E_8: $(M_2M_1W_2W_1)$ E_{14}: $(W_1M_1W_2M_2)$ E_{20}: $(W_2M_1W_1M_2)$
E_3: $(M_1W_1W_2M_2)$ E_9: $(M_2W_1M_1W_2)$ E_{15}: $(W_1M_2M_1W_2)$ E_{21}: $(W_2M_2M_1W_1)$
E_4: $(M_1W_2W_1M_2)$ E_{10}: $(M_2W_1W_2M_1)$ E_{16}: $(W_1M_2W_2M_1)$ E_{22}: $(W_2M_2W_1M_1)$
E_5: $(M_1W_2M_2W_1)$ E_{11}: $(M_2W_2W_1M_1)$ E_{17}: $(W_1W_2M_1M_2)$ E_{23}: $(W_2W_1M_1M_2)$
E_6: $(M_1M_2W_2W_1)$ E_{12}: $(M_2W_2M_1W_1)$ E_{18}: $(W_1W_2M_2M_1)$ E_{24}: $(W_2W_1M_2M_1)$

c. As jobs 3 and 4 are the least desirable (they correspond to positions 3 and 4 of one ordered 4-tuplet), the probability that the two men from the minority group are assigned to these jobs is
$$P(E_{17}) + P(E_{18}) + P(E_{23}) + P(E_{24}) = 4/24 = 1/6.$$

3.12

Each simple event is equally likely, with probability 1/5.

a. S: E_1, E_2, E_3, E_4, E_5 $P(S) = 5/5 = 1$

b. A: E_1, E_3 $P(A) = 2/5$

c. B: E_1, E_2, E_4, E_5 $P(B) = 4/5$

d. C: E_3, E_4 $P(C) = 2/5$

e. $\bar{A}$: E_2, E_4, E_5 $P(\bar{A}) = 3/5$

f. $\bar{B}$: E_3 $P(\bar{B}) = 1/5$

g. AB: E_1 $P(AB) = 1/5$

h. AC: E_3 $P(AC) = 1/5$

i. A|B: E_1 $P(A|B) = 1/4$

j. A ∪ B: S $P(A \cup B) = 1$

k. A ∪ C: E_1, E_3, E_4 $P(A \cup C) = 3/5$

l. A|C: E_3 $P(A|C) = 1/2$

3.13
From Exercise 3.12, $P(AB) = 1/5$, $P(A|B) = 1/4$ and $P(A) = 2/5$. Since $P(AB) \neq 0$, A and B are not mutually exclusive. Since $P(A|B) \neq P(A)$, A and B are not independent.

3.14
The eight simple events and their associated probabilities are given in the exercise.

a. $P(A) = P(E_1) + P(E_4) + P(E_6) = .1 + .05 + .2 = .35$.

b. $P(B) = .05 + .05 + .3 + .2 + .1 = .70$

c. $P(AB) = P(E_4) + P(E_6) = .05 + .2 = .25$

d. $P(A \cup B) = P(B) + P(E_1) = .70 + .1 = .80$

e. Since $P(AB) \neq 0$, A and B are not mutually exclusive.

f. $P(A|B) = P(AB)/P(B) = .25/.70 = .3571$

g. $P(B|A) = P(AB)/P(A) = .25/.35 = .7143$

h. Since $P(A|B) \neq P(A)$, the events A and B are not independent.

3.15
a. The probability table shown below displays the four intersection probabilities $P(AB)$, $P(A\overline{B})$, $P(\overline{A}B)$, and $P(\overline{A}\overline{B})$. Since we are given $P(A)$, $P(B)$ and $P(AB)$, all the other probabilities can be obtained by subtraction.

	B	$\overline{B}$	Totals
A	.15	.05	.2
$\overline{A}$	.45	.35	.8
	.6	.4	1.0

b. From part a, $P(\overline{A}B) = .45$. Then
$$P(A|\overline{B}) = \frac{P(A\overline{B})}{P(\overline{B})} = \frac{.05}{.4} = .125$$
and
$$P(A \cup B) = P(A) + P(B) - P(AB) = .2 + .6 - .15 = .65$$

c. $P(A|B) = \frac{P(AB)}{P(B)} = \frac{.15}{.6} = .25$ while $P(A) = .2$. Since $P(A) \neq P(A|B)$, the events A and B are not independent.

3.16
Define the following events:

A: new order exceeds $1000
B: new order exceeds $2000
C: new order exceeds $3000
D: new order is $2000 or less

a.   P(B) = .25 + .20 + .10 = .55

b.   P(D|A) = P(AD)/P(A) = .35/.90 = .39

c.   P(C|B) = P(BC)/P(B) = .30/.55 = .545

3.17

Define the following events:
- A: adult believes the ad
- B: adult is a college graduate
- C: adult has had some college
- D: adult has not gone to college

a.   If P(B) = .24, P($\bar{A}$B) = P($\bar{A}$|B) P(B) = (1 − .18)(.24) = (.82)(.24) = .1968

b.   P($\bar{A}$|C) = 1 − P(A|C) = 1 − .25 = .75

c.   If P(D) = .4, P(AD) = P(A|D)P(D) = (.27)(.4) = .108

3.18

Define the following events:

- L: company maintains a parental leave program
- S: company pays salary
- H: company pays health care

It is given that P(L) = .27; P(S|L) = 1/3; P(H|L) = 3/4.

a.   P(LS) = P(L)P(S|L) = .27($\frac{1}{3}$) = .09

b.   P(L$\bar{H}$) = P(L)P($\bar{H}$|L) = .27($\frac{1}{4}$) = .0675

3.19

Define the following events:

- A: car has been in an accident in the past year
- B: car has antilock brakes

The four cells of the probability table represent the intersection probabilities AB, A$\bar{B}$, $\bar{A}$B, and $\bar{A}\bar{B}$.

a.   P(A) = P(AB) + P(A$\bar{B}$) = .03 + .12 = .15

b.   P($\bar{A}$B) = .40   (directly from the table)

c.   P(B|A) = $\frac{P(AB)}{P(A)} = \frac{.03}{.15} = .2$

3.20

a.   The probability that the first car has antilock brakes is .43, since that is the proportion of cars in the large group of 1993-model cars having antilock brakes. Whether or not the first car has antilock brakes, the probability that the second car has antilock brakes is still approximately .43, since the number of cars from which we are choosing is so large.

b. Define the following events:

A: first car has antilock brakes
B: second car has antilock brakes

Notice from part a that $P(B|A) = .43$ and $P(B) = .43$, so that events A and B are independent.

c. Define C: third car has antilock brakes. Since events A, B, and C are independent,

$$P(ABC) = P(A)P(B)P(C) = (.43)^3 = .0795$$

d. The event that exactly one of the three cars will have antilock brakes will occur if one of three simple events:

$$A\bar{B}\bar{C} \quad \text{or} \quad \bar{A}B\bar{C} \quad \text{or} \quad \bar{A}\bar{B}C$$

By definition,
$$P(\bar{A}) = P(\bar{B}) = P(\bar{C}) = 1 - .43 = .57$$

and the three events are independent. Hence,

$$P(\text{exactly one car has antilock brakes}) = P(A\bar{B}\bar{C}) + P(\bar{A}B\bar{C}) + P(\bar{A}\bar{B}C)$$

$$= P(A)P(\bar{B})P(\bar{C}) + P(\bar{A})P(B)P(\bar{C}) + P(\bar{A})P(\bar{B})P(C)$$

$$= (.43)(.57)(.57) + (.57)(.43)(.57) + (.57)(.57)(.43) = .4191$$

3.21

Define the event A: executive will use Information Superhighway.

a. $P(A) = .40$

b. If two executives are randomly chosen, a simple event consists of a pair, the first element representing the first executive's action, and the second representing the action of the second executive. Then the event that only one of the two executives will use the Information Superhighway is composed of two simple events, $A\bar{A}$ and $\bar{A}A$. The probability of interest is then

$$P(A\bar{A}) + P(\bar{A}A) = P(A)P(\bar{A}) + P(\bar{A})P(A) = 2(.40)(.60) = .48$$

3.22

There are 35 brands from which to choose, each of which is categorized according to quality.

a. $P(E) = \frac{2}{35}$

b. $P(\text{at least "good"}) = P(G \text{ or } V \text{ or } E) = P(G) + P(V) + P(E) = \frac{11}{35} + \frac{2}{35} + \frac{21}{35} = \frac{34}{35}$

c. $P(\text{not } V \text{ or } E) = P(G \text{ or } F \text{ or } P) = \frac{11 + 1 + 0}{35} = \frac{12}{35}$

d. Define the following events:

A: first brand is "very good"
B: second brand is "very good"

Using the Multiplicative Law of Probability,

$$P(AB) = P(A)P(B|A) = \left(\frac{21}{35}\right)\left(\frac{20}{34}\right) = \frac{420}{1190} = .3529$$

3.23

Define the following events:
 A: device A is activated
 B: device B is activated

It is given that $P(A) = .91$, $P(B) = .95$, and that the two systems (and hence, events A and B) are independent. Then

P(system functions) = P(either or both devices activate)

$$= P(A \cup B) = P(A) + P(B) - P(AB)$$

$$= P(A) + P(B) - P(A)P(B) = .91 + .95 - (.91)(.95) = .9955$$

3.24

Use Bayes' Rule. The denominator given in the formula for $P(s_i|I)$ is

$$P(I) = \sum_{i=1}^{3} P(s_i)P(I|s_i) = .4(.1) + .5(.3) + .1(.2) = .21$$

so that

$$P(s_1|I) = \frac{P(s_1)P(I|s_1)}{P(I)} = \frac{.4(.1)}{.21} = .1905$$

$$P(s_2|I) = \frac{P(s_2)P(I|s_2)}{P(I)} = \frac{.5(.3)}{.21} = .7143$$

$$P(s_3|I) = \frac{P(s_3)P(I|s_3)}{P(I)} = \frac{.1(.2)}{.21} = .0952$$

3.25

This is similar to Exercise 3.24. The denominator given in the formula for $P(s_i|I)$ is

$$P(I) = \sum_{i=1}^{4} P(s_i)P(I|s_i) = .6(.1) + .2(.4) + .2(.3) + .5(.2) = .30$$

so that

$$P(s_1|I) = \frac{P(s_1)P(I|s_1)}{P(I)} = \frac{.6(.1)}{.30} = .2000$$

$$P(s_2|I) = \frac{P(s_2)P(I|s_2)}{P(I)} = \frac{.2(.4)}{.30} = .2667$$

$$P(s_3|I) = \frac{P(s_3)P(I|s_3)}{P(I)} = \frac{.2(.3)}{.30} = .2000$$

$$P(s_4|I) = \frac{P(s_4)P(I|s_4)}{P(I)} = \frac{.5(.2)}{.30} = .3333$$

3.26

Define the following events: D: item is defective
 C: item goes through a complete inspection

It is given that $P(D) = .1$, $P(C|D) = .6$, $P(C|\overline{D}) = .2$. The probability of interest is

$P(D|C)$, the probability that an item is defective given that it goes through a complete inspection. Using Bayes' Rule,

$$P(D|C) = \frac{P(D)P(C|D)}{P(D)\,P(C|D) + P(\overline{D})\,P(C|\overline{D})} = \frac{.1(.6)}{.1(.6) + .9(.2)} = \frac{.06}{.24} = .25$$

3.27

Define the following events:

    A: passenger uses airport A
    B: passenger uses airport B
    C: passenger uses airport C
    D: a weapon is detected

Suppose that a passenger is carrying a weapon. It is given that

$P(D|A) = .99$ $P(A) = .5$
$P(D|B) = .95$ $P(B) = .3$
$P(D|C) = .80$ $P(C) = .2$

The probability of interest is

$$P(A|D) = \frac{P(A)P(D|A)}{P(A)\,P(D|A) + P(B)\,P(D|B) + P(C)\,P(D|C)}$$

$$= \frac{.5(.99)}{.5(.99) + .3(.95) + .2(.80)} = .5266$$

Similarly,

$$P(C|D) = \frac{.2(.80)}{.5(.99) + .3(.95) + .2(.80)} = .1702$$

3.28

Define the following events, under the assumption that an incorrect return has been filed:

    G_1: individual guilty of cheating
    G_2: individual not guilty (filed incorrectly due to lack of knowledge)
    D: individual denies knowledge of error

It is given that $P(G_1) = .05$, $P(G_2) = .02$, $P(D|G_1) = .80$. Note that $P(D|G_2) = 1$, since it follows that, if the individual has incorrectly filed due to lack of knowledge, he will, with probability 1 deny knowledge of the error. Using Bayes' Rule,

$$P(G_1|D) = \frac{P(G_1)P(D|G_1)}{P(G_1)P(D|G_1) + P(G_2)P(D|G_2)} = \frac{.05(.80)}{.05(.80) + .02(1)} = .6667$$

3.29

a. Since one of the requirements of a probability distribution is that $\sum_x p(x) = 1$, we need

$$p(4) = 1 - (.1 + .3 + .4 + .1 + .05) = 1 - .95 = .05$$

b. The probability histogram is shown in Figure 3.1.

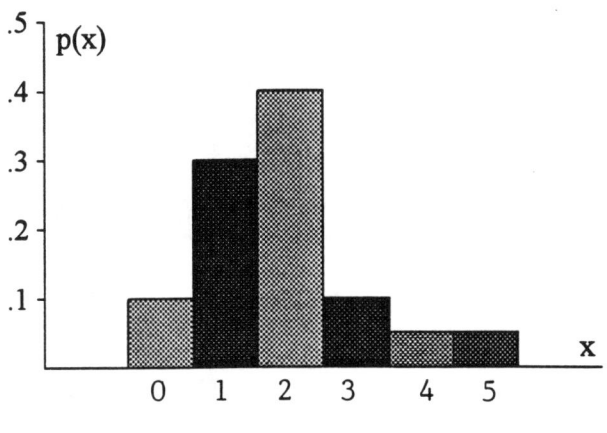

Figure 3.1

3.30

a. $P[x = 2] = p(2) = .4$ and $P[x \geq 2] = p(2) + p(3) + p(4) + p(5) = .6$

b. $P[x \leq 3] = p(0) + p(1) + p(2) + p(3) = .9$

c. $\mu = E(x) = \sum xp(x) = 0(.1) + 1(.3) + 2(.4) + 3(.1) + 4(.05) + 5(.05) = 1.85$

Then the variance of x is defined as

$$\sigma^2 = E[(x - \mu)^2] = (0 - 1.85)^2(.1) + (1 - 1.85)^2(.3) + \cdots + (5 - 1.85)^2(.05)$$

$$= 1.4275$$

3.31

a. Since $0 \leq p(x) \leq 1$ and $\sum_x p(x) = 1$, this is a valid probability distribution.

b. This is not a valid probability distribution, since p(0) is negative.

c. This is not a valid probability distribution, since the sum of all the probabilities is not equal to one.

3.32

a. $\mu = E(x) = \sum xp(x) = 1(.05) + 2(.20) + \cdots + 7(.05) = 3.45$

Then the variance of x is defined as

$$\sigma^2 = E[(x - \mu)^2] = (1 - 3.45)^2(.05) + \cdots + (7 - 3.45)^2(.05)$$

$$= 2.0475$$

and

$$\sigma = \sqrt{2.0475} = 1.4309$$

b-c. The interval $\mu \pm 2\sigma = 3.45 \pm 2.86$ or .59 to 6.31. This interval is shown on the probability histogram below.

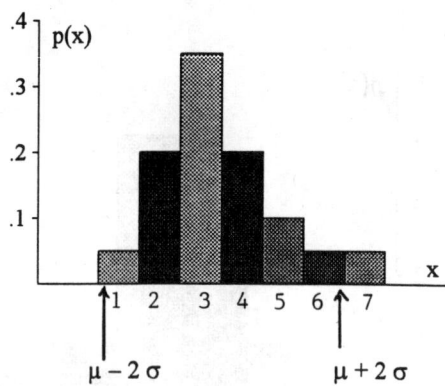

Figure 3.2

Then
$$P[.59 \leq x \leq 6.31] = P[1 \leq x \leq 6] = 1 - p(7) = .95.$$

d. Since the probability that x falls in the interval $\mu \pm 2\sigma$ is .95 from part c, we would expect about 95% of the values to fall in the interval in repeated sampling.

3.33

The random variable G, total gain to the insurance company, will be D if there is no theft, but D−50,000 if there is a theft during a given year. These two events will occur with probability .99 and .01, respectively. Hence, the probability distribution for G is given below.

G	p(G)
D	.99
D−50,000	.01

The expected gain is

$$E(G) = \sum G\, p(G) = .99D + .01(D-50,000)$$
$$= D - 500.$$

In order that $E(G) = 1000$, it is necessary to have

$$1000 = D - 500 \quad \text{or} \quad D = \$1500$$

3.34

a. $E(x) = \sum xp(x) = 3(.03) + 4(.05) + \cdots + 13(.01) = 7.9$

b. $\sigma^2 = \sum(x-\mu)^2 p(x)$

$$= (3-7.9)^2(.03) + (4-7.9)^2(.05) + \cdots + (13-7.9)^2(.01)$$

$$= 4.73 \quad \text{and} \quad \sigma = \sqrt{4.73} = 2.1749$$

c. Calculate $\mu \pm 2\sigma = 7.9 \pm 4.350$ or 3.55 to 12.25. Then, referring to the probability distribution of x,

$$P[3.55 \leq x \leq 12.25] = P[4 \leq x \leq 12] = 1-p(3)-p(13) = 1-.04 = .96$$

3.35

a. Define the event L to be the event that an American agrees with a twelve-year Congressional term limit. Then $P(L) = .61$, and there are three people randomly selected to be interviewed. Let x be the number who agree with the term limit. The simple events in the experiment are:

| LLL | NLL | LNL | LLN |
| NNL | NLN | LNN | NNN |

Then

$$p(0) = P(NNN) = P(N)P(N)P(N) = (.39)^3 = .0593$$

$$p(1) = P(NNL) + P(NLN) + P(LNN) = 3(.61)(.39)^2 = .2783$$

$$p(2) = P(NLL) + P(LNL) + P(LLN) = 3(.61)^2(.39) = .4354$$

$$p(3) = P(LLL) = (.61)^3 = .2270$$

b. The probability histogram for x is shown in Figure 3.3.

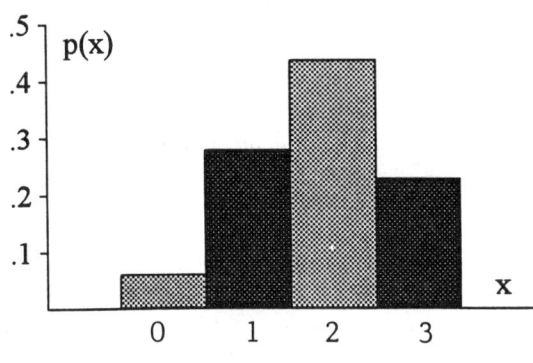

Figure 3.3

c. $P[x \geq 2] = p(2) + p(3) = .4354 + .2270 = .6624$

d. $\mu = E(x) = \sum xp(x) = 0(.0593) + 1(.2783) + 2(.4354) + 3(.2270) = 1.83$

and
$$\sigma^2 = \sum (x-\mu)^2 p(x)$$
$$= \sum x^2 p(x) - \mu^2$$
$$= 0^2(.0593) + 1^2(.2783) + 2^2(.4354) + 3^2(.2270) - (1.83)^2$$
$$= 4.0629 - 3.3489 = .714$$
$$\sigma = \sqrt{.714} = .845$$

3.36

We are asked to find the premium that the insurance company should charge in order to break even. Let c be the unknown value of the premium and x be the gain to the insurance company caused by marketing the new product. There are three possible values for x. If the product is a failure or moderately successful, x will be negative; if the product is a success, the insurance company will gain the amount of the premium and x will be positive. The probability distribution for x follows:

x	p(x)
c	.94
−80,000+c	.01
−25,000+c	.05

In order to break even,

$$E(x) = \sum xp(x) = 0$$

Therefore,

$$.94(c) + (.01)(-80{,}000+c) + (.05)(-25{,}000+c) = 0$$

$$-800 - 1250 + (.01 + .05 + .94)c = 0$$

$$c = 2050$$

Hence, the insurance company should charge a premium of $2050.

3.37

Define the random variable v to be the volume shipped per trailer load. Since there are only two types of trailers, there are only two possible values for v:

(1) $\quad v = (8)(10)(30) = 2400$ with probability $p(2400) = .3$

(2) $\quad v = (8)(10)(40) = 3200$ with probability $p(3200) = .7$

The probability distribution and the calculation of E(v) follow.

v	p(v)
2400	.3
3200	.7

$$E(v) = \sum vp(v)$$
$$= 2400(.3) + 3200(.7)$$
$$= 2960.$$

3.38

Let B_1, B_2, W_1, and W_2 be the two white and two black balls, respectively.

a. The six simple events are

$$\begin{array}{cccccc} B_1B_2 & B_1W_1 & B_1W_2 & B_2W_1 & B_2W_2 & W_1W_2 \\ B_2B_1 & W_1B_1 & W_2B_1 & W_1B_2 & W_2B_2 & W_2W_1 \end{array}$$

Since the balls are randomly selected, each simple event has probability 1/12. Notice that we have listed the **ordered** simple events, since it will be important in the calculation of part b.

b. $\quad P[\text{first ball black}] = 6/12 = 1/2$

c. $\quad P[\text{second ball black}|\text{first ball white}] = \dfrac{P(W_1B_2) + P(W_2B_1) + P(W_1B_1) + P(W_2B_2)}{P[\text{first ball white}]}$

$$= \dfrac{4/12}{6/12} = \dfrac{4}{6} = \dfrac{2}{3}.$$

d. $\quad P[\text{one white and one black ball}] = \dfrac{8}{12} = \dfrac{2}{3}.$

3.39

Refer to Exercise 3.38 and let x be the number of white balls selected.

a. For the twelve simple events given in Exercise 3.38, the appropriate values of x are 0, 1, 1, 1, 1, 2, 0, 1, 1, 1, 1, 2, respectively.

b-c. The values of p(x) are then given as

$$p(0) = \dfrac{2}{12} = \dfrac{1}{6} \qquad p(1) = \dfrac{8}{12} = \dfrac{2}{3} \qquad p(2) = \dfrac{2}{12} = \dfrac{1}{6}$$

and the probability table and histogram are shown below. Notice that $\sum xp(x) = 1$.

x	p(x)
0	1/6
1	2/3
2	1/6

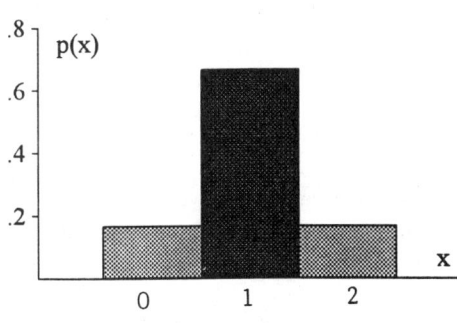

Figure 3.4

3.40

Each student will obtain a different simulation. The relative frequency histogram should be similar to the probability histogram shown in Figure 3.4.

3.41

Refer to Exercise 3.38. The list of simple events is now expanded to include four more:

a.
B_1B_2 B_1W_1 B_1W_2 B_2W_1 B_2W_2 W_1W_2
B_2B_1 W_1B_1 W_2B_1 W_1B_2 W_2B_2 W_2W_1
B_1B_1 B_2B_2 W_1W_1 W_2W_2

Each simple event has probability 1/16.

b. P[second ball black] $= \frac{8}{16} = \frac{1}{2}$

c. P[second ball black|first ball white] $= \frac{4}{8} = \frac{1}{2}$

d. Using the new set of simple events,

$$p(0) = \frac{4}{16} = \frac{1}{4} \qquad p(1) = \frac{8}{16} = \frac{1}{2} \qquad p(2) = \frac{4}{16} = \frac{1}{4}$$

which is different from the probability distribution in Exercise 3.39 b.

3.42

Define the following events:
A: resident is under 25 years of age
B: resident is 65 years or older
C: resident is 25-34
D: resident is 35-44

It is given that P(A) = .15, P(B) = .12, P(C) = .28, and P(D) = .21. Four persons 18 years or older are chosen at random, forming independent events.

a. $P(AAAA) = [P(A)]^4 = (.15)^4 = .0005$

b. $P(BBBB) = [P(B)]^4 = (.12)^4 = .0002$

c. Since the event of interest will occur if either the first, second, third, or fourth person is under 25, and the remaining people are 25 or older, there are four simple events in the event of interest, each of which have probability (.15)(.85)(.85)(.85). Therefore, the probability of interest is
$$4P(A)[P(\bar{A})]^3 = 4(.15)(.85)^3 = .3685$$

3.43

Define the following events:

G: shopper feels government must solve the garbage problem
M: shopper feels the manufacturers must solve the garbage problem
C: shopper feels the consumers must solve the garbage problem

a. $P(GG) = (.18)^2 = .0324$

b. $P(GC) + P(CG) = 2(.18)(.30) = .108$

c. $P(\bar{M}\bar{M}) = (1 - .18)^2 = (.82)^2 = .6724$

3.44

a. The number of bank failures is a discrete random variable, taking the values 0, 1, 2....

b. The floor space area is a continuous random variable, taking values greater than 0.

c. The number of people awaiting treatment is a discrete random variable, taking the values 0, 1, 2....

d. The total number of points scored is a discrete random variable, taking the values 0, 1, 2....

e. The number of claims received is a discrete random variable, taking the values 0, 1, 2....

3.45

a. $\mu = E(x) = \sum xp(x) = .3 + .6 + .6 + .4 + .25 = 2.15$

b. $\sigma^2 = \sum(x-\mu)^2 p(x)$

$= (-2.15)^2(.05) + (-1.15)^2(.3) + \cdots + (2.85)^2(.05) = 1.5275$

c. Calculate $\sigma = \sqrt{1.5275} = 1.236$. Then

$\mu \pm 2\sigma = 2.15 \pm 2.472$ or $-.322$ to 4.622.

The graph of p(x) with the interval $\mu \pm 2\sigma$ superimposed is shown in Figure 3.5.

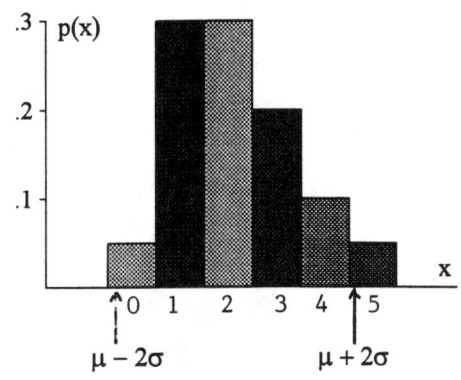

Figure 3.5

d. Referring to the probability distribution of x,

$$P[(\mu - 2\sigma) < x < (\mu + 2\sigma)] = P[-.322 < x < 4.622] = 1 - p(5) = 1 - .05 = .95.$$

3.46 Define the event F to be a patient who fails to pay his bill and is eventually forgiven. Then $P(F) = .30$, and four patients are chosen randomly and independently.

a. $P(FFFF) = (.3)^4 = .0081$

b. P[one of four patients will be forgiven]

$$= P(F\overline{F}\overline{F}\overline{F}) + P(\overline{F}F\overline{F}\overline{F}) + P(\overline{F}\overline{F}F\overline{F}) + P(\overline{F}\overline{F}\overline{F}F)$$

$$= 4P(F)[P(\overline{F})]^3 = 4(.3)(.7)^3 = .4116$$

c. P[none will be forgiven] $= P(\overline{F}\overline{F}\overline{F}\overline{F}) = [P(\overline{F})]^4 = (.7)^4 = .2401$

3.47 Refer to Exercise 3.46. In order to obtain the probability distribution for x, the number of patients whose bills will have to be forgiven, we need p(x) for x = 0, 1, 2, 3, 4. The probabilities associated with x = 0, 1, 4 were calculated in Exercise 3.46. We need only calculate p(2) and p(3).

$$p(2) = P(FF\overline{F}\overline{F}) + P(\overline{F}FF\overline{F}) + P(F\overline{F}F\overline{F}) + P(F\overline{F}\overline{F}F) + P(\overline{F}F\overline{F}F) + P(\overline{F}\overline{F}FF)$$

$$= 6[P(F)]^2[P(\overline{F})]^2 = 6(.3)^2(.7)^2 = .2646$$

$$p(3) = P(FFF\overline{F}) + P(FF\overline{F}F) + P(F\overline{F}FF) + P(\overline{F}FFF)$$

$$= 4[P(F)]^3 P(\overline{F}) = 4(.3)^3(.7) = .0756$$

The probability distribution and the probability histogram follow.

x	p(x)
0	.2401
1	.4116
2	.2646
3	.0756
4	.0081

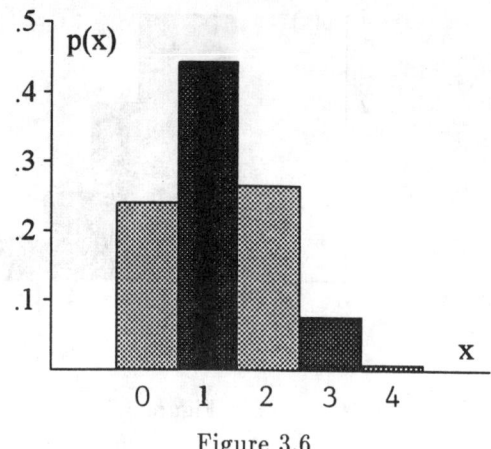

Figure 3.6

3.48

a. Since one of the requirements of a probability distribution is that $\sum_X p(x) = 1$, we need

$$p(3) = 1 - (.1 + .3 + .3 + .1) = 1 - .8 = .2$$

b-c. Each student will obtain a slightly different simulation. The relative frequency histogram should resemble the probability histogram (Figure 3.7) shown below.

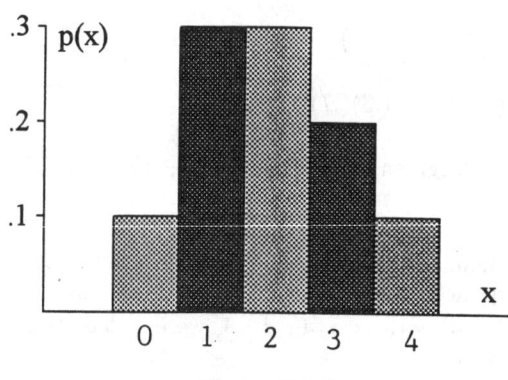

Figure 3.7

3.49

Refer to the solution to Exercise 3.1, where the six simple events in the experiment are given, with $P(E_i) = 1/6$.

a. $S = \{E_1, E_2, E_3, E_4, E_5, E_6\}$ and $P(S) = 1$.

b. $A = \{E_1, E_2, E_3\}$ and $P(A) = 3/6 = 1/2$.

c. $B = \{E_1, E_2\}$ and $P(B) = 2/6 = 1/3$.

d. $C = \{E_4, E_5, E_6\}$ and $P(C) = 3/6 = 1/2$.

e. $AB = \{E_1, E_2\}$ and $P(AB) = 2/6 = 1/3$.

f. AC contains no simple events and $P(AC) = 0$.

g. BC contains no simple events and $P(BC) = 0$.

h. A ∪ B = A and P(A ∪ B) = 1/2.

i. A ∪ C = S and P(A ∪ C) = 1.

j. B ∪ C = {E$_1$, E$_2$, E$_4$, E$_5$, E$_6$} and P(B ∪ C) = 5/6.

3.50

a. From Exercise 3.49, P(AB) = 1/3, while P(AC) = P(BC) = 0. Hence, events AC and BC are mutually exclusive.

b. From Exercise 3.49, P(A) = 1/2, P(B) = 1/3, and P(AB) = 1/3. Then

$$P(A \cup B) = \tfrac{1}{2} + \tfrac{1}{3} - \tfrac{1}{3} = \tfrac{1}{2}$$

c. Using the Multiplicative Rule of Probability,

$$P(A|B) = \frac{P(AB)}{P(B)} = \frac{1/3}{1/3} = 1$$

$$P(A|C) = \frac{P(AC)}{P(C)} = \frac{0}{1/2} = 0$$

d. Since P(A) = $\tfrac{1}{2}$, neither A and B nor A and C are independent events.

3.51

a. If x is the number observed on the upper face of a single die, then p(x) = 1/6 for x = 1, 2, 3, 4, 5, 6.

b.

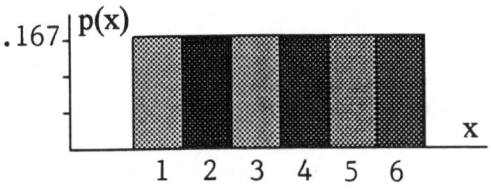

Figure 3.8

c. $\mu = E(x) = \sum x\, p(x) = (1 + \ldots + 5 + 6)\left(\tfrac{1}{6}\right) = 3.5$

$\sigma^2 = \sum (x-\mu)^2 p(x)$

$= \tfrac{1}{6}[(1 - 3.5)^2 + (2 - 3.5)^2 + \cdots + (6 - 3.5)^2]$

$= 2.91667$ and $\sigma = \sqrt{2.91667} = 1.71$

d. Calculate the interval $\mu \pm 2\sigma = 3.5 \pm 3.42$ or $.08$ to 6.92. Since all of the six possible values for x fall in this interval, the probability that x will fall within two standard deviations of the mean is 1.

e. From part d, the exact percentage falling in the interval, 100%, is not too far from the approximate percentage given by the Empirical Rule, 95%.

3.52

In order to find E(N), we must determine the probability distribution for the random variable N. That is, we must know what values N can take and their associated probabilities. It is given that
$$N = 8\pi r^2$$
and the probability distribution for r is given. Hence, since r is positive,

$$P[r = 21] = P[8\pi r^2 = 8\pi(21)^2] = P[N = 11{,}083.539] = .05$$

Similarly, each value of N may be calculated using a particular value of r and

$$P[N = 8\pi r^2] = p(r) \quad \text{for} \quad r = 21, 22, \ldots, 26$$

The probability distribution for N is

N	p(N)
11083.539	.05
12164.247	.20
13295.220	.30
14476.459	.25
15707.963	.15
16989.733	.05

Then
$$E(N) = \sum Np(N) = 554.1769 + 2432.8494 + 3988.5660 + 3619.1147$$
$$+ 2356.1945 + 849.4867$$

$$= 13800.3882$$

3.53

Let y represent the value of the premium that the insurance company charges and let x be the insurance company's gain. There are four possible values for x. If no accident occurs or if an accident results in no damage to the car, the insurance company gains y dollars. If an accident occurs and the car is damaged, the company will gain either $y-15{,}000$ dollars, $y-.6(15{,}000)$ dollars, or $y-.2(15{,}000)$ dollars, depending upon whether the damage to the car is total, 60% of market value, or 20% of market value, respectively. The following probabilities are known.

P[accident occurs] = .15   P[total loss|accident occurs] = .08
P[60% loss|accident occurs] = .12   P[20% loss|accident occurs] = .80

Hence,
$$P[x = y-15{,}000] = P[\text{accident}]P[\text{total loss}|\text{accident}] = .15(.08) = .012$$
Similarly,
$$P[x = y-9000] = .15(.12) = .018 \quad \text{and}$$

$$P[x = y-3000] = .15(.80) = .12$$

The gain x and its associated probability distribution are shown below. Note that p(y) is found by subtraction.

x	p(x)
y − 15000	.012
y − 9000	.018
y − 3000	.12
y	.85

Letting the expected gain equal 0, the value of the premium is obtained.

$$E(x) = \sum xp(x) = .012(y-15,000) + .018(y-9000) + .12(y-3000) + .85y$$

$$E(x) = y - (180 + 162 + 360) = y - 702$$

$$y = \$702$$

3.54

The company will either gain ($15.50−14.80) if the package is delivered on time, or will lose $14.80 if the package is not delivered on time. We assume that, if the package is not delivered within 24 hours, the company does not collect the $15.50 delivery fee. Then the probability distribution for x, the company's gain, is

x	p(x)
.70	.98
−14.80	.02

and

$$E(x) = .70(.98) - 14.80(.02) = .39.$$

The expected gain per package is $0.39.

3.55

Define the following events:

R: shaver repaired S: answering machine repaired
T: shaver thrown away K: answering machine kept unused
 Q: answering machine thrown away

a. $P(\bar{R}) = 1 - P(R) = 1 - .34 = .66$

b. $P(T|\bar{R}) = \dfrac{P(T\bar{R})}{P(\bar{R})} = \dfrac{.34}{.66} = .5152$

c. $P(K) = .2$

d. $P(S|\bar{T}) = \dfrac{P(S\bar{T})}{P(\bar{T})} = \dfrac{.31}{1 - .3} = .4429$

e. $P(R\bar{S}) + P(\bar{R}S) = P(R)P(\bar{S}) + P(\bar{R})P(S) = .34(.69) + .66(.31) = .4392$

3.56

Define the event Y to be a customer who is ready to purchase another PC from Micron. Then $P(Y) = .88$, and four customers are chosen randomly and independently.

a. $P(YYYY) = (.88)^4 = .5997$

b. P[none would purchase another Micron PC] $= P(NNNN) = [P(N)]^4 = (.12)^4 = .0002$

c. P[one of four customers will purchase another Micron PC] =

$$= P(YNNN) + P(NYNN) + P(NNYN) + P(NNNY)$$

$$= 4P(Y)[P(N)]^3 = 4(.88)(.12)^3 = .0061$$

3.57

The numbers in the cells of this two-way table represent the probabilities of the individual simple events involving rating and price range, once they have been divided by 24, the number of brand/models under consideration.

a. $P(E) = \frac{1+3+0+0}{24} = \frac{4}{24} = \frac{1}{6}$

b. P($600 to $1000) = P($600 to $800) + P($800 to $1000)

$$= \frac{1+4+1}{24} + \frac{3+5+1}{24} = \frac{15}{24} = \frac{5}{8}$$

c. $P(\bar{E}) = 1 - \frac{1}{6} = \frac{5}{6}$

d. P($1200 to $1400 and VG) = $\frac{5}{24}$

e. Define E: refrigerator is rated "excellent"
F: refrigerator is $800 to $1000

$$P(E|F) = \frac{P(EF)}{P(F)} = \frac{3/24}{9/24} = \frac{1}{3}$$

f. Define G: refrigerator is $1000 to $1200

$$P(E|G) = \frac{P(EG)}{P(G)} = \frac{0/24}{4/24} = 0$$

CHAPTER 4
Useful Discrete Probability Distributions

4.1 The random variable x is not a binomial random variable since the balls are selected without replacement. For this reason, the probability p of choosing a red ball changes from trial to trial.

4.2 If the sampling in Exercise 4.1 is conducted with replacement, then x is a binomial random variable with n = 2 independent trials, and p = P[red ball] = 3/5, which remains constant from trial to trial.

4.3

a. $3! = 3(2)(1) = 6$

b. $5! = 5(4)(3)(2)(1) = 120$

c. $C_4^6 = \frac{6!}{4!2!} = 15$

d. $C_3^8 = \frac{8!}{3!5!} = 56$

e. $C_2^{10} = \frac{10!}{2!8!} = 45$

f. $C_6^6 = \frac{6!}{6!0!} = 1$

g. $C_0^6 = \frac{6!}{0!6!} = 1$

h. $C_1^9 = \frac{9!}{1!8!} = 9$

i. $C_0^9 = \frac{9!}{0!9!} = 1$

4.4 With n = 5 and three different values of p, the values of $p(x) = C_x^5 p^x q^{5-x}$ are calculated for x = 0, 1, 2, 3, 4, and 5. The values of p(x) are shown in the table below.

	p = .2	p = .5	p = .8
x	p(x)	p(x)	p(x)
0	.32768	.03125	.00032
1	.40960	.15625	.00640
2	.20480	.31250	.05120
3	.05120	.31250	.20480
4	.00640	.15625	.40960
5	.00032	.03125	.32768

The probability histograms are shown in Figure 4.1(a), (b), and (c). Notice that the probability distributions for p = .2 and p = .8 are mirror images.

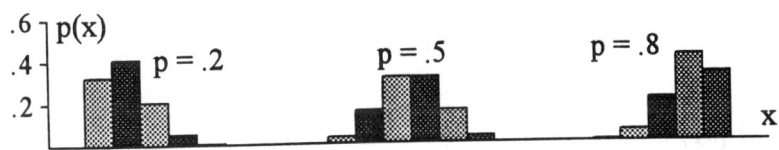

Figure 4.1

4.5 This is similar to Exercise 4.4. With n = 6 and three different values of p, the values of $p(x) = C_x^6 p^x q^{6-x}$ are calculated for x = 0, 1, 2, 3, 4, 5, and 6. The values of p(x) are shown in the table below.

	p = .1	p = .5	p = .9
x	p(x)	p(x)	p(x)
0	.531441	.015625	.000001
1	.352494	.093750	.000054
2	.098415	.234375	.001215
3	.014580	.312500	.014580
4	.001215	.234375	.098415
5	.000054	.093750	.354294
6	.000001	.015625	.531441

The probability histograms are shown in Figure 4.2(a), (b) and (c). Notice that the probability distributions for p = .1 and p = .9 are mirror images. When p is small, the distribution is skewed to the left; when p is large, the distribution is skewed to the right.

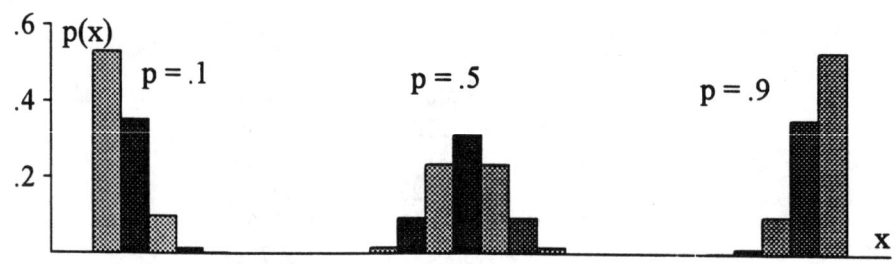

Figure 4.2

4.6 Table 1, Appendix II gives the sum, $\sum_{x=0}^{a} p(x)$, for various values of n and p. The three necessary sums can be found directly in the table, by indexing the proper values of n, p, and a.

a. $\sum_{x=0}^{3} p(x) = .995$ for n = 8 and p = .1

b. $\sum_{x=0}^{7} p(x) = .562$ for n = 12 and p = .6

c. $\sum_{x=0}^{14} p(x) = .788$ for n = 25 and p = .5

4.7

The probability distribution for n = 5 and p = .5 is shown in the following tables (see Exercise 4.4), along with the partial sums, $\sum_{x=0}^{a} p(x)$.

x	p(x)	a	$\sum_{x=0}^{a} p(x)$
0	.03125	0	p(0) = .03125
1	.15625	1	p(0) + p(1) = .18750
2	.31250	2	p(0) + p(1) + p(2) = .50000
3	.31250	3	p(0) + p(1) + p(2) + p(3) = .8125
4	.15625	4	p(0) + p(1) + p(2) + p(3) + p(4) = .96875
5	.03125		

Notice that these sums agree with Table 1, Appendix II, accurate to three decimal places.

4.8

Notice in Exercise 4.7 that the sum for a = 3 contains one more element than does the sum for a = 2, namely, p(3). Hence,

$$p(3) = \sum_{x=0}^{3} p(x) - \sum_{x=0}^{2} p(x) = .812 - .500 = .312 \text{ (from Table 1)}.$$

4.9

The difference between $\sum_{x=0}^{4} p(x)$ and $\sum_{x=0}^{2} p(x)$ is p(3) + p(4) in Table 1. Hence,

$$p(3) + p(4) = .969 - .500 = .469 \text{ (from Table 1)}.$$

From the original probability distribution (see Exercise 4.7),

$$p(3) + p(4) = .3125 + .15625 = .46875.$$

4.10

Use Table 1, Appendix II.

a. P[x < 12] = P[x ≤ 11] = .748

b. P[x ≤ 6] = .610

c. P[x > 4] = 1 − P[x ≤ 4] = 1 − .633 = .367

d. P[x ≥ 6] = 1 − P[x ≤ 5] = 1 − .034 = .966

e. P[3 < x < 7] = P[x ≤ 6] − P[x ≤ 3] = .828 − .172 = .656

4.11

Use Table 1, Appendix II.

a. P[x ≤ 5] = .834   **b.** P[x < 3] = P[x ≤ 2] = .317

c. P[x ≤ 17] = .965   **d.** P[x > 17] = 1 − P[x ≤ 17] = 1 − .965 = .035

e. P[x < 6] = P[x ≤ 5] = .403

4.12

For a binomial random variable x with n identical trials and p = P[success], the mean and standard deviation are $\mu = np$ and $\sigma = \sqrt{npq}$ respectively.

a. $\mu = 1000(.3) = 300;\ \sigma = \sqrt{1000(.3)(.7)} = 14.49$

b. $\mu = 400(.01) = 4;\ \sigma = \sqrt{400(.01)(.99)} = 1.99$

c. $\mu = 500(.5) = 250;\ \sigma = \sqrt{500(.5)(.5)} = 11.18$

d. $\mu = 1600(.8) = 1280;\ \sigma = \sqrt{1600(.8)(.2)} = 16$

4.13

a. $\mu = 100(.01) = 1;\ \sigma = \sqrt{100(.01)(.99)} = .99$

b. $\mu = 100(.9) = 90;\ \sigma = \sqrt{100(.9)(.1)} = 3$

c. $\mu = 100(.3) = 30;\ \sigma = \sqrt{100(.3)(.7)} = 4.58$

d. $\mu = 100(.7) = 70;\ \sigma = \sqrt{100(.7)(.3)} = 4.58$

e. $\mu = 100(.5) = 50;\ \sigma = \sqrt{100(.5)(.5)} = 5$

4.14

The table below shows the values of p and σ (which depends on p) for a given value of n.

p	σ
.01	.99
.30	4.58
.50	5.00
.70	4.58
.90	3.00

The maximum value of p occurs when p = .5.

4.15

The respondent is given four rather than two choices, so that the experiment can result in one of four outcomes: crime, health care, budget deficit, or welfare reform. Hence, each trial <u>does not</u> result in one of two outcomes, but rather four. Therefore, the experiment cannot be a binomial experiment.

4.16

The experiment results in one of two outcomes, satisfied (S) or dissatisfied (F), and there are n = 5000 households (trials) chosen for the experiment. If the n = 5000 households are not randomly chosen, the trials may not be independent. Otherwise, the experiment meets the requirements for a binomial experiment with n = 5000 and x = number of HMO members who are satisfied with their insurance policies.

4.17

Define x to be the number of households that have cable TV. Then p = P[household has cable TV] = .60 and n = 4.

a. $P[x = 4] = C_4^4(.6)^4(.4)^0 = (.6)^4 = .1296$

b. $P[x \geq 1] = 1 - P[x = 0] = 1 - C_0^4(.6)^0(.4)^4 = 1 - .0256 = .9744$

c. $P[x = 1] = C_1^4(.6)^1(.4)^3 = 4(.6)(.064) = .1536$

4.18

The random variable x is approximately binomial, with n = 25 and p = P[CEO is aware of Information Superhighway] = .5. Using Table 1 of Appendix II with n = 25,

a. $P[x = 25] = P[x \leq 25] - P[x \leq 24] = 1 - 1 = 0$

b. $P[x \geq 10] = 1 - P[x \leq 9] = 1 - .115 = .885$

c. $P[x = 10] = P[x \leq 10] - P[x \leq 9] = .212 - .115 = .097$

4.19

Let x be the number of application forms with falsified information and n = 5. Then p = P[application form is falsified] = .35. Since x has a binomial distribution,

$$p(x) = C_x^5(.35)^x(.65)^{5-x}.$$

The probability of interest is

$$P[\text{at least one false application}] = P[x \geq 1] = 1 - P[x = 0]$$

$$= 1 - C_0^5(.35)^0(.65)^5 = 1 - .116 = .884$$

$$P[x \geq 2] = p(2) + p(3) + p(4) + p(5) = 1 - p(0) - p(1)$$

$$= 1 - .116 - C_1^5(.35)^1(.65)^4 = 1 - .116 - .312 = .572.$$

4.20

a. Let x be the number of Japanese who feel that their products are superior to American products. Then x has a binomial distribution with n = 50 and p = P[Japanese feel their products are superior] = .71. That is,

$$p(x) = C_x^{50}(.71)^x(.29)^{50-x}$$

b. If x is defined as the number of Japanese who feel that the U.S. would be the number one economic power in the next century, then x has a binomial distribution with n = 50 and p = P[Japanese feel U.S. will be #1 economic power] = .42. That is,

$$p(x) = C_x^{50}(.42)^x(.58)^{50-x}$$

c. For the random variable x described in part b,

$$\mu = np = 50(.42) = 21 \quad \text{and} \quad \sigma = \sqrt{npq} = \sqrt{50(.42)(.58)} = 3.490$$

d. From part c, we know that $\mu = 21$ and $\sigma = 3.490$. We can determine whether or not the value x = 5 is an unlikely observation by calculating the z-score (Chapter 2) as

$$z = \frac{x - \mu}{\sigma} = \frac{5 - 21}{3.490} = -4.58$$

That is, x = 5 lies more than four standard deviations below the mean. This is a very unlikely occurrence.

4.21

Define x to be the number qualifying for favorable rates. Then p = P[qualify] = .7 and n = 5.

a. $P[x = 5] = p(5) = C_5^5(.7)^5(.3)^0 = (.7)^5 = .16807$ (*Table 1*: $1 - P[x \leq 4] = .168$)

b. $P[x \geq 4] = p(4) + p(5) = C_4^5(.7)^4(.3)^1 + C_5^5(.7)^5(.3)^0$

$= .36015 + .16807 = .52822$ (*Table 1*: $1 - P[x \leq 3] = .528$)

4.22

Define x to be the number of underfilled cans. Then p = P[underfilled can] = .5 and n = number of cans tested.

a. When n = 3, $P[x = 3] = P[x \leq 3] - P[x \leq 2] = 1 - .875 = .125$
When n = 5, $P[x = 5] = P[x \leq 5] - P[x \leq 4] = 1 - .969 = .031$
When n = 10, $P[x = 10] = P[x \leq 10] - P[x \leq 9] = 1 - .999 = .001$

b. The more cans that are tested, and the longer the sequence of underfilled cans becomes, the more unlikely it becomes that the machine is in fact working as expected. It is more probable that p = P[underfilled can] is greater than .5.

c. When the series of underfilled cans (the run) becomes long enough that the production supervisor deems it very unlikely under the assumption that p = .5 (see part a), the process should be stopped and the machine checked for improper settings, broken parts, etc.

4.23

For this binomial experiment, n = 600, p = .033, and x = number of customers ordering hot dogs. Then

$\mu = np = 600(.033) = 19.8$

$\sigma^2 = npq = 600(.033)(.967) = 19.1466$

and $\sigma = \sqrt{npq} = \sqrt{19.1466} = 4.376$

Since p is so close to 0, this binomial distribution will be skewed to the right, and the Empirical Rule is not appropriate. The results of Tchebysheff's Theorem are given in the following table:

k	$\mu \pm k\sigma$	Interval Boundaries	At least
2	19.8 ± 8.752	11.048 to 28.552	.75
3	19.8 ± 13.128	6.672 to 32.928	.89

The number of fast food customers ordering hot dogs in a sample of 600 should be between 7 and 32.

4.24

Using an electronic calculator, we find $e^{-\mu} = e^{-1.2} = .301194$. Then

$$p(x) = \frac{\mu^x e^{-\mu}}{x!} = \frac{(1.2)^x e^{-1.2}}{x!}$$

a. $p(0) = \frac{(1.2)^0 (.301194)}{0!} = .301194$

b. $p(1) = \frac{(1.2)^1(.301194)}{1!} = .3614328$

c. $P[x \leq 2] = p(0) + p(1) + \frac{(1.2)^2(.301194)}{2!} = .879487$

d. $P[x > 1] = 1 - P[x \leq 1] = 1 - p(0) - p(1)$

$= 1 - .301194 - .3614328 = .3373732$

4.25

This is similar to Exercise 4.24, with $p(x) = \frac{2^x e^{-2}}{x!} = \frac{2^x(.135335)}{x!}$. (Table 2 of cumulative Poisson probabilities can be used for this exercise.)

a. $p(0) = .135335$

b. $P[x > 1] = 1 - P[x \leq 1] = 1 - .135335 - \frac{2^1 e^{-2}}{1!} = .593994$

c. $P[x < 2] = P[x \leq 1] = .135335 + .270671 = .406006$

4.26

Table 2, Appendix II, tabulates the partial sums, $P[x \leq a] = \sum_{x=0}^{a} p(x)$, where

$$p(x) = \frac{\mu^x e^{-\mu}}{x!}$$

Hence, to find $p(x)$ for a particular value of x, say $x = b$, we calculate

$$P[x = b] = P[x \leq b] - P[x \leq (b-1)]$$

Subtracting successive entries in the table gives the desired values of $p(x)$ for $\mu = 1$.

x	0	1	2	3	4	5	6	7
p(x)	.368	.368	.184	.061	.015	.003	.001	.000

The graph is shown in Figure 4.3.

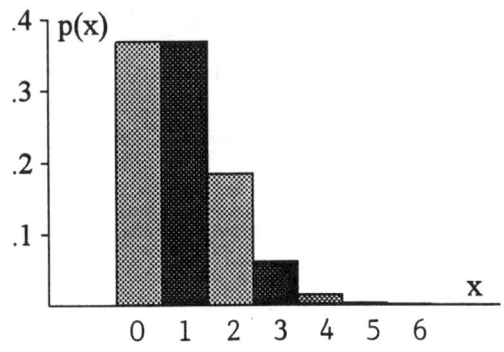

Figure 4.3

4.27

As in Exercise 4.26, successive entries in Table 2 are subtracted from each other to obtain $p(x)$ for $\mu = 3$.

x	p(x)	x	p(x)
0	.050	6	.050
1	.149	7	.022
2	.224	8	.008
3	.224	9	.003
4	.168	10	.001
5	.101		

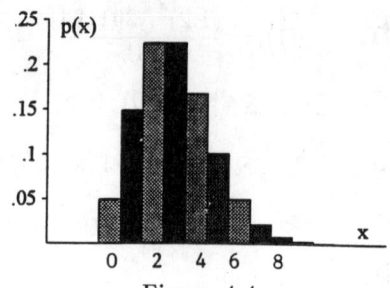

Figure 4.4

4.28

a. Since the probability of interest is of the form $P[x \leq a]$, it can be read directly from Table 2 with $\mu = 3$ and $a = 2$. Then

$$P[x \leq 2] = .423$$

b. $P[x \geq 1] = 1 - P[x = 0] = 1 - .368 = .632$

c. $P[x = 2] = P[x \leq 2] - P[x \leq 1] = .677 - .406 = .271$

4.29

a. From Table 1, with $n = 20$ and $p = .2$,

$$P[x = 2] = P[x \leq 2] - P[x \leq 1] = .206 - .069 = .137$$

b. From Table 2, with $\mu = np = 4$,

$$P[x = 2] \approx .238 - .092 = .146$$

4.30

Consider a random variable x that possesses, approximately, a Poisson distribution, with a population mean of 1. The probability that the patrolman will miss a given location will be

$$p(0) = \frac{1^0 e^{-1}}{0!} = .367879$$

The probabilities of visiting a location once or twice are

$$p(1) = \frac{1^1 e^{-1}}{1!} = .367879 \quad \text{and} \quad p(2) = \frac{1^2 e^{-1}}{2!} = .183940$$

The probability that the patrolman will visit a given location at least once is

$$P[x \geq 1] = 1 - P[x = 0] = 1 - .367879 = .632121$$

4.31

The number of accidents follows an approximate Poisson distribution with $\mu = 3.5$, and

$$p(x) = \frac{(3.5)^x e^{-3.5}}{x!}$$

a. $p(0) = \frac{(3.5)^0 e^{-3.5}}{0!} = .030197$

b. For this Poisson random variable, $\mu = 3.5$ and $\sigma = \sqrt{\mu} = \sqrt{3.5} = 1.871$. The value $x = 7$ lies $(7 - 3.5)/1.871 = 1.871$ standard deviations above the mean. This is not an unlikely occurrence.

c. The value x = 9 lies (9 − 3.5)/1.871 = 2.94 standard deviations above the mean. This is an unlikely occurrence, if in fact $\mu = 3.5$. Perhaps the mean has changed.

4.32

It is given that x has a Poisson distribution with $\mu = 4$. Using Table 2 with $\mu = 4$,

a. $P[x = 1] = P[x \leq 1] - P[x \leq 0] = .092 - .018 = .074$ and $P[x \leq 1] = .092$

b. $P[x > 9] = 1 - P[x \leq 9] = 1 - .992 = .008$. Hence, it is unlikely that x will exceed 9.

4.33

Let x be the number of telephone inquiries during a particular period of time. Then x has a Poisson distribution with μ = average number of inquiries in the designated period of time.

a. For a two-hour time period, $\mu = 4/4 = 1$ (the office averages 4 inquiries per eight-hour day). Then
$$P[x = 0] = \frac{1^1 e^{-1}}{1!} = e^{-1} = .368$$

b. For an eight-hour time period, $\mu = 4$. Then, using Table 2,

$$P[x \geq 5] = 1 - P[x \leq 4] = 1 - .629 = .371$$

4.34

It is given that x has a Poisson distribution with $\mu = 5$. Using Table 2 with $\mu = 5$,

a. $P[x = 2] = P[x \leq 2] - P[x \leq 1] = .125 - .040 = .085$ and $P[x \leq 2] = .125$

b. $P[x > 10] = 1 - P[x \leq 10] = 1 - .986 = .014$. Hence, it is unlikely that x will exceed 10.

4.35

The formula for p(x) is

$$p(x) = \frac{C_x^3 \, C_{2-x}^7}{C_2^{10}} \qquad \text{for } x = 0, 1, 2.$$

a. $p(0) = \frac{C_0^3 \, C_2^7}{C_2^{10}} = \frac{21}{45}$ $\qquad p(1) = \frac{C_1^3 \, C_1^7}{C_2^{10}} = \frac{21}{45}$ $\qquad p(2) = \frac{C_2^3 \, C_0^7}{C_2^{10}} = \frac{3}{45}$

b. The graph of p(x) is shown in Figure 4.5.

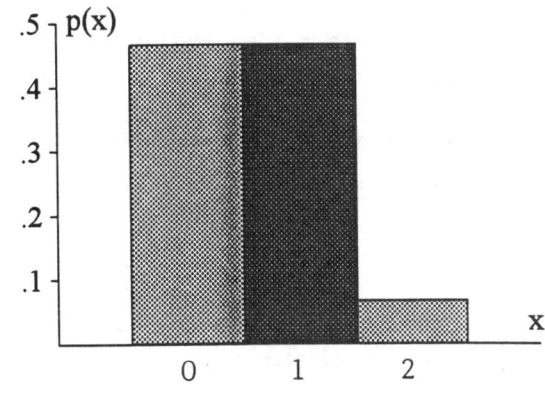

Figure 4.5

4.36

This is similar to Exercise 4.35 with $p(x) = \dfrac{C_x^3 C_{3-x}^{17}}{C_3^{20}}$ for $x = 0, 1, 2$ and 3.

a. $p(0) = \dfrac{C_0^3 C_3^{17}}{C_3^{20}} = \dfrac{680}{1140} = .5965$ $\qquad p(1) = \dfrac{C_1^3 C_2^{17}}{C_3^{20}} = \dfrac{408}{1140} = .3579$

$p(2) = \dfrac{C_2^3 C_1^{17}}{C_3^{20}} = \dfrac{51}{1140} = .0447$ $\qquad p(2) = \dfrac{C_2^3 C_1^{17}}{C_3^{20}} = \dfrac{1}{1140} = .0009$

b. The graph of p(x) is shown in Figure 4.6.

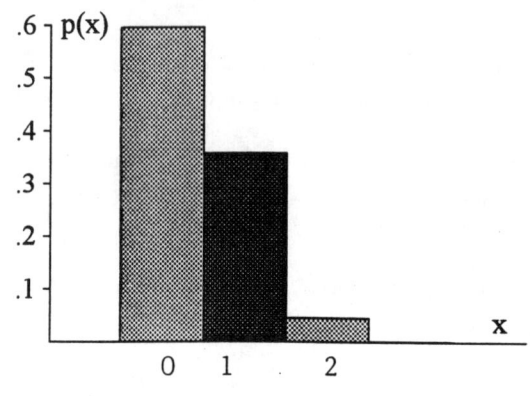

Figure 4.6

4.37

For the random variable described in Exercise 4.36, $N = 20$, $n = 3$, and $r = 3$. Then

$$\mu = n\left(\frac{r}{N}\right) = 3\left(\frac{3}{20}\right) = .45$$

$$\sigma^2 = n\left(\frac{r}{N}\right)\left(\frac{N-r}{N}\right)\left(\frac{N-n}{N-1}\right) = 3\left(\frac{3}{20}\right)\left(\frac{17}{20}\right)\left(\frac{17}{19}\right) = \frac{2601}{7600} = .3422$$

and

$$\sigma = \sqrt{.3422} = .5850$$

Calculate $\mu \pm 2\sigma = .45 \pm 1.17$ or $-.72$ to 1.62. The probability of interest is then

$$P[-.72 < x < 1.62] = p(0) + p(1) = .5965 + .3579 = .9544$$

4.38

For this exercise, $N = 20$, $n = 10$, $r = 5$, and $p(x) = \dfrac{C_x^5 C_{10-x}^{15}}{C_{10}^{20}}$ where x is the number of "best engineers" chosen. In order that all five "best engineers" are chosen in the ten selected, we must select the five best and five others. Hence, the probability of interest is

$$p(5) = \dfrac{C_5^5 C_{10-5}^{15}}{C_{10}^{20}} = \dfrac{10(9)(8)(7)(6)}{20(10)(18)(17)(16)} = \dfrac{21}{1292} = .0163$$

4.39

For this exercise, $N = 20$, $n = 5$, $r = 4$, and $p(x) = \dfrac{C_x^4 C_{5-x}^{16}}{C_5^{20}}$ where x is the number of defectives chosen. The probability of interest is:

$$P[\text{reject lot}] = P[x > 1] = 1 - P[x \leq 1]$$
$$= 1 - \frac{C_0^4 C_5^{16}}{C_5^{20}} - \frac{C_1^4 C_4^{16}}{C_5^{20}} = .2487$$

4.40

For this exercise, $N = 10$, $n = 4$, $r = 3$, and $p(x) = \dfrac{C_x^3 C_{4-x}^7}{C_4^{10}}$ where x is the number of CEOs whose companies recently restructured.

a. $P[x = 3] = \dfrac{C_3^3 C_{4-3}^7}{C_4^{10}} = \dfrac{7}{210}$

b. $P[x = 0] = \dfrac{C_0^3 C_{4-0}^7}{C_4^{10}} = \dfrac{35}{210}$

c. $P[x \geq 1] = 1 - P[x = 0] = 1 - \dfrac{35}{210} = \dfrac{175}{210}$

4.41

See Section 4.2 in the text.

4.42

The random variable x is defined to be the number of heads observed when a coin is flipped three times. Then $p = P[\text{success}] = P[\text{head}] = 1/2$, $q = 1 - p = 1/2$, and $n = 3$. The binomial formula yields the following results.

a.
$P[x = 0] = p(0) = C_0^3 (1/2)^0 (1/2)^3 = 1/8$
$P[x = 1] = p(1) = C_1^3 (1/2)^1 (1/2)^2 = 3/8$
$P[x = 2] = p(2) = C_2^3 (1/2)^2 (1/2)^1 = 3/8$
$P[x = 3] = p(3) = C_3^3 (1/2)^3 (1/2)^0 = 1/8$

b. The associated probability histogram is found in Figure 4.7.

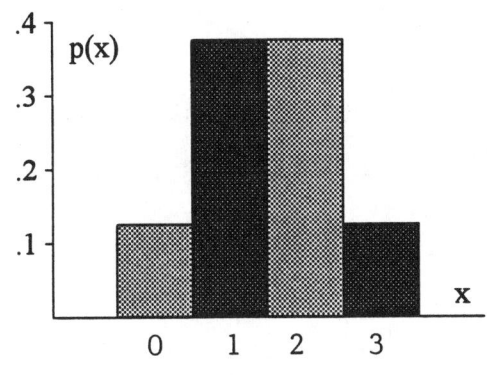

Figure 4.7

c. $\mu = np = 3(1/2) = 1.5$

$\sigma = \sqrt{npq} = \sqrt{3(1/2)(1/2)} = .866$

d. The desired intervals are

$\mu \pm \sigma = 1.5 \pm .866$ or .634 to 2.366

$$\mu \pm 2\sigma = 1.5 \pm 1.732 \quad \text{or} \quad -.232 \text{ to } 3.232$$

The values of x that fall in the first interval are $x = 1$ and $x = 2$, and the fraction of measurements in this interval will be $3/8 + 3/8 = 3/4$. The second interval encloses all four values of x, and thus the fraction of measurements within two standard deviations of the mean will be 1, or 100%. These results are consistent with both Tchebysheff's Theorem and the Empirical Rule.

4.43

The partial sum is read directly from Table 1, indexing the proper values.

a. .851 **b.** .829 **c.** .011

4.44

From Table 1, $p(C) = \sum_{x=0}^{C} p(x) - \sum_{x=0}^{C-1} p(x)$. Hence,

a. $p(6) = .618 - .367 = .251$ **b.** $p(5) = .151 - .059 = .092$

c. $p(3) = .411 - .206 = .205$

4.45

Use Table 1, indexing the proper values.

a. $\sum_{x=1}^{10} p(x) = 1 - .349 = .651$ **b.** $\sum_{x=7}^{9} p(x) = .893 - .121 = .772$

c. $\sum_{x=4}^{15} p(x) = 1 - .091 = .909$

4.46

a. $P[x \leq 1] = .737$

b. $P[x > 1] = 1 - P[x \leq 1] = 1 - .737 = .263$

c. $P[x < 1] = P[x \leq 0] = .328$

d. $P[x = 1] = P[x \leq 1] - P[x \leq 0] = .737 - .238 = .409$

e. $P[x = 1] = P[x \leq 1] - P[x \leq 0] = .376 - .107 = .269$

4.47

a. $P[x \leq 4] = .377$

b. $P[x < 4] = P[x \leq 3] = .172$

c. $P[x > 4] = 1 - P[x \leq 4] = 1 - .377 = .623$

d. $P[x = 4] = P[x \leq 4] - P[x \leq 3] = .377 - .172 = .205$

4.48

Define x to be the number of days on which the train is more than thirty-five minutes late. Then x has a binomial distribution with $p = .5$ and $n = 5$.

$$P[x = 5] = p(5) = C_5^5(.5)^5(.5)^0 = (.5)^5 = .03125$$

The probability that the train is late more than three times out of five is

$$P[x > 3] = p(4) + p(5) = C_4^5(.5)^4(.5)^1 + C_5^5(.5)^5(.5)^0 = .18750$$

4.49

For this exercise, $N = 10$, $n = 5$, $r = 4$, and $p(x) = \dfrac{C_x^4 \, C_{5-x}^6}{C_5^{10}}$ where x is the number of defectives chosen. The probability of interest is:

$$p(0) = \dfrac{C_0^4 \, C_5^6}{C_5^{10}} = \dfrac{6}{252} = .0238.$$

4.50

Let x be the number favoring unionization. Then x has a hypergeometric distribution with $N = 80$, $n = 20$, $r = 45$.

a. $\mu = n\left(\dfrac{r}{N}\right) = 20\left(\dfrac{45}{80}\right) = 11.25$

b. $\sigma^2 = n\left(\dfrac{r}{N}\right)\left(\dfrac{N-r}{N}\right)\left(\dfrac{N-n}{N-1}\right) = 20\left(\dfrac{45}{80}\right)\left(\dfrac{35}{80}\right)\left(\dfrac{60}{79}\right) = 3.73813$

and

$$\sigma = \sqrt{3.73813} = 1.933$$

c. The value $x = 9$ lies $(9 - 11.25)/1.933 = -1.16$ standard deviations below the mean. It is not an unlikely occurrence.

4.51

Use Table 1 with $n = 10$ and $p = .3$.

a. $P[x \le 1] = .149$

b. $P[x > 5] = 1 - P[x \le 5] = 1 - .953 = .047$

c. $P[x = 3] = P[x \le 3] - P[x \le 2] = .650 - .383 = .267$

4.52

If there are 10,000,000 calls per year, then $n = 10{,}000{,}000$ and $p = 1/3$.

a. Let x be the number of nonbusiness calls in the 10,000,000. Then

$$E(x) = np = 10{,}000{,}000(1/3) = 3{,}333{,}333.33$$

b. $\sigma = \sqrt{npq} = \sqrt{10{,}000{,}000(1/3)(2/3)} = 1490.71$

c. The value $x = 3{,}300{,}000$ lies $(3{,}300{,}000 - 3{,}333{,}333.33)/1490.71 = -22.36$ standard deviations below the mean. By either Tchebysheff's Theorem or the Empirical Rule, this is a highly unlikely occurrence.

4.53

Define x to be the number of Hondas that were made in North America. Then x has a binomial distribution with $p = .6$ and $n = 15$. Use Table 1 in Appendix II.

a. $P[x \le 8] = .390$

b. $P[8 \le x \le 12] = P[x \le 12] - P[x \le 7] = .973 - .213 = .760$

c. $P[x = 10] = P[x \leq 10] - P[x \leq 9] = .783 - .597 = .186$

4.54

Let x be the number of employees who have Hondas made in North America. Then x has a hypergeometric distribution with N = 15, n = 5, r = 8. The probability of interest is:

$$P[x = 5] = p(5) = \frac{C_5^8 \, C_0^7}{C_5^{15}} = \frac{56}{3003} = .0186$$

4.55

Assuming that the sample of twenty customers is a random and independent sample, we have a binomial experiment with p = .9 and q = .1. Success is here defined to be "no claim" and failure will be "having a claim". We are asked for the probability that at least two of these customers will have claims against the guarantee, where x is equal to the number of people who have no claim. That is, we are asked for the porbability that there are two or more failures. This is equivalent to having eighteen or less successes. Referring to Table 1, Appendix II, with n = 20, p = .9 and a = 18, we have $P[x \leq 18] = .608$.

4.56

Let x be defined as the number of successes among the ten new textbooks. The probability of success is given to be one out of ten; that is, p = .1 and q = .9.

a. $P[x = 1] = C_1^{10}(.1)^1(.9)^9 = .3874$

b. $P[\text{at least one success}] = 1 - P[x = 0] = 1 - C_0^{10}(.1)^0(.9)^{10} = .6513$

c. $P[\text{at least two successes}] = P[x \geq 2] = 1 - P[x \leq 1]$

$1 - C_0^{10}(.1)^0(.9)^{10} - C_1^{10}(.1)^1(.9)^9 = 1 - .7362 = .2638$

4.57

Assuming that the properties of a binomial experiment are satisfied, we have n = 20, p = .2, and q = .8. Define x as the number of homes heated by natural gas.

a. Using Table 1 in Appendix II with a = 0, $P[x = 0] = .012$.
Alternatively,
$$p(0) = C_0^{20}(.2)^0(.8)^2 = (.8)^{20} = .0115292.$$

b. $P[\text{no more than 4}] = P[x \leq 4] = .630$.

c. The hypothesis of independence may not be satisfied. Even though the city block was chosen randomly, the homes within the block will not necessarily be independent of each other. This violates property number four of the definition of the binomial experiment.

4.58

It is given that n = 2300 and p = .35.

a. $E(x) = np = 2300(.35) = 805$ **b.** $\sigma = \sqrt{npq} = \sqrt{523.25} = 22.875$

c. Calculate $\mu \pm 2\sigma = 805 \pm 45.749$ or 759.21 to 850.749.

d. The value x = 249 does not fall within two standard deviations of the mean. In fact, it lies $(249 - 805)/22.875 = -24.306$ standard deviations below the mean. The company's rate is nowhere as high as the overall rate.

4.59

Use the binomial tables in Appendix II, indexing n = 5, a = 1 in the first case and n = 25, a = 5 in the second (see Figure 4.8).

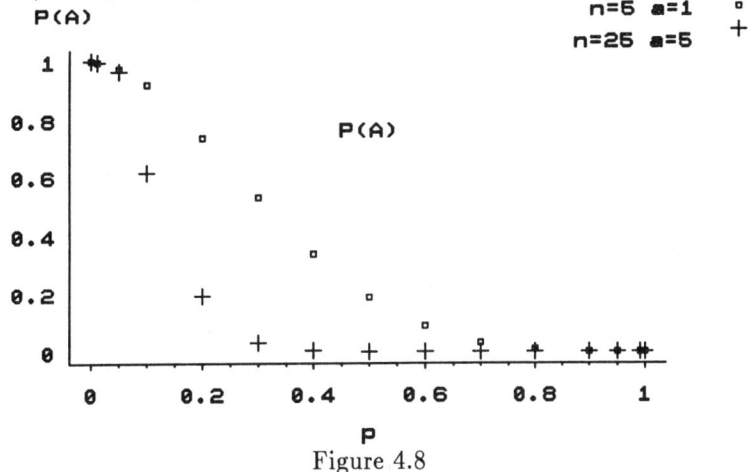

Figure 4.8

b. If the fraction defective in the lot ranges from p = 0 to p = .10, the seller would want the probability of acceptance in this interval to be as high as possible. Hence, he would choose the second plan.

c. If the buyer wishes to be protected against accepting lots with the fraction defective greater than .3, he would want the probability of acceptance when p is greater than .3 to be as small as possible. Thus, he would also choose the second plan.

4.60

Using the binomial tables and indexing n = 20, a = 1 and the appropriate value of p, the probabilities of accepting the lot are obtained.

p	P[acceptance]
.00	1.000
.01	.983
.05	.736
.10	.392
.20	.069
1.00	.000

A sketch of the operating characteristic curve for this plan is shown in Figure 4.9.

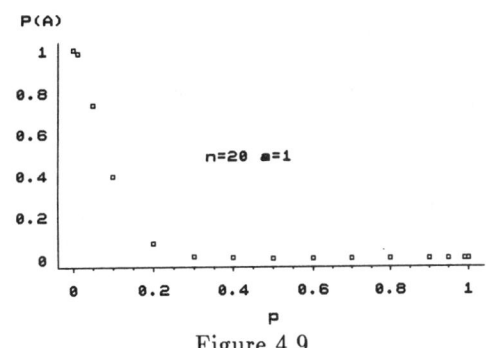

Figure 4.9

4.61

The MINITAB printout given in the exercise gives $P[x \leq k]$, where x, the number of accidents per 100,000 full-time employees, has a Poisson probability distribution with $\mu = 7.4$.

a. $P[x = 0] = .0006$

b. $P[x \leq 7] = .5393$

c.   Since $\mu = 7.4$, $\sigma = \sqrt{\mu} = \sqrt{7.4} = 2.72$ and $\mu \pm 2\sigma = 7.4 \pm 5.44$, which is the interval from 1.96 to 12.84. The probability that x will lie in this interval is

$$P[2 \leq x \leq 12] = P[x \leq 12] - P[x \leq 1] = .9609 - .0051 = .9558$$

This is fairly close to the percentage suggested by the Empirical Rule, which says that, for mound-shaped data, approximately 95% of the measurements should fall in this interval.

4.62

The random variable x has a binomial distribution with n = 20 and p = .5. From Table 1 in Appendix II,

$$P[x \geq 15] = 1 - P[x \leq 14] = 1 - .979 = .021$$

4.63

Refer to Exercise 4.62.

a. $E(x) = np = 500(.5) = 250$ and $\sigma = \sqrt{npq} = \sqrt{500(.5)(.5)} = 11.18034$

b.   The value $x = 280$ lies $(280 - 250)/11.18034 = 2.68$ standard deviations above the mean. It is an unlikely occurrence if in fact the two machines are equally desirable.

4.64

The number x of fatal aircraft accidents in a given year has approximately a Poisson distribution, whose mean is assumed to be $\mu = 3.5$.

a. $P[x \geq 11] = 1 - P[x \leq 10] = 1 - .999 = .001$ from Table 2, Appendix II.

b.   The mean and standard deviation of x are calculated to be

$$\mu = 3.5 \text{ and } \sigma = \sqrt{3.5} = 1.87$$

Hence, the value $x = 11$ lies $(11 - 3.5)/1.87 = 4.01$ standard deviations above the mean. This is a very unlikely occurrence, assuming that the mean is, in fact, still 3.5. Perhaps the mean number of accidents has increased.

4.65

The number x of Americans who think that the top U.S. goal should be preventing the spread of nuclear weapons has a binomial distribution with n = 600 and p = .7.

a.   For this binomial random variable,

$$\mu = np = 600(.7) = 420 \text{ and } \sigma = \sqrt{npq} = \sqrt{600(.7)(.3)} = 11.225$$

b.   The value $x = 365$ lies z-score $= (365 - 420)/11.225 = -4.90$ standard deviations below the mean. This is a very unlikely occurrence, assuming that the 70% figure is correct.

c. Based on the results of part b, we would suspect that the 70% figure is not correct, but may in fact be too high.

4.66

The number x of company audits containing substantial errors has a binomial distribution with n = 10 and p = .9. Use Table 1 in Appendix II.

a. $P[x = 9] = P[x \leq 9] - P[x \leq 8] = .651 - .264 = .387$

b. $P[x \leq 9] = .651$

c. $P[x \geq 9] = 1 - P[x \leq 8] = 1 - .264 = .736$

4.67

The retailer is using a lot acceptance sampling plan with n = 5 and a = 1. Suppose that p = P[observe a defective] = 20/200 = .1.

a. $P[\text{acceptance}] = P[\text{observe 0 or 1 defective}] = C_0^5(.1)^0(.9)^5 + C_1^5(.1)^1(.9)^4$

$= .59049 + .32805 = .91854$

b. It is necessary to determine

P[observing 1 defective when lot is accepted]

= P[observing 1 defective when 0 or 1 defect has been observed]

$$= \frac{P[\text{observing 1 defective}]}{P[\text{observing 0 or 1 defective}]}$$

using the laws of conditional probability. Now,

$P[\text{observing 1 defective}] = C_1^5(.1)^1(.9)^4 = .32805$ and

$P[\text{observe 0 or 1 defective}] = .91854$ from part a. Hence,

$P[\text{observing 1 defective when lot is accepted}] = \frac{.32805}{.91854} = .35714.$

4.68

Let x be the number of errors per cashier per day. Then x has a Poisson distribution with $\mu = 1.5$ and $\sigma = \sqrt{1.5} = 1.225$. The value x = 4 lies $(4 - 1.5)/1.225 = 2.04$ standard deviations above the mean. It is a somewhat unlikely occurrence.

4.69

Since $(q + p)^2 = q^2 + 2pq + p^2$, we can write

$(q + p)^3 = (q^2 + 2pq + p^2)(p + q)$

$= pq^2 + 2p^2q + p^3 + q^3 + 2pq^2 + p^2q$

$= q^3 + 3pq^2 + 3p^2q + p^3$

Using $p(x) = C_x^n p^x q^{n-x}$ for x = 0, 1, 2, 3 and n = 3, we have

$$p(0) = C_0^3 p^0 q^3 = q^3 \qquad p(2) = C_2^3 p^2 q = 3p^2 q$$

$$p(1) = C_1^3 p^1 q^2 = 3pq^2 \qquad p(3) = C_3^3 p^3 q^0 = p^3$$

Note the correspondence in terms for the two.

4.70

a. Let S denote a success and F denote a failure. The experiment consists of triples, the outcome of the first, second and third trials.

E_1: SSS E_5: FSS
E_2: SSF E_6: FSF
E_3: SFS E_7: FFS
E_4: SFF E_8: FFF

b-c. Since the trials are independent, with $P(S) = p$ and $P(F) = 1 - p = q$,

$$P(E_1) = P(SSS) = [P(S)]^3 = p^3$$

$$P(E_2) = P(SSF) = [P(S)]^2 P(F) = p^2 q$$

and so on. The probabilities $P(E_i)$ are shown below along with the appropriate values of x for each simple event.

Simple Event	$P(E_i)$	x
E_1: SSS	p^3	3
E_2: SSF	$p^2 q$	2
E_3: SFS	$p^2 q$	2
E_4: SFF	pq^2	1
E_5: FSS	$p^2 q$	2
E_6: FSF	pq^2	1
E_7: FFS	pq^2	1
E_8: FFF	q^3	0

d. The probability distribution for x is then

$$p(0) = P[x = 0] = P(E_8) = q^3$$

$$p(1) = P[x = 1] = P(E_4) + P(E_6) + P(E_7) = 3pq^2$$

$$p(2) = P[x = 2] = P(E_2) + P(E_3) + P(E_5) = 3p^2 q$$

$$p(3) = P[x = 3] = P(E_1) = p^3$$

e. Refer to Exercise 4.69 and note the correspondence.

4.71

a. Using the binomial tables and indexing $n = 10$, $a = 0$ and the appropriate value of p, the probabilities of accepting the lot are obtained.

p	P[acceptance]
.00	1.000
.05	.599
.10	.349
.20	.107
.30	.028
1.00	.000

b. A sketch of the operating characteristic curve for this plan is shown in Figure 4.10.

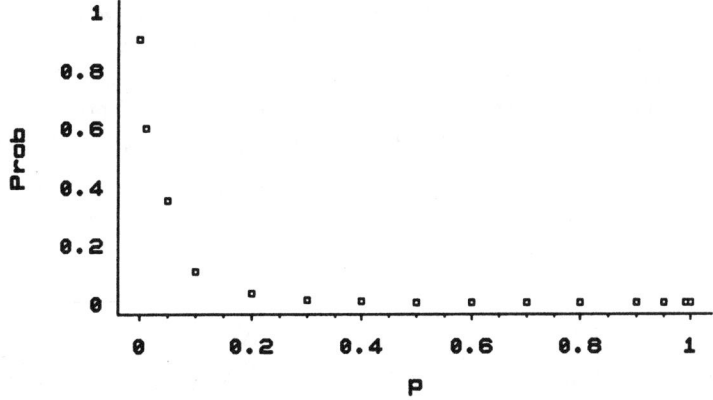

Figure 4.10

c. The probability of accepting lots will not be very high unless the proportion of defective nightstands is very low. Even with p as low as .05, the probability of acceptance is only .599. Unless p is .01 or less (acceptance probability = .904), most of the lots will be sent back.

CHAPTER 5
The Normal and Other Continuous Probability Distributions

5.1

The first few exercises are designed to provide practice for the student in evaluating areas under the normal curve. The following notes may be of some assistance.

(1) Table 3, Appendix II tabulates the area under a standard normal curve from the mean, $z = 0$, to a specified value of z.

(2) Because of the symmetry of the normal distribution, and since the total area under the curve is one, the total area lying on one side of the mean will be .5. Thus, in order to calculate a "tail area," such as the one shown in Figure 5.1, the value $z = z_0$ will be indexed in Table 3, and the area that is obtained will be subtracted from .5. Denote the area obtained by indexing $z = z_0$ in Table 3 by $A(z_0)$ and the desired area by A. Then, in the above example, $A = .5 - A(z_0)$.

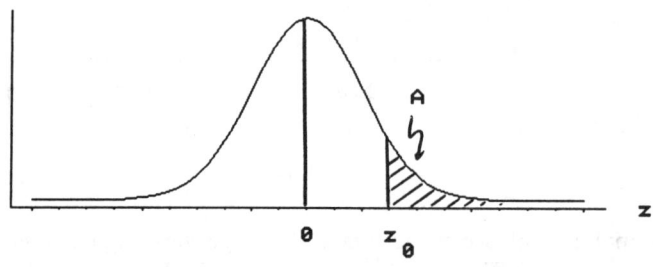

Figure 5.1

(3) Note that z, similar to x, is actually a random variable that may take on an infinite number of values, both positive and negative. However, since the standardized normal curve is symmetric about a mean of zero, a left-hand area (i.e., an area corresponding to a negative value of z) may be evaluated by indexing the positive value in Table 3.

a. It is necessary to find the area between $z = 0$ and $z = 1.6$.

$$A = A(1.6) = .4452$$

b. The area between $z = 0$ and $z = 1.83$ is $A = A(1.83) = .4664$.

5.2

a. $A = A(.90) = .3159$

b. $A = A(-.90) = A(.90) = .3159$

5.3

a. $A(-1.3) + A(1.8) = .4032 + .4641 = .8673$

b. Refer to Figure 5.2. The area between $z = 1.2$ and $z = 0$ is $A(1.2)$, while the area between .6 and 0 is $A(.6)$. The area of interest is the shaded area,

$$A = A(1.2) - A(.6) = .3849 - .2257 = .1592$$

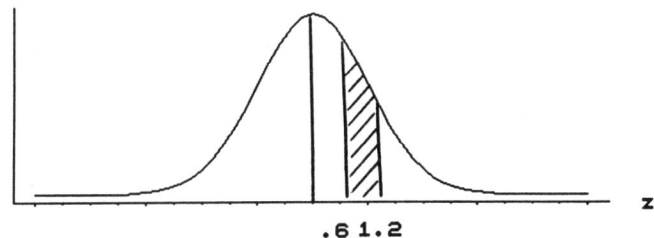

Figure 5.2

5.4

a. $A = A(-1.4) + A(1.4) = 2\,A(1.4) = 2(.4192) = .8384$

b. $A = A(-2.0) + A(2.0) = 2\,A(2.0) = 2(.4772) = .9544$

c. $A = A(-3.0) + A(3.0) = 2\,A(3.0) = 2(.4987) = .9974$

5.5

a. $A = A(-1.43) + A(.68) = .4236 + .2517 = .6753$

b. $A = A(1.74) - A(.58) = .4591 - .2190 = .2401$

c. $A = A(-1.55) - A(-.44) = .4394 - .1700 = .2694$

5.6

Now we are asked to find the z value corresponding to a particular area. We need to find a z_0 such that $P[z < z_0] = .025$. This is equivalent to finding an indexed area of $.5 - .025 = .475$. Using Table 3, we find that $A(1.96) = .475$. Therefore, $z_0 = 1.96$ is the desired z value (see Figure 5.3).

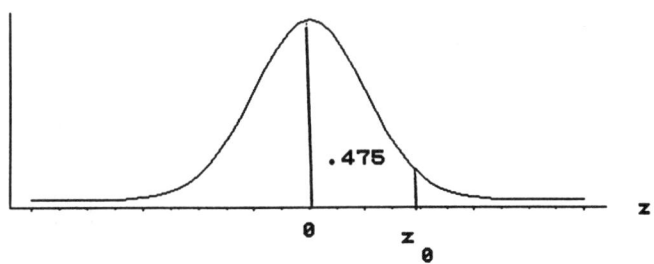

Figure 5.3

5.7

We need to find a z_0 such that $P[z < z_0] = .9251$. Since the area under the entire distribution is equal to 1, we are finding a z value such that $P[z \geq z_0] = 1 - .9251 = .0749$. Using Table 3, we find a value such that the indexed area is $.5 - .0749 = .4251$. Then $z_0 = 1.44$ is the desired value (see Figure 5.4).

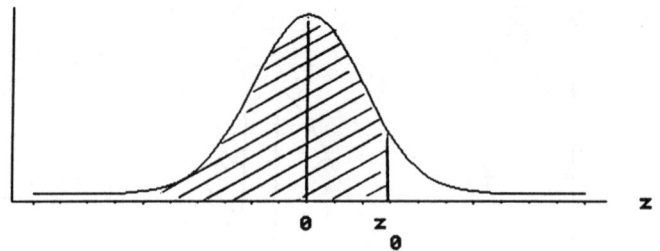

Figure 5.4

5.8

We want to find a z_0 such that $P[z < z_0] = .2981$. Refer to Figure 5.5 where $A_1 = .2981$. Then,

$$A_1 + A_2 = .5$$
$$A_2 = .5 - A_1$$
$$A_2 = .5 - .2981$$
$$A_2 = .2019$$

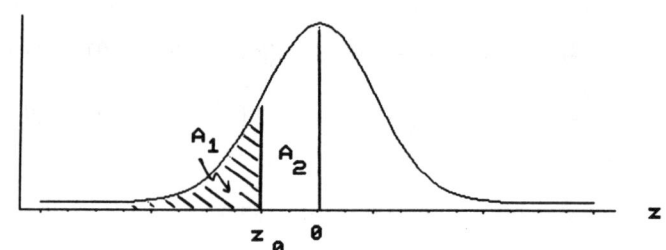

Figure 5.5

Using Table 3, we find that a z value of .53 corresponds to the above area. Since the desired z value is to the left of the mean, $z = -.53$.

5.9

This is similar to previous exercises. The z_0 is in the left-hand side of the curve and corresponds to $A(z_0) = .1985$. Hence, $z_0 = -.52$.

5.10

We want to find a z value such that $P[-z_0 < z < z_0] = .4714$ (see Figure 5.6). That is,

$$A(z_0) + A(-z_0) = .4714$$
$$2A(z_0) = .4714$$
$$A(z_0) = .2357$$

From Table 3, $z_0 = .63$.

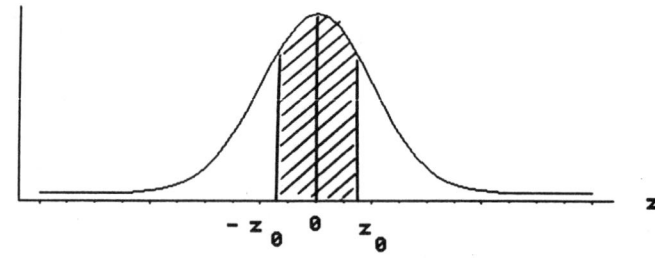

Figure 5.6

5.11

It is given that $P[z < z_0] = A_1 = .0500$. But $A_1 + A_2 = .5000$ by the symmetry of the normal distribution, so that $.0500 + A_1 = .5000$, and $A_2 = A(z_0) = .4500$. The desired value is not tabulated in Table 3 but falls between two tabulated values, .4495 and .4505. Hence, z_0 will lie between -1.64 and -1.65, or $z_0 = -1.645$.

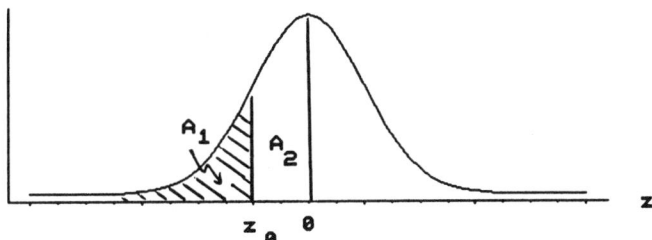

Figure 5.7

This method of evaluation is called "linear interpolation," which is performed as follows:

(1) The difference between two entries in the table is called a "tabular difference." Interpolation is accomplished by taking appropriate portions of this difference.

(2) Let P_0 be the probability associated with z_0 (i.e., $P_0 = A(z_0)$) and let P_1 and P_2 be the two tabulated probabilities with corresponding z values, z_1 and z_2. Consider

$$\frac{P_0 - P_1}{P_2 - P_1}$$

which is the proportion of the distance from P_1 to P_0.

(3) Multiply

$$\frac{P_0 - P_1}{P_2 - P_1}(z_2 - z_1)$$

to obtain a corresponding proportion for the z values and add this value to z_1. This value is the desired z_0. Thus, in this case,

$$\frac{P_0 - P_1}{P_2 - P_1} = \frac{.4500 - .4495}{.4505 - .4495} = \frac{.0005}{.0010} = \frac{1}{2}$$

and

$$z_0 = z_1 + \frac{P_0 - P_1}{P_2 - P_1}(z_2 - z_1) = -1.64 + (\tfrac{1}{2})[-1.65 - (-1.64)]$$

$$= -1.64 + (\tfrac{1}{2})(.01) = -1.645$$

5.12

It is given that $P[-z_0 < z < z_0] = .9000$. That is,

$A(z_0) + A(-z_0) = .9000$

$2A(z_0) = .9000$

$A(z_0) = .4500$

From Exercise 5.11, $z_0 = 1.645$.

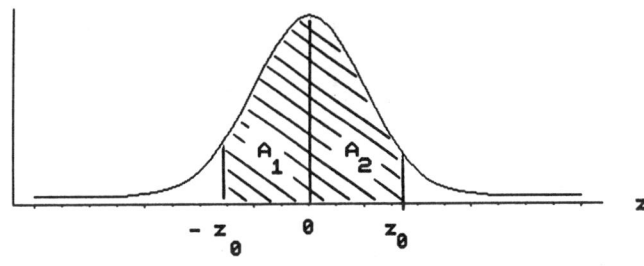

Figure 5.8

5.13

Refer to Figure 5.8 and consider

$$P[-z_0 < z < z_0] = A_1 + A_2 = 2A(z_0) = .9900$$

Then $A(z_0) = .4950$. Linear interpolation must now be used to determine the value of z_0, which will lie between $z_1 = 2.57$ and $z_2 = 2.58$. Hence,

$$z_0 = z_1 + \frac{P_0 - P_1}{P_2 - P_1}(z_2 - z_1) = 2.57 + \left(\frac{.4950 - .4949}{.4951 - .4949}\right)(2.58 - 2.57)$$

$$= 2.57 + (\tfrac{1}{2})(.01) = 2.575$$

5.14

Since $z = (x - \mu)/\sigma$ measures the number of standard deviations an observation lies from its mean, it can be used to standardize any normal random variable x so that Table 3 can be used.

a. Calculate $z = \frac{x - \mu}{\sigma} = \frac{13.5 - 10}{2} = 1.75$. Then

$$P[x > 13.5] = P[z > 1.75] = .5 - .4599 = .0401.$$

This probability is the shaded area in Figure 5.9.

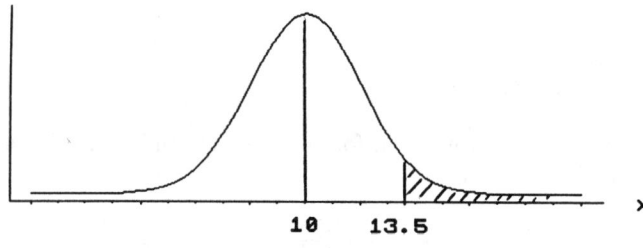

Figure 5.9

b. Calculate $z = \frac{x - \mu}{\sigma} = \frac{8.2 - 10}{2} = -0.9$. Then

$$P[x < 8.2] = P[z < -0.9] = .5 - .3159 = .1841.$$

This probability is the shaded area in Figure 5.10.

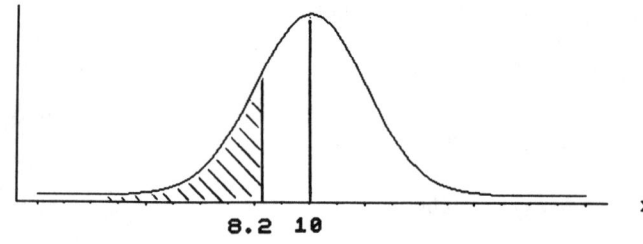

Figure 5.10

c. Calculate $z_1 = \frac{9.4 - 10}{2} = -0.3$ and $z_2 = \frac{10.6 - 10}{2} = 0.3$. Then

$$P[9.4 < x < 10.6] = P[-0.3 < z < 0.3] = 2(.1179) = .2358.$$

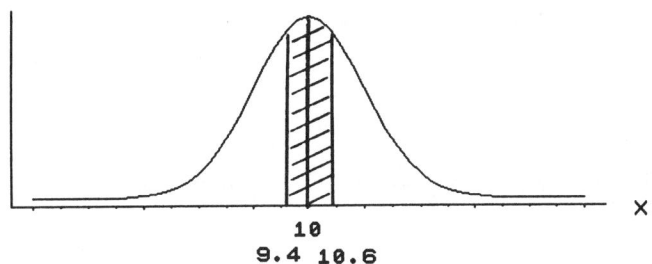

Figure 5.11

5.15

This is similar to Exercise 5.14.

a. Calculate $z_1 = \frac{1.00 - 1.20}{.15} = -1.33$ and $z_2 = \frac{1.10 - 1.20}{.15} = -.67$. Then

$$P[1.00 < x < 1.10] = P[-1.33 < z < -.67] = .4082 - .2486 = .1596$$

b. Calculate $z = \frac{x - \mu}{\sigma} = \frac{1.38 - 1.20}{.15} = 1.2$. Then

$$P[x > 1.38] = P[z > 1.2] = .5 - .3849 = .1151.$$

c. Calculate $z_1 = \frac{1.35 - 1.20}{.15} = 1$ and $z_2 = \frac{1.50 - 1.20}{.15} = 2$. Then

$$P[1.35 < x < 1.50] = P[1 < z < 2] = .4772 - .3413 = .1359$$

5.16

It is given that x is normally distributed with $\sigma = 2$ but with unknown mean μ, and that $P[x > 7.5] = .8023$. This random variable has a normal distribution shown in Figure 5.12.

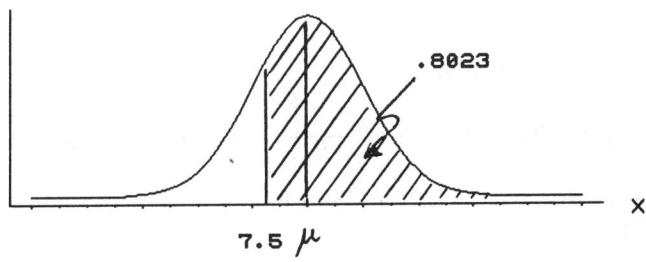

Figure 5.12

Consider the probability $P[x > 7.5]$. In terms of the standard normal random variable z, we can write the z-value corresponding to $x = 7.5$ as

$$z = \frac{7.5 - \mu}{\sigma} = \frac{7.5 - \mu}{2}$$

and $P[x > 7.5] = P[z > \frac{7.5 - \mu}{2}] = .8023$. From Table 3, the value $\frac{7.5 - \mu}{2}$ must be negative, with

$$A\left(\frac{7.5 - \mu}{2}\right) = .3023 \quad \text{or} \quad \frac{7.5 - \mu}{2} = -.85.$$

Solving for μ, $\mu = 7.5 + 2(.85) = 9.2$.

5.17

This is similar to Exercise 5.16. Since $P[x > 14.4] = .3015$, we can write the z-value corresponding to $x = 14.4$ as

$$z = \frac{14.4 - \mu}{1.8}$$

and $P[x > 14.4] = P[z > \frac{14.4 - \mu}{1.8}] = .3015$. From Table 3, the value $\frac{14.4 - \mu}{1.8}$ must be positive, with

$$A\left(\frac{14.4 - \mu}{1.8}\right) = .1985 \quad \text{or} \quad \frac{14.4 - \mu}{1.8} = .52.$$

Solving for μ, $\mu = 14.4 - 1.8(.52) = 13.464$.

5.18

The random variable x is normal with unknown μ and σ. However, it is given that

$$P[x > 4] = P[z > \frac{4 - \mu}{\sigma}] = .9772 \text{ and } P[x > 5] = P[z > \frac{5 - \mu}{\sigma}] = .9332.$$ These probabilities are shown in Figure 5.13.

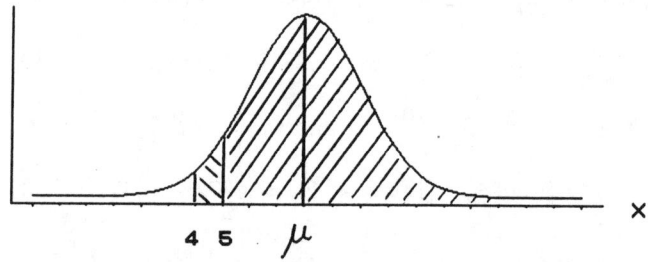

Figure 5.13

The value $\frac{4 - \mu}{\sigma}$ is negative, with

$$A\left(\frac{4 - \mu}{\sigma}\right) = .4772 \quad \text{or} \quad \frac{4 - \mu}{\sigma} = -2 \qquad (i)$$

The value $\frac{5 - \mu}{\sigma}$ is also negative, with

$$A\left(\frac{5 - \mu}{\sigma}\right) = .4332 \quad \text{or} \quad \frac{5 - \mu}{\sigma} = -1.5 \qquad (ii)$$

Equations (i) and (ii) provide two equations in two unknowns that can be solved simultaneously for μ and σ. From (i),

$$\sigma = \frac{\mu - 4}{2}$$

which, when substituted into (ii) yields

$$5 - \mu = -1.5\left(\frac{\mu - 4}{2}\right)$$

$$10 - 2\mu = -1.5\mu + 6$$

$$\mu = 8 \quad \text{and from (i),} \quad \sigma = \frac{8 - 4}{2} = 2.$$

5.19

a. The random variable x, prime interest rate forecasts, is approximately normal with

$\mu = 7.75$ and $\sigma = 1.6$. It is necessary to determine the probability that x exceeds 9. Calculate

$$z = \frac{x - \mu}{\sigma} = \frac{9 - 7.75}{1.6} = .78$$

Then $P[x > 9] = P[z > .78] = .5 - A(.78) = .5 - .2823 = .2177$

b. Calculate $z = \frac{x - \mu}{\sigma} = \frac{6 - 7.75}{1.6} = -1.09$. Then

$$P[x < 6] = P[z < -1.09] = .5 - A(1.09) = .5 - .3621 = .1379$$

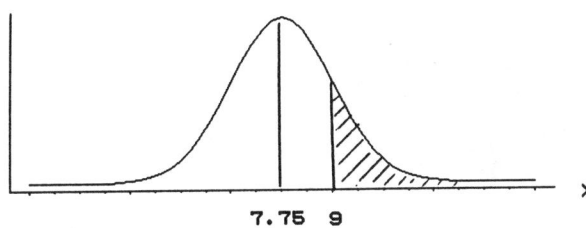
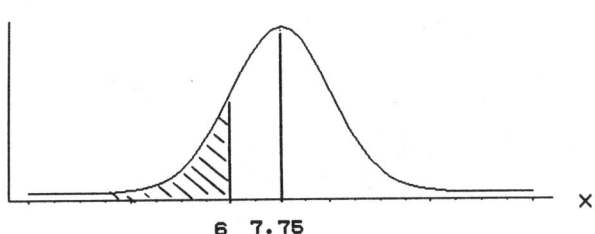

Figure 5.14   Figure 5.15

5.20

The random variable x, total weight of eight people, has a mean of $\mu = 1200$ and a variance $\sigma^2 = 9800$. It is necessary to find $P[x > 1300]$ and $P[x > 1500]$ if the distribution of x is approximately normal. Refer to Figure 5.16. The z-value corresponding to $x_1 = 1300$ is

$$z_1 = \frac{x_1 - \mu}{\sigma} = \frac{1300 - 1200}{\sqrt{9800}} = \frac{100}{98.995} = 1.01$$

Hence,
$$P[x > 1300] = P[z > 1.01] = .5 - A(1.01) = .5 - .3438 = .1562$$

Similarly, the z-value corresponding to $x_2 = 1500$ is

$$z_2 = \frac{x_2 - \mu}{\sigma} = \frac{1500 - 1200}{\sqrt{9800}} = 3.03$$

and $P[x > 1500] = P[z > 3.03] = .5 - A(3.03) = .5 - .4988 = .0012$

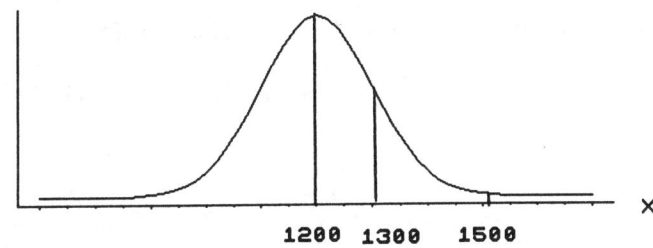

Figure 5.16

5.21

The random variable x, daily discharge, has mean $\mu = 27$ and variance $\sigma^2 = 196$. If x is normally distributed, it is necessary to find $P[x > 50]$. The z-value corresponding to $x = 50$ is

$$z = \frac{x - \mu}{\sigma} = \frac{50 - 27}{14} = 1.64$$

and $P[x > 50] = P[z > 1.64] = .5 - A(1.64) = .5 - .4495 = .0505$

5.22

It is given that x is normally distributed with $\mu = 45$ and $\sigma = 4.5$, where x is the percent of retail price at which the collection sells.

a. If the collection is worth \$30,000 and is selling for more than \$15,000, then x is more than 50 (percent). Calculate

$$z = \frac{x - \mu}{\sigma} = \frac{50 - 45}{4.5} = 1.11$$

Then

$$P[x > 50] = P[z > 1.11] = .5 - .3665 = .1335$$

b. $P[x < 50] = P[z < 1.11] = .5 + .3665 = .8665$ from part a.

c. The value \$12,000 is 40 percent of the collection's worth. Calculate

$$z = \frac{x - \mu}{\sigma} = \frac{40 - 45}{4.5} = -1.11. \quad \text{Then}$$

$$P[x < 40] = P[z < -1.11] = .5 - .3665 = .1335$$

5.23

Define x to be the percentage of returns audited for a particular state. It is given that x is normally distributed with $\mu = 1.55$ and $\sigma = .45$.

a. $P[x > 2.5] = P[z > \frac{2.5 - 1.55}{.45}] = P[z > 2.11] = .5 - .4826 = .0174$

b. $P[x < 1] = P[z < \frac{1 - 1.55}{.45}] = P[z < -1.22] = .5 - .3888 = .1112$

5.24

Define x to be the year-end bonus received by a worker for meeting quality production and profitability targets in 1993 and suppose that x is normally distributed with $\mu = 2800$ and $\sigma = 500$.

a. To find $P[x > 3500]$, calculate $z = \frac{x - \mu}{\sigma} = \frac{3500 - 2800}{500} = 1.4.$ Then

$$P[x > 3500] = P[z > 1.4] = .5 - .4192 = .0808$$

b. It is necessary to find two values of x, say x_1 and x_2, such that

$$P[x_1 < x < x_2] = .95$$

Recall that for the standard normal distribution, 95% of the measurements fall between $z_1 = -1.96$ and $z_2 = 1.96$. That is, the probability statement above will be satisfied if we let

$$z_1 = \frac{x_1 - \mu}{\sigma} = \frac{x_1 - 2800}{500} = -1.96 \quad \text{and}$$

$$z_2 = \frac{x_2 - \mu}{\sigma} = \frac{x_2 - 2800}{500} = 1.96$$

Solving for x_1 and x_2, we have

$$x_1 = 2800 - 1.96(500) = 1820 \text{ and } x_2 = 2800 + 1.96(500) = 3780$$

5.25

Define x to be the cost of natural gas per metric cubic foot and suppose that x is normally distributed with $\mu = 6.00$ and $\sigma = 1.20$.

a. To find $P[7.60 < x < 8.00]$, calculate

$$z_1 = \frac{x_1 - \mu}{\sigma} = \frac{x_1 - 6.00}{1.20} = \frac{7.60 - 6.00}{1.20} = 1.33$$

$$z_2 = \frac{x_2 - \mu}{\sigma} = \frac{x_2 - 6.00}{1.20} = \frac{8.00 - 6.00}{1.20} = 1.67$$

Then $P[7.60 < x < 8.00] = P[1.33 < z < 1.67] = .4525 - .4082 = .0443$

b. The median cost per MCF for natural gas is that value, m, such that $P[x > m] = P[x < m] = .5$. For the standard normal random variable, z, the median value, which has area .50 to its left and right, is $z = 0$. The corresponding value for x, which defines the median, is

$$\frac{m - 6.00}{1.20} = 0$$

or $m = 1.20(0) + 6.00 = 6.00$

c. The lower and upper quartiles of the standard normal distribution are those values, say z_1 and z_2, which have area .25 to their left and right, respectively. Using Table 3, we search for a value of z_0 that has area $A(z_0) = .5 - .25 = .25$. This value is $z_0 = .675$, so that

$$z_1 = -.675 \quad \text{and} \quad z_2 = .675.$$

The corresponding values of x for this particular normal distribution, with $\mu = 6.00$ and $\sigma = 1.20$, are found by solving the equations,

$$z_1 = \frac{x_1 - \mu}{\sigma} = \frac{x_1 - 6.00}{1.20} = -.675 \text{ and}$$

$$z_2 = \frac{x_2 - \mu}{\sigma} = \frac{x_2 - 6.00}{1.20} = .675$$

or $x_1 = 5.19$ and $x_2 = 6.81$

5.26

a. From Table 1, Appendix II,

$$P[8 \leq x \leq 10] = P[x \leq 10] - P[x \leq 7] = .902 - .512 = .390$$

b. Calculate $\mu = np = 7.5$ and $\sigma = \sqrt{25(.3)(.7)} = 2.2913$. The probability of interest is the area under the binomial probability histogram corresponding to the rectangles $x = 8, 9$ and 10 in Figure 5.17. To approximate this area, use the "correction for continuity" and find the area under a normal curve with mean $\mu = 7.5$ and $\sigma = 2.2913$ between $x_1 = 7.5$ and $x_2 = 10.5$. The z-values corresponding to the two values of x are

$$z_1 = \frac{7.5 - 7.5}{2.2913} = 0 \quad \text{and} \quad z_2 = \frac{10.5 - 7.5}{2.2913} = 1.31$$

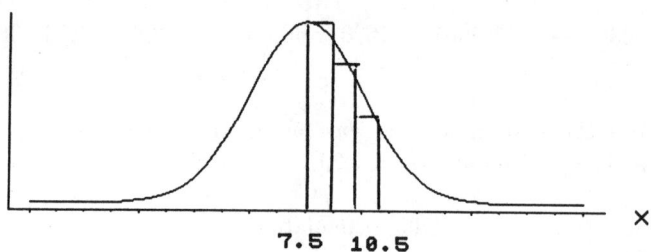

Figure 5.17

The approximating probability is $P[7.5 < x \leq 10.5] = P[0 \leq z \leq 1.31] = .4049$, which is not too far from the actual probability calculated in part a.

5.27

The exact probability is the area under the binomial histogram corresponding to the probability rectangles $x = 6, 7, 8, 9$ and 10. Hence, the approximating probability will be $P[x > 5.5]$ where x has a normal distribution with

$$\mu = np = 10(.5) = 5 \quad \text{and} \quad \sigma = \sqrt{npq} = \sqrt{10(.5)(.5)} = 1.5811$$

as shown in Figure 5.18. Then

$$P[x > 5.5] = P[z > \frac{5.5 - 5}{1.5811}]$$
$$= P[z > .32] = .5 - .1255$$
$$= .3745$$

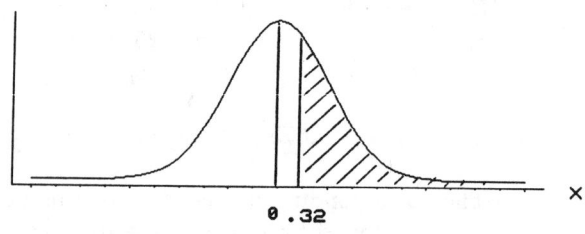

Figure 5.18

5.28

Refer to Exercise 5.27. The probability of interest is now the area under the rectangles $x = 7, 8, 9$, and 10. Hence, the approximating probability will be

$$P[x > 6.5] = P[z > \frac{6.5 - 5}{1.5811}] = P[z > .95] = .5 - .3289 = .1711$$

5.29

This is similar to Exercise 5.27. The approximating probability will be $P[x > 22.5]$ where x has a normal distribution with

$$\mu = 100(.2) = 20 \quad \text{and} \quad \sigma = \sqrt{100(.2)(.8)} = 4$$

Then

$$P[x > 22.5] = P[z > \frac{22.5 - 20}{4}] = P[z > .62] = .5 - .2324 = .2676$$

5.30

Refer to Exercise 5.29. The approximating probability is now $P[x > 21.5]$ since the entire rectangle corresponding to $x = 22$ must be included.

$$P[x > 21.5] = P[z > \frac{21.5 - 20}{4}] = P[z > .38] = .5 - .1480 = .3520$$

5.31

a. Given a binomial random variable, n = 25 and p = .2, use Table 1 to calculate

$$P[4 \leq x \leq 6] = P[x \leq 6] - P[x \leq 3] = .780 - .234 = .546$$

b. The desired area is the shaded area inside the rectangles in Figure 5.19. To approximate this area, use a "correction for continuity" and find the area under the normal curve between 3.5 and 6.5. This is done in order to include the entire area under the rectangles associated with the different values of x. To find the approximate area, first find the mean and standard deviation of the binomial random variable.

$$\mu = np = 25(.2) = 5$$

$$\sigma = \sqrt{npq} = \sqrt{25(.2)(.8)} = 2$$

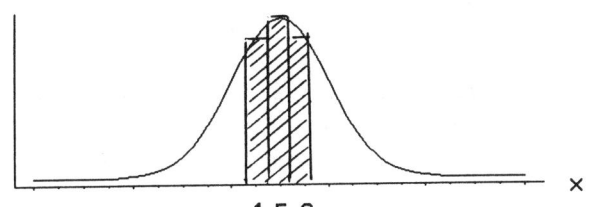

Figure 5.19

The z-values corresponding to the values $x_1 = 3.5$ and $x_2 = 6.5$ are

$$z_1 = \frac{3.5 - 5}{2} = -.75$$

$$z_2 = \frac{6.5 - 5}{2} = .75$$

Then
$$P[4 \leq x \leq 6] = P[-.75 \leq z \leq .75] = 2(.2734) = .5468$$

5.32

This is similar to Exercise 5.31.

a. With n = 20 and p = .4, $P[x \geq 10] = 1 - P[x \leq 9] = 1 - .755 = .245$.

b. To use the normal approximation, find the mean and standard deviation of this binomial random variable.
$$\mu = np = 20(.4) = 8$$

$$\sigma = \sqrt{npq} = \sqrt{20(.4)(.6)} = 2.191$$

Using the continuity correction, it is necessary to find the area to the right of 9.5. The z-value corresponding to x = 9.5 is

$$z = \frac{9.5 - 8}{2.191} = .68 \quad \text{and}$$

$$P[x \geq 10] \approx P[z > .68] = .5 - .2517 = .2483.$$

Note that the normal approximation is very close to the exact binomial probability.

5.33

The normal approximation with "correction for continuity" is $P[354.5 < x < 360.5]$ where x is normally distributed with mean and standard deviation given by

$$\mu = np = 400(.9) = 360 \quad \text{and} \quad \sigma = \sqrt{npq} = \sqrt{400(.9)(.1)} = 6$$

Then
$$P[354.5 < x < 360.5] = P[\frac{354.5 - 360}{6} < z < \frac{360.5 - 360}{6}]$$
$$= P[-.92 < z < .08] = .3212 + .0319 = .3531$$

5.34

Define x to be the number of Maytag electric dryers that have ever needed repair. Then x has a binomial distribution with n = 56 and p = .10. We will use the normal approximation with

$$\mu = 56(.10) = 5.6 \text{ and } \sigma = \sqrt{56(.1)(.9)} = \sqrt{5.04} = 2.24$$

a. $P[x \geq 10] \approx P[z > \frac{9.5 - 5.6}{2.24}] = P[z > 1.74] = .5 - .4591 = .0409$

b. $P[x < 5] \approx P[z < \frac{4.5 - 5.6}{2.24}] = P[z < -.49] = .5 - .1879 = .3121$

c. The sample must have been randomly selected from the population, so that the trials are independent.

d. Calculating the z-score (Chapter 2), the value x = 15 lies z = (15 − 5.6)/2.24 = 4.19 standard deviations above the mean. This is an ususally large z-score; it would suggest that the 10% figure is incorrect.

5.35

Define x to be the number of guests claiming a reservation at the motel. Then p = P[guest claims reservation] = 1 − .1 = .9 and n = 215. The motel has only 200 rooms. Hence, if x > 200, a guest will not receive a room. The probability of interest is then $P[x \leq 200]$. Using the normal approximation, calculate

$$\mu = np = 215(.9) = 193.5 \text{ and } \sigma = \sqrt{215(.9)(.1)} = \sqrt{19.35} = 4.399$$

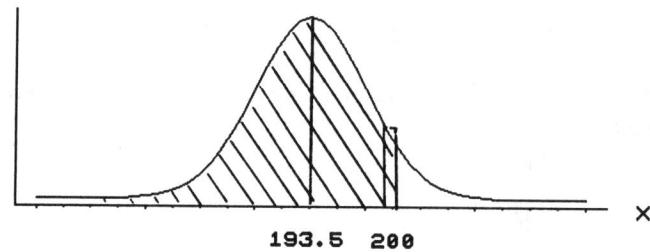

Figure 5.20

The probability $P[x \leq 200]$ is approximated by the area under the appropriate normal curve to the left of 200.5. The z-value corresponding to x = 200.5 is

$$z = \frac{200.5 - 193.5}{\sqrt{19.35}} = 1.59 \text{ and}$$

$$P[x \leq 200] \approx P[z < 1.59] = .5 + .4441 = .9441$$

5.36

Define x to be the number of boards of directors that have an average age of 57 or more. Then x has a binomial distribution with n = 400 and p = .67. Use the normal approximation to the binomial with

$$\mu = np = 400(.67) = 268 \text{ and } \sigma = \sqrt{npq} = \sqrt{400(.67)(.33)} = 9.40425$$

a. $P[x \geq 300] \approx P[z > \frac{299.5 - 268}{9.40425}] = P[z > 3.35] \approx .5 - .5 = 0$

b. This is similar to Exercise 5.24b. It is necessary to find two values of x, say x_1 and x_2, such that
$$P[x_1 < x < x_2] = .95$$

Recall that for the standard normal distribution, 95% of the measurements fall between $z_1 = -1.96$ and $z_2 = 1.96$. That is, the probability statement above will be satisfied if we let

$$z_1 = \frac{x_1 - \mu}{\sigma} = \frac{x_1 - 268}{9.40425} = -1.96 \text{ and}$$

$$z_2 = \frac{x_2 - \mu}{\sigma} = \frac{x_2 - 268}{9.40425} = 1.96$$

Solving for x_1 and x_2, we have

$$x_1 = 268 - 1.96(9.40425) = 249.57 \text{ and } x_2 = 268 + 1.96(9.40425) = 286.43$$

Since x is a binomial random variable that can only take integer values, we conclude that x will fall between 250 and 286 with probability .95.

5.37

This is similar to Exercise 5.36, with x = number who would be willing to buy another Dell PC, n = 200 and p = .82. Use the normal approximation to the binomial with

$$\mu = 200(.82) = 164 \text{ and } \sigma = \sqrt{200(.82)(.18)} = \sqrt{29.52} = 5.433$$

a. $P[x \leq 160] \approx P[z \leq \frac{160.5 - 164}{5.433}] = P[z \leq -.64] = .5 - .2389 = .2611$

b. Most values of x should lie within three standard deviations of the mean, in the interval,

$$\mu \pm 3\sigma \rightarrow 164 \pm 3(5.433) \rightarrow 164 \pm 16.299$$

or between 147.701 and 180.299. Since x is a binomial random variable that can only take integer values, we conclude that most of the values of x will fall between 148 and 180.

5.38

The random variable x, the number of householders between 45 and 64, has a binomial distribution with n = 500 and p = .31. Calculate

$$\mu = 500(.31) = 155 \text{ and } \sigma = \sqrt{500(.31)(.69)} = \sqrt{106.95} = 10.342$$

a. $P[x < 135] \approx P[z < \frac{134.5 - 155}{10.342}] = P[z < -1.98] = .5 - .4761 = .0239$

b. $P[135 \leq x \leq 180] \approx P[\frac{134.5 - 155}{10.342} \leq z \leq \frac{180.5 - 155}{10.342}]$

$= P[-1.98 < z < 2.47] = .4761 + .4932 = .9693$

5.39

Let x be the number of wage earners whose incomes exceed the nationwide median. Then x has a binomial distribution with n = 25 and p = .5. Calculate

$$\mu = 25(.5) = 12.5 \quad \text{and} \quad \sigma = \sqrt{25(.5)(.5)} = 2.5$$

a. From Table 1, Appendix II, $P[x \geq 20] = 1 - .998 = .002$

b. Using the normal approximation,

$$P[x \geq 20] \approx P[z > \frac{19.5 - 12.5}{2.5}] = P[z > 2.8] = .5 - .4974 = .0026$$

c. Observing at least twenty incomes that exceed the median is a very unusual event, given that the sample we have chosen (from this one particular area) is representative of the entire nation. Perhaps this area is not representative of the whole nation, but has a preponderance of incomes which are larger than the nationwide median.

5.40

a. The area between $z = 0$ and $z = 1.2$ is $A(1.2) = .3849$

b. The area between $z = 0$ and $z = -.9$ is $A(-.9) = A(.9) = .3159$

5.41

a. $A(1.6) = .4452$

b. $A(.75) = .2734$. Note that when two decimal places are needed, the value of z must be indexed down the left-hand column as well as across the top row.

5.42

a. $A(1.46) = .4279$

b. $A(-.42) = A(.42) = .1628$

5.43

a. $A(-1.44) = A(1.44) = .4251$

b. $A(2.01) = .4778$

5.44

a. The desired area is A_1, as shown in Figure 5.21. Note that

$$A(.3) = A_2 = .1179 \quad \text{and}$$

$$A(1.56) = A_1 + A_2 = .4406$$

Thus,

$$A_1 = A(1.56) - A(.3) = .4406 - .1179 = .3227$$

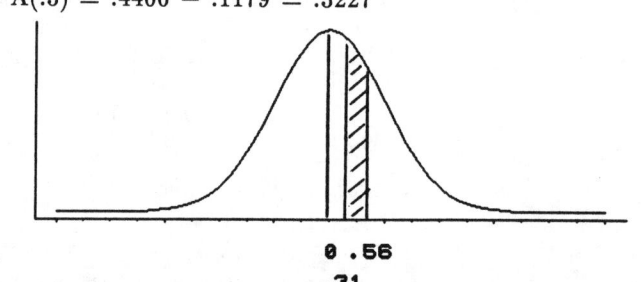

Figure 5.21

b. The desired area is

$$A_1 + A_2 = A(-.2) + A(.2) = .0793 + .0793 = .1586$$

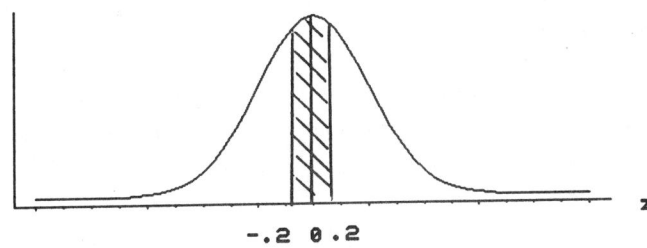

Figure 5.22

5.45

a. $A = A(1.85) - A(.88) = .4678 - .3106 = .1572$

b. $A = A(-.31) + A(1.63) = .4484 + .1217 = .5701$

5.46

a. $A(1.75) - A(1.21) = .4599 - .3869 = .0730$

b. $A(-1.3) + A(1.74) = .4032 + .4591 = .8623$

5.47

$P[z \geq -.75] = .5000 + .2734 = .7734$

5.48

$P[z < 1.35] = .5000 + .4115 = .9115$

5.49

z_0 is negative and $A(z_0) = .4750$. Hence, $z_0 = -1.96$.

5.50

z_0 is positive and $A(z_0) = .1406$. Hence, $z_0 = .36$.

5.51

We want to find a z value such that $P[-z_0 < z < z_0] = .8262$.

$$A(z_0) + A(-z_0) = .8262$$

$$2A(z_0) = .8262$$

$$A(z_0) = .4131$$

From Table 3, $z_0 = 1.36$.

5.52

z_0 is positive and $A(z_0) = .4505$. Hence, $z_0 = 1.65$.

5.53
Since $2A(z_0) = .7458$, $A(z_0) = .3729$ and $z_0 = 1.14$.

5.54
z_0 is negative and $A(z_0) = .4032$. Hence, $z_0 = -1.30$.

5.55
The procedure is reversed now, because the area under the curve is known. The objective is to determine the particular value, z_0, which will yield the given probability. In this exercise, it is necessary to find a z_0 such that $P[z > z_0] = .5000$. By the symmetry of the normal distribution, half of the area falls on each side of the mean. Thus, $P[z > 0] = .5000$, and the desired value of z_0 is 0.

5.56
It is given that $P[z < z_0] = A_1 + A_2 = .8643$ as shown in Figure 5.23. Then

$$.8643 = .5000 + A(z_0)$$
$$.3643 = A(z_0)$$

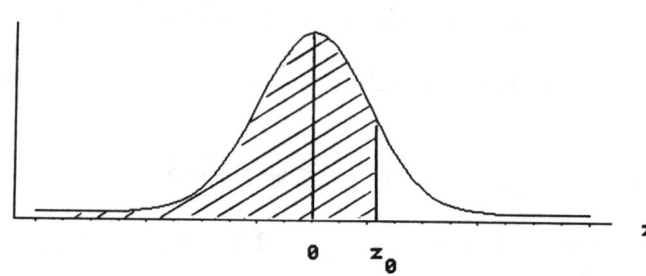

Figure 5.23

The value of z_0 that will satisfy the above equation is $z_0 = 1.10$ from Table 3, Appendix II.

5.57
$P[.7 \leq z \leq 1.63] = A(1.63) - A(.7) = .4484 - .2580 = .1904$

5.58
The random variable x is normally distributed with mean $\mu = 7$ and standard deviation $\sigma = 1.5$. It is necessary to find $P[8 < x < 9]$. The z-values corresponding to $x = 8$ and $x = 9$ are

$$z = \frac{x - \mu}{\sigma} = \frac{8 - 7}{1.5} = .67 \quad \text{and} \quad z = \frac{x - \mu}{\sigma} = \frac{9 - 7}{1.5} = 1.33$$

Then
$$P[8 < x < 9] = P[.67 < z < 1.33] = A(1.33) - A(.67) = .1596$$

5.59
$P[-.2 \leq z \leq 1.83] = A(1.83) + A(-.2) = .4664 + .0793 = .5457$

5.60
$P[-1.48 \leq z \leq 1.48] = 2(.4306) = .8612$

5.61
$P[-z_0 \leq z \leq z_0] = 2A(z_0) = .5000$. Hence, $A(z_0) = .2500$. The desired value, z_0, will be between $z_1 = .67$ and $z_2 = .68$ with associated probabilities $P_1 = .2486$ and $P_2 = .2517$. Performing a linear interpolation, we obtain the desired value, z_0.

$$z_0 = z_1 + \frac{P_0 - P_1}{P_2 - P_1}(z_2 - z_1) = .67 + (\frac{.2500 - .2486}{.2517 - .2486})(.68 - .67)$$

$$= .67 + \frac{.0014}{.0031}(.01) = .6745$$

5.62

It is given that x is approximately normally distributed with $\mu = 75$ and $\sigma = 12$.

a. Calculate

$$z = \frac{x - \mu}{\sigma} = \frac{60 - 75}{12} = -1.25$$

Then $P[x < 60] = P[z < -1.25] = .5 - .3944 = .1056$

b. $P[x > 60] = 1 - P[x < 60] = 1 - .1056 = .8944$

c. If the bit is replaced after more than 90 hours, then $x > 90$. Calculate

$$z = \frac{x - \mu}{\sigma} = \frac{90 - 75}{12} = 1.25$$

Then $P[x > 90] = P[z > 1.25] = .5 - .3944 = .1056$

5.63

The range of faculty ages should be approximately from 25 to 65 (25 for a new PhD). However, the distribution will not be normal (with average value 45 and symmetric) since there will be an overabundance of older tenured faculty. The distribution will probably be skewed to the right.

5.64

For this exercise, it is given that the population of bolt diameters is normally distributed with $\mu = .498$ and $\sigma = .002$. Thus, no correction for continuity is necessary. The fraction of acceptable bolts will be those which lie in the interval from .496 to .504. All others are unacceptable. The desired fraction of acceptable bolts is calculated, and the fraction of unacceptable bolts is obtained by subtracting from the total probability, which is 1.

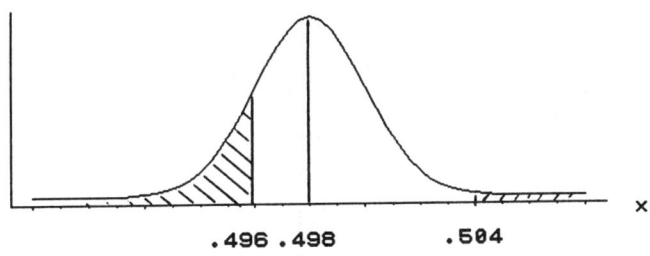

Figure 5.24

The fraction of acceptable bolts is then

$$P[.496 \leq x \leq .504] = P[\frac{.496 - .498}{.002} \leq z \leq \frac{.504 - .498}{.002}]$$

$$= P[-1 \leq z \leq 3] = .3413 + .4987 = .8400$$

and the fraction of unacceptable bolts is $1 - .84 = .16$.

5.65

It is given that x is normally distributed with $\mu = 10$ and $\sigma = 3$. Let t be the guarantee time for the car. It is necessary that only 5% of the cars fail before time t (see Figure 5.25). That is,

$$P[x < t] = .05 \text{ or}$$

$$P[z < \frac{t - 10}{3}] = .05$$

From Table 3, we know that the value of z that satisfies the above probability statement is $z = -1.645$. Hence,

$$\frac{t - 10}{3} = -1.645 \text{ or } t = 5.065 \text{ months}$$

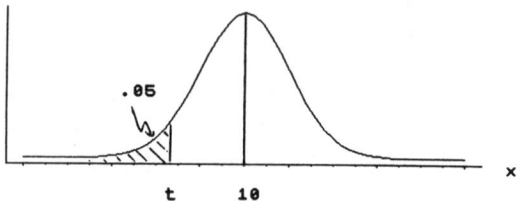

Figure 5.25

5.66

It is given that x is normally distributed with $\mu = 30$ and $\sigma = 11$. The probability of interest is

$$P[x > 50] = P[z > \frac{50 - 30}{11}] = P[z > 1.82] = .5 - .4656 = .0344$$

5.67

a. Using the binomial tables and indexing $n = 25$ and $p = .4$ in Table 1,

$$P[8 \leq x \leq 11] = P[x \leq 11] - P[x \leq 7] = .732 - .154 = .578$$

b. To use the normal approximation to the binomial distribution, calculate

$$\mu = np = 25(.4) = 10 \text{ and } \sigma = \sqrt{npq} = \sqrt{25(.4)(.6)} = 2.449$$

The desired probability is the area inside the rectangles formed by the histogram for $x = 8, 9, 10,$ and 11 in Figure 5.26.

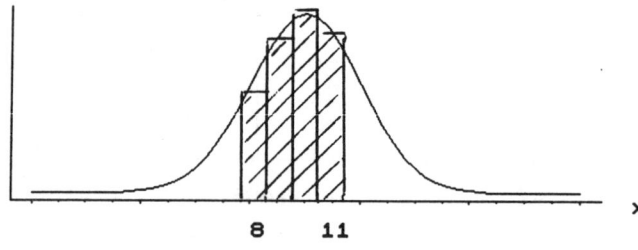

Figure 5.26

Using the correction for continuity to include the entire area under the rectangles, the approximate probability is

$$P[7.5 < x < 11.5] = P[\frac{7.5 - 10}{2.449} < z < \frac{11.5 - 10}{2.449}] = P[-1.02 < z < .61]$$

$$= .3461 + .2291 = .5752$$

5.68

a. Index $n = 25$, $p = .2$, and $a = 4$ in Table 1, Appendix II to obtain $P[x \leq 4] = .421$.

b. To use the normal approximation, calculate

$$\mu = 25(.2) = 5 \quad \text{and} \quad \sigma = \sqrt{25(.2)(.8)} = 2$$

Then $P[x \leq 4] \approx P[z < \frac{4.5 - 5}{2}] = P[z < -.25] = .5 - .0987 = .4013$

5.69

Define $x =$ number of incoming calls that are long distance
         $p =$ P[incoming call is long distance] $= .3$
         $n = 200$

The desired probability is $P[x \geq 50]$, where x is a binomial random variable with

$$\mu = np = 200(.3) = 60 \quad \text{and} \quad \sigma = \sqrt{npq} = \sqrt{200(.3)(.7)} = \sqrt{42} = 6.481$$

A correction for continuity is made to include the entire area under the rectangle corresponding to $x = 50$, and hence the approximation will be

$$P[x \geq 49.5] = P[z \geq \frac{49.5 - 60}{6.481}] = P[z \geq -1.62] = .5 + .4474 = .9474$$

5.70

The random variable of interest is x, the number of persons not showing up for a given flight. This is a binomial random variable with $n = 160$ and $p =$ P[person does not show up] $= .05$. If there is to be a seat available for every person planning to fly, then there must be at least five persons not showing up. Hence, the probability of interest is $P[x \geq 5]$. Calculate

$$\mu = np = 160(.05) = 8 \quad \text{and} \quad \sigma = \sqrt{npq} = \sqrt{160(.05)(.95)} = \sqrt{7.6} = 2.7$$

Referring to Figure 5.27, a correction for continuity is made to include the entire area under the rectangle associated with the value $x = 5$, and the approximation becomes $P[x \geq 4.5]$. The z-value corresponding to $x = 4.5$ is

$$z = \frac{x - \mu}{\sigma} = \frac{4.5 - 8}{\sqrt{7.6}} = -1.27$$

so that

$$P[x \geq 4.5] = P[z \geq -1.27] = .5 + .3980 = .8980$$

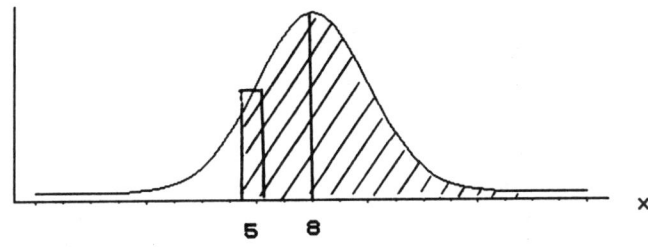

Figure 5.27

5.71

The random variable x is the size of the freshman class. That is, the admissions office will send letters of acceptance to (or accept deposits from) a certain number of qualified students. Of these students, a certain number will actually enter the freshman class. Since the experiment results in one of two outcomes (enter or not enter), the random variable x, the number of students entering the freshman class, has a binomial distribution with

n = number of deposits accepted and
p = P[student, having been accepted, enters freshman class] = .8

a. It is necessary to find a value for n such that $P[x \leq 120] = .95$. Note that

$$\mu = np = .8n \quad \text{and} \quad \sigma = \sqrt{npq} = \sqrt{.16n}$$

Using the normal approximation, we need to find a value of n such that $P[x \leq 120.5] = .95$. The z-value corresponding to $x = 120.5$ is

$$z = \frac{x - \mu}{\sigma} = \frac{120.5 - .8n}{\sqrt{.16n}}$$

From Table 3, the z value corresponding to an area of .05 in the right tail of the normal distribution is 1.645. Then,

$$\frac{120.5 - .8n}{\sqrt{.16n}} = 1.645.$$

Solving for n in the above equation, we obtain the following quadratic equation:

$$.8n + .658\sqrt{n} - 120.5 = 0$$

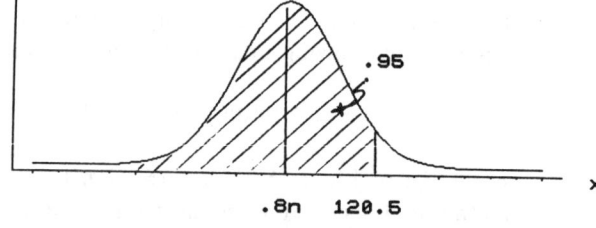

Figure 5.28

Let $x = \sqrt{n}$. Then the equation takes the form

$$ax^2 + bx + c = 0$$

which may be solved using the quadratic formula, $x = \frac{-b \pm \sqrt{b^2 - 4ac}}{2a}$, or

$$x = \frac{-.658 \pm \sqrt{.433 + 4(96.4)}}{1.6} = \frac{-.658 \pm 19.648}{2}$$

Since x must be positive, the desired root is

$$x = \sqrt{n} = \frac{18.990}{1.6} = 11.869 \quad \text{or} \quad n = (11.869)^2 = 140.86$$

Thus, 141 deposits should be accepted.

b. Once $n = 141$ has been determined, the mean and standard deviation of the distribution are

$$\mu = np = 141(.8) = 112.8 \quad \text{and} \quad \sigma = \sqrt{npq} = \sqrt{22.56} = 4.750$$

Then the approximation for P[x < 105] is

$$P[x \leq 104.5] = P[z \leq \frac{104.5 - 112.8}{4.750}] = P[z \leq -1.75] = .5 - .4599 = .0401$$

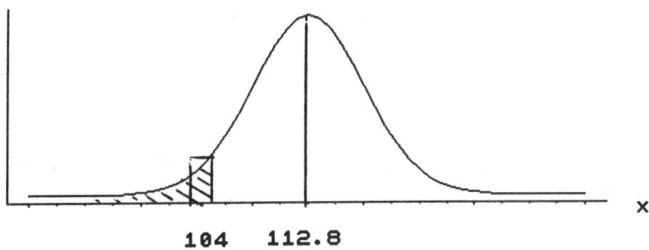

Figure 5.29

5.72

Let x be the number of customers who are in the 35-44 age category. Then x has a binomial distribution with n = 100 and p = .48. Calculate

$$\mu = np = 100(.48) = 48 \quad \text{and} \quad \sigma = \sqrt{npq} = \sqrt{24.96} = 4.996$$

a. $P[x \leq 40] \approx P[z < \frac{40.5 - 48}{4.996}] = P[z < -1.50] = .5 - .4332 = .0668$ and

$P[x \geq 60] \approx P[z > \frac{59.5 - 48}{4.996}] = P[z > 2.30] = .5 - .4893 = .0107$

b. Calculate the z-score (Chapter 2) and notice that the value x = 30 lies

$$z = \frac{x - \mu}{\sigma} = \frac{30 - 48}{4.996} = -3.60$$

standard deviations below the mean. This is a very unlikely occurrence, assuming that the 48% figure is correct. Perhaps the percentage is somewhat less than 48%.

5.73

Refer to Exercise 5.72. Now let x be the number of customers who are in the 25-34 age category. Then x has a binomial distribution with n = 200 and p = .46.

a. $\mu = np = 200(.46) = 92$ and $\sigma = \sqrt{npq} = \sqrt{49.68} = 7.048$

b. Using the normal approximation,

$$P[x < 80] \approx P[z < \frac{79.5 - 92}{7.048}] = P[z < -1.77] = .5 - .4616 = .0384$$

5.74

Let x be the distance traveled on a fall pleasure trip, and suppose that x has a normal distribution with $\mu = 870$ and unknown standard deviation.

a. It is given that $P[570 < x < 1170] = .60$. In terms of the standard normal random variable z, this means that

$$P[\frac{570 - 870}{\sigma} < z < \frac{1170 - 870}{\sigma}] = .60$$

99

$$P[\tfrac{-300}{\sigma} < z < \tfrac{300}{\sigma}] = .60$$

Then, if we let $z_0 = \tfrac{300}{\sigma}$, we need $P[-z_0 < z < z_0] = .60$ or $A(z_0) = .3$. From Table 3, we find that
$$z_0 = \tfrac{300}{\sigma} = .84 \quad \text{so that } \sigma = 300/.84 = 357.14$$

b. $P[x > 1000] = P[z > \tfrac{1000 - 870}{357.14}] = [z > .36] = .5 - .1406 = .3594$

c. $P[x < 500] = P[z < \tfrac{500 - 870}{357.14}] = [z < -1.04] = .5 - .3508 = .1492$

5.75

It is given that x, the yield per share of a group of stocks, has an exponential distribution with $\mu = 4$. Then, since $\mu = 1/\lambda$, we calculate $\lambda = 1/4 = .25$.

a. Using the formula for the right-tail area of an exponential distribution,
$$P[x > 10] = e^{-\lambda(10)} = e^{-.25(10)} = e^{-2.5} = .0821$$

b. It is necessary to find a value of x, say x_0, such that
$$P[x < x_0] = .95 \quad \text{as shown in Figure 5.30.}$$

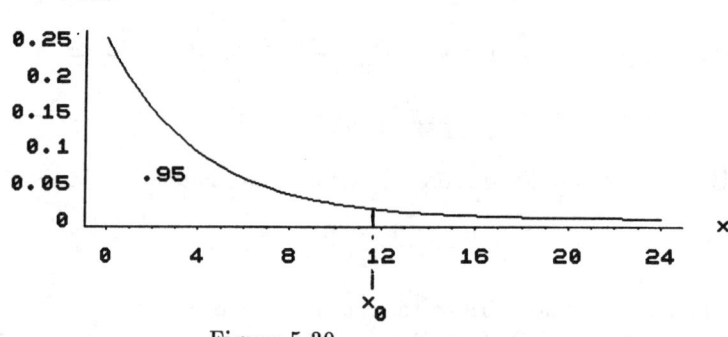

Figure 5.30

That is,
$$P[x \geq x_0] = .05$$
$$e^{-.25(x_0)} = .05$$
$$-.25(x_0) = \ln(.05) \quad \text{or} \quad x_0 = -4\ln(.05) = 11.98\%$$

5.76

The repair time, x, to fix a flat has a uniform distribution on the interval 5 to 15.

a. $P[x > 10] = \tfrac{15 - 10}{15 - 5} = \tfrac{5}{10} = \tfrac{1}{2}$

b. $P[\text{both repairs take less than 10 minutes}] = (\tfrac{1}{2})^2 = \tfrac{1}{4}$

5.77

It is given that x, the time until malfunction of a copying machine, has an exponential distribution with $\mu = 30$. Then, since $\mu = 1/\lambda$, we calculate $\lambda = 1/30$.

a. Using the formula for the right-tail area of an exponential distribution,
$$P[x < 30] = 1 - P[x \geq 30] = 1 - e^{-\lambda(30)}$$

$$= 1 - e^{-\frac{1}{30}(30)} = 1 - e^{-1} = .6321$$

b. It is necessary to find a value of x, say x_0, such that

$$P[x < x_0] = .2 \quad \text{as shown in Figure 5.31.}$$

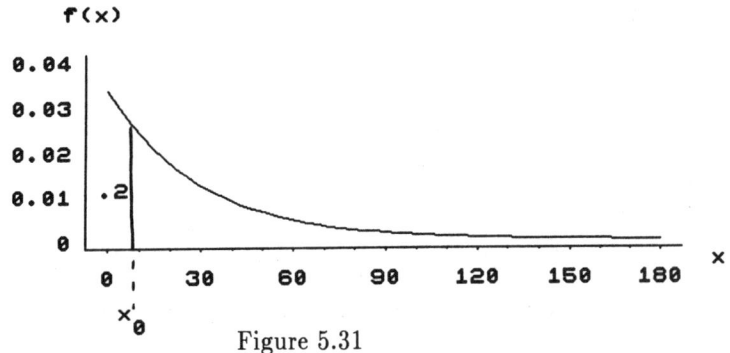

Figure 5.31

That is,
$$P[x \geq x_0] = .8$$

$$e^{-\frac{x_0}{30}} = .8$$

$$-x_0/30 = \ln(.8) \quad \text{or} \quad x_0 = -30 \ln(.8) = 6.694$$

The machine should be serviced every six days.

5.78

The amount of gasoline sold per week has a uniform distribution on the interval 5000 to 15,000.

a. $P[\text{runs out of gas}] = P[x > 12,000] = \frac{15,000 - 12,000}{15,000 - 5000} = \frac{3000}{10,000} = .3$

b. Let C be the amount of gas stored in the tank. Then

$$P[\text{runs out of gas}] = P[\text{demand exceeds amount in storage}] = P[x > C].$$

In order that $P[x > C] = .001$, we must have

$$\frac{15,000 - C}{15,000 - 5,000} = \frac{15,000 - C}{10,000} = .001$$

or $\quad C = 15,000 - 10,000(.001) = 14,990 \quad$ gallons in the tank.

5.79

a. Use the formula given in the exercise for $P[x > 45]$ and solve for λ.

$$e^{-\lambda(45)} = .53$$

$$-45\lambda = \ln(.53) \quad \text{or} \quad \lambda = -\frac{\ln(.53)}{45} = .0141$$

so that $\mu = \frac{1}{\lambda} = 70.88$.

b. $P[x \geq 60] = e^{-\lambda(60)} = e^{-(.0141)(60)} = .4291$

5.80

The random variable x, the number of female purchasers, has a binomial distribution with $n = 400$ and $p = .5$. Calculate

$$\mu = 400(.5) = 200 \quad \text{and} \quad \sigma = \sqrt{400(.5)(.5)} = 10$$

Using the normal approximation to the binomial distribution,

$$P[x > 175] \approx P[z > \tfrac{175.5 - 200}{10}] = P[z > -2.45] = .5 + .4929 = .9929$$

5.81

The random variable x, the number of sales, has a binomial distribution with $n = 50$ and $p = .3$. Calculate

$$\mu = 50(.3) = 15 \quad \text{and} \quad \sigma = \sqrt{50(.3)(.7)} = 3.24$$

Using the normal approximation to the binomial distribution,

$$P[x \geq 10] \approx P[z > \tfrac{9.5 - 15}{3.24}] = P[z > -1.7] = .5 + .4554 = .9554$$

5.82

It is given that the random variable x (ounces of fill) is normally distributed with mean μ and standard deviation $\sigma = .3$. It is necessary to find a value of μ so that $P[x > 8] = .01$. That is, an 8-ounce cup will overflow when $x > 8$, and this should happen only 1% of the time. Then,

$$P[x > 8] = P[z > \tfrac{8 - \mu}{.3}] = .01.$$

From Table 3, the value of z corresponding to an area (in the upper tail of the distribution) of .01 is $z_0 = 2.33$. Hence, the value of μ can be obtained by solving for μ in the following equation:

$$2.33 = \tfrac{8 - \mu}{.3} \quad \text{or} \quad \mu = 7.301$$

5.83

The 3000 light bulbs utilized by the manufacturing plant comprise the entire population (that is, this is not a sample from the population) whose length of life is normally distributed with mean $\mu = 500$ and standard deviation $\sigma = 50$. The objective is to find a particular value, x_0, so that

$$P[x \leq x_0] = .01$$

That is, only 1% of the bulbs will burn out before they are replaced at time x_0. Then

$$P[x \leq x_0] = P[z \leq z_0] = .01 \quad \text{where} \quad z_0 = \tfrac{x_0 - 500}{50}$$

From Table 3, the value of z corresponding to an area (in the left tail of the distribution) of .01 is $z_0 = -2.33$. Solving for x_0 corresponding to $z_0 = -2.33$,

$$-2.33 = \tfrac{x_0 - 500}{50}$$

$$-116.5 = x_0 - 500$$

$$x_0 = 383.5$$

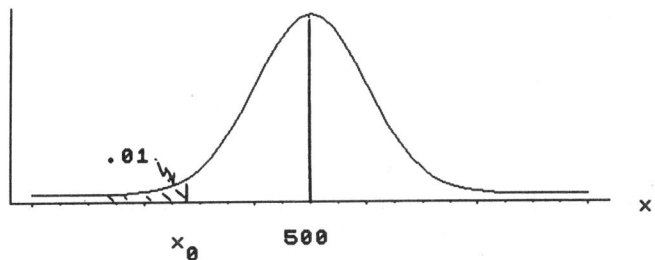

Figure 5.32

5.84

The random variable x is normally distributed with $\mu = 530$ and $\sigma = 120$.

a. The z-value corresponding to x = 700 is $z = \frac{x - \mu}{\sigma} = \frac{700 - 530}{120} = 1.42$ and

$$P[x > 700] = P[z > 1.42] = .5 - .4222 = .0778$$

b. The z-value corresponding to x = 300 is $z = \frac{x - \mu}{\sigma} = \frac{300 - 530}{120} = -1.92$ and

$$P[\text{restaurant does not break even}] = P[x < 300]$$

$$= P[z < -1.92] = .5 - .4726 = .0274$$

5.85

It is given that $\mu = 3.1$ and $\sigma = 1.2$. A washer must be replaced if its lifetime is less than one year, so that the desired fraction is $A = P[x < 1]$. The corresponding z-value is

$$z = \frac{x - \mu}{\sigma} = \frac{1 - 3.1}{1.2} = -1.75 \text{ and}$$

$$P[x < 1] = P[z < -1.75] = .5 - .4599 = .0401$$

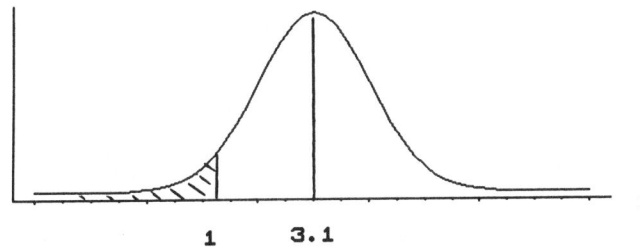

Figure 5.33

5.86

It is given that x is normally distributed with μ unknown and $\sigma = 25.7$. It is necessary to have $P[x > 2000] = .01$. Calculate

$$z = \frac{2000 - \mu}{25.7} \quad \text{and} \quad P[x > 2000] = P[z > \frac{2000 - \mu}{25.7}] = .01$$

Since the value $(2000 - \mu)/25.7$ is a constant (although its value is unknown), it can be treated as such in Table 3. It is necessary to find z_0 such that

$$P[z > z_0] = .01 \quad \text{or} \quad A(z_0) = .4900$$

From Table 3, $z_0 = 2.33$, so that

$$\frac{2000 - \mu}{25.7} = 2.33 \quad \text{and} \quad \mu = 1940.119$$

5.87

Let w be the number of words specified in the contract. Then x, the number of words in the manuscript, is normally distributed with $\mu = w + 20{,}000$ and $\sigma = 10{,}000$. The publisher would like to specify w so that

$$P[x < 100{,}000] = .95$$

As in Exercise 5.86, calculate $z = \dfrac{100{,}000 - (w + 20{,}000)}{10{,}000} = \dfrac{80{,}000 - w}{10{,}000}$
Then
$$P[x < 100{,}000] = P[z < \frac{80{,}000 - w}{10{,}000}] = .95$$

It is necessary that $z_0 = (80{,}000 - w)/10{,}000$ be such that

$$P[z < z_0] = .95 \quad \Rightarrow \quad A(z_0) = .45 \quad \text{or} \quad z_0 = 1.645$$

Hence,
$$\frac{80{,}000 - w}{10{,}000} = 1.645 \quad \text{or } w = 63{,}550$$

CHAPTER 6
Sampling Distributions

6.1
Regardless of the shape of the population from which we are sampling, the sampling distribution of the sample mean will have a mean μ equal to the mean of the population from which we are sampling, and a standard deviation equal to $\sigma/\sqrt{n}$.

a. $E(\bar{x}) = \mu = 10;\quad \sigma_{\bar{x}} = \sigma/\sqrt{n} = 3/\sqrt{25} = .6$

b. $E(\bar{x}) = \mu = 5;\quad \sigma_{\bar{x}} = \sigma/\sqrt{n} = 2/\sqrt{100} = .2$

c. $E(\bar{x}) = \mu = 120;\quad \sigma_{\bar{x}} = \sigma/\sqrt{n} = 1/\sqrt{6} = .4082$

6.2

a. If the sampled populations are normal, the distribution of $\bar{x}$ is also normal *for all values of n*.

b. The Central Limit Theorem states that for sample sizes as small as $n = 25$, the sampling distribution of $\bar{x}$ will be approximately normal. Hence, we can be relatively certain that the sampling distribution of $\bar{x}$ for parts a and b will be approximately normal. However, the sample size in part c, $n = 6$, is too small to assume that the distribution of $\bar{x}$ is approximately normal.

6.3

a. The sketch of the normal distribution with mean $\mu = 5$ and standard deviation $\sigma_{\bar{x}} = .2$ is left to the student. The interval $5 \pm .4$ or 4.6 to 5.4 should be located on the $\bar{x}$ axis.

b. The probability of interest is $P[-.15 < (\bar{x} - \mu) < .15]$ and is shown in Figure 6.1.

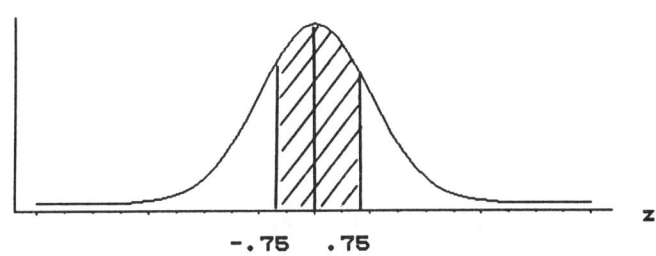

Figure 6.1

c. $P[-.15 < (\bar{x} - \mu) < .15] = P[\frac{-.15}{.2} < \frac{(\bar{x} - \mu)}{\sigma/\sqrt{n}} < \frac{.15}{.2}]$

$= P[-.75 < z < .75] = 2(.2734) = .5468$

6.4

a. The random variable x has a discrete probability distribution given as $p(x) = \frac{1}{6}$ for $x = 1, 2, 3, 4, 5, 6$. Using the formulas given in Chapter 3, calculate

$\mu = \Sigma x p(x) = (1 + 2 + \cdots + 6)(\frac{1}{6}) = \frac{21}{6} = 3.5$

and
$$\sigma^2 = \Sigma x^2 p(x) - \mu^2 = (1^2 + 2^2 + \cdots + 6^2)\left(\tfrac{1}{6}\right) - (3.5)^2$$
$$= \tfrac{91}{6} - 12.25 = 2.9167$$
$$\sigma = \sqrt{2.9167} = 1.71$$

b. The mean of the histogram in Figure 6.3 of the text is the mean for the sample of 100 $\bar{x}$'s. Thus, in Figure 6.3 of the text, the mean is guessed to be near the center of the distribution, or approximately 3.5. Based on the Empirical Rule, approximately 95% of the measurements will lie within two standard deviations of the mean. Hence, we would expect the range to equal approximately four standard deviations. Since R = 5.4 − 1.4 = 4.0,

$$4.0 \approx 4s \qquad \text{or} \qquad s \approx 1.00$$

This is a rough approximation. Actually, s will most likely be somewhat less than 1.00, because the range will probably span more than four standard deviations for 100 measurements.

c. The theoretical mean and standard deviation of the sampling distribution of $\bar{x}$ are

$$\mu = 3.5 \qquad \text{and} \qquad \sigma_{\bar{x}} = \sigma/\sqrt{n} = 1.71/\sqrt{5} = .765$$

These values are not far from the guessed values in part b.

6.5

Refer to Exercise 6.4. As the sample size changes, the mean of the sampling distribution of $\bar{x}$ remains the same—that is, $\mu = 3.5$. However, the standard deviation of the sampling distribution decreases as n increases.

a. $\sigma_{\bar{x}} = \sigma/\sqrt{n} = 1.71/\sqrt{10} = .541$

b. $\sigma_{\bar{x}} = \sigma/\sqrt{n} = 1.71/\sqrt{15} = .442$

c. $\sigma_{\bar{x}} = \sigma/\sqrt{n} = 1.71/\sqrt{25} = .342$

6.6

Refer to Exercise 6.5. The standard error decreases as the sample size (which appears in the denominator of the fraction) increases.

6.7

For a population with $\sigma = 1$, the standard deviation of the sampling distribution of $\bar{x}$ is

$$\sigma/\sqrt{n} = 1/\sqrt{n}.$$

The values of $\sigma_{\bar{x}}$ for various values of n are tabulated and plotted below.

n	1	2	4	9	16	25	100
$\sigma_{\bar{x}}$	1.00	.707	.500	.333	.250	.200	.100

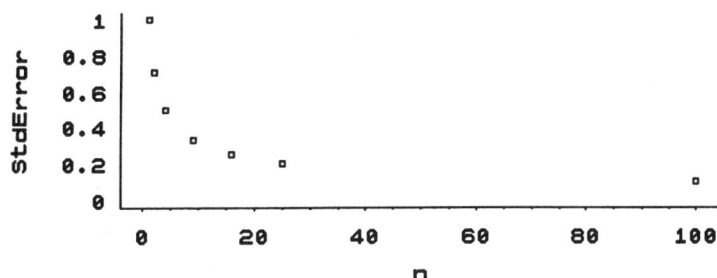

Figure 6.2

6.8

a. If the sampled population is normal, the sampling distribution of $\bar{x}$ will also be normal (regardless of the sample size) with mean $\mu = 1$ and standard deviation (or *standard error*) given as
$$\sigma_{\bar{x}} = \sigma/\sqrt{n} = .36/\sqrt{5} = .161$$

b. Calculate
$$z = \frac{\bar{x} - \mu}{\sigma_{\bar{x}}} = \frac{1.3 - 1}{.161} = 1.86$$

Then
$$P[\bar{x} > 1.3] = P[z > 1.86] = .5 - .4686 = .0314$$

c. Calculate
$$z = \frac{\bar{x} - \mu}{\sigma_{\bar{x}}} = \frac{0.5 - 1}{.161} = -3.11$$

Then
$$P[\bar{x} < 0.5] = P[z < -3.11] \approx 0$$

d. The probability that $\bar{x}$ deviates from $\mu = 1$ by more than .4 is the probability that $\bar{x}$ exceeds 1.4 or is less than .6. Calculate
$$z_1 = \frac{\bar{x}_1 - \mu}{\sigma_{\bar{x}}} = \frac{1.4 - 1}{.161} = 2.48 \quad \text{and} \quad z_2 = \frac{\bar{x}_2 - \mu}{\sigma_{\bar{x}}} = \frac{0.6 - 1}{.161} = -2.48$$

Then
$$P[\bar{x} > 1.4] + P[\bar{x} < 0.6] = P[z > 2.48] + P[z < -2.48]$$
$$= 2(.5 - .4934) = .0132$$

6.9 This is similar to Exercise 6.8.

a. $\mu = 106$, $\sigma_{\bar{x}} = \sigma/\sqrt{n} = 12/\sqrt{25} = 2.4$

b. Calculate
$$z = \frac{\bar{x} - \mu}{\sigma_{\bar{x}}} = \frac{110 - 106}{2.4} = 1.67$$

Then
$$P[\bar{x} > 110] = P[z > 1.67] = .5 - .4525 = .0475$$

c. $P[102 < \bar{x} < 110] = P[-1.67 < z < 1.67] = 2(.4525) = .9050$

6.10

The weight of a truckload of oranges is the sum of the individual orange weights. Hence, since weights and heights are in general normally distributed, so will be the sum of the weights, according to the Central Limit Theorem.

6.11

The number of accidents per year in a manufacturing plant can be thought of as the sum of 365 random variables, each of which is the number of accidents on a particular day. Hence, the Central Limit Theorem insures the approximate normality of the sum.

6.12

The daily catch x is defined to be

$$\sum_{i=1}^{50} x_i$$

where x_i is the daily catch in a single one of fifty traps. Each x_i has mean $\mu = 30$ and variance $\sigma^2 = 25$. According to the Central Limit Theorem, the sum will have a normal distribution with mean $n\mu = 50(30) = 1500$ and standard deviation $\sigma\sqrt{n} = 5\sqrt{50} = 35.355$.

6.13

a. The population from which we are randomly sampling n = 35 measurements is not necessarily normally distributed. However, the sampling distribution of $\bar{x}$ does have an approximately normal distribution, with mean μ and standard deviation $\sigma/\sqrt{n}$. The probability of interest is

$$P[|\bar{x} - \mu| < 1] = P[-1 < (\bar{x} - \mu) < 1]$$

Since

$$z = \frac{\bar{x} - \mu}{\sigma/\sqrt{n}}$$

has a standard normal distribution, we need only find $\sigma/\sqrt{n}$ to approximate the above probability. Though σ is unknown, it can be approximated by $s = 12$ and

$$\sigma/\sqrt{n} \approx 12/\sqrt{35} = 2.028. \text{ Then}$$

$$P[|\bar{x} - \mu| < 1] = P[-1/2.028 < z < 1/2.028]$$

$$= P[-.49 < z < .49] = 2(.1879) = .3758$$

b. No. There are many possible values for x, the actual percent tax savings, as given by the probability distribution for x.

6.14

The small average increases are extremely significant to educators because of the large number of student grades used to calculate the average. That is, n is very large and $\sigma_{\bar{x}}$ is very small.

6.15

The random variable of interest is x, the number of heavy-duty trailers per hour. It is approximately normally distributed with $\mu = 50$ and $\sigma = 7$. Then the sampling distribution of $\bar{x}$ has an approximately normal distribution, with mean $\mu = 50$ and standard deviation $\sigma/\sqrt{n} = 7/\sqrt{25} = 1.4$.

a. The probability of interest is $P[\bar{x} > 55]$. Calculate

$$z = \frac{\overline{x} - \mu}{\sigma/\sqrt{n}} = \frac{55 - 50}{1.4} = 3.57$$

and

$$P[\overline{x} > 55] = P[z > 3.57] = .5 - A(3.57) \approx .5 - .5 = 0$$

b. This is similar to part a, except that $\overline{x}$ is calculated based on n = 4 observations. The z value corresponding to $\overline{x} = 55$ is

$$z = \frac{\overline{x} - \mu}{\sigma/\sqrt{n}} = \frac{55 - 50}{7/\sqrt{4}} = 1.43$$

and

$$P[\overline{x} > 55] = P[z > 1.43] = .5 - A(1.43) = .5 - .4236 = .0764$$

c. The total number of trucks for a four-hour period is the sum of four observations,

$$\sum_{i=1}^{4} x_i$$

each of which is normally distributed with $\mu = 50$ and $\sigma = 7$. The sum will also be normally distributed with mean $n\mu = 4(50) = 200$ and standard deviation $\sigma\sqrt{n} = 7\sqrt{4} = 14$. Hence, the z value corresponding to $\Sigma x_i = 180$ is

$$z = \frac{\Sigma x_i - n\mu}{\sigma\sqrt{n}} = \frac{180 - 200}{14} = -1.43$$

and $P[\Sigma x_i > 180] = P[z > -1.43] = .5 + .4236 = .9236$

6.16

a. The sample mean $\overline{x}$ is normally distributed, since the original strength measurements are normal, with mean μ and standard deviation

$$\sigma/\sqrt{n} \approx 2/\sqrt{10}$$

b. If $\mu = 21$ and n = 10, the z-value corresponding to $\overline{x} = 20$ is

$$z = \frac{\overline{x} - \mu}{\sigma/\sqrt{n}} = \frac{20 - 21}{2/\sqrt{10}} = -1.58$$

and

$$P[\overline{x} < 20] = P[z < -1.58] = .5 - .4429 = .0571$$

c. Refer to part b and assume that μ is unknown. It is necessary to find a value μ_0 so that $P[\overline{x} < 20] = .001$

Recall that if $\mu = \mu_0$, the z-value corresponding to $\overline{x} = 20$ is

$$z = \frac{\overline{x} - \mu_0}{\sigma/\sqrt{n}} = \frac{20 - \mu_0}{2/\sqrt{10}} = \frac{20 - \mu_0}{.632}$$

It is necessary then to have

$$P[\overline{x} < 20] = P[z < \frac{20 - \mu_0}{.632}] = .5 - A\left(\frac{20 - \mu_0}{.632}\right) = .001$$

or

$$A\left(\frac{20 - \mu_0}{.632}\right) = .4990$$

Refer to Figure 6.3. The value of z that cuts off .001 in the left-hand tail of the normal distribution is negative and is such that $A(z_0) = .499$. From Table 3, this value is $z_0 = -3.08$. Hence, the two values, $(20 - \mu_0)/.632$ and $z_0 = -3.08$, must be the same.

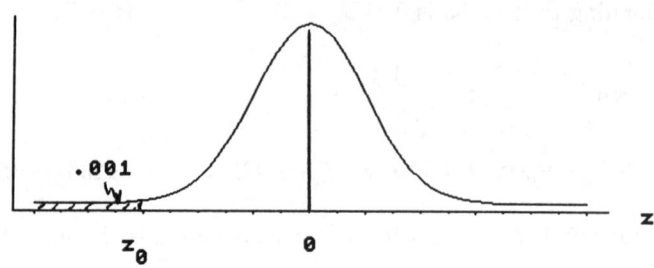

Figure 6.3

Solving for μ_0,

$$\frac{20 - \mu_0}{.632} = -3.08 \quad \text{or} \quad \mu_0 = 21.948$$

6.17

a. The time required to load *Ami Pro 2.0* can be thought of as a combination or average of several random variables, such as amount of software on the PC, type of software on the PC, amount of disk space available, etc. Hence, the Central Limit Theorem insures the approximate normality of the average.

b. If $\mu = 1.33$ and $\sigma = 0.2$, the z-value corresponding to $x = 1.4$ is

$$z = \frac{x - \mu}{\sigma} = \frac{1.4 - 1.33}{0.2} = .35$$

and

$$P[x > 1.4] = P[z > .35] = .5 - .1368 = .3632$$

c. If $n = 5$, the z-value corresponding to $\bar{x} = 1.4$ is

$$z = \frac{\bar{x} - \mu}{\sigma/\sqrt{n}} = \frac{1.4 - 1.33}{0.2/\sqrt{5}} = .78$$

and

$$P[\bar{x} > 1.4] = P[z > .78] = .5 - .2823 = .2177$$

6.18

a. $\mu_{\hat{p}} = p = .3; \quad \sigma_{\hat{p}} = \sqrt{\frac{pq}{n}} = \sqrt{\frac{.3(.7)}{100}} = .0458$

b. $\mu_{\hat{p}} = p = .1; \quad \sigma_{\hat{p}} = \sqrt{\frac{pq}{n}} = \sqrt{\frac{.1(.9)}{400}} = .015$

c. $\mu_{\hat{p}} = p = .6; \quad \sigma_{\hat{p}} = \sqrt{\frac{pq}{n}} = \sqrt{\frac{.6(.4)}{250}} = .0310$

6.19

Since the sample sizes are very large, the sampling distributions in Exercise 6.18 will each be approximately normal, with appropriate means and variances. The interval $p \pm 2\sigma_{\hat{p}}$ should cover 95% of the measurements.

6.20

a-b. The probability of interest is

$$P[\,|\hat{p} - p| \leq .08\,] = P[-.08 \leq (\hat{p} - p) \leq .08]$$

Since $\hat{p}$ is approximately normal, with standard deviation $\sigma_{\hat{p}} = .0458$ from Exercise 6.18,

$$P[-.08 \leq (\hat{p} - p) \leq .08] = P[\tfrac{-.08}{.0458} \leq z \leq \tfrac{.08}{.0458}]$$

$$= P[-1.75 \leq z \leq 1.75] = 2(.4599) = .9198$$

6.21

For n = 1000 and p = .1, the sample proportion $\hat{p}$ is approximately normally distributed with

$$\mu_{\hat{p}} = p = .1; \quad \sigma_{\hat{p}} = \sqrt{\tfrac{pq}{n}} = \sqrt{\tfrac{.1(.9)}{1000}} = .0095$$

a. $P[\hat{p} > .12] = P[z > \tfrac{.12 - .1}{.0095}] = P[z > 2.11] = .5 - .4826 = .0174$

b. $P[\hat{p} < .10] = P[z < \tfrac{.10 - .1}{.0095}] = P[z < 0] = .5$

c. $P[-.02 \leq (\hat{p} - p) \leq .02] = P[-2.11 \leq z \leq 2.11] = 2(.4826) = .9652$

6.22

The values $\sigma_{\hat{p}} = \sqrt{pq/n}$ for n = 100 and various values of p are tabulated and graphed below. Notice that $\sigma_{\hat{p}}$ is maximum for p = .5 and becomes very small for p near 0 and 1.

p	.01	.10	.30	.50	.70	.90	.99
$\sigma_{\hat{p}}$	.0099	.03	.0458	.05	.0458	.03	.0099

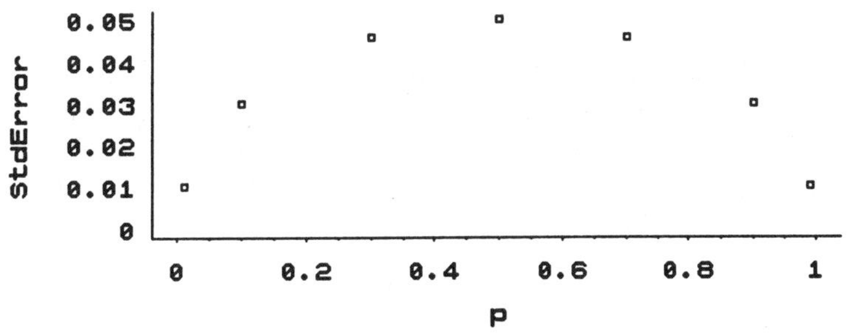

Figure 6.4

6.23

For fixed p, $\sigma_{\hat{p}} = \sqrt{pq/n}$. If n is increased, $\sigma_{\hat{p}}$ decreases. Similarly, since $\sigma_{\overline{x}} = \sigma/\sqrt{n}$, the standard deviation of $\overline{x}$ will decrease as n increases.

6.24

For n = 400 and p = .8, the sample proportion $\hat{p}$ is approximately normally distributed with

$$\mu_{\hat{p}} = p = .8; \quad \sigma_{\hat{p}} = \sqrt{\frac{pq}{n}} = \sqrt{\frac{.8(.2)}{400}} = .02$$

a. $P[\hat{p} > .83] = P[z > \frac{.83 - .8}{.02}] = P[z > 1.5] = .5 - .4332 = .0668$

b. $P[.76 < \hat{p} < .84] = P[\frac{.76 - .8}{.02} < z < \frac{.84 - .8}{.02}] = P[-2 < z < 2]$

$\qquad = 2(.4772) = .9544$

6.25

It is given that n = 1600 and $\hat{p}$ = .71, where p is the proportion of "upscale" men in the population who find shopping by computer convenient.

a. The sample proportion $\hat{p}$ is approximately normally distributed with mean p (unknown) and standard deviation

$$\sigma_{\hat{p}} = \sqrt{\frac{pq}{n}} \approx \sqrt{\frac{.71(.29)}{1600}} = .011344$$

b. $P[-.03 \leq (\hat{p} - p) \leq .03] = P[\frac{-.03}{.011344} \leq z \leq \frac{.03}{.011344}]$

$\qquad = P[-2.64 \leq z \leq 2.64] = 2(.4959) = .9918$

6.26

It is given that n = 250 and p is the proportion of people whose home cost more than the median home price, $112,000. Hence, by definition of the median, p = .5.

a. The sample proportion $\hat{p}$ is approximately normally distributed with mean p = .5 and standard deviation

$$\sigma_{\hat{p}} = \sqrt{\frac{pq}{n}} = \sqrt{\frac{.5(.5)}{250}} = .03162$$

b. $P[\hat{p} > .66] = P[z > \frac{\hat{p} - p}{\sigma_{\hat{p}}}] = P[z > \frac{.66 - .5}{.03162}] = P[z > 5.06] \approx 0$

c. If a sample of size n = 250 produces a sample proportion of $\hat{p} = \frac{165}{250} = .66$, this would be a very unlikely occurrence under the assumption that p = .5 (see part b).
Either we have observed a very unlikely event, or perhaps (1) the sample was not randomly selected from all home buyers in the first quarter of 1994, or (2) the median home price nationally may be incorrect.

6.27

It is given that n = 300 and p = .8, where p is the proportion of married people who feel they have an equal say in major home purchases.

a. The sample proportion $\hat{p}$ is approximately normally distributed with mean p = .8 and standard deviation

$$\sigma_{\hat{p}} = \sqrt{\frac{pq}{n}} = \sqrt{\frac{.8(.2)}{300}} = .02309$$

Then

$P[\hat{p} > .85] = P[z > \frac{\hat{p} - p}{\sigma_{\hat{p}}}] = P[z > \frac{.85 - .8}{.02309}] = P[z > 2.17] = .5 - .4850 = .0150$

b. Since $\hat{p}$ is approximately normally distributed, we would expect approximately 95% of the sample proportions to lie within two standard deviations of the mean. That is,

$$p \pm 2\sigma_{\hat{p}} \Rightarrow .8 \pm 2(.02309) \Rightarrow .8 \pm .046 \quad \text{or } .754 \text{ to } .846$$

c. $P[.75 < \hat{p} < .85] = P[-2.17 < z < 2.17] = 2(.4850) = .9700$

6.28
a. The upper and lower control limits are

$$\text{UCL} = \overline{\overline{x}} + 3\frac{s}{\sqrt{n}} = 20.74 + 3\frac{.87}{\sqrt{10}} = 20.74 + .83 = 21.57$$

$$\text{LCL} = \overline{\overline{x}} - 3\frac{s}{\sqrt{n}} = 20.74 - 3\frac{.87}{\sqrt{10}} = 20.74 - .83 = 19.91$$

b. Control charts are used to monitor the process variable, detecting shifts that might indicate control problems.

c. The control chart is constructed by plotting two horizontal lines, one the upper control limit and one the lower control limit (see Figure 6.11 in the text). Values of $\overline{x}$ are plotted, and should remain within the control limits. If not, the process should be checked.

6.29
This is similar to Exercise 6.28.

a. The upper and lower control limits are

$$\text{UCL} = \overline{\overline{x}} + 3\frac{s}{\sqrt{n}} = 155.9 + 3\frac{4.3}{\sqrt{5}} = 155.9 + 5.77 = 161.67$$

$$\text{LCL} = \overline{\overline{x}} - 3\frac{s}{\sqrt{n}} = 155.9 - 3\frac{4.3}{\sqrt{5}} = 155.9 - 5.77 = 150.13$$

b. The control chart is constructed by plotting two horizontal lines, one the upper control limit and one the lower control limit (see Figure 6.11 in the text). Values of $\overline{x}$ are plotted, and should remain within the control limits. If not, the process should be checked.

6.30
a. The upper and lower control limits are

$$\text{UCL} = \overline{\overline{x}} + 3\frac{s}{\sqrt{n}} = 10{,}752 + 3\frac{1605}{\sqrt{5}} = 10{,}752 + 2153.3 = 12{,}905.3$$

$$\text{LCL} = \overline{\overline{x}} - 3\frac{s}{\sqrt{n}} = 10{,}752 - 3\frac{1605}{\sqrt{5}} = 10{,}752 - 2153.3 = 8598.7$$

b. The $\overline{x}$ chart will allow the manager to monitor daily gains or losses to see whether there is a problem with any particular table.

6.31
The upper and lower control limits are

$$\text{UCL} = \overline{\overline{x}} + 3\frac{s}{\sqrt{n}} = 7.24 + 3\frac{.07}{\sqrt{3}} = 7.24 + .12 = 7.36$$

$$\text{LCL} = \overline{\overline{x}} - 3\frac{s}{\sqrt{n}} = 7.24 - 3\frac{.07}{\sqrt{3}} = 7.24 - .12 = 7.12$$

6.32
Using all 104 measurements, the value of s is calculated to be $s = .006717688$ and $\overline{\overline{x}} = .0256$. Then the upper and lower control limits are

$$UCL = \bar{x} + 3\frac{s}{\sqrt{n}} = .0256 + 3\frac{.006717688}{\sqrt{4}} = .0357$$

$$LCL = \bar{x} - 3\frac{s}{\sqrt{n}} = .0256 - 3\frac{.006717688}{\sqrt{4}} = .0155$$

6.33

a. Each student will obtain a slightly different sampling distribution.

b. Since the Central Limit Theorem should not be used with n = 3, the sample mean will not necessarily be normal, but will have mean $\mu = 3.5$ and standard deviation $\sigma/\sqrt{n} = 1.71/\sqrt{3} = .987$. Note that since n is small, the approximate normality is not too good, but the mean and standard deviation of the sampling distribution are still exact.

c. Answers will vary from student to student, but should be close to 3.5 and .987, respectively.

6.34

Answers will vary from student to student.

6.35

a. $C_2^4 = \frac{4!}{2!2!} = 6$ samples are possible.

b-c. The six samples along with the sample means for each are shown below.

Sample	Observations	$\bar{x}$
1	6, 1	3.5
2	6, 3	4.5
3	6, 2	4.0
4	1, 3	2.0
5	1, 2	1.5
6	3, 2	2.5

d. Since each of the six distinct values of $\bar{x}$ are equally likely (due to random sampling), the sampling distribution of $\bar{x}$ is given as

$$p(\bar{x}) = \frac{1}{6} \quad \text{for } \bar{x} = 1.5, 2, 2.5, 3.5, 4, 4.5$$

The sampling distribution is shown in Figure 6.5.

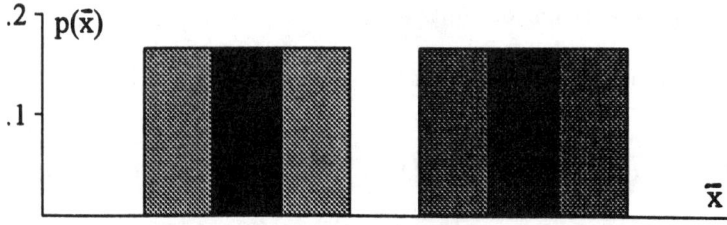

Figure 6.5

e. The population mean is $\mu = (6 + 1 + 3 + 2)/4 = 3$. Notice that none of the samples of size n = 2 produce a value of $\bar{x}$ exactly equal to the population mean.

6.36

Refer to Exercise 6.35. If samples of size n = 3 are drawn without replacement, there are four possible samples with sample means shown below.

Sample	Observations	$\bar{x}$
1	6, 1, 3	3.333
2	6, 1, 2	3
3	6, 3, 2	3.667
4	1, 3, 2	2

The sampling distribution of $\bar{x}$ is now

$$p(\bar{x}) = \tfrac{1}{4} \qquad \text{for } \bar{x} = 2, 3, 3.333 \text{ and } 3.667$$

The graph is shown in Figure 6.6.

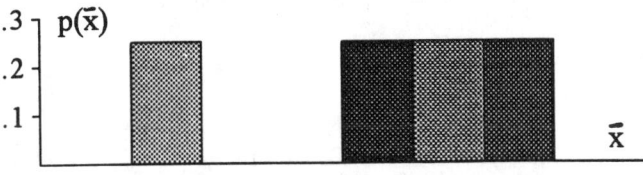

Figure 6.6

6.37

Refer to Exercise 6.35. If samples of size n = 3 are drawn without replacement, there are four possible samples with sample medians shown below.

Sample	Observations	Median (m)
1	6, 1, 3	3
2	6, 1, 2	2
3	6, 3, 2	3
4	1, 3, 2	2

The sampling distribution of the sample median is now

$$p(m) = \tfrac{1}{2} \qquad \text{for } m = 2 \text{ and } 3$$

The graph is shown in Figure 6.7.

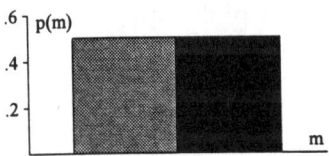

Figure 6.7

6.38

a. $E(\bar{x}) = \mu = 40$

b. $\sigma_{\bar{x}} = \frac{\sigma}{\sqrt{n}} = \frac{4}{\sqrt{100}} = .4$

c. Calculate
$$z = \frac{\bar{x} - \mu}{\sigma_{\bar{x}}} = \frac{41 - 40}{.4} = 2.5$$

Then
$$P[\bar{x} > 41] = P[z > 2.5] = .5 - .4938 = .0062$$

6.39

a. For n = 250, the sample proportion $\hat{p}$ is approximately normally distributed with mean p (unknown) and standard deviation
$$\sigma_{\hat{p}} = \sqrt{\frac{pq}{n}} \approx \sqrt{\frac{(.25)(.75)}{250}} = .02739$$

b. $P[-.01 < (\hat{p} - p) < .01] = P[\frac{-.01}{.02739} < z < \frac{.01}{.02739}]$
$$= P[-.37 < z < .37] = 2(.1443) = .2886$$

6.40

It is given that $\mu = 1.4$ and $\sigma = .7$, although x, the service time for one vehicle, need not be normally distributed. When n = 50, the probability of interest is

$$P[\bar{x} > 1.6] = P[z > \frac{1.6 - 1.4}{.7/\sqrt{50}}] = P[z > 2.02] = .5 - .4783 = .0217$$

6.41

a. It is given that $\mu = 1085$. If the range $R = 1135 - 1005 = 130$ represents three standard deviations on either side of the mean (or 6σ), then $6\sigma = 130$ or $\sigma = 21.67$. The sample mean $\bar{x}$ is approximately normally distributed, according to the Central Limit Theorem, with mean $\mu = 1085$ and standard deviation

$$\sigma/\sqrt{n} \approx 21.67/\sqrt{100} = 2.167$$

b. $P[\bar{x} > 1090] = P[z > \frac{1090 - 1085}{2.167}] = P[z > 2.31] = .5 - .4896 = .0104$

c. $P[\bar{x} < 1078] = P[z < \frac{1078 - 1085}{2.167}] = P[z < -3.23] \approx 0$

If the observed sample mean is $\bar{x} = 1078$, a very unlikely occurrence, we might conclude that one of our original assumptions, either about the mean, the standard deviation, or perhaps the randomness of our sample should be questioned.

6.42

a-b. It is given that $\mu = 31{,}256$, n = 25 and $\sigma = 1550$. The sample mean $\bar{x}$ is approximately normally distributed, according to the Central Limit Theorem, with mean $\mu = 31{,}256$ and standard deviation

$$\sigma/\sqrt{n} \approx 1550/\sqrt{25} = 310$$

c. $P[\bar{x} > 32000] = P[z > \frac{32000 - 31256}{310}] = P[z > 2.40] = .5 - .4918 = .0082$

and $P[\bar{x} > 33000] = P[z > \frac{33000 - 31256}{310}] = P[z > 5.63] \approx 0$

d. Since $\bar{x}$ is approximately normally distributed, approximately 95% of the values of $\bar{x}$ should lie within 1.96 (or 2) standard deviations of the mean $\mu = 31{,}256$. That is, the sample mean $\bar{x}$ should lie between

$$\mu \pm 2 \frac{\sigma}{\sqrt{n}} \;\Rightarrow\; 31256 \pm 2 \frac{1550}{\sqrt{25}} \;\Rightarrow\; 31256 \pm 620$$

or \$30,636 to \$31,876.

6.43

It is given that $n = 1000$ and $p = .86$, where p is the proportion of freshmen receiving financial aid from parents or family.

a. The sample proportion $\hat{p}$ is approximately normally distributed with mean $p = .86$ and standard deviation

$$\sigma_{\hat{p}} = \sqrt{\frac{pq}{n}} = \sqrt{\frac{.86(.14)}{1000}} = .01097$$

b. $P[-.02 < (\hat{p} - p) < .02] = P[\frac{-.02}{.01097} < z < \frac{.02}{.01097}]$

$= P[-1.82 < z < 1.82] = 2(.4656) = .9312$

c. $P[\hat{p} > .90] = P[z > \frac{\hat{p} - p}{\sigma_{\hat{p}}}] = P[z > \frac{.90 - .86}{.01097}] = P[z > 3.65] \approx 0$

This is not a likely event, and we would not expect to see a sample proportion in this range.

6.44

a. The sample proportion $\hat{p}$ has mean $p = .05$ and standard deviation

$$\sigma_{\hat{p}} = \sqrt{\frac{pq}{n}} = \sqrt{\frac{.05(.95)}{500}} = .00975$$

b. The interval $p \pm 3\sqrt{\frac{pq}{n}} \;\Rightarrow\; .05 \pm 3(.00975) \;\Rightarrow\; .05 \pm .02925$ or .02075 to .07925 does lie in the range of $\hat{p}$, that is, between 0 and 1. Therefore, the Central Limit Theorem will apply, and the sampling distribution of $\hat{p}$ will be approximately normal.

c. $P[-.01 < (\hat{p} - p) < .01] = P[\frac{-.01}{.00975} < z < \frac{.01}{.00975}]$

$= P[-1.03 < z < 1.03] = 2(.3485) = .6970$

Hence, the probability that $\hat{p}$ will differ from p by more than .01 is $1 - .6970 = .3030$.

6.45

a. The sample proportion $\hat{p}$ has mean $p = .10$ and standard deviation

$$\sigma_{\hat{p}} = \sqrt{\frac{pq}{n}} = \sqrt{\frac{.10(.90)}{100}} = .03$$

The interval $p \pm 3\sqrt{\frac{pq}{n}} \;\Rightarrow\; .10 \pm 3(.03) \;\Rightarrow\; .10 \pm .09$ or .01 to .19 does lie in the range of $\hat{p}$, that is, between 0 and 1. Therefore, the Central Limit Theorem will apply, and the sampling distribution of $\hat{p}$ will be approximately normal. This approximation will be good.

b. $P[-.05 \leq (\hat{p} - p) \leq .05] = P[\frac{-.05}{.03} \leq z \leq \frac{.05}{.03}]$

$= P[-1.67 \leq z \leq 1.67] = 2(.4525) = .9050$

CHAPTER 7
Estimation of Means and Proportions

7.1
The margin of error in estimation provides a practical upper bound to the difference between a particular estimate and the parameter that it estimates. In this chapter, the margin of error is 1.96 x (standard error of the estimator).

7.2
For the estimate of μ given as $\bar{x}$, the margin of error is $1.96\sigma_{\bar{x}} = 1.96\frac{\sigma}{\sqrt{n}}$.

a. $1.96 \frac{2}{\sqrt{40}} = .620$ **c.** $1.96 \sqrt{\frac{12}{50}} = .960$

b. $1.96 \sqrt{\frac{.9}{100}} = .186$

Notice that, since σ is known, we have no need to use s. Hence, a large sample size is not imperative. The margin of error will still be quite accurate.

7.3
This is similar to Exercise 7.2.

a. $1.96 \frac{.1}{\sqrt{50}} = .028$ **c.** $1.96 \frac{.01}{\sqrt{100}} = .00196$

b. $1.96 \frac{9}{\sqrt{100}} = 1.764$

7.4
A 95% confidence interval for the population mean μ is given by

$$\bar{x} \pm 1.96 \frac{\sigma}{\sqrt{n}}$$

where σ can be estimated by the sample standard deviation s for large values of n.

a. $13.1 \pm 1.96 \sqrt{\frac{3.42}{36}} \Rightarrow 13.1 \pm .604$ or $12.496 < \mu < 13.704$

b. $2.73 \pm 1.96 \sqrt{\frac{.1047}{64}} \Rightarrow 2.73 \pm .079$ or $2.651 < \mu < 2.809$

c. $28.6 \pm 1.96 \sqrt{\frac{1.09}{41}} \Rightarrow 28.6 \pm .320$ or $28.280 < \mu < 28.920$

Intervals constructed in this manner will enclose the true value of μ 95% of the time in repeated sampling. Hence, we are fairly confident that these particular intervals will enclose μ.

7.5
This is similar to Exercise 7.4, with a 90% confidence interval for μ given as

$$\bar{x} \pm 1.645 \frac{\sigma}{\sqrt{n}}$$

where σ can be estimated by the sample standard deviation s for large values of n.

a. $.84 \pm 1.645 \sqrt{\frac{.086}{125}} \Rightarrow .84 \pm .043$ or $.797 < \mu < .883$

b. $21.9 \pm 1.645 \sqrt{\frac{3.44}{50}} \Rightarrow 21.9 \pm .431$ or $21.469 < \mu < 22.331$

c. $907 \pm 1.645 \sqrt{\frac{128}{46}} \Rightarrow 907 \pm 2.744$ or $904.256 < \mu < 909.744$

Intervals constructed in this manner will enclose the true value of μ 90% of the time in repeated sampling. Hence, we are fairly confident that these particular intervals will enclose μ.

7.6

a. $\bar{x} \pm z_{.005} \frac{\sigma}{\sqrt{n}} \Rightarrow \bar{x} \pm 2.58 \frac{\sigma}{\sqrt{n}} \approx 34 \pm 2.58 \sqrt{\frac{12}{38}} \Rightarrow 34 \pm 1.450$

or $32.550 < \mu < 35.450$.

b. $\bar{x} \pm z_{.05} \frac{\sigma}{\sqrt{n}} \Rightarrow \bar{x} \pm 1.645 \frac{\sigma}{\sqrt{n}} \approx 1049 \pm 1.645 \sqrt{\frac{51}{65}} \Rightarrow 1049 \pm 1.457$

or $1047.543 < \mu < 1050.457$.

c. $\bar{x} \pm z_{.025} \frac{\sigma}{\sqrt{n}} \Rightarrow \bar{x} \pm 1.96 \frac{\sigma}{\sqrt{n}} \approx 66.3 \pm 1.96 \sqrt{\frac{2.48}{89}} \Rightarrow 66.3 \pm .327$

or $65.973 < \mu < 66.627$.

7.7

The width of a 95% confidence interval for μ is given as $1.96 \frac{\sigma}{\sqrt{n}}$. Hence,

a. When n = 100, the width is $2(1.96 \frac{10}{\sqrt{100}}) = 2(1.96) = 3.92$.

b. When n = 200, the width is $2(1.96 \frac{10}{\sqrt{200}}) = 2(1.386) = 2.772$.

c. When n = 400, the width is $2(1.96 \frac{10}{\sqrt{400}}) = 2(.98) = 1.96$.

7.8

Refer to Exercise 7.7.
a. When the sample size is doubled, the width is decreased by $1/\sqrt{2}$.

b. When the sample size is quadrupled, the width is decreased by $1/\sqrt{4} = 1/2$.

7.9

a. A 90% confidence interval for μ is

$$\bar{x} \pm 1.645 \frac{\sigma}{\sqrt{n}}$$

Hence, its width is

$$2\left(1.645 \frac{\sigma}{\sqrt{n}}\right) = 2\left(1.645 \frac{10}{\sqrt{100}}\right) = 2(1.645) = 3.29$$

b. A 99% confidence interval for μ is

$$\bar{x} \pm 2.58 \frac{\sigma}{\sqrt{n}}$$

Hence, its width is

$$2\left(2.58 \frac{\sigma}{\sqrt{n}}\right) = 2\left(2.58 \frac{10}{\sqrt{100}}\right) = 2(2.58) = 5.16$$

c. Notice that as the confidence coefficient increases, so does the width of the confidence interval. If we wish to be more confident of enclosing the unknown parameter, we must make the interval wider.

7.10

The point estimate of μ is $\bar{x} = 7.2\%$ and the margin of error in estimation with $s = 5.6\%$ and $n = 200$ is

$$1.96\sigma_{\bar{x}} = 1.96\frac{\sigma}{\sqrt{n}} \approx 1.96\frac{s}{\sqrt{n}} = 1.96\left(\frac{5.6}{\sqrt{200}}\right) = .776$$

7.11

a. It is given that $\sigma = 2000$, $n = 40$, $\bar{x} = 930$, and that the distribution from which we are sampling is not normal. However, since n is large, the sampling distribution of $\bar{x}$ is *approximately* normal. Hence, the margin of error in estimation is

$$1.96\sigma_{\bar{x}} = 1.96\frac{\sigma}{\sqrt{n}} = 1.96\frac{2000}{\sqrt{40}} = 619.806$$

b. The probability that the error of estimation will be less than 619.806 is approximately .95, according to the Central Limit Theorem.

7.12

a. A 95% confidence interval for μ is

$$\bar{x} \pm 1.96\frac{\sigma}{\sqrt{n}} \Rightarrow 32{,}600 \pm 1.96\frac{1000}{\sqrt{100}} \Rightarrow 32{,}600 \pm 196$$

or $32,404 to $32,796.

b. A point estimate for the mean starting salary for Marketing/Sales is $\bar{x} = \$24{,}100$, and the margin of error is

$$1.96\frac{\sigma}{\sqrt{n}} \approx 1.96\frac{800}{\sqrt{100}} = 156.8$$

c. A 98% confidence interval for μ is

$$\bar{x} \pm 2.33\frac{\sigma}{\sqrt{n}} \Rightarrow 22{,}909 \pm 2.33\frac{800}{\sqrt{100}} \Rightarrow 22{,}909 \pm 186.4$$

or $22,722.60 to $23,095.40.

7.13

For this exercise, $n = 32$, $\bar{x} = 11.7$, and $s = 2.1$. The approximate 90% confidence interval for μ, the mean of the population of forecasts of all economic forecasters, is

$$\bar{x} \pm z_{.05}\frac{s}{\sqrt{n}} \Rightarrow 11.7 \pm 1.645\frac{2.1}{\sqrt{32}} \Rightarrow 11.7 \pm .611$$

or $11.089 < \mu < 12.311$.

7.14

For this exercise, $n = 30$, $\bar{x} = 42.1$, and $s = 19.6$. The approximate 90% confidence interval for μ is

$$\bar{x} \pm z_{.05} \tfrac{s}{\sqrt{n}} \Rightarrow 42.1 \pm 1.645 \tfrac{19.6}{\sqrt{30}} \Rightarrow 42.1 \pm 5.887$$

or $36.213 < \mu < 47.987$.

7.15

It is given that $\bar{x} = 5474$, $s = 764$, and $n = 36$. The upper one-sided 90% confidence interval is approximately

$$\bar{x} + 1.28 \tfrac{s}{\sqrt{n}} = 5474 + 1.28 \left(\tfrac{764}{6}\right) = 5636.987$$

or $\mu < 5636.987$.

7.16

Refer to Table 4, Appendix II, indexing d.f. along the left or right margin and t_α across the top.

a. $t_{.05} = 2.015$ with 5 d.f. **b.** $t_{.025} = 2.306$ with 8 d.f.

c. $t_{.10} = 1.330$ with 18 d.f. **d.** $t_{.025} = 1.96$ with 30 d.f.

7.17

The value $P[t > t_a] = a$ is the tabled entry for a particular number of degrees of freedom.

a. $t_{.10} = 1.356$ with 12 d.f. **b.** $t_{.01} = 2.485$ with 25 d.f.

c. $t_{.05} = 1.746$ with 16 d.f.

7.18

Small sample confidence intervals are quite similar to their large sample counterparts; however, these intervals must be based on the *t-distribution*. Thus, the confidence interval for the single population mean described in this exercise will be

$$\bar{x} \pm t_{\alpha/2} \tfrac{s}{\sqrt{n}}$$

where $t_{\alpha/2}$ is a value of t (Table 4) based on $(n-1)$ degrees of freedom that has area $\alpha/2$ to its right (see Figure 7.1).

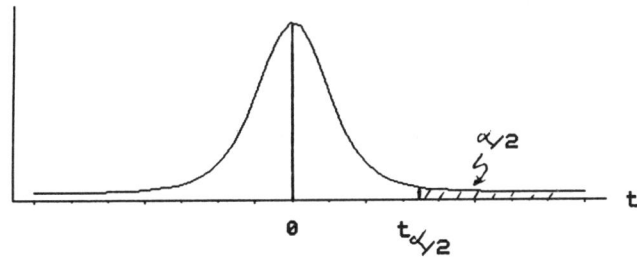

Figure 7.1

a. Using the formulas given in Chapter 2, calculate $\Sigma x_i = 38$ and $\Sigma x_i^2 = 242.28$. Then

$$\bar{x} = \tfrac{\Sigma x_i}{n} = \tfrac{38}{6} = 6.333$$

$$s^2 = \frac{\Sigma x_i^2 - \frac{(\Sigma x_i)^2}{n}}{n-1} = \frac{242.28 - \frac{(38)^2}{6}}{5} = .322667 \text{ and } s = \sqrt{.322667} = .5680$$

b. Indexing $t_{.05}$ with $n - 1 = 5$ degrees of freedom in Table 4, we have $t_{.05} = 2.015$. Hence the 90% confidence interval is

$$\bar{x} \pm t_{.05} \frac{s}{\sqrt{n}} \Rightarrow 6.333 \pm 2.015 \sqrt{\frac{.322667}{6}} \Rightarrow 6.333 \pm .467$$

or $5.866 < \mu < 6.800$.

c. Indexing $t_{.025}$ with $n - 1 = 5$ degrees of freedom in Table 4, we have $t_{.025} = 2.571$. Hence the 95% confidence interval is

$$\bar{x} \pm t_{.025} \frac{s}{\sqrt{n}} \Rightarrow 6.333 \pm 2.571 \sqrt{\frac{.322667}{6}} \Rightarrow 6.333 \pm .596$$

or $5.737 < \mu < 6.929$.

d. Intervals constructed using this procedure will enclose μ 90% (or 95%) of the time in repeated sampling. Hence, we are fairly certain that this particular interval encloses μ.

7.19

For this exercise, $n = 12$, $\bar{x} = 125.12$, and $s = 12.3$.

a. Indexing $t_{.01}$ with $n - 1 = 11$ degrees of freedom in Table 4, we have $t_{.01} = 2.718$. Hence the 98% confidence interval is

$$\bar{x} \pm t_{.01} \frac{s}{\sqrt{n}} \Rightarrow 125.12 \pm 2.718 \frac{12.3}{\sqrt{12}} \Rightarrow 125.12 \pm 9.651$$

or $115.469 < \mu < 134.771$.

b. With $t_{.005} = 3.106$, the 99% confidence interval is

$$\bar{x} \pm t_{.005} \frac{s}{\sqrt{n}} \Rightarrow 125.12 \pm 3.106 \frac{12.3}{\sqrt{12}} \Rightarrow 125.12 \pm 11.028$$

or $114.092 < \mu < 136.148$.

c. Intervals constructed using this procedure will enclose μ 98% (or 99%) of the time in repeated sampling. Hence, we are fairly certain that this particular interval encloses μ.

7.20

a. Indexing $t_{.05}$ with $n - 1 = 9$ degrees of freedom in Table 4, we have $t_{.05} = 1.833$. Hence the 90% confidence interval is

$$\bar{x} \pm t_{.05} \frac{s}{\sqrt{n}} \Rightarrow 8.5 \pm 1.833 \frac{.23}{\sqrt{10}} \Rightarrow 8.5 \pm .133$$

or $8.367 < \mu < 8.633$.

b. The 90% confidence interval does not contain the mean 8.7% rate that prevails in the bank's market area. Hence, we can conclude with a fair amount of certainty that the mean of the population from which we are sampling is not 8.7%. Perhaps the loan applicants at this particular bank do not represent a random sample of all applicants in the market area.

7.21

a. Calculate $\Sigma x_i = 43{,}300$ and $\Sigma x_i^2 = 384{,}221{,}400$. Then

$$\bar{x} = \frac{\Sigma x_i}{n} = \frac{43{,}300}{5} = 8660$$

$$s^2 = \frac{\Sigma x_i^2 - \frac{(\Sigma x_i)^2}{n}}{n-1} = \frac{384{,}221{,}400 - \frac{(43{,}300)^2}{5}}{4} = 2{,}310{,}850$$

Then $t_{.025} = 2.776$ and the 95% confidence interval is

$$\bar{x} \pm t_{.025} \frac{s}{\sqrt{n}} = 8660 \pm 2.776 \sqrt{\frac{2{,}310{,}850}{5}} = 8660 \pm 1887.210$$

or $6772.79 < \mu < 10{,}547.21$.

b. Since $8500 is included in the interval constructed in part a, we cannot be certain that the average profit will exceed $8500.

7.22

a. Calculate $\Sigma x_i = 10{,}854$ and $\Sigma x_i^2 = 5{,}837{,}026$. Then

$$\bar{x} = \frac{\Sigma x_i}{n} = \frac{10{,}854}{21} = 516.857$$

$$s^2 = \frac{\Sigma x_i^2 - \frac{(\Sigma x_i)^2}{n}}{n-1} = \frac{5{,}837{,}026 - \frac{(10{,}854)^2}{21}}{20} = 11{,}352.92857 \text{ and } s = 106.550$$

b. The 99% confidence interval for μ is

$$\bar{x} \pm t_{.005} \frac{s}{\sqrt{n}} \Rightarrow 516.857 \pm 2.845 \sqrt{\frac{11352.92857}{21}} \Rightarrow 516.857 \pm 66.150$$

or $450.707 < \mu < 583.007$.

c. If $x = 1085$ is added to the $n = 21$ observations, then

$$\bar{x} = \frac{\Sigma x_i}{n} = \frac{11939}{22} = 542.682$$

$$s^2 = \frac{\Sigma x_i^2 - \frac{(\Sigma x_i)^2}{n}}{n-1} = \frac{7{,}014{,}251 - \frac{(11939)^2}{22}}{21} = 25{,}484.41775$$

and the 99% confidence interval for μ is

$$\bar{x} \pm t_{.005} \frac{s}{\sqrt{n}} = 542.682 \pm 2.831 \sqrt{\frac{25484.41775}{22}} = 542.682 \pm 96.353$$

or $446.329 < \mu < 639.035$.

d. Yes, since the value $x = 1085$ is not included in the 99% confidence interval calculated in part c.

7.23

a. Calculate $\Sigma x_i = 171.6$ and $\Sigma x_i^2 = 2699.08$. Then

$$\bar{x} = \frac{\Sigma x_i}{n} = \frac{171.6}{11} = 15.6$$

$$s^2 = \frac{\Sigma x_i^2 - \frac{(\Sigma x_i)^2}{n}}{n-1} = \frac{2699.08 - \frac{(171.6)^2}{11}}{10} = 2.212$$

From Table 4, $t_{.01} = 2.764$ and the 98% confidence interval is

$$\bar{x} \pm t_{.01} \frac{s}{\sqrt{n}} \Rightarrow 15.6 \pm 2.764 \sqrt{\frac{2.212}{11}} \Rightarrow 15.6 \pm 1.239$$

or $14.361 < \mu < 16.839$.

b. No. Projections are never completely accurate, and it is very rare that a projected figure equals the actual figure.

7.24

When estimating the difference $\mu_1 - \mu_2$ using the sample estimate $\bar{x}_1 - \bar{x}_2$, the margin of error is

$$1.96 \sqrt{\frac{\sigma_1^2}{n_1} + \frac{\sigma_2^2}{n_2}}$$

Estimating σ_1^2 and σ_2^2 with s_1^2 and s_2^2, the approximate bound is

$$1.96 \sqrt{\frac{1.38}{35} + \frac{4.14}{49}} = .690$$

7.25

The 90% confidence interval for $\mu_1 - \mu_2$ is

$$(\bar{x}_1 - \bar{x}_2) \pm 1.645 \sqrt{\frac{\sigma_1^2}{n_1} + \frac{\sigma_2^2}{n_2}}$$

Estimating σ_1^2 and σ_2^2 with s_1^2 and s_2^2, the approximate interval is

$$(2.9 - 5.1) \pm 1.645 \sqrt{\frac{.83}{64} + \frac{1.67}{64}} = -2.2 \pm .325$$

or $-2.525 < (\mu_1 - \mu_2) < -1.875$.

Intervals constructed in this manner will enclose $(\mu_1 - \mu_2)$ 90% of the time. Hence, we are fairly certain that this particular interval encloses $(\mu_1 - \mu_2)$.

7.26

The degrees of freedom for s^2, the pooled estimator of σ^2, are $n_1 + n_2 - 2$.

a. $n_1 + n_2 - 2 = 16 + 8 - 2 = 22$

b. $n_1 + n_2 - 2 = 10 + 12 - 2 = 20$

c. $n_1 + n_2 - 2 = 15 + 3 - 2 = 16$

7.27

a. $s^2 = \frac{(n_1 - 1)s_1^2 + (n_2 - 1)s_2^2}{n_1 + n_2 - 2} = \frac{9(3.4) + 3(4.9)}{10 + 4 - 2} = 3.775$

b. $s^2 = \dfrac{(n_1 - 1)s_1^2 + (n_2 - 1)s_2^2}{n_1 + n_2 - 2} = \dfrac{11(18) + 20(23)}{12 + 21 - 2} = 21.2258$

7.28

When the actual data are given and s^2 must be calculated, the calculation is done using the shortcut formula, noting that

$$\Sigma(x_i - \bar{x})^2 = \Sigma x_i^2 - \dfrac{(\Sigma x_i)^2}{n}$$

Hence, for the pooled estimator,

$$s^2 = \dfrac{\Sigma x_{1i}^2 - \dfrac{(\Sigma x_{1i})^2}{n_1} + \Sigma x_{2i}^2 - \dfrac{(\Sigma x_{2i})^2}{n_2}}{n_1 + n_2 - 2}$$

The preliminary calculations are shown below:

Sample 1	Sample 2
$\Sigma x_{1i} = 28$	$\Sigma x_{2i} = 43$
$\Sigma x_{1i}^2 = 242$	$\Sigma x_{2i}^2 = 411$
$n_1 = 4$	$n_2 = 5$

Then

$$s^2 = \dfrac{242 - \dfrac{(28)^2}{4} + 411 - \dfrac{(43)^2}{5}}{4 + 5 - 2} = \dfrac{46 + 41.2}{7} = 12.45714286$$

7.29

Refer to Exercise 7.28. A 95% confidence interval for $(\mu_1 - \mu_2)$ is given as

$$(\bar{x}_1 - \bar{x}_2) \pm t_{.025}\sqrt{s^2\left(\dfrac{1}{n_1} + \dfrac{1}{n_2}\right)}$$

$$(7 - 8.6) \pm 2.365\sqrt{12.4571\left(\dfrac{1}{4} + \dfrac{1}{5}\right)}$$

$$-1.6 \pm 5.599 \quad\text{or}\quad -7.199 < (\mu_1 - \mu_2) < 3.999$$

7.30

The pooled estimator of σ^2 is calculated as

$$s^2 = \dfrac{(n_1 - 1)s_1^2 + (n_2 - 1)s_2^2}{n_1 + n_2 - 2} = \dfrac{15(4.8) + 12(5.9)}{16 + 13 - 2} = 5.2889$$

and the 99% confidence interval for $(\mu_1 - \mu_2)$ is given as

$$(\bar{x}_1 - \bar{x}_2) \pm t_{.005}\sqrt{s^2\left(\dfrac{1}{n_1} + \dfrac{1}{n_2}\right)}$$

$$(34.6 - 32.2) \pm 2.771\sqrt{5.2889\left(\dfrac{1}{16} + \dfrac{1}{13}\right)}$$

$$2.4 \pm 2.380 \quad\text{or}\quad 0.020 < (\mu_1 - \mu_2) < 4.780$$

7.31

This is similar to previous exercises. The 90% confidence interval for $\mu_1 - \mu_2$ is approximately

$$(\bar{x}_1 - \bar{x}_2) \pm 1.645 \sqrt{\frac{s_1^2}{n_1} + \frac{s_2^2}{n_2}}$$

$$(2.4 - 3.1) \pm 1.645 \sqrt{\frac{1.44}{100} + \frac{2.64}{100}}$$

$$-0.7 \pm .332 \quad \text{or} \quad -1.032 < (\mu_1 - \mu_2) < -0.368$$

Intervals constructed in this manner will enclose $(\mu_1 - \mu_2)$ 90% of the time. Hence, we are fairly certain that this particular interval encloses $(\mu_1 - \mu_2)$.

7.32

a. The parameter to be estimated is μ, the mean number of ships passing within 10 miles of the proposed power site location per day. The 95% confidence interval is approximately

$$\bar{x} \pm 1.96 \frac{s}{\sqrt{n}} \Rightarrow 7.2 \pm 1.96 \sqrt{\frac{8.8}{60}} \Rightarrow 7.2 \pm .751 \quad \text{or} \quad 6.449 < \mu < 7.951$$

b. Now we are interested in the difference between means, $\mu_1 - \mu_2$, for summer versus winter months. The 90% confidence interval for $\mu_1 - \mu_2$ is approximately

$$(\bar{x}_1 - \bar{x}_2) \pm 1.645 \sqrt{\frac{s_1^2}{n_1} + \frac{s_2^2}{n_2}}$$

$$(7.2 - 4.7) \pm 1.645 \sqrt{\frac{8.8}{60} + \frac{4.9}{90}}$$

$$2.5 \pm .738 \quad \text{or} \quad 1.762 < (\mu_1 - \mu_2) < 3.238$$

c. The population used is the difference between the mean number of ships sighted in summer months and the mean number sighted in winter months for all summers and winters. One possible problem with the samples is that the months were not chosen independently or randomly and all of the months were chosen in the same year. Practically, it would be nearly impossible to choose them independently and randomly.

7.33

a. The pooled estimator of σ^2 is calculated as

$$s^2 = \frac{(n_1 - 1)s_1^2 + (n_2 - 1)s_2^2}{n_1 + n_2 - 2} = \frac{7(24,400)^2 + 7(28,600)^2}{8 + 8 - 2} = 706,660,000$$

and the 99% confidence interval for $(\mu_1 - \mu_2)$ is given as

$$(\bar{x}_1 - \bar{x}_2) \pm t_{.005} \sqrt{s^2 \left(\frac{1}{n_1} + \frac{1}{n_2}\right)}$$

$$(106,200 - 111,900) \pm 2.977 \sqrt{706,660,000 \left(\frac{1}{8} + \frac{1}{8}\right)}$$

$$-5700 \pm 39,568.911 \quad \text{or} \quad -45,268.911 < (\mu_1 - \mu_2) < 33,868.911$$

b. Since the confidence interval contains the value $(\mu_1 - \mu_2) = 0$, there is no evidence to indicate that μ_1 is not equal to μ_2. That is, the value $(\mu_1 - \mu_2) = 0$ or $\mu_1 = \mu_2$ is not unlikely.

7.34

The 98% confidence interval for $\mu_1 - \mu_2$ is

$$(\bar{x}_1 - \bar{x}_2) \pm z_{.01} \sqrt{\frac{\sigma_1^2}{n_1} + \frac{\sigma_2^2}{n_2}}$$

Estimating σ_1^2 and σ_2^2 with s_1^2 and s_2^2, the approximate interval is

$$(78{,}100 - 82{,}700) \pm 2.33 \sqrt{\frac{(6300)^2}{57} + \frac{(7100)^2}{66}} = -4600 \pm 2815.450$$

or $-7415.45 < (\mu_1 - \mu_2) < -1784.55$.

Intervals constructed in this manner will enclose $(\mu_1 - \mu_2)$ 98% of the time. Hence, we are fairly certain that this particular interval encloses $(\mu_1 - \mu_2)$.

7.35

The preliminary calculations are shown below:

Sample 1	Sample 2
$\Sigma x_{1i} = 17.2$	$\Sigma x_{2i} = 18.4$
$\Sigma x_{1i}^2 = 59.86$	$\Sigma x_{2i}^2 = 68.1$
$n_1 = 5$	$n_2 = 5$

The pooled estimator of σ^2 is calculated as,

$$s^2 = \frac{\Sigma x_{1i}^2 - \frac{(\Sigma x_{1i})^2}{n_1} + \Sigma x_{2i}^2 - \frac{(\Sigma x_{2i})^2}{n_2}}{n_1 + n_2 - 2}$$

$$= \frac{59.86 - \frac{(17.2)^2}{5} + 68.1 - \frac{(18.4)^2}{5}}{5 + 5 - 2} = \frac{.692 + .388}{8} = .135$$

Then the 90% confidence interval for $(\mu_1 - \mu_2)$ is given as

$$(\bar{x}_1 - \bar{x}_2) \pm t_{.05}\sqrt{s^2\left(\frac{1}{n_1} + \frac{1}{n_2}\right)}$$

$$(3.44 - 3.68) \pm 1.86\sqrt{.135\left(\frac{1}{5} + \frac{1}{5}\right)}$$

$$-0.24 \pm .432 \quad \text{or} \quad -.672 < (\mu_1 - \mu_2) < .192$$

7.36

When sampling from a binomial distribution with large n, the 99% confidence interval for p is given as

$$\hat{p} \pm 2.58 \sqrt{\frac{pq}{n}}$$

Since p is unknown, we use $\hat{p}$ as an approximation to p and obtain the approximate confidence interval

$$\hat{p} \pm 2.58 \sqrt{\frac{\hat{p}\hat{q}}{n}} \quad \text{where} \quad \hat{p} = \frac{x}{n} = \frac{655}{900} = .728$$

$$.728 \pm 2.58 \sqrt{\frac{.728(.272)}{900}}$$

$$.728 \pm .038 \quad \text{or} \quad .690 < p < .766$$

Intervals constructed in this manner will enclose p 99% of the time. Hence, we are fairly certain that this particular interval encloses p.

7.37

Calculate $\hat{p} = \frac{x}{n} = \frac{263}{300} = .877$. Then an approximate 90% confidence interval for p is

$$\hat{p} \pm 1.645 \sqrt{\frac{\hat{p}\hat{q}}{n}} \Rightarrow .877 \pm 1.645 \sqrt{\frac{.877(.123)}{300}} \Rightarrow .877 \pm .031$$

or $.846 < p < .908$.

7.38

Calculate $\hat{p} = \frac{x}{n} = \frac{140}{500} = .280$. Then an approximate 95% confidence interval for p is

$$\hat{p} \pm 1.96 \sqrt{\frac{\hat{p}\hat{q}}{n}} \Rightarrow .280 \pm 1.96 \sqrt{\frac{.28(.72)}{500}} \Rightarrow .280 \pm .039$$

or $.241 < p < .319$.

7.39

Calculate $\hat{p} = \frac{x}{n} = \frac{27}{500} = .054$. Then an approximate 95% confidence interval for p is

$$\hat{p} \pm 1.96 \sqrt{\frac{\hat{p}\hat{q}}{n}} \Rightarrow .054 \pm 1.96 \sqrt{\frac{.054(.946)}{500}} \Rightarrow .054 \pm .020$$

or $.034 < p < .074$. Notice that the interval is narrower than the one calculated in Exercise 7.38, even though the confidence coefficient and n are the same. This is because the value of p (estimated by $\hat{p}$) is quite close to 0, causing $\sigma_{\hat{p}}$ to be small.

7.40

a. For this sample, $\hat{p} = .80$ and the 95% confidence interval is

$$\hat{p} \pm 1.96 \sqrt{\frac{\hat{p}\hat{q}}{n}} \Rightarrow .80 \pm 1.96 \sqrt{\frac{.80(.20)}{3000}} \Rightarrow .80 \pm .014$$

or $.786 < p < .814$.

b. $\hat{p} \pm 1.96 \sqrt{\frac{\hat{p}\hat{q}}{n}} \Rightarrow .58 \pm 1.96 \sqrt{\frac{.58(.42)}{3000}} \Rightarrow .58 \pm .018$

or $.562 < p < .598$.

c. The point estimate for p is $\hat{p} = .56$ and the margin of error is

$$1.96 \sqrt{\frac{\hat{p}\hat{q}}{n}} = 1.96 \sqrt{\frac{.56(.44)}{3000}} = .018$$

7.41

a. With $n = 7000$ and $\hat{p} = .80$, the point estimate of p is $\hat{p} = .80$ and the approximate margin of error in estimation is

$$1.96 \sqrt{\frac{\hat{p}\hat{q}}{n}} = 1.96 \sqrt{\frac{.80(.20)}{7000}} = .0094$$

The error is far less than three percentage points, or .03.

b. The 95% confidence interval for p is then
$$\hat{p} \pm 1.96\sqrt{\tfrac{\hat{p}\hat{q}}{n}} \Rightarrow .80 \pm 1.96\sqrt{\tfrac{.80(.20)}{7000}} \Rightarrow .80 \pm .0094$$
or $.7906 < p < .8094$.

7.42
a. With $n = 1600$ and $\hat{p} = .24$, the approximate 90% confidence interval for p is
$$\hat{p} \pm 1.645\sqrt{\tfrac{\hat{p}\hat{q}}{n}} \Rightarrow .24 \pm 1.645\sqrt{\tfrac{.24(.76)}{1600}} \Rightarrow .24 \pm .018$$
or $.222 < p < .258$.

b. The point estimate of p is $\hat{p} = .22$ and the approximate margin of error is
$$1.96\sqrt{\tfrac{\hat{p}\hat{q}}{n}} = 1.96\sqrt{\tfrac{.22(.78)}{1600}} = .020$$

7.43
For $n = 1000$, the margins of error for the three population percentages are

(1) $1.96\sqrt{\tfrac{.54(.46)}{1000}} = .031$ (2) $1.96\sqrt{\tfrac{.52(.48)}{1000}} = .031$

(3) $1.96\sqrt{\tfrac{.44(.56)}{1000}} = .031$

7.44
a. The approximate margin of error is
$$1.96\sqrt{\tfrac{\hat{p}\hat{q}}{n}} = 1.96\sqrt{\tfrac{.57(.43)}{1002}} = .031$$

b. The approximate margin of error is
$$1.96\sqrt{\tfrac{\hat{p}\hat{q}}{n}} = 1.96\sqrt{\tfrac{.09(.91)}{1002}} = .018$$

Recall that the quantity $\hat{p}\hat{q}$ attains its maximum value when $\hat{p} = .5$, and decreases as $\hat{p}$ gets close to 0 or 1. Hence, since $\hat{p} = .09$ in part b, the margin of error is smaller than in part a, where $\hat{p} = .57$.

c. The 95% confidence interval for p is
$$\hat{p} \pm 1.96\sqrt{\tfrac{\hat{p}\hat{q}}{n}} \Rightarrow .37 \pm 1.96\sqrt{\tfrac{.37(.63)}{1002}} \Rightarrow .37 \pm .030$$
or $.340 < p < .400$.

7.45
The estimate of $p_1 - p_2$ is $\hat{p}_1 - \hat{p}_2$ where $\hat{p}_1 = \tfrac{x_1}{n_1} = \tfrac{120}{500} = .24$ and $\hat{p}_2 = \tfrac{x_2}{n_2} = \tfrac{147}{500} =$

.294. Then, estimating p_1 and p_2 with $\hat{p}_1$ and $\hat{p}_2$ in the standard error, the approximate margin of error is
$$1.96\sqrt{\tfrac{\hat{p}_1\hat{q}_1}{n_1} + \tfrac{\hat{p}_2\hat{q}_2}{n_2}} = 1.96\sqrt{\tfrac{.24(.76)}{500} + \tfrac{.294(.706)}{500}} = .055$$

7.46

a. Calculate $\hat{p}_1 = \frac{x_1}{n_1} = \frac{337}{800} = .42$ and $\hat{p}_2 = \frac{x_2}{n_2} = \frac{374}{640} = .58$. The approximate 90% confidence interval is

$$(\hat{p}_1 - \hat{p}_2) \pm 1.645 \sqrt{\frac{\hat{p}_1 \hat{q}_1}{n_1} + \frac{\hat{p}_2 \hat{q}_2}{n_2}}$$

$$(.42 - .58) \pm 1.645 \sqrt{\frac{.42(.58)}{800} + \frac{.58(.42)}{640}}$$

$$-.16 \pm .043 \quad \text{or} \quad -.203 < (p_1 - p_2) < -.117$$

b. The two binomial samples must be random and independent and the sample sizes must be large enough that the distributions of $\hat{p}_1$ and $\hat{p}_2$ are approximately normal. Assuming that the samples are random, these conditions are met in this exercise.

7.47

a. Calculate $\hat{p}_1 = \frac{x_1}{n_1} = \frac{849}{1265} = .671$ and $\hat{p}_2 = \frac{x_2}{n_2} = \frac{910}{1688} = .539$. The approximate 99% confidence interval is

$$(\hat{p}_1 - \hat{p}_2) \pm 2.58 \sqrt{\frac{\hat{p}_1 \hat{q}_1}{n_1} + \frac{\hat{p}_2 \hat{q}_2}{n_2}}$$

$$(.671 - .539) \pm 2.58 \sqrt{\frac{.671(.329)}{1265} + \frac{.539(.461)}{1688}}$$

$$.132 \pm .046 \quad \text{or} \quad .086 < (p_1 - p_2) < .178$$

b. The two binomial samples must be random and independent and the sample sizes must be large enough that the distributions of $\hat{p}_1$ and $\hat{p}_2$ are approximately normal. Assuming that the samples are random, these conditions are met in this exercise.

7.48

a. Calculate $\hat{p}_1 = \frac{x_1}{n_1} = \frac{108}{314} = .344$ and $\hat{p}_2 = \frac{x_2}{n_2} = \frac{102}{207} = .493$. The approximate 95% confidence interval is

$$(\hat{p}_1 - \hat{p}_2) \pm 1.96 \sqrt{\frac{\hat{p}_1 \hat{q}_1}{n_1} + \frac{\hat{p}_2 \hat{q}_2}{n_2}}$$

$$(.344 - .493) \pm 1.96 \sqrt{\frac{.344(.656)}{314} + \frac{.493(.507)}{207}}$$

$$-.149 \pm .086 \quad \text{or} \quad -.235 < (p_1 - p_2) < -.063$$

b. The two binomial samples must be random and independent and the sample sizes must be large enough that the distributions of $\hat{p}_1$ and $\hat{p}_2$ are approximately normal. Assuming that the samples are random, these conditions are met in this exercise.

7.49

It is given that $n_1 = 414$, $\hat{p}_1 = .058$, $n_2 = 1029$, $\hat{p}_2 = .036$. The approximate 95% confidence interval is

$$(\hat{p}_1 - \hat{p}_2) \pm 1.96 \sqrt{\frac{\hat{p}_1 \hat{q}_1}{n_1} + \frac{\hat{p}_2 \hat{q}_2}{n_2}}$$

$$(.058 - .036) \pm 1.96 \sqrt{\frac{.058(.942)}{414} + \frac{.036(.964)}{1029}}$$

$$.022 \pm .025 \quad \text{or} \quad -.003 < (p_1 - p_2) < .047$$

7.50

a. Given $n_1 = 5000$, $\hat{p}_1 = .10$, $n_2 = 5000$, $\hat{p}_2 = .08$, the approximate 95% confidence interval is

$$(.10 - .08) \pm 1.96 \sqrt{\frac{.10(.90)}{5000} + \frac{.08(.92)}{5000}}$$

$$.02 \pm .011 \quad \text{or} \quad .009 < (p_1 - p_2) < .031$$

b. The two samples of size $n_1 = n_2 = 5000$ must be random and independent (they cannot be based on the same 5000 households each month).

7.51

Given $n_1 = 3000$, $\hat{p}_1 = .88$, $n_2 = 2500$, $\hat{p}_2 = .58$, the approximate 99% confidence interval is

$$(.88 - .58) \pm 2.58 \sqrt{\frac{.88(.12)}{3000} + \frac{.58(.42)}{2500}}$$

$$.30 \pm .030 \quad \text{or} \quad .270 < (p_1 - p_2) < .330$$

7.52

a. Given $n_1 = 1000$, $\hat{p}_1 = .10$, $n_2 = 1000$, $\hat{p}_2 = .19$, the approximate 95% confidence interval is

$$(.10 - .19) \pm 1.96 \sqrt{\frac{.10(.90)}{1000} + \frac{.19(.81)}{1000}}$$

$$-.09 \pm .031 \quad \text{or} \quad -.121 < (p_1 - p_2) < -.059$$

b. Since there are two age groups combined to form the first sample, $n_1 = 2000$ and $\hat{p}_1 = .95$, while $n_2 = 1000$ and $\hat{p}_2 = .93$. The point estimate of $p_1 - p_2$ is $.95 - .93 = .02$, and the approximate margin of error is

$$1.96 \sqrt{\frac{\hat{p}_1 \hat{q}_1}{n_1} + \frac{\hat{p}_2 \hat{q}_2}{n_2}} = 1.96 \sqrt{\frac{.95(.05)}{2000} + \frac{.93(.07)}{1000}} = .018$$

7.53

It is necessary to find the sample size required to estimate a certain parameter to within a given bound with confidence $(1 - \alpha)$. Recall from Section 7.4 that we may estimate a parameter with $(1 - \alpha)100\%$ confidence within the interval (estimator) $\pm z_{\alpha/2} \times$ (standard error of the estimator). Thus,

$$z_{\alpha/2} \times \text{(standard error)}$$

provides a margin of error with $(1 - \alpha)100\%$ confidence. The experimenter will specify a given bound B. If we let $z_{\alpha/2} \times$ (standard error) $\leq B$, we will be $(1 - \alpha)100\%$ confident that the estimator will lie within B units of the parameter of interest.

For this exercise, the parameter of interest is μ, $B = 1.6$ and $1 - \alpha = .95$. Hence, we must have

$$1.96 \frac{\sigma}{\sqrt{n}} \leq 1.6 \quad \Rightarrow \quad 1.96 \frac{12.7}{\sqrt{n}} \leq 1.6$$

$$\sqrt{n} \geq \frac{1.96(12.7)}{1.6} = 15.5575$$

$$n \geq 242.04 \quad \text{or} \quad n \geq 243$$

7.54

For this exercise, B = .04 for the binomial estimator $\hat{p}$, where $\sigma_{\hat{p}} = \sqrt{\frac{pq}{n}}$. Assuming maximum variation, which occurs if p = .3 (since we suspect that .1 < p < .3) and $z_{.025} = 1.96$, we have

$$1.96 \, \sigma_{\hat{p}} \leq B \quad \Rightarrow \quad 1.96 \sqrt{\frac{pq}{n}} \leq B$$

$$1.96 \sqrt{\frac{.3(.7)}{n}} \leq .04 \quad \Rightarrow \quad \sqrt{n} \geq \frac{1.96\sqrt{.3(.7)}}{.04}$$

$$n \geq 504.21 \quad \text{or} \quad n \geq 505$$

7.55

In this exercise, the parameter of interest is $\mu_1 - \mu_2$, $n_1 = n_2 = n$, and $\sigma_1^2 \approx \sigma_2^2 \approx 27.8$. Then we must have

$$z_{\alpha/2} \sigma_{(\overline{x}_1 - \overline{x}_2)} \leq B$$

$$1.645 \sqrt{\frac{\sigma_1^2}{n_1} + \frac{\sigma_2^2}{n_2}} \leq .17 \quad \Rightarrow \quad 1.645 \sqrt{\frac{27.8}{n} + \frac{27.8}{n}} \leq .17$$

$$\sqrt{n} \geq \frac{1.645\sqrt{55.6}}{.17}$$

$$n \geq 5206.06 \quad \text{or} \quad n_1 = n_2 = 5207$$

7.56

In this exercise, the parameter of interest is $p_1 - p_2$, $n_1 = n_2 = n$, and B = .05. Since no prior knowledge is available about p_1 and p_2, we assume the largest possible variation, which occurs if $p_1 = p_2 = .5$. Then

$$z_{\alpha/2} \times (\text{standard error of } \hat{p}_1 - \hat{p}_2) \leq B$$

$$z_{.01} \sqrt{\frac{p_1 q_1}{n_1} + \frac{p_2 q_2}{n_2}} \leq .05 \quad \Rightarrow \quad 2.33 \sqrt{\frac{(.5)(.5)}{n} + \frac{(.5)(.5)}{n}} \leq .05$$

$$\sqrt{n} \geq \frac{2.33\sqrt{.5}}{.05}$$

$$n \geq 1085.78 \quad \text{or} \quad n_1 = n_2 = 1086$$

7.57

For this exercise, B = 5000 and $\sigma \approx R/4 = 100{,}000/4 = 25{,}000$. Then we must have

$$1.96 \frac{\sigma}{\sqrt{n}} \leq 5000 \quad \Rightarrow \quad 1.96 \frac{25{,}000}{\sqrt{n}} \leq 5000$$

$$\sqrt{n} \geq \frac{1.96(25{,}000)}{5000} = 9.8$$

$$n \geq 96.04 \quad \text{or} \quad n \geq 97$$

7.58

Assume that $n_1 = n_2 = n$ and $B = 1000$. From Exercise 7.34, $\sigma_1 \approx 6300$ and $\sigma_2 \approx 7100$. Hence,

$$n \geq \frac{z_{\alpha/2}^2(\sigma_1^2 + \sigma_2^2)}{B^2}$$

$$n \geq \frac{(1.96)^2(6300^2 + 7100^2)}{1000^2} = 346.128$$

and $n_1 = n_2 = 347$ loans should be included in each sample.

7.59

With $B = 20$ and $\sigma \approx 58$, solve

$$n \geq \frac{z_{\alpha/2}^2 \sigma^2}{B^2} = \frac{(2.58)^2 58^2}{20^2} = 55.98$$

We should sample fifty-six <u>independent</u> accounts.

7.60

With $B = .04$ and $p_1 = p_2 = .5$, solve

$$n \geq \frac{2(.25)z_{\alpha/2}^2}{B^2} = \frac{2(.25)(1.96)^2}{(.04)^2} = 1200.5$$

In each market, randomly sample 1201 consumers.

7.61

With $B = .01$ and p unknown, solve

$$n \geq \frac{(.25)z_{\alpha/2}^2}{B^2} = \frac{(.25)(2.58)^2}{(.01)^2} = 16{,}641$$

Ford Motor Company would have to survey 16,641 professional leaders.

7.62

a. If $.7 < p < .9$, the largest value of $\sigma_{\hat{p}}$ will occur if $p = .7$. Hence, we use this value to estimate p and solve

$$n \geq \frac{z_{\alpha/2}^2 pq}{B^2} = \frac{(2.58)^2(.7)(.3)}{(.01)^2} = 13{,}978.44$$

or $n = 13{,}979$.

b. The sample size required is smaller than in Exercise 7.61, since we have more knowledge of the range of values that p might take. With more information available, a smaller sample size is required.

7.63

If $B = .03$ and we wish to estimate a binomial proportion p, with no additional knowledge of p, we need to have

$$n \geq \frac{(.25)z_{\alpha/2}^2}{B^2} = \frac{(.25)(1.96)^2}{(.03)^2} = 1067.11$$

or $n = 1068$ households.

7.64

See Section 6.3 of the text.

7.65

A Student's t statistic can be employed to construct a confidence interval for a single population mean when the sample has been randomly selected from a normal population. It will work quite satisfactorily for populations that possess mound-shaped frequency distributions resembling the normal distribution.

7.66

As in the case of the single population mean, random samples must be independently drawn from two populations that possess normal distributions with a common variance, σ^2. Consequently, it is logical that information in the two sample variances, s_1^2 and s_2^2, should be pooled in order to give the best estimate of the common variance, σ^2. In this way, all of the sample information is being utilized to its best advantage.

7.67

The point estimate of μ is $\bar{x} = 29.1$ and the margin of error in estimation with $s = 3.9$ and $n = 64$ is

$$1.96\sigma_{\bar{x}} = 1.96\frac{\sigma}{\sqrt{n}} \approx 1.96\frac{s}{\sqrt{n}} = 1.96\left(\frac{3.9}{\sqrt{64}}\right) = .9555$$

7.68

Refer to Exercise 7.67. The approximate 90% confidence interval is

$$\bar{x} \pm 1.645\frac{s}{\sqrt{n}} = 29.1 \pm 1.645\frac{3.9}{\sqrt{64}} = 29.1 \pm .802 \quad \text{or} \quad 28.298 < \mu < 29.902$$

Intervals constructed in this manner enclose the true value of μ 90% of the time in repeated sampling. Therefore, we are fairly certain that this particular interval encloses μ.

7.69

Refer to Exercise 7.67, with $B = .5$, $\sigma \approx 3.9$, and $1 - \alpha = .95$. Then we must solve for n in the following inequality:

$$1.96\frac{\sigma}{\sqrt{n}} \leq B \quad \Rightarrow \quad 1.96\frac{3.9}{\sqrt{n}} \leq .5$$

$$\sqrt{n} \geq 15.288$$

$$n \geq 233.723 \quad \text{or} \quad n \geq 234$$

Alternatively, use the formula $n \geq \dfrac{z_{\alpha/2}^2 \sigma^2}{B^2} = \dfrac{(1.96)^2(3.9)^2}{(.5)^2} = 233.723$.

7.70

The 90% confidence interval for $\mu_1 - \mu_2$ is approximately

$$(\bar{x}_1 - \bar{x}_2) \pm 1.645\sqrt{\frac{s_1^2}{n_1} + \frac{s_2^2}{n_2}}$$

$$(100.4 - 96.2) \pm 1.645\sqrt{\frac{(.8)^2}{50} + \frac{(1.3)^2}{60}}$$

$$4.2 \pm .333 \qquad \text{or} \quad 3.867 < (\mu_1 - \mu_2) < 4.533$$

7.71

Refer to Exercise 7.70, with $B = .2$, $1 - \alpha = .95$, $n_1 = n_2 = n$, $s_1 = .8$, and $s_2 = 1.3$. Using these values to estimate σ_1 and σ_2, the following inequality must be solved:

$$1.96 \sqrt{\frac{\sigma_1^2}{n_1} + \frac{\sigma_2^2}{n_2}} \leq .2 \quad \Rightarrow \quad 1.96 \sqrt{\frac{(.8)^2}{n} + \frac{(1.3)^2}{n}} \leq .2$$

$$\sqrt{n} \geq 14.959$$

$$n \geq 223.77 \qquad \text{or} \quad n_1 = n_2 = 224$$

Alternatively, solve

$$n \geq \frac{z_{\alpha/2}^2 (\sigma_1^2 + \sigma_2^2)}{B^2}$$

$$n \geq \frac{(1.96)^2 [(.8)^2 + (1.3)^2]}{(.2)^2} = 223.77$$

7.72

a. The point estimate for p is $\hat{p} = \frac{x}{n} = \frac{240}{500} = .48$ with approximate margin of error

$$1.96 \sqrt{\frac{\hat{p}\hat{q}}{n}} = 1.96 \sqrt{\frac{.48(.52)}{500}} = .044$$

b. An approximate 90% confidence interval for p is

$$\hat{p} \pm 1.645 \sqrt{\frac{\hat{p}\hat{q}}{n}} = .48 \pm 1.645 \sqrt{\frac{.48(.52)}{500}} = .48 \pm .037$$

or $.443 < p < .517$. Intervals constructed in this manner enclose the true value of p 90% of the time. Hence, we are fairly certain that this particular interval encloses p.

7.73

Assuming maximum variation with $p = .5$, solve

$$1.645 \sqrt{\frac{pq}{n}} \leq .025$$

$$\sqrt{n} \geq \frac{1.645 \sqrt{.5(.5)}}{.025} = 32.9$$

$$n \geq 1082.41 \qquad \text{or } n \geq 1083$$

Alternatively, solve $n \geq \frac{(.25) z_{\alpha/2}^2}{B^2} = \frac{(.25)(1.645)^2}{(.025)^2} = 1082.41$.

7.74

Calculate
$$\hat{p}_1 = \frac{17}{40} = .425 \qquad \hat{p}_2 = \frac{23}{80} = .2875$$

The approximate 90% confidence interval for $p_1 - p_2$ is

$$(\hat{p}_1 - \hat{p}_2) \pm 1.645 \sqrt{\frac{\hat{p}_1\hat{q}_1}{n_1} + \frac{\hat{p}_2\hat{q}_2}{n_2}}$$

$$(.425 - .2875) \pm 1.645 \sqrt{\frac{.425(.575)}{40} + \frac{.7125(.2875)}{80}}$$

$$.1375 \pm .1532 \qquad \text{or} \qquad -.0157 < (p_1 - p_2) < .2907$$

Intervals constructed in this manner enclose the true value of $p_1 - p_2$ 90% of the time in repeated sampling. Hence, we are fairly certain that this particular interval encloses $p_1 - p_2$.

7.75

Assuming maximum variation ($p_1 = p_2 = .5$) and $n_1 = n_2 = n$, the inequality to be solved is

$$z_{.05} \sqrt{\frac{p_1 q_1}{n_1} + \frac{p_2 q_2}{n_2}} \leq .06$$

$$1.645 \sqrt{\frac{(.5)(.5)}{n} + \frac{(.5)(.5)}{n}} \leq .06$$

$$\sqrt{n} \geq 19.387$$

$$n \geq 375.837 \qquad \text{or} \qquad n_1 = n_2 = 376$$

Alternatively, solve $n \geq \dfrac{2(.25)z^2_{\alpha/2}}{B^2} = \dfrac{2(.25)(1.645)^2}{(.06)^2} = 375.837$.

7.76

Calculate $\Sigma x_i = 3557$ and $\Sigma x_i^2 = 1{,}415{,}919$. Then

$$\bar{x} = \frac{\Sigma x_i}{n} = \frac{3557}{16} = 222.3125$$

$$s^2 = \frac{\Sigma x_i^2 - \frac{(\Sigma x_i)^2}{n}}{n-1} = \frac{1{,}415{,}919 - \frac{(3557)^2}{16}}{15} = 41{,}676.896$$

and the 90% confidence interval is

$$\bar{x} \pm t_{.05} \frac{s}{\sqrt{n}} \Rightarrow 222.3125 \pm 1.753 \sqrt{\frac{41{,}676.896}{16}} \Rightarrow 222.3125 \pm 89.468$$

or $132.8445 < \mu < 311.7805$.

7.77

It is given that $n = 10$, $\bar{x} = 7.1$ and $s = .12$. The 99% confidence interval for μ is

$$\bar{x} \pm t_{.005} \frac{s}{\sqrt{n}} \Rightarrow 7.1 \pm 3.250 \frac{.12}{\sqrt{10}} \Rightarrow 7.1 \pm .123$$

or $6.977 < \mu < 7.223$.

7.78

The pooled estimator of σ^2 is calculated as

$$s^2 = \frac{(n_1 - 1)s_1^2 + (n_2 - 1)s_2^2}{n_1 + n_2 - 2} = \frac{10(31.4) + 10(44.82)}{11 + 11 - 2} = 38.11$$

and the 90% confidence interval for $(\mu_1 - \mu_2)$ is given as

$$(\bar{x}_1 - \bar{x}_2) \pm t_{.05}\sqrt{s^2\left(\frac{1}{n_1} + \frac{1}{n_2}\right)}$$

$$(60.4 - 65.3) \pm 1.725\sqrt{38.11\left(\frac{1}{11} + \frac{1}{11}\right)}$$

$$-4.9 \pm 4.541 \quad \text{or} \quad -9.441 < (\mu_1 - \mu_2) < -.359$$

7.79

Given $n = 391$ and $x = 22$, the point estimate of p is $\hat{p} = \frac{x}{n} = \frac{22}{391} = .056$ and the margin of error is approximately

$$1.96\sqrt{\frac{.056(.944)}{391}} = .023$$

7.80

a. It is given that $n_1 = n_2 = 1200$ with $\hat{p}_1 = .45$ and $\hat{p}_2 = .66$. The individual margins of error for the point estimates of p_1 and p_2 are

$$1.96\sqrt{\frac{p_1 q_1}{n_1}} \approx 1.96\sqrt{\frac{(.45)(.55)}{1200}} = .028 \text{ and}$$

$$1.96\sqrt{\frac{p_2 q_2}{n_2}} \approx 1.96\sqrt{\frac{(.66)(.34)}{1200}} = .027$$

b. The point estimate of the difference $p_1 - p_2$ is $\hat{p}_1 - \hat{p}_2 = .45 - .66 = -.21$ and the approximate margin of error is

$$1.96\sqrt{\frac{p_1 q_1}{n_1} + \frac{p_2 q_2}{n_2}} \approx 1.96\sqrt{\frac{.45(.55)}{1200} + \frac{.66(.34)}{1200}} = .039$$

7.81

Calculate $\hat{p} = \frac{x}{n} = \frac{40}{400} = .1$. Then an approximate 90% confidence interval for p is

$$\hat{p} \pm 1.645\sqrt{\frac{\hat{p}\hat{q}}{n}} \Rightarrow .1 \pm 1.645\sqrt{\frac{.1(.9)}{400}} \Rightarrow .1 \pm .025$$

or $.075 < p < .125$.

7.82

With $B = 100$ and $\sigma \approx 1000$, n is obtained by solving for n in the inequality,

$$n \geq \frac{z_{\alpha/2}^2 \sigma^2}{B^2} = \frac{(1.96)^2 1000^2}{100^2} = 384.16 \quad \text{or } n \geq 385$$

Then, given that the mean of this sample of 385 measurements is $14,800, a 95% confidence interval for μ is

$$\bar{x} \pm 1.96 \frac{\sigma}{\sqrt{n}} \approx 14,800 \pm 1.96 \frac{1000}{\sqrt{385}} \Rightarrow 14,800 \pm 99.891$$

or $14,700.109 < \mu < 14,899.891$.

7.83

The object is to estimate the population mean μ to within $B = .5$ with probability $.90$. Hence, solve for n in the following inequality:

$$n \geq \frac{z_{\alpha/2}^2 \sigma^2}{B^2} = \frac{(1.645)^2 2^2}{(.5)^2} = 43.296 \quad \text{or } n \geq 44$$

That is, a sample of size $n = 44$ must be used in order to achieve the desired bound.

7.84

With $B = 4$, $\sigma = 16$ and $z_{\alpha/2} = 1.645$, the inequality to be solved is

$$n \geq \frac{z_{\alpha/2}^2 \sigma^2}{B^2} = \frac{(1.645)^2 16^2}{4^2} = 43.296 \quad \text{or } n \geq 44$$

7.85

a. With $x = 25$ and $n = 400$, the best estimate of p, the proportion of unemployed workers, is

$$\hat{p} = \frac{x}{n} = \frac{25}{400} = .0625$$

with a margin of error approximately

$$1.96 \sqrt{\frac{\hat{p}\hat{q}}{n}} = 1.96 \sqrt{\frac{.0625(.9375)}{400}} = .0237$$

b. Estimating p by $\hat{p}$ and assuming that the margin of error is .02, we have

$$1.96 \sqrt{\frac{\hat{p}\hat{q}}{n}} \leq .02$$

$$1.96 \sqrt{\frac{.0625(.9375)}{n}} \leq .02$$

$$\sqrt{n} \geq 23.722$$

$$n \geq 562.73 \quad \text{or} \quad n \geq 563$$

7.86

Calculate $\hat{p} = \frac{x}{n} = \frac{60}{87} = .69$. Then an approximate 90% confidence interval for p is

$$\hat{p} \pm 1.645 \sqrt{\frac{\hat{p}\hat{q}}{n}} \Rightarrow .69 \pm 1.645 \sqrt{\frac{.69(.31)}{87}} \Rightarrow .69 \pm .082$$

or $.608 < p < .772$.

7.87

Using the range approximation to obtain an estimate of σ, we have

$$\sigma \approx \frac{R}{4} = \frac{13{,}000 - 4800}{4} = 2050$$

and the desired value of n is obtained:

$$1.96 \frac{\sigma}{\sqrt{n}} \leq B$$

$$1.96 \frac{2050}{\sqrt{n}} \leq 500$$

n ≥ 64.58   or   n ≥ 65

7.88

The following sample information is available:

$$n_1 = 200, \ n_2 = 180, \ \hat{p}_1 = .32 \ \hat{p}_2 = .21$$

The best estimate for $p_1 - p_2$ is

$$(\hat{p}_1 - \hat{p}_2) = .32 - .21 = .11$$

and the approximate margin of error is

$$1.96\sqrt{\frac{\hat{p}_1\hat{q}_1}{n_1} + \frac{\hat{p}_2\hat{q}_2}{n_2}} = 1.96\sqrt{\frac{.32(.68)}{200} + \frac{.21(.79)}{180}} = .0879$$

7.89

The approximate 95% confidence interval for μ is

$$\bar{x} \pm 1.96\frac{s}{\sqrt{n}} = 34 \pm 1.96\frac{3}{\sqrt{100}} = 34 \pm .59 \text{ or } \quad 33.41 < \mu < 34.59$$

7.90

It is assumed that $p = .2$ and that the desired bound is .01. Hence,

$$1.96\sqrt{\frac{pq}{n}} \leq .01$$

$$\sqrt{n} \geq \frac{1.96\sqrt{.2(.8)}}{.01} = 78.4$$

n ≥ 6146.56   or   n ≥ 6147

Alternatively, solve

$$n \geq \frac{z_{\alpha/2}^2 pq}{B^2} = \frac{(1.96)^2(.2)(.8)}{(.01)^2} = 6146.56$$

7.91

The approximate 90% confidence interval for $p_1 - p_2$ is

$$(\hat{p}_1 - \hat{p}_2) \pm 1.645\sqrt{\frac{\hat{p}_1\hat{q}_1}{n_1} + \frac{\hat{p}_2\hat{q}_2}{n_2}}$$

$$(.2 - .1) \pm 1.645\sqrt{\frac{.2(.8)}{400} + \frac{.1(.9)}{400}}$$

.1 ± .041   or $.059 < (p_1 - p_2) < .141$

7.92

a. The preliminary calculations are shown below:

Sample 1	Sample 2
$\Sigma x_{1i} = 585$	$\Sigma x_{2i} = 466$
$\Sigma x_{1i}^2 = 42{,}845$	$\Sigma x_{2i}^2 = 36{,}246$
$n_1 = 8$	$n_2 = 6$

The pooled estimator of σ^2 is calculated as

$$s^2 = \frac{\Sigma x_{1i}^2 - \frac{(\Sigma x_{1i})^2}{n_1} + \Sigma x_{2i}^2 - \frac{(\Sigma x_{2i})^2}{n_2}}{n_1 + n_2 - 2}$$

$$= \frac{42{,}845 - \frac{(585)^2}{8} + 36{,}246 - \frac{(466)^2}{6}}{8 + 6 - 2} = \frac{66.875 + 53.3333}{12} = 10.017$$

Then the 90% confidence interval for $(\mu_1 - \mu_2)$ is given as

$$(\bar{x}_1 - \bar{x}_2) \pm t_{.05}\sqrt{s^2\left(\frac{1}{n_1} + \frac{1}{n_2}\right)}$$

$$(73.125 - 77.667) \pm 1.782\sqrt{10.017\left(\frac{1}{8} + \frac{1}{6}\right)}$$

$$-4.542 \pm 3.046 \quad \text{or} \quad -7.588 < (\mu_1 - \mu_2) < -1.496$$

b. Since the value $\mu_1 - \mu_2 = 0$ is not in the interval calculated in part a, it is an unlikely value. Hence, it is not likely that $\mu_1 = \mu_2$. We would conclude that there is a difference in the mean efficiency ratings for the two types of heaters.

7.93

The 95% confidence interval for $\mu_1 - \mu_2$ is approximately

$$(\bar{x}_1 - \bar{x}_2) \pm 1.96\sqrt{\frac{s_1^2}{n_1} + \frac{s_2^2}{n_2}}$$

$$(5.3 - 6.4) \pm 1.96\sqrt{\frac{(2.7)^2}{100} + \frac{(2.9)^2}{100}}$$

$$-1.1 \pm .777 \quad \text{or} \quad -1.877 < (\mu_1 - \mu_2) < -0.323$$

Since the value $\mu_1 - \mu_2 = 0$ is not in the interval calculated in part a, it is an unlikely value. Hence, it is not likely that $\mu_1 = \mu_2$. We would conclude that there is a difference in the mean percentage shrinkage for the two stores.

7.94

a. Calculate $\Sigma x_i = 29{,}210$ and $\Sigma x_i^2 = 142{,}237{,}728$. Then

$$\bar{x} = \frac{\Sigma x_i}{n} = \frac{29{,}210}{7} = 4172.857$$

$$s^2 = \frac{\Sigma x_i^2 - \frac{(\Sigma x_i)^2}{n}}{n - 1} = \frac{142{,}237{,}728 - \frac{(29{,}210)^2}{7}}{6} = 3{,}391{,}428.476$$

$$s = \sqrt{s^2} = 1841.5831$$

b. The 95% confidence interval is

$$\bar{x} \pm t_{.025} \frac{s}{\sqrt{n}} = 4172.857 \pm 2.447 \frac{1841.5831}{\sqrt{7}} = 4172.857 \pm 1703.242$$

or $2469.615 < \mu < 5876.099$.

c. The MINITAB printout agrees with the hand calculation.

7.95

The preliminary calculations are shown below:

Sample 1	Sample 2
$\Sigma x_{1i} = 41.6$	$\Sigma x_{2i} = 112.9$
$\Sigma x_{1i}^2 = 255.96$	$\Sigma x_{2i}^2 = 1832.11$
$n_1 = 7$	$n_2 = 7$

The pooled estimator of σ^2 is calculated as

$$s^2 = \frac{\Sigma x_{1i}^2 - \frac{(\Sigma x_{1i})^2}{n_1} + \Sigma x_{2i}^2 - \frac{(\Sigma x_{2i})^2}{n_2}}{n_1 + n_2 - 2}$$

$$= \frac{255.96 - \frac{(41.6)^2}{7} + 1832.11 - \frac{(112.9)^2}{7}}{7 + 7 - 2} = \frac{8.737 + 11.194}{12} = 1.66095$$

Then the 90% confidence interval for $(\mu_1 - \mu_2)$ is given as

$$(\bar{x}_1 - \bar{x}_2) \pm t_{.05} \sqrt{s^2 \left(\frac{1}{n_1} + \frac{1}{n_2}\right)}$$

$$(5.943 - 16.129) \pm 1.782 \sqrt{1.66095 \left(\frac{1}{7} + \frac{1}{7}\right)}$$

$$-10.186 \pm 1.228 \quad \text{or} \quad -11.414 < (\mu_1 - \mu_2) < -8.958$$

7.96

It is given that $n = 100$, $\bar{x} = 39.1$, and $s = 17.3$. The approximate 90% confidence interval for μ is

$$\bar{x} \pm 1.645 \frac{s}{\sqrt{n}} \Rightarrow 39.1 \pm 1.645 \frac{17.3}{\sqrt{100}} \Rightarrow 39.1 \pm 2.846$$

or $36.254 < \mu < 41.946$.

7.97

This is similar to previous exercises. The 95% confidence interval for $\mu_1 - \mu_2$ is approximately

$$(\bar{x}_1 - \bar{x}_2) \pm 1.96 \sqrt{\frac{s_1^2}{n_1} + \frac{s_2^2}{n_2}}$$

$$(28.3 - 47.7) \pm 1.96\sqrt{\frac{(15.5)^2}{30} + \frac{(18.9)^2}{30}}$$

$$-19.4 \pm 8.747 \quad \text{or} \quad -28.147 < (\mu_1 - \mu_2) < -10.653$$

7.98

Refer to Exercise 7.97. With $B = 5$, $n_1 = n_2 = n$, and $\sigma_1 \approx 15.5$, $\sigma_2 \approx 18.9$, the inequality to be solved is

$$n \geq \frac{z_{\alpha/2}^2 (\sigma_1^2 + \sigma_2^2)}{B^2}$$

$$n \geq \frac{(1.96)^2[(15.5)^2 + (18.9)^2]}{(5)^2} = 91.808$$

There should be ninety-two clients of each type in the sample.

7.99

Calculate $\hat{p} = \frac{x}{n} = \frac{1914}{2300} = .832$. Then an approximate 95% confidence interval for p is

$$\hat{p} \pm 1.96 \sqrt{\frac{\hat{p}\hat{q}}{n}} = .832 \pm 1.96 \sqrt{\frac{.832(.168)}{2300}} = .832 \pm .015$$

or $.817 < p < .847$.

7.100

The point estimate of μ is $\bar{x} = 45.273$ and the margin of error is approximately

$$1.96 \frac{s}{\sqrt{n}} = 1.96 \frac{1.199}{\sqrt{30}} = .429$$

Therefore, using this margin of error, we would not expect the true mean μ to differ from the estimate, 45.273, by more than .429. This means that the minimum value for the true mean μ is $45.273 - .429 = 44.844$. The 44.7 driving range reported by the California Air Resources Board is probably incorrect.

7.101

We choose to construct a 95% confidence interval, and then check to see if the value $\mu_1 - \mu_2 = 0$ is contained in this interval. The preliminary calculations are shown below:

Standard	Enhanced
$\Sigma x_{1i} = 13.9$	$\Sigma x_{2i} = 12.78$
$\Sigma x_{1i}^2 = 19.4708$	$\Sigma x_{2i}^2 = 16.9368$
$n_1 = 10$	$n_2 = 10$

The pooled estimator of σ^2 is calculated as

$$s^2 = \frac{\Sigma x_{1i}^2 - \frac{(\Sigma x_{1i})^2}{n_1} + \Sigma x_{2i}^2 - \frac{(\Sigma x_{2i})^2}{n_2}}{n_1 + n_2 - 2}$$

$$= \frac{19.4708 - \frac{(13.9)^2}{10} + 16.9368 - \frac{(12.78)^2}{10}}{10 + 10 - 2} = \frac{.14980 + .60396}{18} = .041876$$

Then the 95% confidence interval for $(\mu_1 - \mu_2)$ is given as

$$(\bar{x}_1 - \bar{x}_2) \pm t_{.025}\sqrt{s^2\left(\frac{1}{n_1} + \frac{1}{n_2}\right)}$$

$$(1.39 - 1.278) \pm 2.101\sqrt{.041876\left(\frac{1}{10} + \frac{1}{10}\right)}$$

$$.112 \pm .192 \quad \text{or} \quad -.080 < (\mu_1 - \mu_2) < .304$$

Since the value $\mu_1 - \mu_2 = 0$ is contained in this interval, this is not an unlikely value. Hence, we cannot conclude that there is a difference between the two means.

CHAPTER 8
Tests of Hypotheses for Means and Proportions

8.1 In this exercise, the parameter of interest is μ, the population mean. The objective of the experiment is to show that the mean exceeds 2.3.

a. We want to prove the alternative hypothesis that μ is, in fact, greater than 2.3. Hence, the alternative hypothesis is

$$H_a: \mu > 2.3$$

and the null hypothesis is

$$H_0: \mu = 2.3$$

b. The probability of a Type I error is defined as

$$\alpha = P[\text{reject } H_0 \text{ when } H_0 \text{ true}] = P[\text{decide } \mu > 2.3 \text{ when } \mu = 2.3]$$

c. The best estimator for μ is the sample average $\bar{x}$, and the test statistic is

$$z = \frac{\bar{x} - \mu}{\sigma/\sqrt{n}}$$

which represents the distance (measured in units of standard deviations) from $\bar{x}$ to the hypothesized mean μ. Hence, if this value is large in absolute value, one of two conclusions may be drawn. Either a very unlikely event has occurred, or the hypothesized mean is incorrect. Refer to part b. If $\alpha = .05$, the critical value of z that separates the rejection and nonrejection regions will be a value (denoted by z_0) such that

$$P[z > z_0] = \alpha = .05$$

That is, $z_0 = 1.645$ (see Figure 8.1). Hence, H_0 will be rejected if $z > 1.645$.

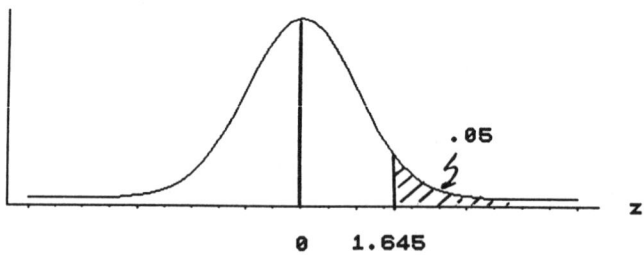

Figure 8.1

d. To conduct the test, calculate the value of the test statistic using the information contained in the sample. Note that the value of the true standard deviation, σ, is approximated using the sample standard deviation s.

$$z = \frac{\bar{x} - \mu}{\sigma/\sqrt{n}} \approx \frac{\bar{x} - \mu}{s/\sqrt{n}} = \frac{2.4 - 2.3}{.29/\sqrt{35}} = 2.04$$

The observed value of the test statistic, z = 2.04, falls in the rejection region and the null hypothesis is rejected. There is sufficient evidence to indicate that $\mu > 2.3$.

8.2

If the experimenter wishes to prove that $\mu < 2.9$, this is the alternative hypothesis. Hence, the hypothesis to be tested is

$$H_0: \mu = 2.9 \qquad H_a: \mu < 2.9$$

and the test is one-tailed, since only values of $\bar{x}$ smaller than 2.9 (and hence in the lower tail of the sampling distribution) would tend to disprove H_0.

8.3

If the experimenter wishes to prove that $\mu \neq 2.9$, this is the alternative hypothesis. Hence, the hypothesis to be tested is

$$H_0: \mu = 2.9 \qquad H_a: \mu \neq 2.9,$$

and the test is two-tailed, since values of $\bar{x}$ much smaller or much larger than 2.9, in both tails of the sampling distribution) would tend to disprove H_0.

8.4

a-b. This is similar to Exercise 8.1. The hypothesis to be tested is

$$H_0: \mu = 84 \qquad H_a: \mu < 84$$

c. Since $\alpha = $ P[reject H_0 when H_0 true] = P[decide $\mu < 84$ when $\mu = 84$], we must have $\alpha = .05$.

d. The critical value of z that separates the lower-tailed rejection region from the non-rejection region will be z_0 where

$P[z < z_0] = .05$ (see Figure 8.2).

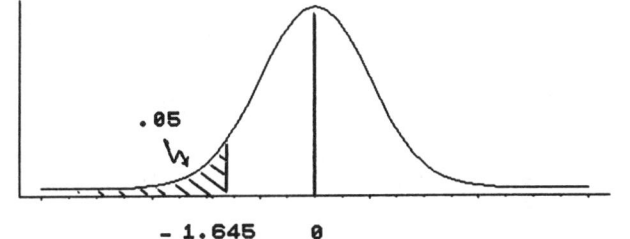

Figure 8.2

That is, $z_0 = -1.645$. Hence, H_0 will be rejected if $z < -1.645$. The observed value of the test statistic is
$$z = \frac{\bar{x} - \mu}{\sigma/\sqrt{n}} \approx \frac{\bar{x} - \mu}{s/\sqrt{n}} = \frac{83.8 - 84}{2.9/\sqrt{40}} = -.436$$

which does not fall in the rejection region. The null hypothesis is not rejected. There is insufficient evidence to conclude that $\mu < 84$.

8.5

a. Refer to Exercise 8.4, in which the rejection region was given as $z < -1.645$ where

$$z = \frac{\bar{x} - \mu}{s/\sqrt{n}} = \frac{\bar{x} - 84}{2.9/\sqrt{40}}$$

Solving for $\bar{x}$, we obtain the critical value of $\bar{x}$ necessary for rejection of H_0.

$$\frac{\bar{x} - 84}{2.9/\sqrt{40}} < -1.645$$

$$\bar{x} < 84 - 1.645\frac{2.9}{\sqrt{40}} = 83.246$$

b. The probability of a Type II error is defined as

$$\beta = P[\text{accept } H_0 \text{ when } H_0 \text{ is false}]$$

Since the acceptance region is $\bar{x} \geq 83.246$ from part b, β can be rewritten as

$$\beta = P[\bar{x} \geq 83.246 \text{ when } H_0 \text{ is false}] = P[\bar{x} \geq 83.246 \text{ when } \mu < 84]$$

Several alternative values of μ are given in this exercise. For $\mu = 82.8$,

$$\beta = P[\bar{x} \geq 83.246 \text{ when } \mu = 82.8]$$

$$= P[z \geq \frac{83.246 - 82.8}{2.9/\sqrt{40}}] = P[z \geq .97] = .5 - .3340 = .1660$$

For $\mu = 82.6$,

$$\beta = P[\bar{x} \geq 83.246 \text{ when } \mu = 82.6]$$

$$= P[z \geq \frac{83.246 - 82.6}{2.9/\sqrt{40}}] = P[z \geq 1.41] = .5 - .4207 = .0793$$

For $\mu = 82.4$,

$$\beta = P[\bar{x} \geq 83.246 \text{ when } \mu = 82.4]$$

$$= P[z \geq \frac{83.246 - 82.4}{2.9/\sqrt{40}}] = P[z \geq 1.85] = .5 - .4678 = .0322$$

For $\mu = 83.4$,

$$\beta = P[\bar{x} \geq 83.246 \text{ when } \mu = 83.4]$$

$$= P[z \geq \frac{83.246 - 83.4}{2.9/\sqrt{40}}] = P[z \geq -.34] = .5 + .1331 = .6331$$

c. The power curve is graphed using the values calculated in part b and is shown in Figure 8.3.

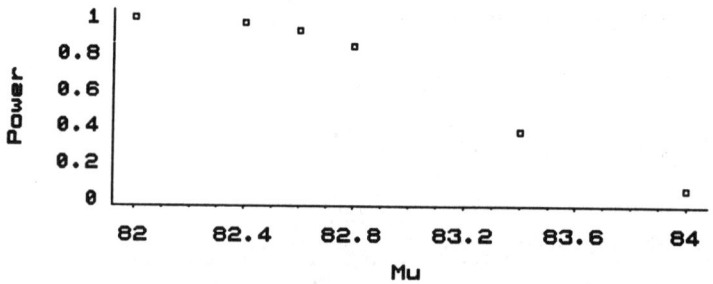

Figure 8.3

8.6

a-b. The hypothesis to be tested is

$$H_0: \mu = 10 \qquad H_a: \mu > 10$$

However, there are only four measurements on which to base the test, and a t statistic must be employed.

c. The rejection region will no longer be determined by using the standard normal distribution. Rather, it will be determined based on the *t distribution* with $(n-1)$ degrees of freedom. For $\alpha = .10$, index $t_{.10}$ in Table 4, Appendix II and the rejection region will be $t > 1.638$ (See Figure 8.4).

d. The test statistic is identical to the test statistic used earlier in this section, except that its sampling distribution *is not* normal, but has a t distribution. It is calculated as

$$t = \frac{\bar{x} - \mu}{s/\sqrt{n}},$$

where
$$\bar{x} = \frac{\Sigma x_i}{n} = \frac{43.9}{4} = 10.975$$

and
$$s^2 = \frac{\Sigma x_i^2 - \frac{(\Sigma x_i)^2}{n}}{n-1} = \frac{486.25 - \frac{(43.9)^2}{4}}{3} = 1.4825$$

Then
$$t = \frac{\bar{x} - \mu}{s/\sqrt{n}} = \frac{10.975 - 10}{\sqrt{\frac{1.4825}{10}}} = 1.602.$$

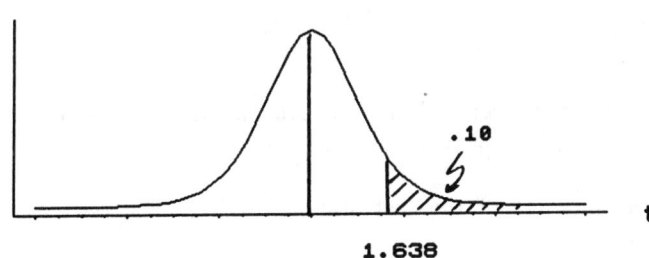

Figure 8.4

The observed value of the test statistic does not fall in the rejection region, and H_0 is not rejected. There is insufficient evidence to conclude that μ is greater than 10.

8.7

a-b. The hypothesis of interest is

$$H_0: \mu = 7 \qquad H_a: \mu < 7$$

c. The rejection region, with $\alpha = .10$ and $n - 1 = 5$ degrees of freedom is $t < -t_{.10} = -1.476$.

d. Calculate $\Sigma x_i = 36.9$ and $\Sigma x_i^2 = 246.71$. Then

$$\bar{x} = \frac{\Sigma x_i}{n} = \frac{36.9}{6} = 6.15$$

$$s^2 = \frac{\Sigma x_i^2 - \frac{(\Sigma x_i)^2}{n}}{n-1} = \frac{246.71 - \frac{(36.9)^2}{6}}{5} = 3.955$$

The observed value of the test statistic is

$$t = \frac{\bar{x} - \mu}{s/\sqrt{n}} = \frac{6.15 - 7}{\sqrt{\frac{3.955}{6}}} = -1.047$$

and H_0 is not rejected. There is insufficient evidence to indicate that μ is less than 7.

8.8

The rejection region, for $\alpha = .05$, $n - 1 = 11$ degrees of freedom is $t > t_{.05} = 1.796$ and the test statistic is

$$t = \frac{\bar{x} - \mu}{s/\sqrt{n}} = \frac{3.18 - 3}{\sqrt{\frac{.21}{12}}} = 1.361$$

The observed value of the test statistic does not fall in the rejection region, and H_0 is not rejected. There is insufficient evidence to conclude that μ is greater than 3.

8.9

The hypothesis of interest is

$$H_0: \mu = 48 \qquad H_a: \mu \neq 48$$

and the test statistic is

$$t = \frac{\bar{x} - \mu}{s/\sqrt{n}} = \frac{47.1 - 48}{\sqrt{\frac{4.7}{25}}} = -2.076$$

The two-tailed rejection region with $\alpha = .10$ and d.f. $= n - 1 = 24$ is found in Table 4 by indexing $t_{.05}$ to be $|t| > t_{.05} = 1.711$.

Since the observed value of the test statistic falls in the rejection region, we reject H_0 and conclude that $\mu \neq 48$.

8.10

1. The hypothesis of interest is

$$H_0: \mu = 5 \qquad H_a: \mu > 5$$

2. The test statistic is $t = \frac{\bar{x} - \mu}{s/\sqrt{n}} = 0.80$

3. The rejection region, with $\alpha = .01$ is found in Table 4 with $n - 1 = 10$ degrees of freedom: H_0 will be rejected if $t > 2.764$.

4. <u>Conclusion:</u> Do not reject H_0. There is insufficient evidence to indicate that $\mu > 5$.

8.11

a-b. The hypothesis of interest is

$$H_0: \mu = 19 \qquad H_a: \mu > 19$$

c. The rejection region, with $\alpha = .05$ and $n - 1 = 2$ degrees of freedom is $t > t_{.05} = 2.920$.

d. Calculate $\Sigma x_i = 65$ and $\Sigma x_i^2 = 1589$. Then

$$\bar{x} = \frac{\Sigma x_i}{n} = \frac{65}{3} = 21.667$$

$$s^2 = \frac{\Sigma x_i^2 - \frac{(\Sigma x_i)^2}{n}}{n-1} = \frac{1589 - \frac{(65)^2}{3}}{2} = 90.333$$

The observed value of the test statistic is

$$t = \frac{\bar{x} - \mu}{s/\sqrt{n}} = \frac{21.667 - 19}{\sqrt{\frac{90.333}{3}}} = .486$$

and H_0 is not rejected. There is insufficient evidence to indicate that μ is greater than 19.

8.12

a-c. The test is one-tailed with

$$H_0: \mu = 4.8 \qquad H_a: \mu > 4.8$$

d. The probability that the owner describes is

$$P[\text{reject } H_0 \text{ when } H_a \text{ true}] = 1 - \beta$$

The value $\alpha = .01$ is the probability of falsely rejecting H_0, of showing that the company is operating at an acceptable level when, in fact, $\mu = 4.8$.

e. The one-tailed rejection region with $\alpha = .01$ is $z > 2.33$.

f. The observed value of the test statistic is

$$z = \frac{\bar{x} - \mu}{\sigma/\sqrt{n}} \approx \frac{4.87 - 4.8}{\frac{3.9}{\sqrt{80}}} = .16$$

and H_0 is not rejected. There is insufficient evidence to indicate that the company averages more than 4.8% profit on new cars.

8.13

a-c. The test is one-tailed with

$$H_0: \mu = 1200 \qquad H_a: \mu < 1200$$

d. The one-tailed rejection region with $\alpha = .10$ is $z < -1.28$.

e. The observed value of the test statistic is

$$z = \frac{\bar{x} - \mu}{\sigma/\sqrt{n}} \approx \frac{1186 - 1200}{\sqrt{\frac{2480}{30}}} = -1.54$$

and H_0 is rejected. We conclude that $\mu < 1200$.

8.14

The hypothesis to be tested is

$$H_0: \mu = 30.31 \qquad H_a: \mu < 30.31$$

Calculate $\Sigma x_i = 578.7$ and $\Sigma x_i^2 = 18{,}462.09$. Then

$$\bar{x} = \frac{\Sigma x_i}{n} = \frac{578.7}{20} = 28.935$$

$$s^2 = \frac{\Sigma x_i^2 - \frac{(\Sigma x_i)^2}{n}}{n-1} = \frac{18{,}462.09 - \frac{(578.7)^2}{20}}{19} = 90.3898$$

The observed value of the test statistic is

$$t = \frac{\bar{x} - \mu}{s/\sqrt{n}} = \frac{28.935 - 30.31}{\sqrt{\frac{90.3898}{20}}} = -.647$$

The rejection region, with $\alpha = .05$ and $n - 1 = 19$ degrees of freedom is

$$t < -t_{.05} = -1.729$$

as shown in Figure 8.5. Since the observed value of t does not fall in the rejection region, H_0 is not rejected. There is no reason to believe that the mean has decreased.

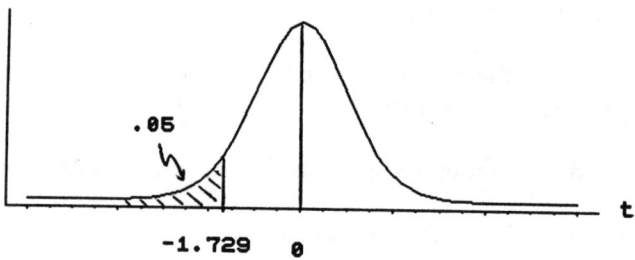

Figure 8.5

8.15

a. The hypothesis to be tested is

$$H_0: \mu = 44.7 \qquad H_a: \mu \neq 44.7$$

The observed value of the test statistic is

$$t = \frac{\bar{x} - \mu}{s/\sqrt{n}} = 2.62$$

and the rejection region, with $\alpha = .05$ and $n - 1 = 29$ degrees of freedom, is

$$|t| > t_{.025} = 2.045$$

Since the observed value of t does fall in the rejection region, H_0 is rejected. There is reason to believe that the mean is not 44.7.

b. The results of Exercise 7.100 are confirmed by the test in part a.

8.16

a. In order to test the hypothesis

$$H_0: \mu = 94 \qquad H_a: \mu < 94,$$

the test statistic is

$$z \approx \frac{\overline{x} - \mu}{s/\sqrt{n}} = \frac{92.9 - 94}{4.1/\sqrt{52}} = -1.93$$

Since the test is one-tailed, the p-value is

$$P[z < -1.93] = .5 - .4732 = .0268.$$

b. If the test is conducted with $\alpha = .05$, H_0 can be rejected since any value of α larger than .0268 allows rejection of H_0. We conclude that $\mu < 94$.

8.17

a. In order to test the hypothesis

$$H_0: \mu = 94 \qquad H_a: \mu \neq 94,$$

the test statistic is

$$z \approx \frac{\overline{x} - \mu}{s/\sqrt{n}} = \frac{92.1 - 94}{4.1/\sqrt{52}} = -3.34$$

Since the test is two-tailed, the p-value is

$$\text{p-value} = P[|z| > 3.34] = P[z > 3.34] + P[z < -3.34]$$

Since the largest tabled entry in Table 3 is $A(3.09) = .4990$, we can say that

$$P[z > 3.34] < P[z > 3.09] = .5 - .4990 = .001$$

so that the p-value is less than $.001 + .001 = .002$.

b. The null hypothesis can be rejected for any value of α greater than .002. Hence, for $\alpha = .05$, H_0 is rejected.

8.18

a. The hypothesis to be tested is

$$H_0: \mu = 15 \qquad H_a: \mu \neq 15$$

and the test statistic is

$$t = \frac{\overline{x} - \mu}{s/\sqrt{n}} = \frac{15.7 - 15}{2.4/\sqrt{18}} = 1.237$$

Since the test is two-tailed, the p-value is

$$\text{p-value} = P[|t| > 1.237] = P[t > 1.237] + P[t < -1.237]$$

Refer to Table 4, Appendix II, with $n - 1 = 17$. The exact probability $P[t > 1.237]$ is unavailable; however, it is evident that $t = 1.237$ is smaller than $t_{.10} = 1.333$. Therefore, the area to the right of t = 1.237 must be greater than .10. Then

$$P[|t| > 1.237] = 2\,P[t > 1.237] > 2(.10) = .20$$

and the p-value can be approximated as p-value > .20.

b. The null hypothesis can be rejected for any value of α greater than .20. Hence, for α = .05, H_0 cannot be rejected.

8.19

From Exercise 8.12, z = −1.54 for a one-tailed test. Hence,

$$\text{p-value} = P[z < -1.54] = .5 - .4382 = .0618$$

8.20

From Exercise 8.19, p-value = .0618. Since this value is the smallest value of α for which test results are significant, the test results will not be significant for α = .05.

8.21

From Exercise 8.11, t = .486 for a one-tailed test. Refer to Table 4, Appendix II with 2 degrees of freedom. The observed value of t is smaller than $t_{.10}$ = 1.886. Therefore,

$$\text{p-value} = P[t > .486] > .10$$

8.22

From Exercise 8.12, z = .16 for a one-tailed test. Then

$$\text{p-value} = P[z > .16] = .5 - .0636 = .4364$$

8.23

From Exercise 8.14, t = −.647 for a one-tailed test. Refer to Table 4, Appendix II with 9 degrees of freedom. The observed value of t is smaller than $t_{.10}$ = 1.328. Therefore,

$$\text{p-value} = P[t < -.647] > .10$$

8.24

a. The hypothesis to be tested is

$$H_0: \mu = 8.7 \qquad H_a: \mu < 8.7$$

and the test statistic is

$$t = \frac{\bar{x} - \mu}{s/\sqrt{n}} = \frac{8.5 - 8.7}{.23/\sqrt{10}} = -2.750$$

The rejection region with α = .05 and 9 degrees of freedom is t < −1.833 and H_0 is rejected. There is evidence to indicate that $\mu < 8.7$.

b. Refer to Table 4, Appendix II. The observed value of t = −2.75 falls between $-t_{.025}$ = −2.262 and $-t_{.01}$ = −2.821. Therefore,

$$\text{p-value} = P[t < -2.750]$$

must lie between .01 and .025, or .01 < p-value < .025.

8.25

a. The hypothesis to be tested is

$$H_0: \mu = 800 \qquad H_a: \mu > 800$$

and the test statistic is

$$z \approx \frac{\bar{x} - \mu}{s/\sqrt{n}} = \frac{930 - 800}{2000/\sqrt{40}} = .41$$

The rejection region with $\alpha = .05$ is $z > 1.645$, and H_0 is not rejected.

b. For this one-tailed test,

$$\text{p-value} = P[z > .41] = .5 - .1591 = .3409$$

This p-value is very large. We would not reject H_0 for such a large observed significance level.

8.26

a-b. The hypothesis of interest is one-tailed:

$$H_0: \mu_1 - \mu_2 = 0 \qquad H_a: \mu_1 - \mu_2 > 0$$

c. The test statistic, calculated under the assumption that $\mu_1 - \mu_2 = 0$, is

$$z = \frac{(\bar{x}_1 - \bar{x}_2) - (\mu_1 - \mu_2)}{\sqrt{\frac{\sigma_1^2}{n_1} + \frac{\sigma_2^2}{n_2}}}$$

with σ_1^2 and σ_2^2 known, or estimated by s_1^2 and s_2^2, respectively. The rejection region, with $\alpha = .10$, is $z > 1.28$.

e. For this exercise,

$$z \approx \frac{(\bar{x}_1 - \bar{x}_2) - 0}{\sqrt{\frac{s_1^2}{n_1} + \frac{s_2^2}{n_2}}} = \frac{11.6 - 9.7}{\sqrt{\frac{27.9}{80} + \frac{38.4}{80}}} = 2.087$$

Since the observed value of z falls in the rejection region, H_0 is rejected. There is evidence to indicate that $\mu_1 - \mu_2 > 0$, or $\mu_1 > \mu_2$.

8.27

Refer to Exercise 8.26. For this one-tailed test, $z = 2.09$ and

$$\text{p-value} = P[z > 2.09] = .5 - .4817 = .0183$$

8.28

a. If your research objective were to show that μ_1 is different from μ_2 (whether larger or smaller), the test would be two-tailed.

b. The hypothesis of interest would be

$$H_0: \mu_1 - \mu_2 = 0 \qquad H_a: \mu_1 - \mu_2 \neq 0$$

c. For a two-tailed test with $\alpha = .05$, the rejection region is $|z| > 1.96$ (that is, $z > 1.96$ or $z < -1.96$). Since the observed value of z is $z = 2.087$, H_0 is rejected.

d. There is reason to believe that μ_1 is different from μ_2.

8.29

The probability that you are making an incorrect decision is influenced by the fact that if $\mu_1 - \mu_2 = 0$, it is just as likely that $\bar{x}_1 - \bar{x}_2$ will be positive as that it will be negative. Hence, a two-tailed rejection region *must* be used. Choosing a one-tailed region after determining the sign of $\bar{x}_1 - \bar{x}_2$ simply tells us which of the two pieces of the rejection region is being used. Hence,

$$\alpha = P[\text{reject } H_0 \text{ when } H_0 \text{ true}] = P[z > 1.28 \text{ or } z < -1.28 \text{ when } H_0 \text{ true}]$$

$$= \alpha_1 + \alpha_2 = .20$$

which is twice what the experimenter thinks it is. Hence, one cannot choose the rejection region after the test is performed.

8.30

a-b. The hypothesis to be tested is one-tailed:

$$H_0: \mu_1 - \mu_2 = 0 \qquad H_a: \mu_1 - \mu_2 > 0$$

c. The rejection region with $\alpha = .10$ and $n_1 + n_2 - 2 = 8 - 2 = 6$ degrees of freedom is $t > t_{.10} = 1.440$.

d. The preliminary calculations are shown below:

Sample 1	Sample 2
$\Sigma x_{1i} = 49$	$\Sigma x_{2i} = 38$
$\Sigma x_{1i}^2 = 617$	$\Sigma x_{2i}^2 = 366$
$\bar{x}_1 = 12.25$	$\bar{x}_2 = 9.5$
$n_1 = 4$	$n_2 = 4$

Then

$$s^2 = \frac{617 - \frac{(49)^2}{4} + 366 - \frac{(38)^2}{4}}{4 + 4 - 2} = \frac{16.75 + 5.00}{6} = 3.625$$

The test statistic, under the assumption that $\sigma_1^2 = \sigma_2^2 = \sigma^2$, is calculated using the pooled value of s^2 in the t statistic shown below:

$$t = \frac{(\bar{x}_1 - \bar{x}_2) - 0}{\sqrt{s^2\left(\frac{1}{n_1} + \frac{1}{n_2}\right)}} = \frac{12.25 - 9.5}{\sqrt{3.625\left(\frac{1}{4} + \frac{1}{4}\right)}} = 2.043$$

and H_0 is rejected.

e. Since the observed t falls between $t_{.05}$ and $t_{.025}$, $.025 < \text{p-value} < .05$.

8.31

a-b. The hypothesis to be tested is two-tailed:

$$H_0: \mu_1 - \mu_2 = 0 \qquad H_a: \mu_1 - \mu_2 \neq 0$$

c. The rejection region with $\alpha = .01$ is $|z| > 2.58$.

d. The observed value of the test statistic is

$$z \approx \frac{(\bar{x}_1 - \bar{x}_2) - 0}{\sqrt{\frac{s_1^2}{n_1} + \frac{s_2^2}{n_2}}} = \frac{264 - 199}{\sqrt{\frac{157^2}{30} + \frac{111^2}{30}}} = 1.85$$

and H_0 is not rejected. There is insufficient evidence to indicate that $\mu_1 - \mu_2 \neq 0$.

8.32

a. The hypothesis to be tested is two-tailed:

$$H_0: \mu_1 - \mu_2 = 0 \qquad H_a: \mu_1 - \mu_2 \neq 0$$

and the rejection region with $\alpha = .10$ is $|z| > 1.645$. The observed value of the test statistic is

$$z \approx \frac{(\bar{x}_1 - \bar{x}_2) - 0}{\sqrt{\frac{s_1^2}{n_1} + \frac{s_2^2}{n_2}}} = \frac{78{,}100 - 82{,}700}{\sqrt{\frac{6300^2}{57} + \frac{7100^2}{66}}} = -3.807$$

and H_0 is rejected. There is sufficient evidence to indicate that the means differ from April to May.

b. In this exercise, H_0 would be rejected in either case.

8.33

a-b. The hypothesis to be tested is one-tailed:

$$H_0: \mu_1 - \mu_2 = 0 \qquad H_a: \mu_1 - \mu_2 < 0$$

since increased scores imply $\mu_2 > \mu_1$.

c. The rejection region with $\alpha = .05$ and 18 d.f. is $t < -1.734$.

d. The observed value of the test statistic with $s^2 = \frac{9(.95)^2 + 9(.56)^2}{18} = .60805$, is

$$t = \frac{(\bar{x}_1 - \bar{x}_2) - 0}{\sqrt{s^2 \left(\frac{1}{n_1} + \frac{1}{n_2}\right)}} = \frac{6.82 - 8.17}{\sqrt{.60805 \left(\frac{1}{10} + \frac{1}{10}\right)}} = -3.871$$

and H_0 is rejected. The training course was effective in increasing customer service scores.

e. Since H_0 is rejected, the risk is

$$\alpha = P[\text{reject } H_0 \text{ when } H_0 \text{ true}] = .05$$

8.34

The hypothesis to be tested is two-tailed:

$$H_0: \mu_1 - \mu_2 = 0 \qquad H_a: \mu_1 - \mu_2 \neq 0$$

and the preliminary calculations are shown below:

Sample 1	Sample 2
$\Sigma x_{1i} = 13.9$	$\Sigma x_{2i} = 12.78$
$\Sigma x_{1i}^2 = 19.4708$	$\Sigma x_{2i}^2 = 16.9368$
$\bar{x}_1 = 1.39$	$\bar{x}_2 = 1.278$
$n_1 = 10$	$n_2 = 10$

Then

$$s_1^2 = \frac{19.4708 - \frac{(13.9)^2}{10}}{9} = .016644 \qquad s_2^2 = \frac{16.9368 - \frac{(12.78)^2}{10}}{9} = .067107$$

The test statistic, under the assumption that $\sigma_1^2 \neq \sigma_2^2$, is calculated using the t statistic shown below:

$$t = \frac{(\bar{x}_1 - \bar{x}_2) - 0}{\sqrt{\frac{s_1^2}{n_1} + \frac{s_2^2}{n_2}}} = \frac{1.39 - 1.278}{\sqrt{\frac{.016644}{10} + \frac{.067107}{10}}} = 1.22$$

The degrees of freedom for this test statistic are approximated as

$$df \approx \frac{\left(\frac{s_1^2}{n_1} + \frac{s_2^2}{n_2}\right)^2}{\frac{\left(\frac{s_1^2}{n_1}\right)^2}{n_1 - 1} + \frac{\left(\frac{s_2^2}{n_2}\right)^2}{n_2 - 1}} = \frac{(.016644 + .067107)^2}{\frac{(.016644)^2}{9} + \frac{(.067107)^2}{9}} = 13.2$$

The rejection region with $\alpha = .01$ and a conservative 13 degrees of freedom is $|t| > t_{.005} = 3.012$ and H_0 is not rejected. There is insufficient evidence to indicate a difference in the two average times.

8.35

a. As in Exercise 8.34, $H_0: \mu_1 - \mu_2 = 0$ and $H_a: \mu_1 - \mu_2 \neq 0$.

b. From the last line of printout, p-value = .24.

c. Since the p-value is quite large, there is insufficient evidence for rejection of H_0. We cannot conclude that the means are different.

8.36

The test statistic is

$$t = \frac{\bar{d} - \mu_d}{s_d/\sqrt{n}} = \frac{.3 - 0}{\sqrt{\frac{.16}{10}}} = 2.372$$

with n − 1 = 9 degrees of freedom. The p-value is then

$$P[|t| > 2.372] = 2\, P[t > 2.372]$$

Since the value t = 2.372 falls between two tabled entries for 9 d.f., $t_{.025} = 2.262$ and $t_{.01} = 2.821$, we can conclude that

$$2(.01) < \text{p-value} < 2(.025)$$

$$.02 < \text{p-value} < .05.$$

8.37

A 95% confidence interval for $\mu_1 - \mu_2 = \mu_d$ is

$$\overline{d} \pm t_{.025} \frac{s_d}{\sqrt{n}} \Rightarrow .3 \pm 2.262 \sqrt{\frac{.16}{10}} \Rightarrow .3 \pm .286$$

or $.014 < (\mu_1 - \mu_2) < .586$.

8.38

Using $s_d^2 = .16$ from Exercise 8.36 and B = .1, the inequality to be solved is approximately

$$1.96\, \frac{s_d}{\sqrt{n}} \leq .1$$

$$\sqrt{n} \geq \frac{1.96\sqrt{.16}}{.1} = 7.84 \quad \Rightarrow \quad n \geq 61.47 \quad \text{or } n = 62$$

Since this value of n is greater than 30, the sample size, n = 62 pairs, will be valid.

8.39

a. The hypothesis of interest is

$$H_0: \mu_1 - \mu_2 = 0 \quad \text{or} \quad H_0: \mu_d = 0$$

$$H_a: \mu_1 - \mu_2 > 0 \quad \text{or} \quad H_a: \mu_d > 0$$

b. The test statistic is

$$t = \frac{\overline{d} - \mu_d}{s_d/\sqrt{n}} = \frac{5.7 - 0}{\sqrt{\frac{256}{18}}} = 1.511$$

The rejection region with $\alpha = .05$ and n − 1 = 17 d.f. is $t > t_{.05} = 1.740$, and H_0 is not rejected. We cannot conclude that $\mu_d > 0$.

8.40

The 90% confidence interval for $\mu_1 - \mu_2 = \mu_d$ is

$$\overline{d} \pm t_{.05} \frac{s_d}{\sqrt{n}} \Rightarrow .13 \pm 1.796 \sqrt{\frac{.001}{12}} \Rightarrow .13 \pm .016$$

or $.114 < (\mu_1 - \mu_2) < .146$.

8.41

a. It is necessary to use a paired-difference test, since the two samples are not random and independent. The hypothesis of interest is

$$H_0: \mu_1 - \mu_2 = 0 \quad \text{or} \quad H_0: \mu_d = 0$$

$$H_a: \mu_1 - \mu_2 \neq 0 \quad \text{or} \quad H_a: \mu_d \neq 0$$

The table of differences, along with the calculation of $\bar{d}$ and s_d^2, is presented below.

d_i	d_i^2
.1	.01
.1	.01
0	.00
.2	.04
−.1	.01
.3	.07

$$\bar{d} = \frac{\Sigma d_i}{n} = \frac{.3}{5} = .06$$

$$s_d^2 = \frac{\Sigma d_i^2 - \frac{(\Sigma d_i)^2}{n}}{n-1} = \frac{.07 - \frac{(.3)^2}{5}}{4} = .013$$

The test statistic is

$$t = \frac{\bar{d} - \mu_d}{s_d/\sqrt{n}} = \frac{.06 - 0}{\sqrt{\frac{.013}{5}}} = 1.177$$

The rejection region with $\alpha = .05$ and $n - 1 = 4$ d.f. is $|t| > t_{.025} = 2.776$, and H_0 is not rejected. We cannot conclude that the means are different.

b. The observed significance level is

$$P[|t| > 1.177] = 2\,P[t > 1.177] > 2(.10) = .20.$$

c. A 95% confidence interval for $\mu_1 - \mu_2 = \mu_d$ is

$$\bar{d} \pm t_{.025} \frac{s_d}{\sqrt{n}} \Rightarrow .06 \pm 2.776 \sqrt{\frac{.013}{5}} \Rightarrow .06 \pm .142$$

or $-.082 < (\mu_1 - \mu_2) < .202$.

d. In order to use the paired-difference test, it is necessary that the n paired observations be randomly selected from normally distributed populations.

8.42

The hypothesis of interest is

$$H_0: \mu_1 - \mu_2 = 0 \quad \text{or} \quad H_0: \mu_d = 0$$

$$H_a: \mu_1 - \mu_2 \neq 0 \quad \text{or} \quad H_a: \mu_d \neq 0$$

The table of differences, along with the calculation of $\bar{d}$ and s_d^2, is presented below.

d_i	d_i^2
.1	.01
.7	.49
.3	.09
−.1	.01
.5	.25
.2	.04
.5	.25

$$\bar{d} = \frac{\Sigma d_i}{n} = \frac{2.2}{7} = .3143$$

$$s_d^2 = \frac{\Sigma d_i^2 - \frac{(\Sigma d_i)^2}{n}}{n-1} = \frac{1.14 - \frac{(2.2)^2}{7}}{6} = .07476$$

The test statistic is

$$t = \frac{\bar{d} - \mu_d}{s_d/\sqrt{n}} = \frac{.3143 - 0}{\sqrt{\frac{.07476}{7}}} = 3.041$$

The rejection region with $\alpha = .01$ and $n - 1 = 6$ d.f. is $|t| > t_{.005} = 3.707$, and H_0 is not rejected. We cannot conclude that the means are different.

b. The observed significance level is

$$P[|t| > 3.041] = 2\,P[t > 3.041]$$

Since $P[t > 3.041]$ falls between .01 and .025,

$$2(.01) < \text{p-value} < 2(.025)$$

$$.02 < \text{p-value} < .05$$

c. A 95% confidence interval for $\mu_1 - \mu_2 = \mu_d$ is

$$\bar{d} \pm t_{.025} \frac{s_d}{\sqrt{n}} \Rightarrow .314 \pm 2.447 \sqrt{\frac{.07476}{7}} \Rightarrow .314 \pm .253$$

or $.061 < (\mu_1 - \mu_2) < .567$.

d. In order to use the paired-difference test, it is necessary that the n paired observations be randomly selected from normally distributed populations.

8.43

a. The hypothesis of interest is

$$H_0: \mu_A - \mu_B = 0 \quad \text{or} \quad H_0: \mu_d = 0$$

$$H_a: \mu_A - \mu_B > 0 \quad \text{or} \quad H_a: \mu_d > 0$$

The table of differences, along with the calculation of $\bar{d}$ and s_d^2, is presented below.

d_i	d_i^2
1.2	1.44
1.6	2.56
2.9	8.41
4.1	16.81
−0.4	0.16
1.8	3.24
0.8	0.64
−0.1	0.01

$$\bar{d} = \frac{\Sigma d_i}{n} = \frac{11.9}{8} = 1.4875$$

$$s_d^2 = \frac{\Sigma d_i^2 - \frac{(\Sigma d_i)^2}{n}}{n-1} = \frac{33.27 - \frac{(11.9)^2}{8}}{7} = 2.2241$$

The test statistic is

$$t = \frac{\bar{d} - \mu_d}{s_d/\sqrt{n}} = \frac{1.4875 - 0}{\sqrt{\frac{2.2241}{8}}} = 2.821$$

The rejection region with $\alpha = .05$ and $n - 1 = 7$ d.f. is $t > t_{.05} = 1.895$, and H_0 is rejected. We can conclude that assessor A gives higher assessments than assessor B.

b. A 90% confidence interval for $\mu_A - \mu_B = \mu_d$ is

$$\bar{d} \pm t_{.025} \frac{s_d}{\sqrt{n}} \Rightarrow 1.4875 \pm 2.365 \sqrt{\frac{2.2241}{8}} \Rightarrow 1.4875 \pm 1.247$$

or $.2405 < (\mu_A - \mu_B) < 2.7345$.

c. In order to use the paired-difference test, it is necessary that the n paired observations be randomly selected from normally distributed populations.

d. Yes. If the individual assessments are normally distributed, then the mean of four assessments will be normally distributed. Hence, the difference $x_A - \bar{x}$ will be normally distributed and the t test on the differences is valid as in part c.

8.44

a. The hypothesis of interest is

$$H_0: \mu_1 - \mu_2 = 0 \quad \text{or} \quad H_0: \mu_d = 0$$

$$H_a: \mu_1 - \mu_2 > 0 \quad \text{or} \quad H_a: \mu_d > 0$$

where μ_1 is the mean for the current year and μ_2 is the mean for last year. The differences, along with necessary calculations, are given below:

d_i
.54
-.94
-.23
.30
.28
.76

$$\bar{d} = \frac{\Sigma d_i}{n} = \frac{.71}{6} = .118333$$

$$s_d^2 = \frac{\Sigma d_i^2 - \frac{(\Sigma d_i)^2}{n}}{n-1} = \frac{1.9741 - \frac{(.71)^2}{6}}{5} = .3780167$$

The test statistic is

$$t = \frac{\bar{d} - \mu_d}{s_d/\sqrt{n}} = \frac{.118333 - 0}{\sqrt{\frac{.3780167}{6}}} = .471$$

The rejection region with $\alpha = .05$ and $n - 1 = 5$ d.f. is $t > t_{.05} = 2.015$, and H_0 is not rejected. We cannot conclude that the mean has increased.

b. Since the observed t is less than $t_{.10}$, p-value > .10.

c. A 95% confidence interval for $\mu_1 - \mu_2 = \mu_d$ is

$$\bar{d} \pm t_{.025} \frac{s_d}{\sqrt{n}} \Rightarrow .118 \pm 2.571 \sqrt{\frac{.3780167}{6}} \Rightarrow .118 \pm .645$$

or $-.527 < (\mu_1 - \mu_2) < .763$.

8.45

a. The hypothesis of interest is
$$H_0: \mu_1 - \mu_2 = 0 \quad \text{or} \quad H_0: \mu_d = 0$$

$$H_a: \mu_1 - \mu_2 > 0 \quad \text{or} \quad H_a: \mu_d > 0$$

where μ_1 is the mean for the old system and μ_2 is the mean for the incentive system. The differences along with necessary calculations are given below:

d_i		
64	39	57
58	33	-2
36	2	51
0	59	
66	-5	
36	45	

$$\bar{d} = \frac{\Sigma d_i}{n} = \frac{539}{15} = 35.933$$

$$s_d^2 = \frac{\Sigma d_i^2 - \frac{(\Sigma d_i)^2}{n}}{n-1} = \frac{28{,}407 - \frac{(539)^2}{15}}{14} = 645.638$$

The test statistic is

$$t = \frac{\bar{d} - \mu_d}{s_d/\sqrt{n}} = \frac{35.933 - 0}{\sqrt{\frac{645.638}{15}}} = 5.477$$

and the observed level of significance is p-value = $P[t > 5.477]$, which is less than .005 from Table 4. Hence, H_0 is rejected, and we conclude that the average payment time is reduced under the incentive system.

b. A 95% confidence interval for $\mu_1 - \mu_2 = \mu_d$ is

$$\bar{d} \pm t_{.025} \frac{s_d}{\sqrt{n}} \Rightarrow 35.933 \pm 2.145 \sqrt{\frac{645.638}{15}} \Rightarrow 35.933 \pm 14.073$$

or $21.860 < (\mu_1 - \mu_2) < 50.006$.

8.46

a. The hypothesis of interest is

$$H_0: \mu_1 - \mu_2 = 0 \quad \text{or} \quad H_0: \mu_d = 0$$

$$H_a: \mu_1 - \mu_2 \neq 0 \quad \text{or} \quad H_a: \mu_d \neq 0$$

Using the MINITAB printout, the test statistic is

$$t = \frac{\bar{d} - \mu_d}{s_d/\sqrt{n}} = -1.90$$

and the rejection region with $n - 1 = 11$ d.f. is $|t| > 1.796$. H_0 is rejected, and we conclude that there is a difference in mean daily sales for "no music" versus "slow tempo."

b. A 90% confidence interval for $\mu_1 - \mu_2 = \mu_d$ is

$$\bar{d} \pm t_{.05} \frac{s_d}{\sqrt{n}} \Rightarrow -635 \pm 1.796\,(333) \Rightarrow -635 \pm 598$$

or $-1233 < (\mu_1 - \mu_2) < -37$. Since this interval is calculated using rounded values, it does not exactly match the interval given in the printout as $(-1233, -36)$.

8.47

a-b. The hypothesis of interest concerns the binomial parameter p and is one-tailed:

$$H_0: p = .3 \qquad H_a: p < .3$$

c. It is given that x = 279 and n = 1000, so that $\hat{p} = \frac{x}{n} = \frac{279}{1000} = .279$. The test statistic is then

$$z = \frac{\hat{p} - p_0}{\sqrt{\frac{p_0 q_0}{n}}} = \frac{.279 - .3}{\sqrt{\frac{.3(.7)}{1000}}} = -1.449$$

The rejection region is one-tailed, with $\alpha = .05$ or $z < -1.645$ and H_0 is not rejected. We cannot conclude that $p < .3$.

8.48

a-b. The hypothesis of interest is one-tailed:

$$H_0: p = .6 \qquad H_a: p > .6$$

c. It is given that x = 1238 and n = 2000, so that $\hat{p} = \frac{x}{n} = \frac{1238}{2000} = .619$. The test statistic is then

$$z = \frac{\hat{p} - p_0}{\sqrt{\frac{p_0 q_0}{n}}} = \frac{.619 - .6}{\sqrt{\frac{.6(.4)}{2000}}} = 1.734$$

The rejection region is one-tailed, with $\alpha = .05$ or $z > 1.645$ and H_0 is rejected. We can conclude that $p > .6$.

8.49

The hypothesis of interest is one-tailed:

$$H_0: p = .5 \qquad H_a: p > .5$$

With x = 72 and n = 120, so that $\hat{p} = \frac{x}{n} = \frac{72}{120} = .6$, the test statistic is

$$z = \frac{\hat{p} - p_0}{\sqrt{\frac{p_0 q_0}{n}}} = \frac{.6 - .5}{\sqrt{\frac{.5(.5)}{120}}} = 2.191$$

The rejection region is one-tailed, with $\alpha = .05$ or $z > 1.645$ and H_0 is rejected. We conclude that $p > .5$.

8.50

From Exercise 8.49, the observed value of the test statistic is $z = 2.19$, so that

$$\text{p-value} = P[z > 2.19] = .5 - .4857 = .0143$$

Since the p-value is less than $\alpha = .05$, H_0 is rejected. This confirms the conclusion drawn in Exercise 8.49.

8.51

a. The hypothesis of interest is two-tailed:

$$H_0: p = .86 \qquad H_a: p \neq .86$$

With $\hat{p} = .89$, the test statistic is

$$z = \frac{\hat{p} - p_0}{\sqrt{\frac{p_0 q_0}{n}}} = \frac{.89 - .86}{\sqrt{\frac{.86(.14)}{1000}}} = 2.73$$

The rejection region is two-tailed, with $\alpha = .05$ or $|z| > 1.96$ and H_0 is rejected. We conclude that $p \neq .86$.

b. From part a, $z = 2.73$, and the p-value is

$$\text{p-value} = P[|z| \geq 2.73] = 2(.5 - .4968) = .0064$$

Since this p-value is less than $\alpha = .05$, H_0 is rejected, as in part a.

8.52

a-b. If the check verification system reduces the percentage of bad checks, then with this system in operation, we should have $p < .05$. Hence, the hypothesis of interest should be one-tailed:

H_0: $p = .05$ $\qquad$ H_a: $p < .05$

d. The test statistic is calculated using $x = 45$, $n = 1124$, $\hat{p} = 45/1124 = .04$.

$$z = \frac{\hat{p} - p_0}{\sqrt{\frac{p_0 q_0}{n}}} = \frac{.04 - .05}{\sqrt{\frac{.05(.95)}{1124}}} = -1.538$$

The rejection region is one-tailed, with $\alpha = .05$ or $z < -1.645$ and H_0 is not rejected. There is no reason to believe the check verification system is effective.

8.53

a-b. The hypothesis of interest should be one-tailed:

H_0: $p = .10$ $\qquad$ H_a: $p > .10$

c. The rejection region is one-tailed, with $\alpha = .05$ or $z > 1.645$.

d. The observed value of the test statistic is calculated with $\hat{p} = 33/200 = .165$.

$$z = \frac{\hat{p} - p_0}{\sqrt{\frac{p_0 q_0}{n}}} = \frac{.165 - .10}{\sqrt{\frac{.10(.90)}{200}}} = 3.064$$

H_0 is rejected. We conclude that p exceeds .10.

8.54

a. Since it is of interest to determine whether or not p has decreased from the known past value of 0.6, the hypothesis of interest is

H_0: $p = .6$ $\qquad$ H_a: $p < .6$

b. Calculate $\hat{p} = x/n = 108/200 = .54$. The test statistic is

$$z = \frac{\hat{p} - p_0}{\sqrt{\frac{p_0 q_0}{n}}} = \frac{.54 - .6}{\sqrt{\frac{.6(.4)}{200}}} = -1.732$$

The rejection region is one-tailed, with $\alpha = .05$ or $z < -1.645$ and H_0 is rejected. There is evidence of a reduction in the proportion of all subscribers who will renew.

c. The approximate 95% confidence interval for p is

$$\hat{p} \pm 1.96\sqrt{\frac{\hat{p}\hat{q}}{n}} \Rightarrow .54 \pm 1.96\sqrt{\frac{(.54)(.46)}{200}} \Rightarrow .54 \pm .069$$

or $.471 < p < .609$.

d. With $B = .01$ and p known from past experience to be approximately .6, the inequality to be solved is

$$1.96\sqrt{\frac{.6(.4)}{n}} \leq .01$$

$$\sqrt{n} \geq 96.02 \quad \text{or } n \geq 9220$$

8.55

a-b. Since it is necessary to detect either $p_1 > p_2$ or $p_1 < p_2$, a two-tailed test is necessary:

$$H_0: p_1 - p_2 = 0 \qquad H_a: p_1 - p_2 \neq 0$$

c. The test statistic, based on the sample data, will be

$$z = \frac{\hat{p}_1 - \hat{p}_2 - (p_1 - p_2)}{\sqrt{\frac{p_1 q_1}{n_1} + \frac{p_2 q_2}{n_2}}}$$

In order to evaluate the denominator, estimates for p_1 and p_2 must be obtained, using the assumption that $p_1 - p_2 = 0$. Because we are assuming that $p_1 = p_2$, the best estimate for this common value will be

$$\hat{p} = \frac{x_1 + x_2}{n_1 + n_2} = \frac{74 + 81}{140 + 140} = .554$$

Also,

$$\hat{p}_1 = \frac{74}{140} = .529, \qquad \hat{p}_2 = \frac{81}{140} = .579$$

The test statistic is then

$$z = \frac{\hat{p}_1 - \hat{p}_2}{\sqrt{\hat{p}\hat{q}\left(\frac{1}{n_1} + \frac{1}{n_2}\right)}} = \frac{.529 - .579}{\sqrt{.554(.446)(2/140)}} = -.842$$

The rejection region with $\alpha = .05$ is $|z| > 1.96$ and H_0 is not rejected. There is no evidence of a difference in the two population parameters.

8.56

a-b. If p_1 cannot be larger than p_2, the only alternative to $H_0: p_1 - p_2 = 0$ is that $p_1 < p_2$, and the one-tailed alternative is $H_a: p_1 - p_2 < 0$.

c. The rejection region with $\alpha = .10$ is $z < -1.28$ and the observed value of the test statistic is $z = -.842$. The null hypothesis is not rejected. There is insufficient evidence to indicate that p_1 is less than p_2.

8.57

a. The hypothesis of interest is one-tailed:

$$H_0: p_1 - p_2 = 0 \qquad H_a: p_1 - p_2 < 0$$

b. Calculate

$$\hat{p}_1 = \frac{132}{280} = .471, \quad \hat{p}_2 = \frac{178}{350} = .509, \text{ and } \hat{p} = \frac{x_1 + x_2}{n_1 + n_2} = \frac{132 + 178}{280 + 350} = .492$$

The test statistic is then

$$z = \frac{\hat{p}_1 - \hat{p}_2}{\sqrt{\hat{p}\hat{q}\left(\frac{1}{n_1} + \frac{1}{n_2}\right)}} = \frac{.471 - .509}{\sqrt{.492(.508)(1/280 + 1/350)}} = -.948$$

The rejection region with $\alpha = .05$ is $z < -1.645$ and H_0 is not rejected. There is insufficient evidence to indicate that $p_1 < p_2$.

8.58

a. Since the modification can only decrease the fraction defective, p_2 can only be less than p_1 if the modification is effective. Hence, the hypothesis to be tested is

$$H_0: p_1 - p_2 = 0 \qquad H_a: p_1 - p_2 > 0$$

b. It is given that $\hat{p}_1 = .0525$, $\hat{p}_2 = .035$, and $n_1 = n_2 = 400$. Hence, $x_1 = n_1\hat{p}_1 = 21$ and $x_2 = n_2\hat{p}_2 = 14$. Then

$$\hat{p} = \frac{x_1 + x_2}{n_1 + n_2} = \frac{35}{800} = .04375$$

The test statistic is

$$z = \frac{\hat{p}_1 - \hat{p}_2}{\sqrt{\hat{p}\hat{q}\left(\frac{1}{n_1} + \frac{1}{n_2}\right)}} = \frac{.0525 - .035}{\sqrt{.04375(.95625)(1/400 + 1/400)}} = 1.210$$

The rejection region with $\alpha = .05$ is $z > 1.645$ and H_0 is not rejected. There is insufficient evidence to show that the modification is effective.

8.59

a. The hypothesis of interest is one-tailed:

$$H_0: p_1 - p_2 = 0 \qquad H_a: p_1 - p_2 > 0$$

where p_1 is the probability that a business executive is rated high or very high in 1993. Calculate

$$\hat{p}_1 = .20, \quad \hat{p}_2 = .18, \text{ and } \hat{p} = \frac{n_1\hat{p}_1 + n_2\hat{p}_2}{n_1 + n_2} = \frac{200 + 180}{2000} = .19$$

The test statistic is then

$$z = \frac{\hat{p}_1 - \hat{p}_2}{\sqrt{\hat{p}\hat{q}\left(\frac{1}{n_1} + \frac{1}{n_2}\right)}} = \frac{.20 - .18}{\sqrt{.19(.81)(1/1000 + 1/1000)}} = 1.14$$

The rejection region with $\alpha = .05$ is $z > 1.645$ and H_0 is not rejected. There is insufficient evidence to indicate that the rating of business executives has increased from 1992 to 1993.

b. From part a, p-value = $P[z > 1.14] = .5 - .3729 = .1271$.

With $\alpha = .05$, this p-value is too large to allow rejection of H_0.

8.60

a. The hypothesis is two-tailed:

$$H_0: p_1 - p_2 = 0 \qquad H_a: p_1 - p_2 \neq 0$$

Calculate

$$\hat{p}_1 = .60, \quad \hat{p}_2 = .49, \quad \text{and } \hat{p} = \frac{n_1\hat{p}_1 + n_2\hat{p}_2}{n_1 + n_2} = \frac{300 + 245}{1000} = .545$$

The test statistic is then

$$z = \frac{\hat{p}_1 - \hat{p}_2}{\sqrt{\hat{p}\hat{q}\left(\frac{1}{n_1} + \frac{1}{n_2}\right)}} = \frac{.60 - .49}{\sqrt{.545(.455)(1/500 + 1/500)}} = 3.49$$

The rejection region with $\alpha = .01$ is $|z| > 2.58$ and the null hypothesis is rejected. There is a difference in the proportions for men and women.

b. From part a, p-value $= P[|z| \geq 3.49] = 2P[z \geq 3.49]$. Since $z = 3.49$ is larger than the largest value, $z = 3.09$, in Table 3, $P[z \geq 3.49] < P[z \geq 3.09] = .5 - .4990 = .001$. Then

$$\text{p-value} = 2P[z \geq 3.49] < 2(.001) = .002$$

Since this p-value is less than $\alpha = .01$, H_0 is rejected, as in part a.

8.61

a. The hypothesis to be tested is

$$H_0: p_1 - p_2 = 0 \qquad H_a: p_1 - p_2 \neq 0$$

It is given that $\hat{p}_1 = .45$, $\hat{p}_2 = .76$, $n_1 = 56$ and $n_2 = 146$. Hence, $x_1 = n_1\hat{p}_1 \approx 25$ and $x_2 = n_2\hat{p}_2 \approx 111$. Then

$$\hat{p} = \frac{x_1 + x_2}{n_1 + n_2} \approx \frac{25 + 111}{56 + 146} = .67$$

The test statistic is

$$z = \frac{\hat{p}_1 - \hat{p}_2}{\sqrt{\hat{p}\hat{q}\left(\frac{1}{n_1} + \frac{1}{n_2}\right)}} = \frac{.45 - .76}{\sqrt{.67(.33)(1/56 + 1/146)}} = -4.19$$

The rejection region with $\alpha = .01$ is $|z| > 2.58$ and H_0 is rejected. There is evidence to show that there is a difference in the two proportions.

b. $(\hat{p}_1 - \hat{p}_2) \pm 2.58\sqrt{\frac{\hat{p}_1\hat{q}_1}{n_1} + \frac{\hat{p}_2\hat{q}_2}{n_2}}$

$$(.45 - .76) \pm 2.58\sqrt{\frac{.45(.55)}{56} + \frac{.76(.24)}{146}}$$

$$-.31 \pm .194$$

or $-.504 < (p_1 - p_2) < -.116$. Notice that the value $(p_1 - p_2) = 0$ does not fall in the interval, confirming the results of part a.

8.62

a. The hypothesis to be tested is

$$H_0: p_1 - p_2 = 0 \qquad H_a: p_1 - p_2 \neq 0$$

Calculate

$$\hat{p}_1 = \frac{62}{200} = .31, \quad \hat{p}_2 = \frac{54}{200} = .27, \quad \text{and } \hat{p} = \frac{x_1 + x_2}{n_1 + n_2} = \frac{116}{400} = .29$$

The test statistic is then

$$z = \frac{\hat{p}_1 - \hat{p}_2}{\sqrt{\hat{p}\hat{q}\left(\frac{1}{n_1} + \frac{1}{n_2}\right)}} = \frac{.31 - .27}{\sqrt{.29(.71)(1/200 + 1/200)}} = .88$$

The rejection region with $\alpha = .05$ is $|z| > 1.96$ and H_0 is not rejected. There is insufficient evidence to indicate a difference in the two proportions.

b. p-value $= P[|z| > .88] = 2P[z > .88] = 2(.5 - .3106) = .3788$

8.63

It is necessary to test

$$H_0: \sigma^2 = 15 \qquad H_a: \sigma^2 > 15$$

This will be done by using s^2, the sample variance, which is a good estimate for σ^2. Refer to Section 8.9 of the text and notice that the quantity

$$\chi^2 = \frac{(n-1)s^2}{\sigma^2}$$

possesses a chi-square distribution in repeated sampling. This distribution is shown in Figure 8.6. Notice that the distribution is nonsymmetrical and that the random variable

$$\frac{(n-1)s^2}{\sigma^2}$$

takes on values commencing at 0 (since s^2, σ^2 and $(n-1)$ are never negative).

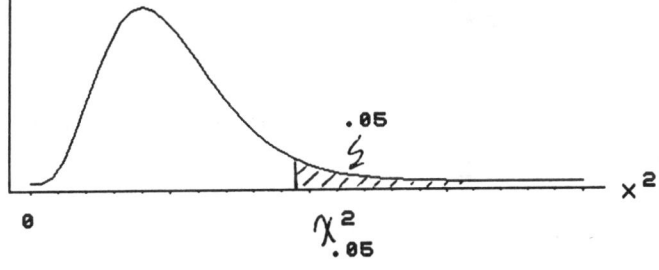

Figure 8.6

The test statistic is

$$\chi^2 = \frac{(n-1)s^2}{\sigma_0^2} = \frac{24(21.4)}{15} = 34.24$$

A one-tailed test of an hypothesis is required. Hence, a critical value of χ^2 (denoted by χ_C^2) must be found such that

$$P[\chi^2 > \chi_C^2] = .05$$

Indexing $\chi_{.05}^2$ with $(n-1) = 24$ degrees of freedom (see Table 5), the critical value is found to be $\chi_{.05}^2 = 36.4151$ (see Figure 8.6). The value of the test statistic does not fall in the rejection region. Hence, H_0 is not rejected. We cannot conclude that the variance exceeds 15.

8.64

For this exercise, $s^2 = .3214$ and $n = 15$. A 90% confidence interval for σ^2 will be

$$\frac{(n-1)s^2}{\chi_{\alpha/2}^2} < \sigma^2 < \frac{(n-1)s^2}{\chi_{(1-\alpha/2)}^2}$$

where $\chi_{\alpha/2}^2$ represents the value of χ^2 such that 5% of the area under the curve (shown in Figure 8.7) lies to its right. Similarly, $\chi_{(1-\alpha/2)}^2$ will be the χ^2 value such that an area .95 lies to its right.

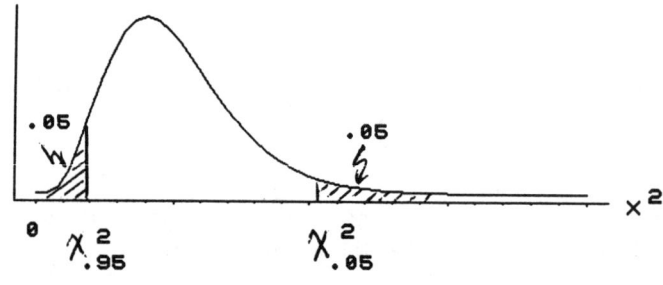

Figure 8.7

Hence, we have located one-half of α in each tail of the distribution. Indexing $\chi_{.05}^2$ and $\chi_{.95}^2$ with $(n-1) = 14$ degrees of freedom in Table 5 yields

$$\chi_{.05}^2 = 23.6848 \quad \text{and} \quad \chi_{.95}^2 = 6.57063$$

and the confidence interval is

$$\frac{14(.3214)}{23.6848} < \sigma^2 < \frac{14(.3214)}{6.57063} \quad \text{or} \quad .190 < \sigma^2 < .685$$

8.65

It is necessary to test

$$H_0: \sigma^2 = 25 \qquad H_a: \sigma^2 < 25$$

and the test statistic is

$$\chi^2 = \frac{(n-1)s^2}{\sigma_0^2} = \frac{21(14.14)}{25} = 11.8776$$

The one-tailed rejection region with $\alpha = .05$ and $(n - 1) = 21$ degrees of freedom is

$$\chi^2 < \chi^2_{.95} = 11.5913$$

and H_0 is not rejected. There is insufficient evidence to indicate that σ^2 is less than 25.

8.66

From Exercise 8.65, $n = 22$ and $s^2 = 14.14$. Then

$$\chi^2_{.05} = 32.6705 \quad \text{and} \quad \chi^2_{.95} = 11.5913$$

and the 90% confidence interval is

$$\frac{21(14.14)}{32.6705} < \sigma^2 < \frac{21(14.14)}{11.5913} \quad \text{or} \quad 9.089 < \sigma^2 < 25.617$$

8.67

When the assumptions for the F distribution are met, then s_1^2/s_2^2 possesses an F distribution with $(n_1 - 1)$ and $(n_2 - 1)$ degrees of freedom. Note that $(n_1 - 1)$ and $(n_2 - 1)$ are the degrees of freedom associated with s_1^2 and s_2^2, respectively. The F distribution is non-symmetrical with the degree of skewness dependent on the above-mentioned degrees of freedom. Table 6 presents the critical values of F (depending on the degrees of freedom) such that

$$P[F > F_a] = a$$

for $a = .10, .05, .025, .01$ and $.005$, respectively. Because right-hand tail areas correspond to an upper-tailed test of an hypothesis, we will always identify the larger sample variance as s_1^2 (that is, we will always place the larger sample variance in the numerator of $F = s_1^2/s_2^2$). Hence, an upper-tailed test is implied and the critical values of F will determine the rejection region. If we wish to test the hypothesis

$$H_0: \sigma_1^2 = \sigma_2^2 \qquad H_a: \sigma_1^2 \neq \sigma_2^2$$

there will be another portion of the rejection region in the lower tail of the distribution. The area to the right of the critical value will represent only $\alpha/2$, and the probability of a Type I error is $2(\alpha/2) = \alpha$.

a. In this exercise, the hypothesis of interest is

$$H_0: \sigma_1^2 = \sigma_2^2 \qquad H_a: \sigma_1^2 \neq \sigma_2^2$$

and the test statistic is

$$F = \frac{s_1^2}{s_2^2} = \frac{55.7}{31.4} = 1.774$$

The rejection region (two-tailed) will be determined by a critical value of F based on $\nu_1 = (n_1 - 1) = 15$ and $\nu_2 = (n_2 - 1) = 19$ degrees of freedom, with area .025 to its right. That is, from Table 6, $F > 2.62$. The observed value of F does not fall in the rejection region, and we cannot conclude that the variances are different.

b. The student will need to find critical values of F for various levels of a in order to find the approximate p-value. The critical values with $\nu_1 = 15$ and $\nu_2 = 19$ are shown below from Table 6.

a	F_a
.10	1.86
.05	2.23
.025	2.62
.01	3.15
.005	3.59

Hence,

p-value = $2 P[F > 1.774] > 2(.10) = .20$

8.68

a. The hypothesis of interest is

$$H_0: \sigma_1^2 = \sigma_2^2 \qquad H_a: \sigma_1^2 > \sigma_2^2$$

and the test statistic is

$$F = \frac{s_1^2}{s_2^2} = \frac{18.3}{7.9} = 2.316$$

The rejection region with $\alpha = .05$ and $\nu_1 = \nu_2 = 12$ degrees of freedom is $F > 2.69$ and H_0 is not rejected.

b. The critical values of F for various values of a are given below.

a	F_a
.10	2.15
.05	2.69
.025	3.28
.01	4.16
.005	4.91

Hence,

p-value = $P[F > 2.316]$ lies between .05 and .10.

8.69

a. The hypothesis of interest is

$$H_0: \sigma_1^2 = \sigma_2^2 \qquad H_a: \sigma_1^2 > \sigma_2^2$$

and the test statistic is

$$F = \frac{s_1^2}{s_2^2} = \frac{92{,}000}{37{,}000} = 2.486$$

The one-tailed rejection region with $\alpha = .05$ and $\nu_1 = \nu_2 = 49$ degrees of freedom is found by interpolation in Table 6. The value $F_{49,49}$ is roughly halfway between $F_{40,40} = 1.69$ and $F_{60,60} = 1.53$; therefore, we reject H_0 if $F > F_{49,49} \approx 1.61$. The observed value of the test statistic falls in the rejection region and we conclude that the "suspect line" possesses a larger variance.

b. The student must obtain various critical levels of F from Table 6. We "roughly" interpolate $F_{49,49}$ as halfway between $F_{40,40}$ and $F_{60,60}$.

a	F_a
.05	1.61
.025	1.775
.01	1.975
.005	2.13

In any event,

p-value = $P[F > 2.486] < .005$

8.70

It is necessary to place a 90% confidence interval on σ^2, the variance of the truck noise emission readings. Indexing $\chi^2_{.05}$ and $\chi^2_{.95}$ with $(n-1) = 5$ degrees of freedom in Table 5 yields

$$\chi^2_{.05} = 11.0705 \quad \text{and} \quad \chi^2_{.95} = 1.145476$$

Calculate
$$s^2 = \frac{\Sigma x_i^2 - \frac{(\Sigma x_i)^2}{n}}{n-1} = \frac{44{,}103.74 - \frac{(514.4)^2}{6}}{5} = .502667$$

and the confidence interval is

$$\frac{5(.502667)}{11.0705} < \sigma^2 < \frac{5(.502667)}{1.145476} \quad \text{or} \quad .227 < \sigma^2 < 2.194$$

Intervals constructed in this manner will enclose σ^2 90% of the time in repeated sampling. Hence, we are fairly certain that σ^2 is between .227 and 2.194.

8.71

It is necessary to test

$$H_0: \sigma = .7 \qquad H_a: \sigma > .7$$

which is equivalent to

$$H_0: \sigma^2 = .49 \qquad H_a: \sigma^2 > .49$$

Calculate
$$s^2 = \frac{\Sigma x_i^2 - \frac{(\Sigma x_i)^2}{n}}{n-1} = \frac{36 - \frac{(10)^2}{4}}{3} = 3.6667$$

and the test statistic is

$$\chi^2 = \frac{(n-1)s^2}{\sigma_0^2} = \frac{3(3.6667)}{.49} = 22.449$$

The one-tailed rejection region with $\alpha = .05$ and $(n-1) = 3$ degrees of freedom is

$$\chi^2 > \chi^2_{.05} = 7.81$$

and H_0 is rejected. There is sufficient evidence to indicate that σ^2 is greater than .49.

8.72

From Exercise 8.71, $n = 4$ and $s^2 = 3.6667$. Then

$$\chi^2_{.05} = 7.81473 \quad \text{and} \quad \chi^2_{.95} = .351846$$

and the 90% confidence interval is

$$\frac{3(3.6667)}{7.81473} < \sigma^2 < \frac{3(3.6667)}{.351846} \quad \text{or} \quad 1.408 < \sigma^2 < 31.264$$

8.73

a. The force transmitted to a wearer, x, is known to be normally distributed with $\mu = 800$ and $\sigma = 40$. Hence,
$$P[x > 1000] = P\left[z > \frac{1000 - 8000}{40}\right] = P[z > 5] \approx 0$$

It is highly improbable that any particular helmet will transmit a force in excess of 1000 pounds.

b. Since n = 40, a large sample test will be used to test

$$H_0: \mu = 800 \qquad H_a: \mu > 800$$

The test statistic is

$$z = \frac{\bar{x} - \mu}{s/\sqrt{n}} = \frac{825 - 800}{\sqrt{\frac{2350}{40}}} = 3.262$$

and the rejection region with $\alpha = .05$ is $z > 1.645$. H_0 is rejected and we conclude that $\mu > 800$.

c. The hypothesis of interest is

$$H_0: \sigma = 40 \qquad H_a: \sigma > 40$$

and the test statistic is

$$\chi^2 = \frac{(n-1)s^2}{\sigma_0^2} = \frac{39(2350)}{40^2} = 57.281$$

The one-tailed rejection region with $\alpha = .05$ and $(n-1) = 39$ degrees of freedom (approximated with 40 degrees of freedom) is

$$\chi^2 > \chi^2_{.05} = 55.7585$$

and H_0 is rejected. There is sufficient evidence to indicate that σ is greater than 40.

8.74

a. The officer was concerned that the variances σ_1^2 and σ_2^2 were different.

b. The hypothesis of interest is

$$H_0: \sigma_1^2 = \sigma_2^2 \qquad H_a: \sigma_1^2 \neq \sigma_2^2$$

Calculate

$$s_1^2 = \frac{\Sigma x_i^2 - \frac{(\Sigma x_i)^2}{n}}{n-1} = \frac{7087 - \frac{(183)^2}{5}}{4} = 97.3$$

and

$$s_2^2 = \frac{\Sigma x_i^2 - \frac{(\Sigma x_i)^2}{n}}{n-1} = \frac{2066 - \frac{(100)^2}{5}}{4} = 16.5$$

The test statistic is

$$F = \frac{s_1^2}{s_2^2} = \frac{97.3}{16.5} = 5.897$$

The upper portion of the rejection region with $\alpha = .10$ is found in Table 6 as $F > F_{.05} = 6.39$ and H_0 is not rejected. There is no reason to believe that the assumption has been violated. The Student's t test is appropriate assuming that the other assumptions have been met.

8.75

The hypothesis of interest is

$$H_0: \sigma_1^2 = \sigma_2^2 \qquad H_a: \sigma_1^2 \neq \sigma_2^2$$

and the test statistic is

$$F = \frac{s_1^2}{s_2^2} = \frac{.273}{.094} = 2.904$$

The upper portion of the rejection region with $\alpha = 2(.05) = .10$ is $F > F_{.05} = 3.18$ (from Table 6) and H_0 is not rejected. The supplier's shipments are identical in variability.

8.76

See Section 8.2 of the text.

8.77

See Section 8.4 of the text.

8.78

The following conditions must hold in order that the z statistic be appropriate:

(1) $\bar{x}$ is a normally distributed random variable (the Central Limit Theorem insures this as long as $n \geq 30$ and the sample is random).

(2) σ is a known quantity, or n is large, so that a good approximation for $\sigma_{\bar{x}}$ can be obtained from the sample observations.

8.79

A Student's t test can be employed to test an hypothesis about a single population mean when the sample has been randomly selected from a normal population. It will work quite satisfactorily for populations that possess mound-shaped frequency distributions resembling the normal distribution.

8.80

As in the case of the single population mean, random samples must be independently drawn from two populations that possess normal distributions with a common variance, σ^2. Consequently, it is logical that information in the two sample variances, s_1^2 and s_2^2, should be pooled in order to give the best estimate of the common variance, σ^2. In this way, all of the sample information is being utilized to its best advantage.

8.81

The parameter of interest is μ, the average daily wage of workers in a given industry. A sample of $n = 40$ workers has been drawn from a particular company within this industry and $\bar{x}$, the sample average, has been calculated. The objective is to determine whether this company pays inferior wages in comparison to the total industry. That is, assume that this sample of forty workers has been drawn from a hypothetical population of workers. Does this population have as an average wage $\mu = 23.20$, or is μ less than 23.20? Thus, the hypothesis to be tested is

$$H_0: \mu = 23.20 \qquad H_a: \mu < 23.20$$

a. The test statistic is

$$z \approx \frac{\bar{x} - \mu}{s/\sqrt{n}} = \frac{21.20 - 23.20}{4.5/\sqrt{40}} = -2.811$$

and the p-value is

$$\text{p-value} = P[z < -2.81] = .5 - .4975 = .0025$$

b. Since $\alpha = .01$ is larger than the p-value, .0025, H_0 can be rejected and we conclude that the company is paying inferior wages.

8.82

a. If the airline is to determine whether or not the flight is <u>unprofitable</u>, they are interested in finding out whether or not $\mu < 60$ (since a flight is <u>profitable</u> if μ is at least 60). Hence, the hypothesis to be tested is

$$H_0: \mu = 60 \qquad H_a: \mu < 60$$

b. Since only small values of $\overline{x}$ (and hence negative values of z) would tend to disprove H_0 in favor of H_a, this is a one-tailed test.

c. For this exercise, n = 120, $\overline{x}$ = 58, and s = 11. Hence, the test statistic is

$$z = \frac{\overline{x} - \mu}{\sigma/\sqrt{n}} \approx \frac{\overline{x} - \mu}{s/\sqrt{n}} = \frac{58 - 60}{11/\sqrt{120}} = -1.992$$

The rejection region with $\alpha = .10$ is $z < -1.28$. The observed value, $z = -1.992$, falls in the rejection region and H_0 is rejected. We conclude that the flight is unprofitable.

8.83

The objective of this experiment is to make a decision about the binomial parameter p, which is the probability that a customer prefers model A. Hence, the null hypothesis will be that a customer has no preference for A, and the alternative will be that he does have a preference. If the null hypothesis is true, then

$$H_0: p = P[\text{customer prefers A}] = 1/3$$

If the customer actually has a preference for A, then

$$H_a: p > 1/3$$

a. The test statistic is calculated with $\hat{p} = \frac{400}{1000} = .4$ as

$$z = \frac{\hat{p} - p_0}{\sqrt{\frac{p_0 q_0}{n}}} = \frac{.4 - 1/3}{\sqrt{\frac{(1/3)(2/3)}{1000}}} = 4.472$$

and the p-value is

$$\text{p-value} = P[z > 4.47] < .5 - .4990 = .001$$

since $P[z > 4.47]$ is surely less than $P[z > 3.09]$, the largest value in Table 3.

b. Since $\alpha = .01$ is larger than the p-value, which is less than .001, H_0 can be rejected. We conclude that customers have a preference for model A.

8.84

The manufacturer claims that at least 20% of the public prefers his product. Hence, in order to test his claim, the following hypothesis is used:

$$H_0: p = .20 \qquad H_a: p < .20$$

Rejection of the null hypothesis would imply that the manufacturer's claim is invalid. The critical value of z that separates the acceptance and rejection regions will be $z = -1.645$, since values of $\hat{p}$ in the lower tail of the distribution will tend to disprove the null hypothesis (see Figure 8.8). The objective is then to determine a value for $\hat{p}$ such that the corresponding test statistic, z, will be less than or equal to -1.645. Under the assumption of the null hypothesis, $\mu_{\hat{p}} = p_0 = .2$ and

$$\sigma_{\hat{p}} = \sqrt{\frac{pq}{n}} = \sqrt{\frac{(.2)(.8)}{100}} = .04$$

A value for $\hat{p}$ must be found so that

$$z = \frac{\hat{p} - .2}{.04} \leq -1.645$$

Solving for $\hat{p}$ yields $\hat{p} \leq .1342$.

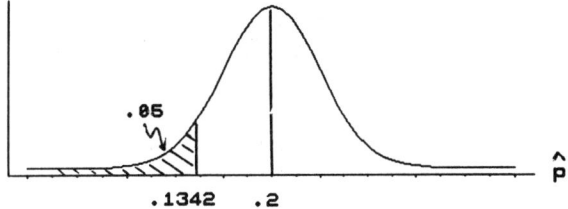

Figure 8.8

8.85

Refer to Exercise 8.84. The observed value of the test statistic is

$$z = \frac{\hat{p} - p_0}{\sqrt{\frac{p_0 q_0}{n}}} = \frac{.16 - .2}{\sqrt{\frac{(.2)(.8)}{100}}} = -1$$

The rejection region for this test with $\alpha = .10$ is $z < -1.28$ and the null hypothesis is not rejected. The manufacturer's claim cannot be refuted.

8.86

The hypothesis to be tested is

$$H_0: \mu = .05 \qquad H_a: \mu > .05$$

and the test statistic is

$$t = \frac{\overline{x} - \mu}{s/\sqrt{n}} = \frac{.058 - .05}{\frac{.012}{\sqrt{10}}} = 2.108$$

The rejection region with $\alpha = .05$ and $n - 1 = 9$ degrees of freedom is located in the upper tail of the t distribution and is found from Table 4 as $t > t_{.05} = 1.833$. Since the observed value of the test statistic falls in the rejection region, H_0 is rejected and we conclude that μ is greater than .05. The observed level of significance, p-value = $P[t > 2.108]$ is between .025 and .05 since $t = 2.108$ falls between $t_{.025}$ and $t_{.05}$.

8.87

The hypothesis to be tested is

$$H_0: \sigma_1^2 = \sigma_2^2 \qquad H_a: \sigma_1^2 \neq \sigma_2^2$$

and the test statistic is

$$F = \frac{s_2^2}{s_1^2} = \frac{172}{81} = 2.123$$

The rejection region (two-tailed) will be determined by critical values of F based on $\nu_1 = (n_2 - 1) = 19$ and $\nu_2 = (n_1 - 1) = 19$ degrees of freedom, such that

$$P[F > F_{.05}] = .05 \text{ and } P[F < F_{.95}] = .95$$

We need only to be concerned with the upper-tail value of F, with area .05 to its right. That is, from Table 6, $F > F_{.05} \approx 2.16$. The observed value of F does not fall in the rejection region, and H_0 is not rejected. We cannot conclude that the variances are significantly different.

Note that 2.16, the F value corresponding to 20 and 19 degrees of freedom, will be slightly smaller than the desired critical F. However, the value of the test statistic is $F = 2.123$. Hence, if we cannot reject for $F_{.05}$ with 20 and 19 degrees of freedom, we will not be able to reject for $F_{.05}$ with 19 and 19 degrees of freedom.

8.88

The hypothesis to be tested is

$$H_0: \sigma_1^2 = \sigma_2^2 \qquad H_a: \sigma_1^2 < \sigma_2^2$$

and the test statistic is

$$F = \frac{s_2^2}{s_1^2} = \frac{(28.2)^2}{(15.6)^2} = 3.268$$

The rejection region (one-tailed) will be determined by a critical value of F based on $\nu_1 = (n_2 - 1) = 29$ and $\nu_2 = (n_1 - 1) = 29$ degrees of freedom, with area .05 to its right. That is, from Table 6, $F > F_{.05} \approx 1.85$. The observed value of F falls in the rejection region, and H_0 is rejected. We conclude that increased maintenance of the older system is needed.

8.89

Define p to be the percentage of defective pieces of equipment in the population. Then the hypothesis to be tested is

$$H_0: p = .05 \qquad H_a: p > .05$$

Note that if the percentage of defectives is greater than 5%, the percentages conforming to specifications will be less than 95%, thus contradicting the manufacturer's claim. The test statistic is calculated to be

$$z = \frac{\hat{p} - p_0}{\sqrt{\frac{p_0 q_0}{n}}} = \frac{\frac{53}{700} - .05}{\sqrt{\frac{(.05)(.95)}{700}}} = 3.122$$

The rejection region will be $z \geq 1.645$. Since the test statistic falls in the rejection region, H_0 is rejected.

8.90

a. The hypothesis to be tested is

$$H_0: \mu_1 - \mu_2 = 0 \qquad H_a: \mu_1 - \mu_2 \neq 0$$

and the observed value of the test statistic is

$$z \approx \frac{(\bar{x}_1 - \bar{x}_2) - 0}{\sqrt{\frac{s_1^2}{n_1} + \frac{s_2^2}{n_2}}} = \frac{1.30 - 1.37}{\sqrt{\frac{.53}{200} + \frac{.64}{200}}} = -.915$$

The rejection region with $\alpha = .05$ is $|z| > 1.96$ and H_0 is not rejected. There is insufficient evidence to indicate a difference in means for the two regions.

b. For each of the two regions, the random variable of interest is x, the number of automobiles owned by a family. Hence, $x = 0, 1, 2,...$ and its probability distribution is discrete. However, the Central Limit Theorem assures us that if n is large, the distribution of $\bar{x}$ will be approximately normal regardless of the distribution of x. Hence, the z test performed in part a is still valid.

c. The approximate 95% confidence interval for μ_2 is

$$\bar{x}_2 \pm 1.96 \frac{s_2}{\sqrt{n_2}} \Rightarrow 1.37 \pm 1.96\sqrt{\frac{.64}{200}} \Rightarrow 1.37 \pm .111$$

or $1.259 < \mu_2 < 1.481$. Intervals constructed in this manner will contain μ_2 95% of the time in repeated sampling. Hence, we are fairly certain that this particular interval contains μ_2.

d. The approximate 90% confidence interval for $\mu_1 - \mu_2$ is

$$(\bar{x}_1 - \bar{x}_2) \pm 1.645\sqrt{\frac{s_1^2}{n_1} + \frac{s_2^2}{n_2}}$$

$$(1.30 - 1.37) \pm 1.645\sqrt{.00585}$$

$$-.07 \pm .126 \qquad \text{or} \qquad -.196 < \mu_1 - \mu_2 < .056$$

8.91

The test is performed as follows:

$$H_0: \mu = 1600 \qquad H_a: \mu < 1600$$

and the test statistic is

$$z \approx \frac{\bar{x} - \mu}{s/\sqrt{n}} = \frac{1570 - 1600}{\frac{120}{\sqrt{100}}} = -2.5$$

a. The p-value is

$$P[z < -2.5] = .5 - .4938 = .0062$$

b. Since $\alpha = .05$ is greater than the observed p-value, .0062, the null hypothesis is rejected. We conclude that $\mu < 1600$.

8.92

a. Let $p = P[\text{customer prefers brand A}]$. Then the hypothesis of interest is

$$H_0: p = .2 \qquad H_a: p > .2$$

b. It is decided to reject H_0 if $x \geq 92$. Hence,

$$\alpha = P[\text{reject } H_0 \text{ when } H_0 \text{ true}] = P[x \geq 92 \text{ when } p = .2]$$

If $p = .2$, then $E(x) = np = 400(.2) = 80$ and $\sigma = \sqrt{npq} = \sqrt{64} = 8$. Using the normal approximation to the binomial distribution, the approximation becomes

$$\alpha \approx P[x > 91.5] = P[z > \frac{91.5 - 80}{8}] = P[z > 1.44] = .5 - .4251 = .0749$$

8.93

a. The hypothesis to be tested is

$$H_0: \mu_1 - \mu_2 = 0 \qquad H_a: \mu_1 - \mu_2 \neq 0$$

and the pooled estimate of σ^2 is calculated as

$$s^2 = \frac{(n_1 - 1)s_1^2 + (n_2 - 1)s_2^2}{n_1 + n_2 - 2} = \frac{9(16.36) + 9(18.92)}{18} = 17.64$$

The test statistic is then

$$t = \frac{\overline{x}_1 - \overline{x}_2 - 0}{\sqrt{s^2\left(\frac{1}{n_1} + \frac{1}{n_2}\right)}} = \frac{22.2 - 28.5}{\sqrt{17.64\left(\frac{2}{10}\right)}} = -3.354$$

With $\alpha = .10$, the rejection region based on 18 degrees of freedom is $|t| > 1.734$ and H_0 is rejected. We conclude that there is a difference in mean servicing times between the two employees.

b. The observed level of significance is p-value $= 2P[t > 3.354]$ for a two-tailed test with 18 degrees of freedom. Since $t = 3.354$ exceeds the largest tabulated value, $t_{.005} = 2.878$, we have p-value $< 2(.005) = .01$.

8.94

The 95% confidence interval is

$$(\overline{x}_1 - \overline{x}_2) \pm t_{\alpha/2}\sqrt{s^2\left(\frac{1}{n_1} + \frac{1}{n_2}\right)}$$

$$(22.2 - 28.5) \pm 2.101\sqrt{17.64\left(\frac{2}{10}\right)}$$

$$-6.3 \pm 3.946 \qquad \text{or} \qquad -10.246 < \mu_1 - \mu_2 < -2.354$$

8.95

The problem of selecting a proper sample size to achieve a given bound is now complicated by the fact that the t value, which is used in calculating the correct sample size, changes as the value of n changes. In Chapter 7, the procedure was to choose n so that the half-width of the $100(1 - \alpha)\%$ confidence interval was less than some given bound, B. That is,

$$z_{\alpha/2}\sigma_{\hat{\theta}} \leq B$$

Now, the inequality to be solved is

$$t_{\alpha/2}\sigma_{\hat{\theta}} \leq B$$

and the $t_{\alpha/2}$ value must be based on $n - 1$ degrees of freedom. Since n is unknown, the procedure is as follows:

(1) Calculate n using $z_{\alpha/2}$ instead of $t_{\alpha/2}$. If the value for n is large (that is, $n \geq 30$), this sample size will be valid.

(2) If the value for n is small, we are not justified in using the value $z_{\alpha/2}$. This value must be replaced by the appropriate t value with $(n - 1)$ degrees of freedom. If the inequality holds, the sample size is valid; if not, it is necessary to pick larger values of n until the inequality will hold. This repetitive procedure is usually not necessary, because the $z_{\alpha/2}$ value will usually yield a satisfactory approximation to the required sample size.

In this exercise, we want to estimate $\mu_1 - \mu_2$ to within 1 with probability .95. Hence, the following inequality must hold:

$$t_{.025} \sqrt{s^2 \left(\frac{1}{n_1} + \frac{1}{n_2}\right)} \leq 1$$

With $n_1 = n_2 = n$, consider the sample size obtained by replacing $t_{.025}$ by $z_{.025}$.

$$1.96 \sqrt{17.64 \left(\frac{2}{n}\right)} \leq 1 \quad \Rightarrow \quad 1.96 \sqrt{35.28} \leq \sqrt{n} \quad \Rightarrow \quad n \geq 135.53 \text{ or } n \geq 136$$

Since this sample size is greater than 30, the sample size $n_1 = n_2 = 136$ is valid.

8.96

No. To gain information, pairing must occur in the experimental design. That is, some real link must exist between paired responses.

8.97

A paired-difference analysis is employed when the two samples (one from population I and one from population II) are physically paired and hence are not independent. See Section 8.6 of the text for a thorough discussion of the paired-difference test.

8.98

The samples are not independent because the paired reactions for a single person will not be independent. However, if the difference between processing times for the two processes is calculated, an independent random sample of differences is generated and a paired-difference analysis can be conducted. The hypothesis of interest is

$$H_0: \mu_2 - \mu_1 = 0 \quad \text{or} \quad H_0: \mu_d = 0$$

$$H_a: \mu_2 - \mu_1 \neq 0 \quad \text{or} \quad H_a: \mu_d \neq 0$$

The differences, $x_2 - x_1$, along with necessary calculations, are given below:

d_i

1 0
1 2
2
−1
1
1

$$\bar{d} = \frac{\Sigma d_i}{n} = \frac{7}{8} = .875$$

$$s_d^2 = \frac{\Sigma d_i^2 - \frac{(\Sigma d_i)^2}{n}}{n - 1} = \frac{13 - \frac{(7)^2}{8}}{7} = .9821$$

The test statistic is

$$t = \frac{\bar{d} - \mu_d}{s_d/\sqrt{n}} = \frac{.875 - 0}{\sqrt{\frac{.9821}{8}}} = 2.497$$

and the rejection region is $|t| > t_{.025} = 2.365$. H_0 is rejected, and we conclude that there is a difference in means between the two processes.

The observed value of t, $t = 2.497$, falls between $t_{.025} = 2.365$ and $t_{.01} = 2.998$, so that

$$2(.01) < \text{p-value} < 2(.025)$$

$$.02 < \text{p-value} < .05$$

8.99

Refer to Exercise 8.98. A 95% confidence interval for $\mu_2 - \mu_1 = \mu_d$ is

$$\bar{d} \pm t_{.025} \frac{s_d}{\sqrt{n}} \Rightarrow .875 \pm 2.365 \sqrt{\frac{.9821}{8}} \Rightarrow .875 \pm .829$$

or $.046 < (\mu_2 - \mu_1) < 1.704$.

8.100

The hypothesis of interest is

$$H_0: \mu = 48{,}000 \qquad H_a: \mu > 48{,}000$$

and the test statistic is

$$t = \frac{\bar{x} - \mu}{s/\sqrt{n}} = \frac{51{,}102 - 48{,}000}{\frac{5127}{\sqrt{12}}} = 2.096$$

The rejection region, with $\alpha = .05$ and $n - 1 = 11$ degrees of freedom, is $t > 1.796$ and H_0 is rejected. There is an increase in water consumption.

8.101

A paired-difference analysis is conducted. The hypothesis of interest is

$$H_0: \mu_A - \mu_B = 0 \quad \text{or} \quad H_0: \mu_d = 0$$

$$H_a: \mu_A - \mu_B \neq 0 \quad \text{or} \quad H_a: \mu_d \neq 0$$

The differences, $x_A - x_B$, along with necessary calculations, are given below:

d_i
2
2
1
1
0
1

$$\bar{d} = \frac{\Sigma d_i}{n} = \frac{12}{10} = 1.2$$

$$s_d^2 = \frac{\Sigma d_i^2 - \frac{(\Sigma d_i)^2}{n}}{n - 1} = \frac{18 - \frac{(12)^2}{10}}{9} = .4$$

The test statistic is

$$t = \frac{\bar{d} - \mu_d}{s_d/\sqrt{n}} = \frac{1.2 - 0}{\sqrt{\frac{.4}{10}}} = 6.00$$

and the rejection region is $|t| > t_{.025} = 2.262$. H_0 is rejected, and we conclude that there is a difference in mean scores for the two methods.

b. A 98% confidence interval for $\mu_A - \mu_B = \mu_d$ is

$$\bar{d} \pm t_{.01} \frac{s_d}{\sqrt{n}} \Rightarrow 1.2 \pm 2.821 \sqrt{\frac{.4}{10}} \Rightarrow 1.2 \pm .564$$

or $.636 < (\mu_A - \mu_B) < 1.764$.

8.102

The hypothesis to be tested is

$$H_0: \sigma^2 = .01 \qquad H_a: \sigma^2 > .01$$

and the test statistic is

$$\chi^2 = \frac{(n-1)s^2}{\sigma_0^2} = \frac{7(.018)}{.01} = 12.6$$

The one-tailed rejection region with $\alpha = .05$ and $(n-1) = 7$ degrees of freedom is

$$\chi^2 > \chi^2_{.05} = 14.07$$

and H_0 is not rejected. There is insufficient evidence to indicate that σ^2 is greater than .01.

8.103

From Exercise 8.102, $n = 8$ and $s^2 = .018$. Then

$$\chi^2_{.05} = 14.0671 \qquad \text{and} \qquad \chi^2_{.95} = 2.16735$$

and the 90% confidence interval is

$$\frac{7(.018)}{14.0671} < \sigma^2 < \frac{7(.018)}{2.16735} \qquad \text{or} \qquad .00896 < \sigma^2 < .05814$$

8.104

In order to employ the F statistic to test an hypothesis concerning the equivalence of two population variances, we must assume that independent random samples have been drawn from two normal populations.

8.105

The hypothesis of interest is

$$H_0: \sigma_1^2 = \sigma_2^2 \qquad H_a: \sigma_1^2 \neq \sigma_2^2$$

and the test statistic is

$$F = \frac{s_2^2}{s_1^2} = \frac{2.96}{1.54} = 1.922$$

The critical values of F for various values of a with $n_2 - 1 = 14$ and $n_1 - 1 = 14$ degrees of freedom cannot be found directly from Table 6. However, they can be approximated as $F_{15,14}$ and are shown in the following table.

a	F_a
.10	2.01
.05	2.46
.025	2.95
.01	3.66
.005	4.25

Hence,

$$\tfrac{1}{2}(\text{p-value}) = P[F > 1.922] \text{ lies below } F_{.10}$$

and

$$\text{p-value} > 2(.10) = .20$$

This is too large to reject H_0.

8.106

The manufacturer claims that the range of the random variable x (purity of his product) is no more than 2%. In terms of the standard deviation, σ, he is claiming that $\sigma \le .5$, since

$$\text{Range} \approx 4\sigma = 2 \quad \text{or} \quad \sigma \approx .5$$

Hence, the hypothesis to be tested is

$$H_0: \sigma = .5 \qquad H_a: \sigma > .5$$

or

$$H_0: \sigma^2 = .25 \qquad H_a: \sigma^2 > .25$$

Calculate $s^2 = \dfrac{\Sigma x_i^2 - \dfrac{(\Sigma x_i)^2}{n}}{n-1} = \dfrac{47{,}982.56 - \dfrac{(489.8)^2}{5}}{4} = .438$

The test statistic is

$$\chi^2 = \frac{(n-1)s^2}{\sigma_0^2} = \frac{4(.438)}{.25} = 7.008$$

The one-tailed rejection region with $\alpha = .05$ and $(n-1) = 4$ degrees of freedom is

$$\chi^2 > \chi^2_{.05} = 9.48773$$

and H_0 is not rejected. There is insufficient evidence to contradict the manufacturer's claim.

8.107

From Exercise 8.106, $n = 5$ and $s^2 = .438$. Then

$$\chi^2_{.05} = 9.48773 \quad \text{and} \quad \chi^2_{.95} = .710721$$

and the 90% confidence interval is

$$\frac{4(.438)}{9.48773} < \sigma^2 < \frac{4(.438)}{.710721} \quad \text{or} \quad .185 < \sigma^2 < 2.465$$

8.108

The hypothesis of interest is

$$H_0: \mu = 16 \qquad H_a: \mu < 16$$

and the test statistic is

$$t = \frac{\bar{x} - \mu}{s/\sqrt{n}} = \frac{15.7 - 16}{\frac{.5}{\sqrt{9}}} = -1.8$$

The rejection region, with $\alpha = .05$ and $n - 1 = 8$ degrees of freedom, is $t < -2.306$ and H_0 is not rejected. There is insufficient evidence to indicate that the mean is less than 16.

8.109

a-b. A paired-difference analysis is conducted. The hypothesis of interest is

$$H_0: \mu_1 - \mu_2 = 0 \quad \text{or} \quad H_0: \mu_d = 0$$

$$H_a: \mu_1 - \mu_2 \neq 0 \quad \text{or} \quad H_a: \mu_d \neq 0$$

Using the MINITAB printout, the test statistic is

$$t = \frac{\bar{d} - \mu_d}{s_d/\sqrt{n}} = 4.97$$

with p-value $= .0004$, and the rejection region is $|t| > t_{.025} = 2.201$. H_0 is rejected, and we conclude that there is a difference in means between the two salespersons.

c. A 95% confidence interval for $\mu_1 - \mu_2 = \mu_d$ is

$$\bar{d} \pm t_{.025} \frac{s_d}{\sqrt{n}} \Rightarrow 15.67 \pm 2.201 \frac{10.92}{\sqrt{12}} \Rightarrow 15.67 \pm 6.94$$

or $8.73 < (\mu_1 - \mu_2) < 22.61$. There is a slight difference in the interval shown on the printout, since rounded values were used in the calculations above.

8.110

a. The hypothesis of interest is

$$H_0: \mu = 1100 \qquad H_a: \mu < 1100$$

b. The rejection region, with $\alpha = .05$, is $z < -1.645$.

c. The test statistic is

$$z \approx \frac{\bar{x} - \mu}{s/\sqrt{n}} = \frac{1060 - 1100}{\frac{340}{\sqrt{260}}} = -1.897$$

and H_0 is rejected. There is sufficient evidence to indicate that the mean is less than 1100; there has been a drop in average daily production.

8.111

The hypothesis to be tested is

$$H_0: p_1 - p_2 = 0 \qquad H_a: p_1 - p_2 \neq 0$$

Calculate

$$\hat{p} = \frac{n_1 \hat{p}_1 + n_2 \hat{p}_2}{n_1 + n_2} = \frac{500(.54) + 400(.46)}{500 + 400} = .504$$

The test statistic is then
$$z = \frac{\hat{p}_1 - \hat{p}_2}{\sqrt{\hat{p}\hat{q}\left(\frac{1}{n_1} + \frac{1}{n_2}\right)}} = \frac{.54 - .46}{\sqrt{.504(.496)(1/500 + 1/400)}} = 2.39$$

The p-value is calculated as
$$\text{p-value} = P[|z| > 2.39] = 2(.5 - .4916) = .0168$$

Since the p-value is very small, the null hypothesis is rejected, indicating a difference in the percentages for the two surveys. There is possibly something wrong with the survey methods, the data analysis, or the sampling plans that were used.

8.112

The hypothesis of interest is
$$H_0: \mu = 60 \qquad H_a: \mu > 60$$

and the test statistic is
$$z \approx \frac{\overline{x} - \mu}{s/\sqrt{n}} = \frac{62 - 60}{\frac{8}{\sqrt{50}}} = 1.768$$

The rejection region, with $\alpha = .10$, is $z > 1.28$ and H_0 is rejected. There is sufficient evidence to indicate that the mean occupancy rate exceeds 60%.

8.113

Refer to Exercise 8.110, in which the rejection region was given as $z < -1.645$ where
$$z = \frac{\overline{x} - \mu}{s/\sqrt{n}} = \frac{\overline{x} - 1100}{340/\sqrt{260}}$$

Solving for $\overline{x}$, we obtain the critical value of $\overline{x}$ necessary for rejection of H_0.
$$\frac{\overline{x} - 1100}{340/\sqrt{260}} < -1.645$$

$$\overline{x} < 1100 - 1.645 \frac{340}{\sqrt{260}} = 1065.314$$

Since the acceptance region is then $\overline{x} \geq 1065.314$, β can be written as
$$\beta = P[\overline{x} \geq 1065.314 \text{ when } H_0 \text{ is false}] = P[\overline{x} \geq 1065.314 \text{ when } \mu < 1100]$$

Several alternative values of μ are given in this exercise. For $\mu = 1040$,
$$\beta = P[\overline{x} \geq 1065.314 \text{ when } \mu = 1040]$$
$$= P[z \geq \frac{1065.314 - 1040}{340/\sqrt{260}}] = P[z \geq 1.2] = .5 - .3849 = .1151$$

For $\mu = 1030$,
$$\beta = P[\overline{x} \geq 1065.314 \text{ when } \mu = 1030]$$

$$= P[z \geq \frac{1065.314 - 1030}{340/\sqrt{260}}] = P[z \geq 1.67] = .5 - .4525 = .0475$$

For $\mu = 1020$,

$$\beta = P[\overline{x} \geq 1065.314 \text{ when } \mu = 1020]$$

$$= P[z \geq \frac{1065.314 - 1020}{340/\sqrt{260}}] = P[z \geq 2.15] = .5 - .4842 = .0158$$

The power curve is graphed using the values calculated above and is shown in Figure 8.9.

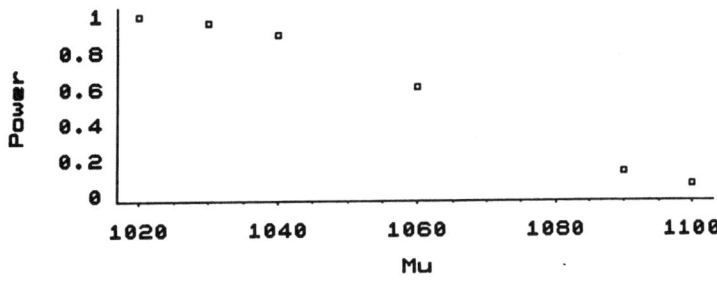

Figure 8.9

8.114

a. The number of bird hits is a discrete random variable, which can be modeled by the Poisson distribution.

b. Birds generally fly in groups; hence, one bird hit might imply several hits and thus the hits would not be independent.

8.115

a. Yes. The aircraft, and hence the values of x, are randomly selected.

b. The hypothesis to be tested is

$$H_0: \mu = 9 \qquad H_a: \mu < 9$$

where μ is the mean number of hits for painted engines. The rejection region for $\alpha = .05$ is $z < -1.645$ and the observed value of the test statistic is

$$z = \frac{\overline{x} - \mu}{\sigma/\sqrt{n}}$$

Since $\hat{\sigma} = \sqrt{\overline{x}} = \sqrt{1} = 1$, the test statistic is

$$z = \frac{1 - 9}{1/\sqrt{40}} = -50.60$$

The null hypothesis is rejected and we conclude that painting produced a marked reduction in mean bird hits per engine.

8.116

The hypothesis to be tested is one-tailed:

$$H_0: \mu_1 - \mu_2 = 0 \qquad H_a: \mu_1 - \mu_2 < 0$$

and the rejection region is $z < -1.645$. With $n_1 = n_2 = 40$ and $\hat{\sigma}_1 = \sqrt{\bar{x}_1} = 1$, $\hat{\sigma}_2 = \sqrt{\bar{x}_2} = 3$, the observed value of the test statistic is

$$z = \frac{(\bar{x}_1 - \bar{x}_2) - 0}{\sqrt{\frac{\sigma_1^2}{n_1} + \frac{\sigma_2^2}{n_2}}} \approx \frac{1-9}{\sqrt{\frac{1}{40} + \frac{9}{40}}} = -16$$

and H_0 is rejected. The mean number of hits is less for planes with painted air intakes.

8.117

a. The hypothesis of interest is

$$H_0: \sigma_1^2 = \sigma_2^2 \qquad H_a: \sigma_1^2 \neq \sigma_2^2$$

and the preliminary calculations are shown below:

Sample 1	Sample 2
$\Sigma x_{1i} = 127$	$\Sigma x_{2i} = 137$
$\Sigma x_{1i}^2 = 1653$	$\Sigma x_{2i}^2 = 2509$
$\bar{x}_1 = 12.7$	$\bar{x}_2 = 17.125$
$n_1 = 10$	$n_2 = 8$

Calculate

$$s_1^2 = \frac{\Sigma x_{1i}^2 - \frac{(\Sigma x_{1i})^2}{n_1}}{n_1 - 1} = \frac{1653 - \frac{(127)^2}{10}}{9} = 4.45556$$

$$s_2^2 = \frac{\Sigma x_{2i}^2 - \frac{(\Sigma x_{2i})^2}{n_2}}{n_2 - 1} = \frac{2509 - \frac{(137)^2}{8}}{7} = 23.26786$$

and the test statistic is

$$F = \frac{s_2^2}{s_1^2} = \frac{23.26786}{4.45556} = 5.22$$

The rejection region with $\alpha = .10$ is $F > 3.29$ and H_0 is rejected. There is a difference in the population variances.

b. The critical values of F for various values of a with $n_2 - 1 = 7$ and $n_1 - 1 = 9$ degrees of freedom can be found directly from Table 6.

a	F_a
.20	2.51
.10	3.29
.05	4.20
.02	5.61
.01	6.88

Hence,

p-value = $P[F > 5.22]$ lies between $F_{.05}$ and $F_{.02}$ so that

$.02 < \text{p-value} < .05$

8.118

a. The hypothesis to be tested is

$$H_0: \mu_1 - \mu_2 = 0 \qquad H_a: \mu_1 - \mu_2 \neq 0$$

The test statistic, under the assumption that $\sigma_1^2 \neq \sigma_2^2$ is calculated using the t statistic shown below:

$$t = \frac{(\bar{x}_1 - \bar{x}_2) - 0}{\sqrt{\frac{s_1^2}{n_1} + \frac{s_2^2}{n_2}}} = \frac{12.7 - 17.125}{\sqrt{\frac{4.455556}{10} + \frac{23.26786}{8}}} = -2.42$$

The degrees of freedom for this test statistic are approximated as

$$df \approx \frac{\left(\frac{s_1^2}{n_1} + \frac{s_2^2}{n_2}\right)^2}{\frac{\left(\frac{s_1^2}{n_1}\right)^2}{n_1 - 1} + \frac{\left(\frac{s_2^2}{n_2}\right)^2}{n_2 - 1}} = \frac{(.4455556 + 2.9084825)^2}{\frac{(.4455556)^2}{9} + \frac{(2.9084825)^2}{7}} = 9.14$$

The rejection region with $\alpha = .05$ and a conservative 9 degrees of freedom is $|t| > t_{.025} = 2.262$ and H_0 is not rejected.

b. From Table 4 with 9 degrees of freedom, the observed value of t is between $t_{.025} = 2.262$ and $t_{.01} = 2.821$. Hence, for a two-tailed test,

$$2(.01) < \text{p-value} < 2(.025) \quad \text{or} \quad .02 < \text{p-value} < .05$$

8.119

The hypothesis of interest is

$$H_0: \mu = 7.8 \qquad H_a: \mu < 7.8$$

Calculate $\Sigma x_i = 42.3$ and $\Sigma x_i^2 = 314.03$. Then

$$\bar{x} = \frac{\Sigma x_i}{n} = \frac{42.3}{6} = 7.05$$

$$s^2 = \frac{\Sigma x_i^2 - \frac{(\Sigma x_i)^2}{n}}{n-1} = \frac{314.03 - \frac{(42.3)^2}{6}}{5} = 3.163$$

The observed value of the test statistic is

$$t = \frac{\bar{x} - \mu}{s/\sqrt{n}} = \frac{7.05 - 7.8}{\sqrt{\frac{3.163}{6}}} = -1.033$$

The rejection region, with $\alpha = .05$ and $n - 1 = 5$ degrees of freedom, is $t < -t_{.05} = -2.015$, and H_0 is not rejected. There is insufficient evidence to indicate that μ is less than 7.8.

8.120

a. The hypothesis of interest is

$$H_0: \mu = 3 \qquad H_a: \mu < 3$$

Calculate

$$\bar{x} = \frac{\Sigma x_i}{n} = \frac{7.3}{3} = 2.433$$

$$s^2 = \frac{\Sigma x_i^2 - \frac{(\Sigma x_i)^2}{n}}{n-1} = \frac{18.27 - \frac{(7.3)^2}{3}}{2} = .2533$$

The observed value of the test statistic is

$$t = \frac{\bar{x} - \mu}{s/\sqrt{n}} = \frac{2.433 - 3}{\sqrt{\frac{.2533}{3}}} = -1.95$$

The rejection region, with $\alpha = .05$ and $n - 1 = 2$ degrees of freedom, is $t < -t_{.05} = -2.92$, and H_0 is not rejected. There is insufficient evidence to indicate that μ is less than 3.

b. The 90% confidence interval for the population mean is

$$\bar{x} \pm t_{.05} \frac{s}{\sqrt{n}} \Rightarrow 2.433 \pm 2.92 \sqrt{\frac{.2533}{3}} \Rightarrow 2.433 \pm .849$$

or $1.584 < \mu < 3.282$. Intervals constructed in this manner will enclose μ 90% of the time in repeated sampling.

8.121

It is necessary to estimate the mean life to within .2, with probability .90. Hence, the following inequality must hold:

$$t_{.05} \frac{s}{\sqrt{n}} \leq .2$$

Consider the sample size obtained by replacing $t_{.05}$ by $z_{.05}$.

$$1.645\sqrt{.2533}/\sqrt{n} \leq .2 \Rightarrow \sqrt{n} \geq 4.140 \quad \text{or} \quad n \geq 17.138$$

Because the sample size is less than 30, this value of n will not necessarily achieve the desired bound when $z_{.05}$ is replaced by $t_{.05}$. The reader may verify that a solution to the inequality can be obtained only by trying a larger value of n. Try $n = 19$.

$$t_{.05} \frac{s}{\sqrt{n}} = 1.734 \sqrt{\frac{.2533}{19}} = .200225$$

which is close to the desired bound. Try $n = 20$.

$$t_{.05} \frac{s}{\sqrt{n}} = 1.729 \sqrt{\frac{.2533}{20}} = .195$$

which actually achieves the necessary bound. Hence, the required sample size is $n = 20$.

8.122

The underlying populations are ratings and can only take on the finite number of values, 1, 2, ..., 9, 10. Neither population has a normal distribution, but both are discrete. Further, the samples are not independent, since the same person is asked to rank each car design. Hence, two of the assumptions required for the Student's t test have been violated.

8.123

A paired-difference test is used. To test $H_0: \mu_1 - \mu_2 = 0$ versus $H_a: \mu_1 - \mu_2 > 0$, where μ_1

is the mean before the safety program and μ_2 is the mean after the program, calculate the differences:

$$7, \ 6, \ -1, \ 5, \ 6, \ 1$$

and $\Sigma d_i = 24$, $\Sigma d_i^2 = 148$, $n = 6$. Then

$$\bar{d} = \frac{\Sigma d_i}{n} = \frac{24}{6} = 4$$

$$s_d^2 = \frac{\Sigma d_i^2 - \frac{(\Sigma d_i)^2}{n}}{n-1} = \frac{148 - \frac{(24)^2}{6}}{5} = 10.4$$

The test statistic is

$$t = \frac{\bar{d} - \mu_d}{s_d/\sqrt{n}} = \frac{4 - 0}{\sqrt{\frac{10.4}{6}}} = 3.038$$

The rejection region with $\alpha = .10$ and $n - 1 = 5$ d.f. is $t > t_{.10} = 1.476$, and H_0 is rejected. We conclude that the program was effective.

8.124

A paired-difference test is used. To test $H_0: \mu_1 - \mu_2 = 0$ versus $H_a: \mu_1 - \mu_2 \neq 0$, calculate the differences:

$$29, \ 31, \ -35, \ -17, \ 99, \ 73, \ 54$$

and $\Sigma d_i = 234$, $\Sigma d_i^2 = 21{,}362$, $n = 7$. Then

$$\bar{d} = \frac{\Sigma d_i}{n} = \frac{234}{7} = 33.4286$$

$$s_d^2 = \frac{\Sigma d_i^2 - \frac{(\Sigma d_i)^2}{n}}{n-1} = \frac{21{,}362 - \frac{(234)^2}{7}}{6} = 2256.619$$

The test statistic is

$$t = \frac{\bar{d} - \mu_d}{s_d/\sqrt{n}} = \frac{33.4286 - 0}{\sqrt{\frac{2256.619}{7}}} = 1.862$$

The rejection region with $\alpha = .05$ and $n - 1 = 6$ d.f. is $|t| > t_{.025} = 2.447$, and H_0 is not rejected. There is insufficient evidence to indicate a greater mean demand for one of the entrees.

8.125

Indexing $\chi^2_{.05}$ and $\chi^2_{.95}$ with $(n - 1) = 19$ degrees of freedom in Table 5 yields

$$\chi^2_{.05} = 30.1435 \quad \text{and} \quad \chi^2_{.95} = 10.117$$

and the 90% confidence interval is

$$\frac{(n-1)s^2}{\chi^2_{.05}} < \sigma^2 < \frac{(n-1)s^2}{\chi^2_{.95}}$$

$$\frac{19(39)}{30.1435} < \sigma^2 < \frac{19(39)}{10.117} \quad \text{or} \quad 24.582 < \sigma^2 < 73.243$$

8.126

a. The hypothesis to be tested is

$$H_0: \mu_1 - \mu_2 = 0 \qquad H_a: \mu_1 - \mu_2 \neq 0$$

and the pooled estimate of σ^2 is calculated as

$$s^2 = \frac{(n_1 - 1)s_1^2 + (n_2 - 1)s_2^2}{n_1 + n_2 - 2} = \frac{19(4.28) + 19(3.89)}{38} = 4.085$$

The test statistic is then

$$t = \frac{\bar{x}_1 - \bar{x}_2 - 0}{\sqrt{s^2\left(\frac{1}{n_1} + \frac{1}{n_2}\right)}} = \frac{43.1 - 44.6}{\sqrt{4.085\left(\frac{2}{20}\right)}} = -2.347$$

With $\alpha = .05$, the rejection region based on 38 degrees of freedom is $|t| > 1.96$ and H_0 is rejected. We conclude that there is a difference in the mean productivity for the two work schedules.

b. The observed level of significance is p-value $= 2P[t > 2.347]$ for a two-tailed test with 38 degrees of freedom. Since $t = 2.347$ lies between $t_{.005} = 2.576$ and $t_{.01} = 2.326$, we have

$$2(.005) < \text{p-value} < 2(.01) \quad \text{or} \quad .01 < \text{p-value} < .02$$

c. The 95% confidence interval is

$$(\bar{x}_1 - \bar{x}_2) \pm t_{\alpha/2}\sqrt{s^2\left(\frac{1}{n_1} + \frac{1}{n_2}\right)}$$

$$(43.1 - 44.6) \pm 1.96\sqrt{4.085\left(\frac{2}{20}\right)}$$

$$-1.5 \pm 1.253 \quad \text{or} \quad -2.753 < \mu_1 - \mu_2 < -.247$$

8.127

With $B = 1$ and $\sigma_1^2 = \sigma_2^2 \approx 4.085$, solve

$$t_{.025}\sqrt{\frac{\sigma_1^2}{n} + \frac{\sigma_2^2}{n}} = B$$

$$1.96\sqrt{4.085(2/n)} = 1 \quad \Rightarrow \quad \sqrt{n} = 5.602$$

$$n = 31.386 \quad \text{or} \quad n = 32$$

8.128

a. Since the range is 6σ, we have

$$\sigma = \frac{\text{Range}}{6} = \frac{25}{6} = 4.167$$

b. The hypothesis of interest is

$$H_0: \sigma = 4.167 \qquad H_a: \sigma > 4.167$$

and the rejection region with $\alpha = .05$ is $\chi^2 > \chi^2_{.05} = 36.4151$. The observed test statistic is

$$\chi^2 = \frac{(n-1)s^2}{\sigma_0^2} = \frac{24(5.2)^2}{(4.167)^2} = 37.374$$

and H_0 is rejected. The variation exceeds $\sigma = 4.167$.

8.129

From Exercise 8.128, n = 25 and $s^2 = 27.04$. Then

$$\chi^2_{.025} = 39.3641 \quad \text{and} \quad \chi^2_{.975} = 12.4011$$

and the 95% confidence interval is

$$\frac{648.96}{39.3641} < \sigma^2 < \frac{648.96}{12.4011} \quad \text{or} \quad 16.486 < \sigma^2 < 52.331$$

CHAPTER 9
The Analysis of Variance

9.1

In comparing six populations, there are $k - 1 = 5$ degrees of freedom for treatments and $n = 6(10) = 60$. The ANOVA table is shown below.

Source	d.f.
Treatments	5
Error	54
Total	59

9.2

a. Refer to Exercise 9.1. The given sums of squares are inserted and missing entries found by subtraction. The mean squares are found as MS = SS/df.

Source	d.f.	SS	MS	F
Treatments	5	5.2	1.04	3.467
Error	54	16.2	0.30	
Total	59	21.4		

b. The F statistic, $F = MST/MSE$, has $\nu_1 = 5$ and $\nu_2 = 54$ d.f.

c. With $\alpha = .05$ and degrees of freedom from part b, H_0 is rejected if $F > F_{.05} \approx 2.37$.

d. Since $F = 3.467$ falls in the rejection region, the null hypothesis is rejected. There is a difference among the means.

9.3

Refer to Exercise 9.2.

a. $\bar{x}_1 \pm t_{.025}\sqrt{\frac{MSE}{n_1}} = 3.07 \pm 1.96\sqrt{\frac{0.3}{10}} = 3.07 \pm .339$

or $2.731 < \mu_1 < 3.409$.

b. $(\bar{x}_1 - \bar{x}_2) \pm t_{.025}\sqrt{MSE\left(\frac{1}{n_1} + \frac{1}{n_2}\right)}$

$(3.07 - 2.52) \pm 1.96\sqrt{0.3\left(\frac{2}{10}\right)}$

$.55 \pm .480$ or $.07 < \mu_1 - \mu_2 < 1.03$

9.4

This is similar to Exercise 9.1. With $n = 4(6) = 24$ and $k = 4$, the sources of variation and associated d.f. are shown below.

Source	d.f.
Treatments	3
Error	20
Total	23

9.5

a. Refer to Exercise 9.4. The given sums of squares are inserted and missing entries found by subtraction. The mean squares are found as MS = SS/df.

Source	d.f.	SS	MS	F
Treatments	3	339.8	113.267	16.98
Error	20	133.4	6.67	
Total	23	473.2		

b. The F statistic, F = MST/MSE, has $\nu_1 = 3$ and $\nu_2 = 20$ d.f.

c. With $\alpha = .10$ and degrees of freedom from part b, H_0 is rejected if $F > F_{.10} = 2.38$.

d. Since F = 16.98 falls in the rejection region, the null hypothesis is rejected. There is a difference among the means.

9.6

Refer to Exercise 9.5.

a. $\bar{x}_1 \pm t_{.05}\sqrt{\frac{MSE}{n_1}} \Rightarrow 88.0 \pm 1.725\sqrt{\frac{6.67}{6}} \Rightarrow 88.0 \pm 1.819$

or $86.181 < \mu_1 < 89.819$.

b. $(\bar{x}_1 - \bar{x}_2) \pm t_{.05}\sqrt{MSE\left(\frac{1}{n_1} + \frac{1}{n_2}\right)}$

$(88.0 - 83.9) \pm 1.725\sqrt{6.67\left(\frac{2}{6}\right)}$

4.1 ± 2.572 or $1.528 < \mu_1 - \mu_2 < 6.672$

9.7

a. Since there are 4 d.f. for treatments, we must have k − 1 = 4 or k = 5 random samples.

b-c. No. The only available information involves the total sample size, $\Sigma n_i = 29 + 1 = 30$.

d. The ANOVA table can be completed using the additivity property of the sum of squares and degrees of freedom.

Source	d.f.	SS	MS	F
Treatments	4	26.3	6.575	3.11
Error	25	52.8	2.112	
Total	29	79.1		

e-f. To test for a difference among the means, the test statistic is $F = \frac{MST}{MSE} = 3.11$ and the rejection region with $\alpha = .05$ and 4 and 25 d.f. is $F > 2.76$. The null hypothesis is rejected and we conclude that there is a difference among the means.

9.8

The following preliminary calculations are necessary:

$T_1 = 14$ $T_2 = 19$ $T_3 = 5$ grand total = 38

a. $CM = \frac{(\Sigma\Sigma x_{ij})^2}{n} = \frac{(38)^2}{14} = 103.142857$

Total SS $= \Sigma\Sigma x_{ij}^2 - CM = 3^2 + 2^2 + \cdots + 2^2 + 1^2 - CM$

$= 130 - 103.142857 = 26.8571$

b. $SST = \Sigma \frac{T_i^2}{n_i} - CM = \frac{14^2}{5} + \frac{19^2}{5} + \frac{5^2}{4} - CM = 117.65 - 103.142857 = 14.5071$

and $MST = \frac{SST}{k-1} = \frac{14.5071}{2} = 7.2536$

c. By subtraction, $SSE = \text{Total SS} - SST = 26.8571 - 14.5071 = 12.3500$ and the degrees of freedom, by subtraction, are $13 - 2 = 11$. Then

$MSE = \frac{SSE}{11} = \frac{12.3500}{11} = 1.1227$

d. The information obtained in parts a-c is consolidated in an ANOVA table.

Source	d.f.	SS	MS
Treatments	2	14.5071	7.2536
Error	11	12.3500	1.1227
Total	13	26.8571	

e. The hypothesis to be tested is

$H_0: \mu_1 = \mu_2 = \mu_3$ H_a: at least one pair of means is different

f. The rejection region for the test statistic $F = \frac{MST}{MSE}$ is based on an F distribution with 2 and 11 degrees of freedom. That is, H_0 is rejected if $F > F_{.05} = 3.98$.

g. The observed value of the test statistic is $F = \frac{MST}{MSE} = \frac{7.2536}{1.1227} = 6.46$ and H_0 is rejected. There is a difference among the means.

9.9

The hypothesis to be tested is $H_0: \mu_2 = \mu_3$ versus $H_a: \mu_2 \neq \mu_3$ and the test statistic is

$t = \frac{\bar{x}_2 - \bar{x}_3}{\sqrt{s^2\left(\frac{1}{n_2} + \frac{1}{n_3}\right)}} = \frac{3.8 - 1.25}{\sqrt{1.1227\left(\frac{1}{5} + \frac{1}{4}\right)}} = 3.59$

Notice that the best estimator of σ^2 is $s^2 = MSE$, which is used in the calculation. The rejection region with $\alpha = .05$ and 11 degrees of freedom is $|t| > t_{.025} = 2.201$ and the null hypothesis is rejected. We conclude that there is a difference between the means.

9.10

a. The 90% confidence interval for μ_1 is

$\bar{x}_1 \pm t_{.05}\sqrt{\frac{MSE}{n_1}} \Rightarrow 2.8 \pm 1.796\sqrt{\frac{1.1227}{5}} \Rightarrow 2.8 \pm .85$

or $1.95 < \mu_1 < 3.65$.

b. The 90% confidence interval for $\mu_1 - \mu_3$ is

$(\bar{x}_1 - \bar{x}_3) \pm t_{.05}\sqrt{MSE\left(\frac{1}{n_1} + \frac{1}{n_3}\right)}$

$$(2.8 - 1.25) \pm 1.796\sqrt{1.1227\left(\tfrac{1}{5} + \tfrac{1}{4}\right)}$$

$$1.55 \pm 1.28 \quad\text{or}\quad .27 < (\mu_1 - \mu_3) < 2.83$$

9.11

The following preliminary calculations are necessary:

$$T_1 = 6 \quad T_2 = 17 \quad T_3 = 13 \quad T_4 = 13 \quad \text{grand total} = 49$$

a. $\text{CM} = \dfrac{(\Sigma\Sigma x_{ij})^2}{n} = \dfrac{(49)^2}{15} = 160.0667$

$\text{Total SS} = \Sigma\Sigma x_{ij}^2 - \text{CM} = 2^2 + 4^2 + \cdots + 6^2 + 4^2 - \text{CM}$

$= 209 - 160.0667 = 48.9333$

b. $\text{SST} = \Sigma \dfrac{T_i^2}{n_i} - \text{CM} = \dfrac{6^2}{3} + \dfrac{17^2}{4} + \dfrac{13^2}{5} + \dfrac{13^2}{3} - \text{CM}$

$= 174.3833 - 160.0667 = 14.3167$

and $\text{MST} = \dfrac{\text{SST}}{k-1} = \dfrac{14.3167}{3} = 4.7722$

c. By subtraction, $\text{SSE} = \text{Total SS} - \text{SST} = 34.6167$ and $\text{MSE} = \dfrac{\text{SSE}}{15-4} = 3.1470$.

d. The information obtained in parts a-c is consolidated in an ANOVA table.

Source	d.f.	SS	MS
Treatments	3	14.3167	4.7722
Error	11	34.6167	3.1470
Total	14	48.9333	

e-g. The hypothesis to be tested is

$$H_0: \mu_1 = \mu_2 = \mu_3 = \mu_4 \qquad H_a: \text{at least one pair of means is different}$$

and the observed value of the test statistic is

$$F = \dfrac{\text{MST}}{\text{MSE}} = \dfrac{4.7722}{3.1470} = 1.52$$

The rejection region for the test is based on an F distribution with 3 and 11 degrees of freedom. That is, H_0 is rejected if $F > F_{.05} = 3.59$. H_0 is not rejected. There is no evidence of a difference among the means.

9.12

The hypothesis to be tested is $H_0: \mu_1 = \mu_2$ versus $H_a: \mu_1 \neq \mu_2$ and the test statistic is

$$t = \dfrac{\bar{x}_1 - \bar{x}_2}{\sqrt{\text{MSE}\left(\tfrac{1}{n_1} + \tfrac{1}{n_2}\right)}} = \dfrac{2 - 4.25}{\sqrt{3.146969\left(\tfrac{1}{3} + \tfrac{1}{4}\right)}} = -1.66$$

The rejection region with $\alpha = .05$ and 11 degrees of freedom is $|t| > t_{.025} = 2.201$ and the null hypothesis is not rejected. We cannot conclude that there is a difference between the means.

9.13

a. The 90% confidence interval for μ_3 is

$$\bar{x}_3 \pm t_{.05}\sqrt{\frac{MSE}{n_3}} \Rightarrow 2.6 \pm 1.796\sqrt{\frac{3.1470}{5}} \Rightarrow 2.6 \pm 1.42$$

or $1.18 < \mu_3 < 4.02$.

b. The 90% confidence interval for $\mu_1 - \mu_3$ is

$$(\bar{x}_1 - \bar{x}_3) \pm t_{.05}\sqrt{MSE\left(\frac{1}{n_1} + \frac{1}{n_3}\right)}$$

$$(2.00 - 2.6) \pm 1.796\sqrt{3.1470\left(\frac{1}{3} + \frac{1}{5}\right)}$$

$-.6 \pm 2.33$ or $-2.93 < (\mu_1 - \mu_3) < 1.73$

9.14

a. The hypothesis to be tested is

$$H_0: \mu_1 = \mu_2 = \mu_3 = \mu_4 \quad\quad H_a: \text{at least one pair of means is different}$$

where μ_i is the average price for a particular brand of bread at location i, i = 1, 2, 3, 4. The analysis of variance F test will be used, and a completely randomized design has been used. Refer to the computer printout. The ANOVA table is

Source	d.f.	SS	MS	F
Treatments	3	.022871	.007624	13.0
Error	10	.005850	.000585	
Total	13	.028721		

The F test is

$$F = \frac{MST}{MSE} = \frac{.007624}{.000585} = 13.0$$

The rejection region with $\alpha = .05$ and 3 and 10 d.f. is $F > 3.71$ and H_0 is rejected. There is a significant difference in the mean price of bread in the different store locations.

b. The 95% confidence interval for $\mu_1 - \mu_4$ is

$$(\bar{x}_1 - \bar{x}_4) \pm t_{.025}\sqrt{MSE\left(\frac{1}{n_1} + \frac{1}{n_4}\right)}$$

$$(1.62 - 1.695) \pm 2.228\sqrt{.000585\left(\frac{1}{4} + \frac{1}{2}\right)}$$

$-.075 \pm .047$ or $-.122 < (\mu_1 - \mu_4) < -.028$

9.15

a. From the computer printout, the test statistic is

$$F = \frac{MST}{MSE} = 5.70$$

and the rejection region with $\alpha = .05$ and 2 and 9 degrees of freedom is $F > F_{.05} = 4.26$. The null hypothesis of equality of means is rejected, and we conclude that there is a difference in mean assembly times for the three programs. The p-value is given on the printout as PR > F = 0.0251. H_0 can be rejected for any value of α greater than 0.0251.

b. From the printout, MSE = 14.9407 and the 90% confidence interval for $\mu_A - \mu_B$ is then

$$(\bar{x}_A - \bar{x}_B) \pm t_{.05}\sqrt{MSE\left(\frac{1}{n_A} + \frac{1}{n_B}\right)}$$

$$(60.5 - 54.667) \pm 1.833\sqrt{14.9407\left(\frac{1}{4} + \frac{1}{3}\right)}$$

$$5.833 \pm 5.411 \quad \text{or} \quad .422 < (\mu_A - \mu_B) < 11.244$$

c. The 90% confidence interval for μ_A is

$$\bar{x}_A \pm t_{.05}\sqrt{\frac{MSE}{n_A}} \Rightarrow 60.5 \pm 1.833\sqrt{\frac{14.9407}{4}} \Rightarrow 60.5 \pm 3.543$$

or $56.957 < \mu_A < 64.043$.

d. Since the measurements represent averages of four assembly times and since time itself is a continuous random variable, the Central Limit Theorem assures us that even for small values of n, the average assembly times will have a fairly mound-shaped distribution.

9.16

a. "No instruction on listening" implies that no treatment has been applied to the experimental unit (the subject); hence, this is called a control group. The effect of the other three treatments is judged in comparison to the control.

b. To test $H_0: \mu_1 = \mu_2 = \mu_3 = \mu_4$, the test statistic is F = MST/MSE = 8.11 and the rejection region with 3 and 95 d.f. is $F > F_{.05} \approx 2.76$. The null hypothesis of no difference is rejected.

c. To test $H_0: \mu_1 - \mu_4 = 0$ versus $H_a: \mu_1 - \mu_4 \neq 0$, the test statistic is

$$t = \frac{\bar{x}_1 - \bar{x}_4}{\sqrt{MSE\left(\frac{1}{n_1} + \frac{1}{n_4}\right)}} = \frac{3.18 - 1.36}{\sqrt{3.69\left(\frac{1}{19} + \frac{1}{22}\right)}} = 3.025$$

The rejection region with $\alpha = .05$ and 95 degrees of freedom is $|t| > t_{.025} = 1.96$ and the null hypothesis is rejected. We can conclude that there is a difference between the means.

9.17

Refer to Exercise 9.14. The following preliminary calculations are necessary:

$$T_1 = 6.48 \quad T_2 = 6.46 \quad T_3 = 6.26 \quad T_4 = 3.39 \quad \text{grand total} = 22.59$$

$$CM = \frac{(\Sigma\Sigma x_{ij})^2}{n} = \frac{(22.59)^2}{14} = 36.45057857$$

$$\text{Total SS} = \Sigma\Sigma x_{ij}^2 - CM = (1.59)^2 + (1.58)^2 + \cdots + (1.63)^2 + (1.58)^2 - CM$$

$$= 36.4793 - CM = .028721429$$

$$SST = \Sigma\frac{T_i^2}{n_i} - CM = \frac{(6.48)^2}{4} + \frac{(6.46)^2}{4} + \frac{(6.26)^2}{4} + \frac{(3.39)^2}{2} - CM$$

$$= 36.47345 - CM = .02287143$$

and $\quad MST = \dfrac{SST}{k-1} = \dfrac{.02287143}{3} = .00762381$

By subtraction, SSE = Total SS − SST = .00585 and MSE = $\dfrac{SSE}{14-4}$ = .000585. The results agree with the computer printout.

Source	d.f.	SS	MS
Treatments	3	.02287143	.0076238
Error	10	.00585	.000585
Total	13	.028721429	

9.18

Refer to Exercise 9.15. The following preliminary calculations are necessary:

$T_1 = 242 \quad T_2 = 164 \quad T_3 = 321 \quad$ grand total = 727

$$CM = \dfrac{(\Sigma\Sigma x_{ij})^2}{n} = \dfrac{(727)^2}{12} = 44{,}044.08333$$

Total SS $= \Sigma\Sigma x_{ij}^2 - CM = 59^2 + 52^2 + \cdots + 63^2 + 64^2 - CM$

$$= 44{,}349 - CM = 304.9167$$

$$SST = \Sigma \dfrac{T_i^2}{n_i} - CM = \dfrac{242^2}{4} + \dfrac{164^2}{3} + \dfrac{321^2}{5} - CM$$

$$= 44{,}214.53333 - CM = 170.45$$

By subtraction, SSE = Total SS − SST = 134.46667. The results agree with the computer printout.

Source	d.f.	SS	MS	F
Treatments	2	170.4500	85.22500	5.70
Error	9	134.4667	14.94074	
Total	11	304.9167		

9.19

In comparing three treatments within six blocks, there are $k-1 = 2$ treatment degrees of freedom and $b-1 = 5$ block d.f. The ANOVA table is shown below.

Source	d.f.
Treatments	2
Blocks	5
Error	10
Total	17

9.20

Refer to Exercise 9.19. The given sums of squares are inserted and missing entries found by subtraction. The mean squares are found as MS = SS/df.

Source	d.f.	SS	MS	F
Treatments	2	11.4	5.70	4.01
Blocks	5	17.1	3.42	2.41
Error	10	14.2	1.42	
Total	17	42.7		

9.21

To compare the treatment means, the test statistic is $F = MST/MSE = 4.01$ and the rejection region with 2 and 10 d.f. is $F > F_{.05} = 4.10$. The null hypothesis is not rejected. There is insufficient evidence to indicate a difference between treatment means.

9.22

The 95% confidence interval for $\mu_A - \mu_B$ is then

$$(\bar{x}_A - \bar{x}_B) \pm t_{.025}\sqrt{MSE\left(\frac{2}{b}\right)}$$

$$(21.9 - 24.2) \pm 2.228\sqrt{1.42\left(\frac{2}{6}\right)}$$

$$-2.3 \pm 1.533 \qquad \text{or} \qquad -3.833 < (\mu_A - \mu_B) < -.767$$

9.23

To test for differences among block means, the test statistic is $F = MSB/MSE = 2.41$ and the rejection region with 5 and 10 d.f. is $F > F_{.05} = 3.33$. The null hypothesis is not rejected. There is insufficient evidence to indicate differences among block means.

9.24

This is similar to Exercise 9.19. Sources of variation and associated d.f. are shown below.

Source	d.f.
Treatments	5
Blocks	3
Error	15
Total	23

9.25

Refer to Exercise 9.24. The given sums of squares are inserted and missing entries found by subtraction. The mean squares are found as $MS = SS/df$.

Source	d.f.	SS	MS	F
Treatments	5	6.1	1.22	4.69
Blocks	3	2.2	0.733	2.82
Error	15	3.9	0.26	
Total	23	12.2		

9.26

To compare the treatment means, the test statistic is $F = MST/MSE = 4.69$ and the rejection region with 5 and 15 d.f. is $F > F_{.10} = 2.27$. The null hypothesis is rejected. There is sufficient evidence to indicate a difference between treatment means.

9.27

The 90% confidence interval for $\mu_A - \mu_B$ is then

$$(\bar{x}_A - \bar{x}_B) \pm t_{.05}\sqrt{MSE\left(\frac{2}{b}\right)}$$

$$(291.2 - 289.7) \pm 1.753\sqrt{.26\left(\frac{2}{4}\right)}$$

$$1.5 \pm .632 \qquad \text{or} \qquad .868 < (\mu_A - \mu_B) < 2.132$$

9.28

To test for differences among block means, the test statistic is F = MSB/MSE = 2.82 and the rejection region with 3 and 15 d.f. is $F > F_{.05} = 3.29$. The null hypothesis is not rejected. There is insufficient evidence to indicate differences among block means.

9.29

a. By subtraction, the degrees of freedom for blocks is $b - 1 = 34 - 28 = 6$. Hence, there are $b = 7$ blocks.

b. There are always $b = 7$ observations in a treatment total.

c. There are $k = 4 + 1 = 5$ observations in a block total.

d.

Source	d.f.	SS	MS	F
Treatments	4	14.2	3.55	9.68
Blocks	6	18.9	3.15	8.59
Error	24	8.8	0.3667	
Total	34	41.9		

e. To test the difference among treatment means, the test statistic is

$$F = \frac{MST}{MSE} = \frac{3.55}{.3667} = 9.68$$

and the rejection region with $\alpha = .10$ and 4 and 24 d.f. is $F > 2.19$. There is a significant difference among the treatment means.

f. To test the difference among block means, the test statistic is

$$F = \frac{MSB}{MSE} = \frac{3.15}{.3667} = 8.59$$

and the rejection region with $\alpha = .10$ and 6 and 24 d.f. is $F > 2.04$. There is a significant difference among the block means.

9.30

a. $CM = \frac{(\Sigma\Sigma x_{ij})^2}{n} = \frac{(113)^2}{12} = 1064.08333$

Total SS $= \Sigma\Sigma x_{ij}^2 - CM = 6^2 + 10^2 + \cdots + 14^2 - CM = 1213 - CM = 148.91667$

b. $SST = \frac{\Sigma T_j^2}{3} - CM = \frac{22^2 + 34^2 + 27^2 + 30^2}{3} - CM = 25.58333$ and

$MST = \frac{SST}{k-1} = \frac{25.58333}{3} = 8.52778$

c. $SSB = \frac{\Sigma B_i^2}{4} - CM = \frac{33^2 + 25^2 + 55^2}{4} - CM = 120.66667$ and

$MSB = \frac{SSB}{b-1} = \frac{120.6667}{2} = 60.33333$

d. $SSE = TSS - SST - SSB = 2.66667$ and

$MSE = \frac{SSE}{n-k-b+1} = \frac{2.66667}{6} = .44444$

e. The ANOVA table is

Source	d.f.	SS	MS	F
Treatments	3	25.5833	8.5278	19.19
Blocks	2	120.6667	60.3333	135.75
Error	6	2.6667	0.4444	
Total	11	148.9167		

f. To test the difference among treatment means, the test statistic is

$$F = \frac{MST}{MSE} = \frac{8.5278}{.4444} = 19.19$$

and the rejection region with $\alpha = .05$ and 3 and 6 d.f. is $F > 4.76$. There is a significant difference among the treatment means.

g. To test the difference among block means, the test statistic is

$$F = \frac{MSB}{MSE} = \frac{60.3333}{.4444} = 135.75$$

and the rejection region with $\alpha = .05$ and 2 and 6 d.f. is $F > 5.14$. There is a significant difference among the block means.

h. Since there is a significant difference among the block means, blocking has been effective. The variation due to block differences can be isolated using the randomized block design.

9.31

Refer to Exercise 9.30. The 90% confidence interval is

$$(\bar{x}_A - \bar{x}_B) \pm t_{.05} \sqrt{MSE\left(\frac{2}{b}\right)}$$

$$(7.3333 - 11.3333) \pm 1.943\sqrt{.4444\left(\frac{2}{3}\right)}$$

$$-4 \pm 1.058 \quad \text{or} \quad -5.058 < (\mu_A - \mu_B) < -2.942$$

9.32

Similar to Exercise 9.30.

a. $CM = \frac{(\Sigma\Sigma x_{ij})^2}{n} = \frac{(48.9)^2}{15} = 159.414$

Total SS $= \Sigma\Sigma x_{ij}^2 - CM = (2.1)^2 + \cdots + (3.9)^2 - CM = 165.91 - CM = 6.496$

b. $SST = \frac{\Sigma T_j^2}{5} - CM = \frac{(12.5)^2 + (18.8)^2 + (17.6)^2}{5} - CM = 4.476$ and

$MST = \frac{SST}{k-1} = \frac{4.476}{2} = 2.238$

c. $SSB = \frac{\Sigma B_i^2}{3} - CM = \frac{(8.5)^2 + (10.0)^2 + \cdots + (10.5)^2}{3} - CM = 1.796$ and

$MSB = \frac{SSB}{b-1} = \frac{1.796}{4} = .449$

d. $SSE = TSS - SST - SSB = .224$ and $MSE = \frac{SSE}{n-k-b+1} = \frac{.224}{8} = .028$

e. The ANOVA table is

Source	d.f.	SS	MS	F
Treatments	2	4.476	2.238	79.93
Blocks	4	1.796	0.449	16.04
Error	8	0.224	0.028	
Total	11	6.496		

f. To test the difference among treatment means, the test statistic is

$$F = \frac{MST}{MSE} = 79.93$$

and the rejection region with $\alpha = .05$ and 2 and 8 d.f. is $F > 4.46$. There is a significant difference among the treatment means.

g. To test the difference among block means, the test statistic is

$$F = \frac{MSB}{MSE} = 16.04$$

and the rejection region with $\alpha = .05$ and 4 and 8 d.f. is $F > 3.84$. There is a significant difference among the block means.

h. Since there is a significant difference among the block means, blocking has been effective. The variation due to block differences can be isolated using the randomized block design.

9.33

Refer to Exercise 9.32. The 99% confidence interval is

$$(\bar{x}_C - \bar{x}_A) \pm t_{.005} \sqrt{MSE\left(\frac{2}{b}\right)}$$

$$(3.52 - 2.50) \pm 3.355 \sqrt{.028\left(\frac{2}{5}\right)}$$

$$1.02 \pm .355 \quad \text{or} \quad .665 < (\mu_C - \mu_A) < 1.375$$

9.34

a. Since variation in sales is expected from week to week, blocking helps to isolate this variation.

b. The test statistic is $F = MST/MSE = 22.03$ and the rejection region is $F > 5.14$. H_0 is rejected; there is a difference due to treatments.

c. Calculate $\bar{x}_S = 14{,}846.75$ and $\bar{x}_F = 12{,}606$. The test is

$$H_0: \mu_S - \mu_F = 0 \qquad H_a: \mu_S - \mu_F \neq 0$$

and the test statistic is

$$t = \frac{\bar{x}_S - \bar{x}_F}{\sqrt{MSE\left(\frac{2}{b}\right)}} = \frac{14{,}846.75 - 12{,}606}{\sqrt{233{,}975\left(\frac{2}{4}\right)}} = 6.551$$

The rejection region is $|t| > t_{.025} = 2.447$ and H_0 is rejected. There is a difference in mean daily sales for the two treatments.

9.35

a-b. Cities (blocks) are used to isolate unwanted regional variation. Promotions (treatments) were randomly assigned to two-week periods within each city.

c. Since substantial variation is expected from city to city, isolating this source of variation will reduce SSE.

d. From the printout, F = MST/MSE = 51.15 is significant with p-value = .0014. Hence, there is a difference among treatment means.

e. The 90% confidence interval is

$$(\bar{x}_A - \bar{x}_C) \pm t_{.05} \sqrt{MSE\left(\frac{2}{b}\right)}$$

$$(4.37 - 4.38) \pm 2.132\sqrt{.0046\left(\frac{2}{3}\right)}$$

$$-.01 \pm .118 \quad \text{or} \quad -.128 < (\mu_A - \mu_C) < .108$$

f. See footnote in Example 9.2.

9.36

a. A randomized block design has been used. Since customers arrive randomly at the station, the treatments can be thought of as being randomly applied within a block (week).

b. No. The design is blocked by week within the same service station. The samples are <u>not</u> independent.

c. The 95% confidence interval is

$$(\bar{x}_2 - \bar{x}_3) \pm t_{.025} \sqrt{MSE\left(\frac{2}{b}\right)}$$

$$(19.95 - 4.8125) \pm 2.145\sqrt{.71\left(\frac{2}{8}\right)}$$

$$15.1375 \pm .904 \quad \text{or} \quad 14.2335 < (\mu_2 - \mu_3) < 16.0415$$

9.37

Calculate

$$CM = \frac{(165{,}980)^2}{12} = 2{,}295{,}780{,}033$$

$$\text{Total SS} = 2{,}309{,}918{,}034 - CM = 14{,}138{,}000.67$$

$$SST = \frac{(56{,}169)^2 + (59{,}387)^2 + (50{,}424)^2}{4} - CM = 10{,}307{,}993.17$$

$$SSB = \frac{(42{,}043)^2 + (41{,}702)^2 + (42{,}941)^2 + (39{,}294)^2}{3} - CM = 2{,}426{,}156.667$$

$$SSE = \text{Total SS} - SST - SSB = 1{,}403{,}850.8333$$

The results agree with the computer printout.

Source	d.f.	SS	MS	F
Treatments	2	10,307,993.17	5,153,996.59	22.03
Blocks	3	2,426,156.67	808,718.89	3.46
Error	6	1,403,850.83	233,975.1389	
Total	11	14,138,000.67		

9.38

Calculate

$$CM = \frac{(40.83)^2}{9} = 185.2321$$

Total SS $= 186.1069 - CM = .8748$

$$SST = \frac{(13.11)^2 + (14.58)^2 + (13.14)^2}{3} - CM = .4706$$

$$SSB = \frac{(14.48)^2 + (13.28)^2 + (13.07)^2}{3} - CM = .3858$$

$$SSE = \text{Total SS} - SST - SSB = .0184$$

The results agree with the computer printout.

Source	d.f.	SS	MS	F
Treatments	2	.4706	.2353	51.15
Blocks	2	.3858	.1929	41.93
Error	4	.0184	.0046	
Total	8	.8748		

9.39

Calculate

$$CM = \frac{(219.3)^2}{24} = 2003.85375$$

Total SS $= 3445.21 - CM = 1441.35625$

$$SST = \frac{(21.2)^2 + (159.6)^2 + (38.5)^2}{8} - CM = 3425.48125 - CM = 1421.6275$$

$$SSB = \frac{(24.9)^2 + (26.9)^2 + \cdots + (26.2)^2}{3} - CM = 2013.663333 - CM = 9.809583$$

$$SSE = \text{Total SS} - SST - SSB = 9.919167$$

The results agree with the computer printout.

Source	d.f.	SS	MS	F
Treatments	2	1421.627500	710.813750	1003.25
Blocks	7	9.809583	1.401369	1.98
Error	14	9.919167	0.708512	
Total	23	1441.356250		

9.40

a-b. There are $4 \times 5 = 20$ treatments and $4 \times 5 \times 3 = 60$ total observations.

c. In a factorial experiment, variation due to the interaction A × B is isolated from SSE. The sources of variation and associated degrees of freedom are given below.

Source	d.f.
A	3
B	4
A × B	12
Error	40
Total	59

9.41

This is similar to Exercise 9.40.

a-b. There are $4 \times 2 = 8$ treatments and $4 \times 2 \times r = 8r$ total observations.

c. The sources of variation and associated degrees of freedom are given below.

Source	d.f.
A	3
B	1
A × B	3
Error	$8r-8$
Total	$8r-1$

9.42

a. The complete ANOVA table is shown below. Since factor A is run at 3 levels, it must have 2 d.f. Other entries are found by similar reasoning.

Source	d.f.	SS	MS	F
A	2	5.3	2.6500	1.30
B	3	9.1	3.0333	1.49
A × B	6	4.8	0.8000	0.39
Error	12	24.5	2.0417	
Total	23	43.7		

b. The test statistic is $F = MS(AB)/MSE = 0.39$ and the rejection region is $F > 3.00$. Hence, H_0 is not rejected. There is insufficient evidence to indicate interaction between A and B.

c. The test statistic for testing factor A is $F = 1.30$ with $F_{.05} = 3.89$. The test statistic for factor B is $F = 1.49$ with $F_{.05} = 3.49$. Neither A nor B are significant.

9.43

Refer to Exercise 9.42. The 95% confidence interval is

$$(\bar{x}_1 - \bar{x}_2) \pm t_{.025}\sqrt{MSE\left(\frac{1}{n_1} + \frac{1}{n_2}\right)}$$

$$(12.4 - 6.3) \pm 2.179\sqrt{2.0417\left(\frac{2}{2}\right)}$$

$$6.1 \pm 3.114 \quad \text{or} \quad 2.986 < \mu_1 - \mu_2 < 9.214$$

9.44

This is similar to previous exercises. The completed ANOVA table is shown below.

Source	d.f.	SS	MS	F
A	1	1.14	1.14	6.51
B	2	2.58	1.29	7.37
A × B	2	0.49	0.245	1.40
Error	24	4.20	0.175	
Total	29	8.41		

a. The test statistic is $F = MS(AB)/MSE = 1.40$ and the rejection region is $F > 3.40$. There is insufficient evidence to indicate an interaction.

b. Using Table 6 with $\nu_1 = 2$ and $\nu_2 = 24$, the following values are obtained.

a	F_a
.10	2.54
.05	3.40
.025	4.32
.010	5.61
.005	6.66

The observed value of F is less than $F_{.10}$, so that p-value $> .10$.

c. The test statistic for testing factor A is $F = 6.51$ with $F_{.05} = 4.26$. There is evidence that factor A affects the response.

d. The test statistic for factor B is $F = 7.37$ with $F_{.05} = 3.40$. Factor B also affects the response.

9.45

Refer to Exercise 9.44. The 95% confidence interval is

$$(\bar{x}_1 - \bar{x}_2) \pm t_{.025}\sqrt{MSE\left(\frac{1}{n_1} + \frac{1}{n_2}\right)}$$

$$(3.7 - 1.4) \pm 2.064\sqrt{.175\left(\frac{2}{15}\right)}$$

$$2.3 \pm .315 \quad \text{or} \quad 1.985 < \mu_1 - \mu_2 < 2.615$$

9.46

a. The nine treatment (cell) totals needed for calculation are shown in the table.

		Factor A		
Factor B	1	2	3	Total
1	12	16	10	38
2	15	25	17	57
3	25	17	27	69
Total	52	58	54	164

$CM = \frac{164^2}{18} = 1494.2222$

Total SS $= 1662 - CM = 167.7778$

$SSA = \frac{52^2 + 58^2 + 54^2}{6} - CM = 3.1111$

$$SSB = \frac{38^2 + 57^2 + 69^2}{6} - CM = 81.4444$$

$$SS(AB) = \frac{12^2 + 16^2 + \cdots + 27^2}{2} - SSA - SSB - CM = 62.2222$$

Source	d.f.	SS	MS	F
A	2	3.1111	1.5556	
B	2	81.4444	40.7222	
A × B	4	62.2222	15.5556	6.67
Error	9	21.0000	2.3333	
Total	17	167.7778		

b-c. The test statistic is F = MS(AB)/MSE = 6.67 and the rejection region is F > 3.63. There is evidence of a significant interaction. That is, the effect of factor A depends upon the level of factor B at which A is measured.

d. Since F = 6.67 lies between $F_{.01}$ and $F_{.005}$, .005 < p-value < .01.

e. $\bar{x}_{22} \pm t_{.025}\sqrt{\frac{MSE}{2}} \Rightarrow 12.5 \pm 2.262\sqrt{\frac{2.3333}{2}} \Rightarrow 12.5 \pm 2.443$

or $10.057 < \mu_{22} < 14.943$.

f. $(\bar{x}_{13} - \bar{x}_{31}) \pm t_{.025}\sqrt{MSE\left(\frac{1}{2} + \frac{1}{2}\right)}$

$(12.5 - 5) \pm 2.262\sqrt{2.3333\left(\frac{2}{2}\right)}$

$7.5 \pm 3.455 \quad$ or $4.045 < \mu_{13} - \mu_{31} < 10.955$

9.47

This is similar to Exercise 9.46.

a. The four treatment (cell) totals needed for calculation are shown in the table.

	Factor A		
Factor B	1	2	Total
1	9.7	13.4	23.1
2	14.0	10.3	24.3
Total	23.7	23.7	47.4

$CM = \frac{(47.4)^2}{16} = 140.4225$

Total SS = 145.02 − CM = 4.5975

$SSA = \frac{(23.7)^2 + (23.7)^2}{8} - CM = 0$

$SSB = \frac{(23.1)^2 + (24.3)^2}{8} - CM = .09$

$SS(AB) = \frac{(9.7)^2 + (14)^2 + (13.4)^2 + (10.3)^2}{4} - SSA - SSB - CM = 3.4225$

Source	d.f.	SS	MS	F
A	1	0.0000	0.0000	
B	1	0.0900	0.0900	
A × B	1	3.4225	3.4225	37.85
Error	12	1.0850	0.090417	
Total	15	4.5975		

b-d. The test statistic is $F = MS(AB)/MSE = 37.85$ and $F_{.005} = 11.75$, so that the p-value $< .005$. There is evidence of a significant interaction. That is, the effect of factor A depends upon the level of factor B at which A is measured.

e. $\bar{x}_{22} \pm t_{.025}\sqrt{\frac{MSE}{4}} = 2.575 \pm 2.179\sqrt{\frac{.090417}{4}} = 2.575 \pm .328$

or $2.247 < \mu_{22} < 2.903$.

f. $(\bar{x}_{12} - \bar{x}_{21}) \pm t_{.025}\sqrt{MSE\left(\frac{1}{4} + \frac{1}{4}\right)}$

$(3.5 - 3.35) \pm 2.179\sqrt{.090417\left(\frac{2}{4}\right)}$

$.15 \pm .463$ or $-.313 < \mu_{12} - \mu_{21} < .613$

9.48

a. Interaction would imply that profits per design would react differently, depending on which foreman was in charge.

b. From the printout, $F = MS(AB)/MSE = 8.48$ with $F_{.05} = 2.51$. There is significant interaction.

c. From the printout, p-value $= .0001$.

d. $(\bar{x}_{11} - \bar{x}_{21}) \pm t_{.025}\sqrt{MSE\left(\frac{1}{3} + \frac{1}{3}\right)}$

$(10.833 - 9.3) \pm 2.064\sqrt{1.3675\left(\frac{2}{3}\right)}$

1.533 ± 1.971 or $-.438 < \mu_{11} - \mu_{21} < 3.504$

9.49

a. From the printout, $F = 1.21$ with $\Pr > F = .3616$. Hence, at the $\alpha = .05$ level, H_0 is not rejected. There is insufficient evidence to indicate interaction.

b. Since no interaction is found, the effects of A and B can be tested individually. Both A and B are significant.

c. $(\bar{x}_{31} - \bar{x}_{32}) \pm t_{.025}\sqrt{MSE\left(\frac{1}{2} + \frac{1}{2}\right)}$

$(-17 + .5) \pm 2.447\sqrt{35.17\left(\frac{2}{2}\right)}$

-16.5 ± 14.51 or $-31.01 < \mu_{31} - \mu_{32} < -1.99$

9.50

a. The total number of participants was sixty, twenty in each of three categories. Hence, the total degrees of freedom is fifty-nine. Factor T was run at two levels, factor A at three levels, resulting in the given degrees of freedom.

b. $F = \dfrac{\text{MST}}{\text{MSE}} = \dfrac{103.7009}{28.3015} = 3.66$ $\qquad F = \dfrac{\text{MSA}}{\text{MSE}} = \dfrac{760.5889}{28.3015} = 26.87$

$F = \dfrac{\text{MS(TA)}}{\text{MSE}} = \dfrac{124.9905}{28.3015} = 4.42$

c. Since interaction is significant, the main effects need not be tested individually. Attention should be focused on the individual cell means.

d. The tabled values for the approximate d.f. are shown below.

a	F(1, 60)	F(2, 60)
.10	2.79	2.39
.05	4.00	3.15
.025	5.29	3.93
.010	7.08	4.98
.005	8.49	5.79

For T, $.05 < \text{p-value} < .10$

For A, $\text{p-value} < .005$

For TA, $.01 < \text{p-value} < .025$

9.51

The individual cell totals are given below.

	Foreman				
Design	A_1	A_2	A_3	A_4	Total
B_1	32.5	27.9	34.1	24.6	119.1
B_2	25.2	29.3	25.2	32.9	112.6
B_3	40.3	32.0	27.9	22.8	123.0
Total	98.0	89.2	87.2	80.3	354.7

$\text{CM} = \dfrac{(354.7)^2}{36} = 3494.780278$

Total SS $= 3619.47 - \text{CM} = 124.689722$

$\text{SSA} = \dfrac{(98.0)^2 + (89.2)^2 + (87.2)^2 + (80.3)^2}{9} - \text{CM} = 17.7275$

$\text{SSB} = \dfrac{(119.1)^2 + (112.6)^2 + (123.0)^2}{12} - \text{CM} = 4.600556$

$\text{SS(AB)} = \dfrac{(32.5)^2 + (27.9)^2 + \cdots + (22.8)^2}{3} - \text{SSA} - \text{SSB} - \text{CM} = 69.54167$

The results agree with the computer printout.

Source	d.f.	SS	MS	F
A	3	17.7275	5.90917	4.32
B	2	4.60056	2.30028	1.68
A × B	6	69.54167	11.59028	8.48
Error	24	32.82000		
Total	35	124.68972		

9.52

The individual cell totals are given below.

Location	Markup A_1	A_2	A_3	Total
B_1	14	4	-34	-16
B_2	32	11	-1	42
Total	46	15	-35	26

$$CM = \frac{(26)^2}{12} = 56.3333$$

Total SS $= 1468 - CM = 1411.6667$

$$SSA = \frac{(46)^2 + (15)^2 + (-35)^2}{4} - CM = 835.1667$$

$$SSB = \frac{(-16)^2 + (42)^2}{6} - CM = 280.3333$$

$$SS(AB) = \frac{(14)^2 + (4)^2 + \cdots + (-1)^2}{2} - SSA - SSB - CM = 85.1667$$

The results agree with the computer printout.

Source	d.f.	SS	MS	F
A	1	835.1667	835.1667	11.87
B	2	280.3333	140.1667	7.97
A × B	2	85.1667	42.5833	1.21
Error	6	211.0000	35.1667	
Total	11	1411.6667		

9.53

Sample means must be independent and based upon samples of equal size.

9.54

Use Tables 7 and 8.

a. $q_{.05}(5,7) = 5.06$ **b.** $q_{.05}(3,10) = 3.88$

c. $q_{.01}(4,8) = 6.20$ **d.** $q_{.01}(7,5) = 9.32$

9.55

a. $\omega = q_{.05}(4,5) \frac{s}{\sqrt{5}} = 4.20 \frac{s}{\sqrt{5}} = 1.878s$

b. $\omega = q_{.01}(6,8) \frac{s}{\sqrt{8}} = 6.10 \frac{s}{\sqrt{8}} = 2.1567s$

9.56

a. $\omega = q_{.05}(6,8) \sqrt{\frac{9.12}{4}} = 4.49 \sqrt{\frac{9.12}{4}} = 6.78$

b. The ranked means are shown below. A line under two or more means indicates a difference <u>less</u> than ω and hence no differences between that group of means.

$\overline{x}_4$	$\overline{x}_2$	$\overline{x}_1$	$\overline{x}_5$	$\overline{x}_3$	$\overline{x}_6$
92.9	98.4	101.6	104.2	112.3	113.8

9.57

a. With $k = 9$, $\nu = 9$, $n_t = 2$, $s = \sqrt{MSE} = \sqrt{2.3333}$,

$$\omega = q_{.05}(9,9) \frac{s}{\sqrt{n_t}} = 5.59 \sqrt{\frac{2.3333}{2}} = 6.038$$

The nine cell means are ranked below. Note the two groups that are not significantly different.

5	6	7.5	8.5	8.5	8.5	12.5	12.5	13.5
A_3B_1	A_1B_1	A_2B_1	A_1B_2	A_3B_2	A_2B_3	A_1B_3	A_2B_2	A_3B_3

9.58

With $k = 4$, $\nu = 12$, $n_t = 4$,

$$\omega = q_{.05}(4,12) \frac{\sqrt{MSE}}{\sqrt{n_t}} = 4.20 \sqrt{\frac{.0904167}{4}} = .631$$

The ranked means are shown below.

2.425	2.575	3.35	3.5
$\overline{x}_{11}$	$\overline{x}_{22}$	$\overline{x}_{12}$	$\overline{x}_{21}$

9.59

With $k = 4$, $\nu = 24$, $n_t = 3$, calculate

$$\omega = q_{.05}(4,24) \frac{\sqrt{MSE}}{\sqrt{n_t}} = 3.90 \sqrt{\frac{1.3675}{3}} = 2.633$$

The ranked means for Design B_1 are shown below.

8.2	9.3	10.833	11.367
A_4	A_2	A_1	A_3

9.60

With $k = 4$, $\nu = 24$, $n_t = 3$, and $\omega = 2.633$ as in Exercise 9.59. The ranked means for Design B_2 are shown below.

8.4	8.4	9.77	10.97
A_1B_2	A_3B_2	A_2B_2	A_4B_2

The four means are not significantly different.

9.61

As in Exercise 9.59, $\omega = 2.633$. The ranked means for Design B_3 are shown below.

7.6	9.3	10.97	13.43
A_4B_3	A_3B_3	A_2B_3	A_1B_3

9.62

With $k = 6$, $\nu = 6$, $n_t = 2$, calculate

211

$$\omega = q_{.05}(6,6) \frac{\sqrt{MSE}}{\sqrt{n_t}} = 5.63\sqrt{\frac{35.1667}{2}} = 23.61$$

The ranked means are shown below.

−17.0	−.05	2.0	5.5	7.0	16.0
A_3B_1	A_3B_2	A_2B_1	A_2B_2	A_1B_1	A_1B_2

9.63

With $k = 6$, $\nu = 54$, $n_t = 20$, calculate

$$\omega = q_{.05}(6,54) \frac{\sqrt{MSE}}{\sqrt{n_t}} = 4.23\sqrt{\frac{28.3015}{20}} = 5.03$$

The ranked means are shown below. If we had used $q(6,60)$ rather than $q(6,40)$ to approximate $q(6,54)$, the results would be the same.

1.610	1.728	5.031	5.648	9.508	17.895
$\bar{x}_{32}$	$\bar{x}_{31}$	$\bar{x}_{21}$	$\bar{x}_{22}$	$\bar{x}_{12}$	$\bar{x}_{11}$

9.64

a. To test the difference in treatment means, use

$$F = \frac{MST}{MSE} = 5.20$$

The rejection region with $\alpha = .05$ and 3 and 16 d.f. is $F > 3.24$ and H_0 is rejected. There is evidence to suggest a difference in mean discharge for the four plants.

b. The hypothesis to be tested is $H_0: \mu_A = 1.5$ versus $H_a: \mu_A > 1.5$ and the test statistic is

$$t = \frac{\bar{x}_A - \mu_A}{\sqrt{\frac{MSE}{n_A}}} = \frac{1.568 - 1.5}{\sqrt{\frac{.0298}{5}}} = .88$$

The rejection region with $\alpha = .05$ and 16 d.f. is $t > t_{.05} = 1.746$ and the null hypothesis is not rejected. We cannot conclude that the limit is exceeded at plant A.

c. The 95% confidence interval for $\mu_A - \mu_D$ is

$$(\bar{x}_A - \bar{x}_D) \pm t_{.025}\sqrt{MSE\left(\frac{1}{n_A} + \frac{1}{n_D}\right)}$$

$$(1.568 - 1.916) \pm 2.12\sqrt{.0298\left(\frac{1}{5} + \frac{1}{5}\right)}$$

$$-.348 \pm .231 \quad \text{or} \quad -.579 < (\mu_A - \mu_D) < -.117$$

9.65

a. The F test for treatments is $F = .07$, which is nonsignificant.

b. The 95% confidence interval for $\mu_1 - \mu_3$ is

$$(\bar{x}_1 - \bar{x}_3) \pm t_{.025}\sqrt{MSE\left(\frac{1}{n_1} + \frac{1}{n_3}\right)}$$

$$(26.5 - 27.88) \pm 2.080\sqrt{1494\left(\frac{1}{8} + \frac{1}{8}\right)}$$

$$-1.38 \pm 40.20 \quad \text{or} \quad -41.58 < (\mu_1 - \mu_3) < 38.82$$

9.66

a-b. The design is completely randomized with three treatments, containing four, two, and one measurements, respectively. The analysis is as follows:

$$T_1 = 102.3 \quad T_2 = 47.2 \quad T_3 = 26.0 \quad \text{grand total} = 175.5$$

(1) $\quad CM = \dfrac{(\Sigma\Sigma x_{ij})^2}{n} = \dfrac{(175.5)^2}{7} = 4400.0357$

(2) $\quad \text{Total SS} = \Sigma\Sigma x_{ij}^2 - CM = 4414.33 - CM = 14.2943$

(3) $\quad SST = \Sigma \dfrac{T_i^2}{n_i} - CM = \dfrac{(102.3)^2}{4} + \dfrac{(47.2)^2}{2} + (26.0)^2 - CM = 4406.2425 - CM$

$\quad \quad = 6.2068$

(4) $\quad SSE = \text{Total SS} - SST = 8.0875$

The ANOVA table is

Source	d.f.	SS	MS
Treatments	2	6.2068	3.1034
Error	4	8.8075	2.0219
Total	6	14.2943	

The F test to detect a difference between treatments is $F = MST/MSE = 1.53$ and the rejection region with 2 and 4 d.f. is $F > F_{.05} = 6.94$. We cannot conclude that there is a significant difference between treatment means A, B, and C.

c. The 90% confidence interval for μ_B is

$$\overline{x}_B \pm t_{.05} \sqrt{\dfrac{MSE}{n_B}} \Rightarrow 23.6 \pm 2.132 \sqrt{\dfrac{2.0219}{2}} \Rightarrow 23.6 \pm 2.144$$

or $21.456 < \mu_B < 25.744$.

d. The 90% confidence interval for $\mu_A - \mu_C$ is

$$(\overline{x}_A - \overline{x}_C) \pm t_{.05} \sqrt{MSE\left(\dfrac{1}{n_A} + \dfrac{1}{n_C}\right)}$$

$$(25.575 - 26.0) \pm 2.132 \sqrt{2.0219\left(\dfrac{1}{4} + \dfrac{1}{1}\right)}$$

$$-.425 \pm 3.389 \quad \text{or} \quad -3.814 < (\mu_A - \mu_C) < 2.964$$

9.67

a. The F test to detect a difference between treatments is

$$F = \dfrac{MST}{MSE} = \dfrac{461.93}{46.93} = 9.84$$

and the rejection region with 3 and 23 d.f. is $F > F_{.05} = 3.03$. We can conclude that there is a significant difference between the training programs.

b. The 90% confidence interval for $\mu_1 - \mu_4$ is

$$(\bar{x}_1 - \bar{x}_4) \pm t_{.05}\sqrt{MSE\left(\frac{1}{n_1} + \frac{1}{n_4}\right)}$$

$$(80.333 - 73.5) \pm 1.714\sqrt{46.93\left(\frac{1}{6} + \frac{1}{8}\right)}$$

$$6.833 \pm 6.341 \quad \text{or} \quad .492 < (\mu_1 - \mu_4) < 13.174$$

c. The 90% confidence interval for μ_2 is

$$\bar{x}_2 \pm t_{.05}\sqrt{\frac{MSE}{n_2}} \Rightarrow 91.875 \pm 1.714\sqrt{\frac{46.93}{8}} \Rightarrow 91.875 \pm 4.151$$

or $87.724 < \mu_2 < 96.026$.

9.68

a. To test for a difference in treatment means, the test statistic is $F = MST/MSE = 19.44$. The rejection region with $\alpha = .05$ and 3 and 6 d.f. is $F > 4.76$, and the null hypothesis is rejected. We conclude that there is a difference due to treatments.

b. To test for a difference in block means, the test statistic is $F = MST/MSE = 40.21$ and the rejection region with $\alpha = .05$ and 2 and 6 d.f. is $F > 5.14$. There is a difference in block means.

c. The 95% confidence interval for $\mu_A - \mu_D$ is

$$(\bar{x}_A - \bar{x}_D) \pm t_{.025}\sqrt{MSE\left(\frac{1}{n_A} + \frac{1}{n_D}\right)}$$

$$(11.4 - 12.8) \pm 2.447\sqrt{.089167\left(\frac{1}{3} + \frac{1}{3}\right)}$$

$$-1.4 \pm .597 \quad \text{or} \quad -1.997 < (\mu_A - \mu_D) < -.803$$

9.69

a. The F statistic to detect a difference due to treatments is

$$F = \frac{MST}{MSE} = 7.196$$

and the rejection region with $\alpha = .05$ and 2 and 6 d.f. is $F > 5.14$. There is a significant difference among the treatment means.

b. The 90% confidence interval is

$$(\bar{x}_A - \bar{x}_B) \pm t_{.05}\sqrt{MSE\left(\frac{2}{b}\right)}$$

$$(32.6125 - 34.8875) \pm 1.943\sqrt{.75472\left(\frac{2}{4}\right)}$$

$$-2.275 \pm 1.194 \quad \text{or} \quad -3.469 < (\mu_A - \mu_B) < -1.081$$

c. The F statistic to detect a difference due to blocks is

$$F \doteq \frac{MSB}{MSE} = 16.610$$

and the rejection region with $\alpha = .05$ and 3 and 6 d.f. is $F > 4.76$. There is evidence that the mean estimate of cost varies from job to job.

9.70

a-b. The experiment is run in a randomized block design, thus allowing the experimenter to isolate unwanted variation due to the particular assembler.

c. The test statistic for treatments is $F = 5.917$ and $F_{.05} = 4.10$. There is a significant difference between sequence means.

9.71

a. The individual cell totals are given below.

	A_1	A_2	A_3	Total
B_1	8	2	20	30
B_2	12	13	29	54
Total	20	15	49	84

$$CM = \frac{(84)^2}{18} = 392$$

Total SS $= 684 - CM = 292$

$$SSA = \frac{(20)^2 + (15)^2 + (49)^2}{6} - CM = 112.3333$$

$$SSB = \frac{(30)^2 + (54)^2}{9} - CM = 32$$

$$SS(AB) = \frac{(8)^2 + (2)^2 + \cdots + (29)^2}{3} - SSA - SSB - CM = 4.333$$

Source	d.f.	SS	MS	F
A	2	112.3333	56.1667	4.70
B	1	32.0000	32.0000	2.68
A × B	2	4.3333	2.1667	0.18
Error	12	143.3333	11.9444	
Total	17	292.0000		

b. For AB, p-value $> .10$. For A, $.025 <$ p-value $< .05$. For B, p-value $> .10$.

c. Only Factor A is significant.

9.72

a. From the printout,

$$F = \frac{MST}{MSE} = \frac{1.4475}{.2242} = 6.46$$

and $F_{.05} = 5.14$. There is a significant difference between brands.

b. $$F = \frac{MSB}{MSE} = \frac{.8400}{.2242} = 3.75$$

and $F_{.05} = 4.76$. There is no significant difference between auto types.

c. $(\bar{x}_A - \bar{x}_B) \pm t_{.05} \sqrt{MSE\left(\frac{2}{b}\right)}$

$\frac{106.1 - 110.9}{4} \pm 1.943\sqrt{.2242\left(\frac{2}{4}\right)}$

$-1.2 \pm .65$ or $-1.85 < (\mu_A - \mu_B) < -.85$

9.73

a. For the transformed data,

$CM = \frac{(8.67)^2}{15} = 5.01126$

Total SS $= 5.0619 - CM = .05064$

$SST = \frac{(2.78)^2 + (3.09)^2 + (2.80)^2}{5} - CM = .01204$

Source	d.f.	SS	MS	F
Treatments	2	.01204	.00602	1.87
Error	12	.03860	.00320	
Total	14	.05064		

b. Since $F_{.05} = 3.89$, the observed value of F is nonsignificant. There is no significant difference in the three advertising plans.

9.74

a. Calculate

$CM = \frac{(4.50)^2}{15} = 1.35$

Total SS $= 1.3958 - CM = .0458$

$SST = \frac{(1.40)^2 + (1.69)^2 + (1.41)^2}{5} - CM = .01084$

$SSE = $ Total SS $- SST = .03496$

Source	d.f.	SS	MS	F
Treatments	2	.01084	.00542	1.86
Error	12	.03496	.002913	
Total	14	.04580		

b. Since $F_{.05} = 3.89$, the observed $F = 1.86$ is nonsignificant.

c. There was no difference in the conclusion using transformed or untransformed data.

CHAPTER 10
Quality Control

10.1

a. The range estimate is calculated as

$$\hat{\sigma} = \frac{\overline{R}}{d_2} = \frac{16.80}{3.078} = 5.458$$

where d_2 is found in Table 9, Appendix II, with $n = 10$.

b. The estimate s calculated for 300 observations will be a better estimate of σ than $\hat{\sigma}$ calculated in part a; however, it is more difficult to calculate.

c. The upper and lower control limits are

$$\text{UCL} = \overline{\overline{x}} + A_2\overline{R} = 50.25 + .308(16.80) = 55.4244$$

$$\text{LCL} = \overline{\overline{x}} - A_2\overline{R} = 50.25 - .308(16.80) = 45.0756$$

where A_2 is found in Table 9, Appendix II.

d-e. The control chart is constructed by plotting two horizontal lines, one the upper control limit and one the lower control limit (see Figure 10.3 in the text). Values of $\overline{x}$ are plotted, and should remain within the control limits. If not, the process should be checked.

10.2

This is similar to Exercise 10.1.

a. The range estimate is calculated as

$$\hat{\sigma} = \frac{\overline{R}}{d_2} = \frac{12.45}{2.326} = 5.353$$

where d_2 is found in Table 9, Appendix II.

b. $3\hat{\sigma}_{\overline{x}} = A_2\overline{R} = .577(12.45) = 7.184$

Then $\text{UCL} = \overline{\overline{x}} + A_2\overline{R} = 70.38 + 7.184 = 77.564$

$\text{LCL} = \overline{\overline{x}} - A_2\overline{R} = 70.38 - 7.184 = 63.196$

c. See part c, Exercise 10.1.

10.3

From Exercise 10.1, $\overline{R} = 16.80$ and $n = 10$. The upper and lower control limits for the R chart, using Table 9, Appendix II, are

$$\text{UCL} = D_4\overline{R} = 1.777(16.80) = 29.8536$$

$$\text{LCL} = D_3\overline{R} = .223(16.80) = 3.7464$$

The R chart is constructed as in Figure 10.5 of the text. As sample ranges are plotted, they should stay within the control limits. If not, the process should be checked.

10.4

This is similar to Exercise 10.3. With $\bar{R} = 12.45$ and $n = 5$,

$$UCL = D_4\bar{R} = 2.115(12.45) = 26.332$$

$$LCL = D_3\bar{R} = 0(12.45) = 0$$

10.5

a. This is similar to Exercise 10.1. The upper and lower control limits, with $n = 5$, are

$$UCL = \bar{\bar{x}} + A_2\bar{R} = 10{,}752 + .577(6425) = 14{,}459.225$$

$$LCL = \bar{\bar{x}} - A_2\bar{R} = 10{,}752 - .577(6425) = 7044.775$$

where A_2 is found in Table 9, Appendix II.

b. See part c, Exercise 10.1.

10.6

This is similar to Exercise 10.3. The control limits are

$$UCL = D_4\bar{R} = 2.115(6425) = 13{,}588.875$$

$$LCL = D_3\bar{R} = 0(6425) = 0$$

The R chart is constructed and interpreted as in Exercise 10.3.

10.7

a. Using Tchebysheff's Theorem or the Empirical Rule (if the distribution is mound-shaped), x should fall within 3σ of its mean.

b. For this special case, the control limits are

$$\bar{x} \pm 3s \Rightarrow 10{,}940 \pm 3(5130) \Rightarrow 10{,}940 \pm 15{,}390$$

or $-4450 < x < 26{,}330$.

c. The manager could use the chart to check the honesty of each dealer.

10.8

In order to calculate $\bar{\bar{x}}$ and $\bar{R}$, values of $\bar{x}$ and R are calculated for each sample of size $n = 4$. The twenty-six values of $\bar{x}$ and R are shown in the table.

Week	$\bar{x}$	R	Week	$\bar{x}$	R
1	.03110	.002	14	.02875	.001
2	.02525	.001	15	.03025	.002
3	.02975	.002	16	.01575	.003
4	.03525	.003	17	.01975	.002
5	.02275	.002	18	.02425	.001
6	.02975	.001	19	.02800	.002
7	.01875	.001	20	.03075	.002
8	.02775	.001	21	.04000	.004
9	.03300	.002	22	.03525	.002
10	.01725	.002	23	.02225	.003
11	.02075	.002	24	.02925	.001
12	.01700	.002	25	.01650	.001
13	.01675	.003	26	.02075	.002

Then $\bar{\bar{x}} = .66650/26 = .0256$ and $\bar{R} = .050/26 = .001923$. The upper and lower control limits are

$$\text{UCL} = \bar{\bar{x}} + A_2\bar{R} = .0256 + .729(.001923) = .0270$$

$$\text{LCL} = \bar{\bar{x}} - A_2\bar{R} = .0256 - .729(.001923) = .0242$$

10.9

Refer to Exercise 10.8. The control limits are

$$\text{UCL} = D_4\bar{R} = 2.282(.001923) = .0044$$

$$\text{LCL} = D_3\bar{R} = 0(.001923) = 0$$

10.10

a. This is similar to previous exercises. With n = 3, the control limits are

$$\text{UCL} = \bar{\bar{x}} + A_2\bar{R} = 7.24 + 1.023(.27) = 7.516$$

$$\text{LCL} = \bar{\bar{x}} - A_2\bar{R} = 7.24 - 1.023(.27) = 6.964$$

b. For the R chart,

$$\text{UCL} = D_4\bar{R} = 2.575(.27) = .69525$$

$$\text{LCL} = D_3\bar{R} = 0(.27) = 0$$

10.11

The p chart plots the proportion of defectives produced by a process in which an item can be either defective or nondefective. A c chart plots the number of defects <u>within</u> an item. The random variable $\hat{p}$ has a binomial distribution, while c has a Poisson distribution.

10.12

a. It is given that n = 100 and $\bar{p} = .035$. Then

$$\text{UCL} = \bar{p} + 3\sqrt{\frac{\bar{p}(1-\bar{p})}{n}} = .035 + 3\sqrt{\frac{.035(.965)}{100}} = .090$$

and $\quad \text{LCL} = \bar{p} - 3\sqrt{\frac{\bar{p}(1-\bar{p})}{n}} = .035 - 3\sqrt{\frac{.035(.965)}{100}} = -.020$

b. Since the proportion of defects cannot be negative, the control limits will be 0 and .090. If subsequent samples do not stay within these limits, the process should be checked.

10.13

This is similar to Exercise 10.12.

a. The upper and lower control limits are

$$\text{UCL} = \bar{\bar{p}} + 3\sqrt{\frac{\bar{\bar{p}}(1-\bar{\bar{p}})}{n}} = .041 + 3\sqrt{\frac{.041(.959)}{200}} = .0831$$

and

$$\text{LCL} = \bar{\bar{p}} - 3\sqrt{\frac{\bar{\bar{p}}(1-\bar{\bar{p}})}{n}} = .041 - 3\sqrt{\frac{.041(.959)}{200}} = -.0011$$

b. The p chart is constructed by drawing horizontal lines at UCL = .0831 and LCL = 0. The subsequent samples should stay within these limits. If not, the process should be checked.

10.14

a. It is given that $\bar{c} = .7$ and the control limits are

$$\text{UCL} = \bar{c} + 3\sqrt{\bar{c}} = .7 + 3\sqrt{.7} = 3.210$$

$$\text{LCL} = \bar{c} - 3\sqrt{\bar{c}} = .7 - 3\sqrt{.7} = -1.810$$

b. Since the number of defects cannot be negative, we set LCL = 0, UCL = 3.210 and the centerline at $\bar{c} = .7$. The subsequent samples should stay within these limits. If not, the process should be checked.

10.15

This is similar to Exercise 10.14. The control limits are

$$\text{UCL} = \bar{c} + 3\sqrt{\bar{c}} = 1.3 + 3\sqrt{1.3} = 4.72$$

$$\text{LCL} = \bar{c} - 3\sqrt{\bar{c}} = 1.3 - 3\sqrt{1.3} = -2.12$$

Since the number of defects cannot be negative, we set LCL = 0 and UCL = 4.72. The subsequent samples should stay within these limits. If not, the process should be checked.

10.16

This is similar to Exercise 10.12. The control limits, with n = 400 and $\bar{\bar{p}} = .021$, are

$$\text{UCL} = \bar{\bar{p}} + 3\sqrt{\frac{\bar{\bar{p}}(1-\bar{\bar{p}})}{n}} = .021 + 3\sqrt{\frac{.021(.979)}{400}} = .0425$$

and

$$\text{LCL} = \bar{\bar{p}} - 3\sqrt{\frac{\bar{\bar{p}}(1-\bar{\bar{p}})}{n}} = .021 - 3\sqrt{\frac{.021(.979)}{400}} = -.0005$$

Since the proportion of defects cannot be negative, the control limits will be 0 and .0425. If subsequent samples do not stay within these limits, the process should be checked.

10.17

This is similar to Exercise 10.14. With $\bar{c} = 3.7$, the control limits are

$$\text{UCL} = \overline{c} + 3\sqrt{\overline{c}} = 3.7 + 3\sqrt{3.7} = 9.47$$

$$\text{LCL} = \overline{c} - 3\sqrt{\overline{c}} = 3.7 - 3\sqrt{3.7} = -2.07$$

Since the number of defects cannot be negative, we set LCL = 0 and UCL = 9.47. The subsequent samples should stay within these limits. If not, the process should be checked.

10.18

A c chart is used, with control limits

$$\text{UCL} = \overline{c} + 3\sqrt{\overline{c}} = 4.9 + 3\sqrt{4.9} = 11.54$$

$$\text{LCL} = \overline{c} - 3\sqrt{\overline{c}} = 4.9 - 3\sqrt{4.9} = -1.74$$

Since the number of defects cannot be negative, we set LCL = 0 and UCL = 11.54. The subsequent samples should stay within these limits. If not, the process should be checked.

10.19

A p chart is used. The value of $\overline{\overline{p}}$ is calculated by averaging the 30 values of $\hat{p}$ as

$$\overline{\overline{p}} = \frac{.14 + .21 + \cdots + .26}{30} = .1967$$

and the control limits are

$$\text{UCL} = \overline{\overline{p}} + 3\sqrt{\frac{\overline{\overline{p}}(1-\overline{\overline{p}})}{n}} = .1967 + 3\sqrt{\frac{.1967(.8033)}{100}} = .3160$$

and

$$\text{LCL} = \overline{\overline{p}} - 3\sqrt{\frac{\overline{\overline{p}}(1-\overline{\overline{p}})}{n}} = .1967 - 3\sqrt{\frac{.1967(.8033)}{100}} = .0774$$

If subsequent samples do not stay within these limits, the process should be checked.

10.20

Use Table 10 in Appendix II, indexing n_1, n_2 and r_0.

a. $P[r \leq 4] = .076$ **b.** $P[r \leq 4] = .043$

c. $P[r \leq 7] = .117$ **d.** $P[r \leq 5] = .004$

10.21

Use Table 10, Appendix II, finding a tabled probability close to .05.

a. $P[r \leq 4] = .068$; hence, $r = 4$ **b.** $P[r \leq 5] = .063$; hence, $r = 5$

c. $P[r \leq 5] = .051$; hence, $r = 5$ **d.** $P[r \leq 2] = .057$; hence, $r = 2$

10.22

This is similar to Exercise 10.21.

a. $P[r \leq 5] = .095$; hence, $r = 5$ **b.** $P[r \leq 3] = .071$; hence, $r = 3$

c. $P[r \leq 6] = .100$; hence, $r = 6$ **d.** $P[r \leq 8] = .128$; hence, $r = 8$

10.23

Use the formulas given in Section 10.7 of the text.

$$\mu_r = \frac{2n_1 n_2}{n_1 + n_2} + 1 = \frac{2(20)(15)}{35} + 1 = 18.1429$$

$$\sigma_r^2 = \frac{2n_1 n_2 (2n_1 n_2 - n_1 - n_2)}{(n_1 + n_2)^2 (n_1 + n_2 - 1)} = \frac{2(20)(15)(600 - 35)}{35^2 (34)} = 8.13926$$

and $\sigma_r = \sqrt{8.13926} = 2.8529$.

10.24

This is similar to Exercise 10.23.

$$\mu_r = \frac{2n_1 n_2}{n_1 + n_2} + 1 = \frac{2(25)(30)}{55} + 1 = 28.27$$

$$\sigma_r^2 = \frac{2n_1 n_2 (2n_1 n_2 - n_1 - n_2)}{(n_1 + n_2)^2 (n_1 + n_2 - 1)} = \frac{2(25)(30)(1500 - 55)}{55^2 (54)} = 13.26905$$

and $\sigma_r = \sqrt{13.26905} = 3.643$.

10.25

As in Exercise 10.23, calculate

$$\mu_r = \frac{2n_1 n_2}{n_1 + n_2} + 1 = \frac{2(10)(10)}{20} + 1 = 11$$

$$\sigma_r^2 = \frac{2n_1 n_2 (2n_1 n_2 - n_1 - n_2)}{(n_1 + n_2)^2 (n_1 + n_2 - 1)} = \frac{2(10)(10)(200 - 20)}{20^2 (19)} = 4.73684$$

and $\sigma_r = \sqrt{4.73684} = 2.176$. Then the approximate z value corresponding to $r = 8$ is

$$z = \frac{r - \mu_r}{\sigma_r} = \frac{8 - 11}{2.176} = -1.38$$

The large sample approximation is

$$P[r \leq 8] \approx P[z \leq -1.38] = .5 - .4162 = .0838$$

The exact probability from Table 10 with $n_1 = n_2 = 10$ is

$$P[r \leq 8] = .128.$$

Since $n_1 = n_2 = 10$ is just barely large enough for the normal approximation, the approximation is not too good.

10.26

The twenty-six values of $\bar{x}$ are classified as A (above) or B (below) according to whether the value falls above or below the centerline value, $\bar{\bar{x}} = .0256$. The sequence of runs is shown below.

A B A A B A B A A B B B B
A A B B B A A A A B A B B

and $r = 14$. Since $n_1 = n_2 = 13$ are both larger than 10, a large sample runs test is used to test the hypothesis

H_0: the deviations about the centerline are random

H_a: the deviations are nonrandom

The rejection region will be two-tailed (since no knowledge of the process is available to rule out a large number of runs) and the rejection region is $|z| > 1.96$ for $\alpha = .05$. Calculate

$$\mu_r = \frac{2n_1n_2}{n_1 + n_2} + 1 = \frac{2(13)(13)}{26} + 1 = 14$$

and

$$z = \frac{r - \mu_r}{\sigma_r} = \frac{14 - 14}{\sigma_r} = 0$$

Hence, without calculating σ_r, we know that H_0 is not rejected. There is no evidence to suggest nonrandomness.

10.27

This is similar to Exercise 10.26. The $\hat{p}$-values are measured A (above) or B (below) the value of $\overline{\hat{p}} = .1967$. The sequence of runs produces $n_1 = 17$, $n_2 = 13$, and $r = 22$.

B A B B A A A B A B A B A B B
A B A A A B A B A A B A A B A

A two-tailed rejection region for the large sample test is $|z| > 1.96$ for $\alpha = .05$. Calculate

$$\mu_r = \frac{2n_1n_2}{n_1 + n_2} + 1 = \frac{2(17)(13)}{30} + 1 = 15.733$$

$$\sigma_r^2 = \frac{2n_1n_2(2n_1n_2 - n_1 - n_2)}{(n_1 + n_2)^2(n_1 + n_2 - 1)} = \frac{2(17)(13)(442 - 30)}{30^2(29)} = 6.97716$$

The test statistic is

$$z = \frac{r - \mu_r}{\sigma_r} = \frac{22 - 15.733}{\sqrt{6.97716}} = 2.37$$

and H_0 is rejected. There is evidence of nonrandomness. The number of runs is too large.

10.28

With $n_1 = 3$ and $n_2 = 6$, we find $r = 5$. A two-tailed test should be used, since nonrandomness could imply either a large or a small number of runs. From Table 10, the lower tail of the rejection region is $r \le 2$ with $P[r \le 2] = .024$, while the upper tail, $r \ge 7$, has $P[r \ge 7] = 1 - .811 = .119$. Hence, with $\alpha = .143$, H_0 is not rejected. There is no evidence of nonrandomness.

10.29

a. Calculate $\overline{x} = 1082.7/16 = 67.66875$. The observations are classified as A or B, above or below the mean.

A A A A A B B B B B A B A B A

and $r = 7$. A two-tailed rejection region with $n_1 = n_2 = 8$ is found in Table 10. We have

223

$$P[r \leq 5] = .032 \quad \text{and} \quad P[r \geq 13] = 1 - P[r \leq 12] = .032$$

so that $\alpha = .064$. The observed value of $r = 7$ does not fall in the rejection region. There is no evidence of nonrandomness.

b. If the time period is split, and the first eight measurements constitute sample 1, a two-sample t can be used to compare the means of the following two samples:

Sample 1		Sample 2	
68.2	70.4	65.3	66.8
71.6	65.0	64.2	68.9
69.3	63.6	67.6	66.8
71.6	64.7	68.6	70.1

Calculate $\bar{x}_1 = 68.05$, $\bar{x}_2 = 67.2875$.

$$\Sigma(x_{1j} - \bar{x}_1)^2 = 37{,}119.06 - \frac{(544.4)^2}{8} = 72.64$$

$$\Sigma(x_{2j} - \bar{x}_2)^2 = 36{,}247.15 - \frac{(538.3)^2}{8} = 26.28875$$

$$s^2 = \frac{98.92875}{14} = 7.066339$$

and the test statistic is

$$t = \frac{\bar{x}_1 - \bar{x}_2}{\sqrt{s^2\left(\frac{1}{n_1} + \frac{1}{n_2}\right)}} = \frac{.7625}{\sqrt{7.066339\left(\frac{2}{8}\right)}} = .574$$

The rejection region with $\alpha = .05$ is $|t| > 2.145$ and H_0 is not rejected. There is no evidence of a shift in the mean.

10.30

This exercise asks for the probability of accepting a lot of items when the following sampling plan is used: draw a sample of five items and accept the lot if no defectives are observed. Thus,

$$P[\text{acceptance}] = P[\text{no defectives}] = C_0^5 p^0 q^5$$

where p is the probability of observing a defective. By substituting the five specific values of p in the formula, or by using the binomial tables in Appendix II, the appropriate probabilities are obtained.

p	P[acceptance]
.1	.590
.3	.168
.5	.031
0.0	1.000
1.0	0.000

10.31

The sample size is $n = 5$ and $P[\text{acceptance}] = p(0) + p(1) = C_0^5 p^0 q^5 + C_1^5 p^1 q^4$ where p is the probability of observing a defective. Using the binomial tables with $n = 5$ and $a = 1$, the appropriate probabilities are found.

p	P[acceptance]
.1	.919
.3	.528
.5	.188
0.0	1.000
1.0	0.000

10.32

This is similar to Exercise 10.30. For n = 10, a = 0, and the appropriate values of p, the probability of acceptance is found.

p	P[acceptance]
.1	.349
.3	.028
.5	.001
0.0	1.000
1.0	0.000

10.33

This is similar to Exercise 10.30. For n = 10, a = 1, and the appropriate values of p, the probability of acceptance is found.

p	P[acceptance]
.1	.736
.3	.149
.5	.011
0.0	1.000
1.0	0.000

10.34

The four operating characteristic curves are shown in Figure 10.1.

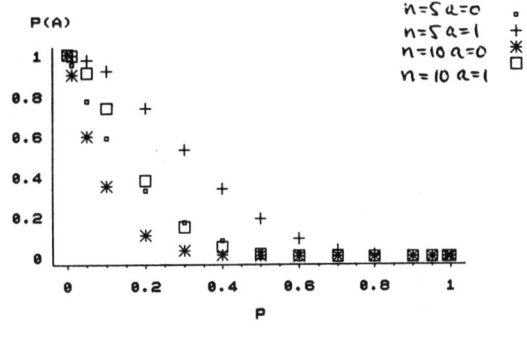

Figure 10.1

With n fixed and a increased, P[acceptance] increases, while with a fixed and n increased, P[acceptance] decreases.

10.35

The producer's risk is the probability of rejecting a lot when p = AQL = .05. For this plan,

$$P[\text{reject lot when } p = .05] = P[x \geq 1 \text{ when } p = .05]$$

$$= 1 - P[x = 0 \text{ when } p = .05] = 1 - .277 = .723$$

10.36

The consumer's risk is

$$P[\text{accept lot when } p = .20] = P[x = 0 \text{ when } p = .20] = .004$$

10.37

Refer to Tables 11 and 12 in Appendix II. The lot size, $N = 2000$, implies code letter K for a normal inspection level. Then, indexing K in Table 12, with AQL $= .015$, the sample size is n $= 125$ and the acceptance number is a $= 0$.

10.38

From Table 11 with $N = 400$ and Level III, we find the code letter J. Hence, from Table 12, we use $n = 80$ and accept the lot for $a = 0$ defectives.

10.39

From Table 11 with $N = 5000$ and Level II, we find the code letter L. Hence, from Table 12, we use $n = 200$ and $a = 0$.

10.40-10.41

See Section 10.2 of the text.

10.42

Answers will vary from student to student. See the appropriate section in Chapter 10 for examples.

10.43

The bottle manufacturer knows that when his process was in control, $\mu = 5.2$ and $\sigma = .3$. The observed fifty samples, with $n = 6$, provide $\bar{R} = .6$ and $s_R = .2$.

a. Using the procedures given in Section 10.3, the centerline of the $\bar{x}$ control chart is taken as $\mu = 5.2$, and the upper and lower control limits are

$$\text{UCL} = 5.2 + A_2\bar{R} = 5.2 + .483(.6) = 5.4898$$

$$\text{LCL} = 5.2 - A_2\bar{R} = 5.2 - .483(.6) = 4.9102$$

The centerline of the R chart is $\bar{R} = .6$ and the control limits are

$$\text{UCL} = D_4\bar{R} = 2.004(.6) = 1.2024$$

$$\text{LCL} = D_3\bar{R} = 0(.6) = 0$$

Notice that the exercise gives information that is not used. The plot is similar to those in the text.

b-c. For the next five days, only the day 2 mean falls within the limits; all but the day 3 range fall within the limits. The process should be stopped and adjusted.

10.44

If the process is in control, then $\mu = 5.2$ and $\sigma = .3$.

a. $P[4.8 < x < 5.5] = P[\frac{4.8 - 5.2}{.3} < z < \frac{5.5 - 5.2}{.3}]$

$= P[-1.33 < z < 1] = .3413 + .4082 = .7495$

b. Since the probability that a bottle does not meet specifications is $1 - .7495 = .2505$, the number of defectives in 10,000 bottles is $10,000(.2505) = 2505$.

10.45

a. This is similar to Exercise 10.19. The average proportion of defectives is

$$\bar{p} = \frac{.04 + .02 + \cdots + .03}{25} = .032$$

and the control limits are

$$UCL = \bar{p} + 3\sqrt{\frac{\bar{p}(1-\bar{p})}{n}} = .032 + 3\sqrt{\frac{.032(.968)}{100}} = .0848$$

and $$LCL = \bar{p} - 3\sqrt{\frac{\bar{p}(1-\bar{p})}{n}} = .032 - 3\sqrt{\frac{.032(.968)}{100}} = -.0208$$

If subsequent samples do not stay within the limits, $UCL = .0848$ and $LCL = 0$, the process should be checked.

b. From part a, we must have $\hat{p} > .0848$.

c. An erroneous conclusion will have occurred if in fact $p < .0848$ and the sample has produced $\hat{p} = .15$ by chance. One can obtain an upper bound on the probability of this particular type of error by calculating $P[\hat{p} \geq .15$ when $p = .0848]$.

10.46

The probability that a shipment of 100 bulbs will have no more than 4% defective is $P[\hat{p} \leq .04]$. However, since n is large, p is small, and $np = .032(100) = 3.2$, the Poisson approximation is more appropriate than the normal approximation. Hence, with $\mu = 3.2$,

$$P[x \leq 4] = \sum_{i=0}^{4} \frac{(3.2)^x e^{-3.2}}{x!} = e^{-3.2}[1 + 3.2 + 5.12 + 5.461 + 4.369] = .7806$$

10.47

Refer to Exercise 10.45, in which $UCL = .0848$ and $LCL = 0$. For the next five samples, the values of $\hat{p}$ are .02, .04, .09, .07, .11. Hence, samples 3 and 5 are producing excess defectives. The process should be checked.

10.48

a. It is given that $p = .03$ when the process is in control. Hence, the control limits are

$$p \pm 3\sqrt{\frac{p(1-p)}{n}}$$

$.03 \pm 3\sqrt{\frac{.03(.97)}{100}} \Rightarrow .03 \pm .051$ or $-.021$ to $.081$

Hence, we set $UCL = .081$ and $LCL = 0$.

b. If p is in fact p = .06, then the probability that a sample proportion will exceed UCL = .081 is

$$P[\hat{p} > .081] = P\left[z > \frac{.081 - .06}{\sqrt{.06(.94)/100}}\right] = P[z > .88] = .5 - .3106 = .1894$$

10.49

a. This is similar to Exercise 10.14. The control limits are calculated with $\bar{c} = 375/75 = 5$.

$$UCL = \bar{c} + 3\sqrt{\bar{c}} = 5 + 3\sqrt{5} = 11.708$$

$$LCL = \bar{c} - 3\sqrt{\bar{c}} = 5 - 3\sqrt{5} = -1.708$$

Since the number of defects cannot be negative, we set LCL = 0 and UCL = 11.708. For the twenty-five panels given here, the process is still in control.

b. The process might be producing one of the unusual observations that exceeds $3\sigma_c$. If it is expensive to stop the process, he should wait; however, he might check to see if there is an obvious and easily corrected problem.

10.50

Use the binomial tables in Appendix II, indexing n = 5, a = 1 in the first case and n = 25, a = 5 in the second (see Figure 10.2).

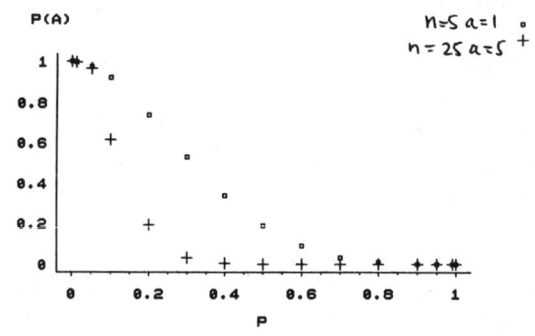

Figure 10.2

a. If the fraction defective in the lot ranges from p = 0 to p = .10, the seller would want the probability of acceptance in this interval to be as high as possible. Hence, he would choose the second plan.

b. If the buyer wishes to be protected against accepting lots with the fraction defective greater than .3, he would want the probability of acceptance when p is greater than .3 to be as small as possible. Thus, he would also choose the second plan.

10.51

a. The operating characteristic curves for n = 25 and a = 1, 2, 3 are graphed using the binomial tables found in Table 1, Appendix II. The graphs are shown in Figure 10.3.

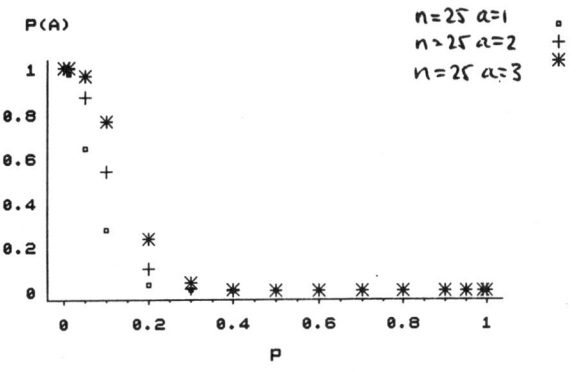

Figure 10.3

b. The plan that best protects the supplier from having acceptable lots returned is the plan for which the probability of accepting is high for small values of p. This plan is (25,3) as shown in Figure 10.3.

c. The plan that best protects the manufacturer from accepting the lot when p is large is the plan for which the probability of accepting is low for large values of p. This plan is (25,1) as shown in Figure 10.3.

d. The compromise plan, acceptable to both the manufacturer and the supplier, would be (25,2).

10.52

Refer to Exercise 10.51. The following specifications must be met:

(i) P[accept lot given p = .01] $\geq$ .90
(ii) P[reject lot given p > .10] $\simeq$ .90

Equation (ii) may be restated as follows:

(iii) P[accept lot given p > .10] $\simeq$ .10

Refer to the binomial tables for n = 25, p = .01. Equation (i) will be satisfied for any acceptance level greater than one (i.e., a $\geq$ 1). Now it is necessary to choose an acceptance level so that equation (iii) is also satisfied. Assume p = .10. Then a = 1 provides the best plan, since

P[accept lot given p = .10] = .271

Notice that for a > 1, the probability of accepting the lot when p = .10 is much larger. Similarly, if p = .20, .30, and so forth, a = 1 still provides the best plan.

10.53

This is similar to Exercise 10.37. From Table 11 with N = 800 and level II, we find code letter J. Hence, for AQL = .01 in Table 12, n = 80 and a = 0.

10.54

This is similar to Exercise 10.37. From Table 11 with N = 3000 and level III, we find code letter L. Hence, for AQL = .065 in Table 12, n = 200 and a = 0.

CHAPTER 11
Linear Regression and Correlation

11.1

The line corresponding to the equation $y = 2x + 1$ can be graphed by locating the y values corresponding to $x = 0, 1,$ and 2.

When $x = 0, y = 2(0) + 1 = 1$

When $x = 1, y = 2(1) + 1 = 3$

When $x = 2, y = 2(2) + 1 = 5$

The graph is shown in Figure 11.1.

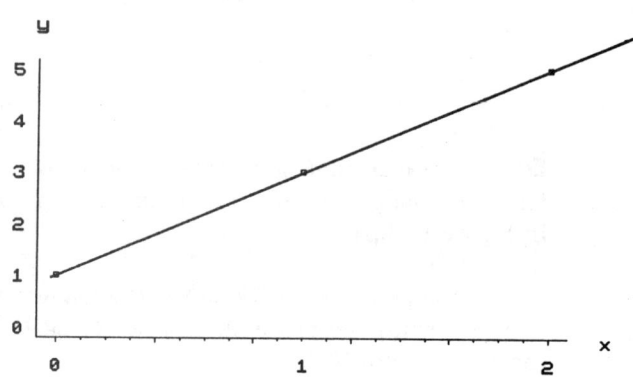

Figure 11.1

Note that the equation is in the form

$$y = \beta_0 + \beta_1 x$$

Thus, the slope of the line is $\beta_1 = 2$ and the y intercept is $\beta_0 = 1$.

11.2

This is similar to Exercise 11.1. When $x = 0, y = -2(0) + 1 = 1$. When $x = 1, y = -2(1) + 1 = -1$, and when $x = 2, y = -2(2) + 1 = -3$. The graph is shown in Figure 11.2.

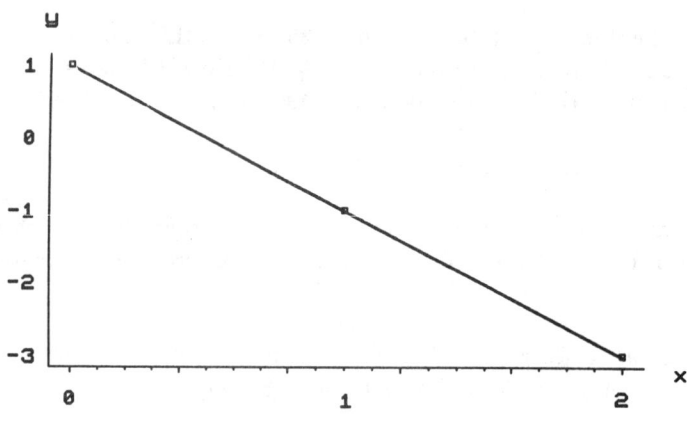

Figure 11.2

Since the line is in the form $y = \beta_0 + \beta_1 x$, the slope of the line is $\beta_1 = -2$ and the y intercept is $\beta_0 = 1$. Notice that the line slopes downward to the right and crosses the y axis at the same point as does the line $y = 2x + 1$. The lines are mirror images.

11.3

Two points are needed to graph a straight line. When x = 0, y = 2. When x = 1, y = 1/2. The line is shown in Figure 11.3.

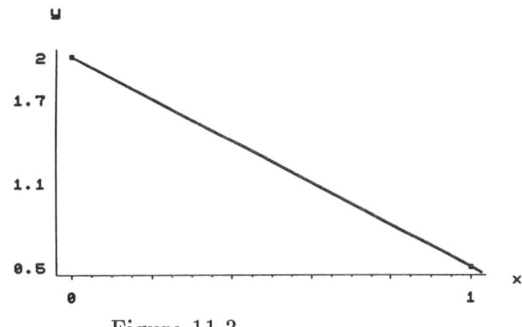

Figure 11.3

11.4

The line 2y = 3x + 4 can be written in the form $y = \frac{3}{2}x + 2$, while the line in Exercise 11.3 can be written in the form $y = -\frac{3}{2}x + 2$. Hence, the y intercepts are identical while the slopes have opposite signs. The two lines are perpendicular.

11.5

If $\beta_0 = 3$ and $\beta_1 = -1$, the straight line is y = 3 − x.

11.6

If $\beta_0 = -3$ and $\beta_1 = 1$, the straight line is y = −3 + x = x − 3.

11.7

A deterministic mathematical model is a model in which the value of a response y is exactly predicted from values of the variables that affect the response. On the other hand, a probabilistic mathematical model is one that contains random elements with specific probability distributions. The value of the response y in this model is not exactly determined.

11.8

a. It is necessary to obtain a prediction equation relating y to x that provides the "best fit" to the data. The "best-fitting" line is one that minimizes the sum of squares of the deviations of the observed y values from the prediction equation. This line, called the "least squares" line, is denoted by

$$\hat{y} = \hat{\beta}_0 + \hat{\beta}_1 x.$$

The equations for calculating the quantities $\hat{\beta}_0$ and $\hat{\beta}_1$ are found in Section 11.6 of the text and the resulting values are given in the computer printout as

$$\hat{\beta}_1 = 1.2 \quad \text{and} \quad \hat{\beta}_0 = 3$$

The least-squares line is

$$\hat{y} = \hat{\beta}_0 + \hat{\beta}_1 x = 3 + 1.2x.$$

b. The least-squares line is graphed in Figure 11.4 along with the five points. Since the line provides a good fit, the above calculations are probably correct.

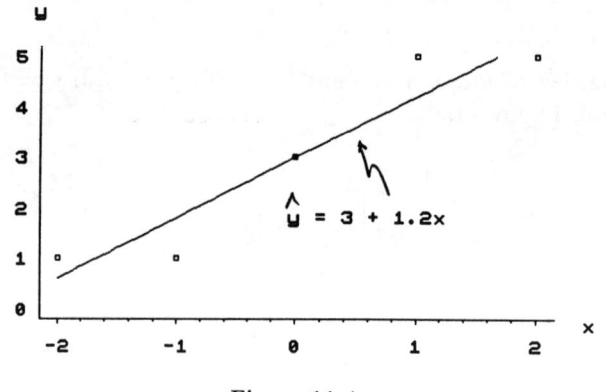

Figure 11.4

11.9

a. The values for the slope and intercept are found in the computer printout to be

$$\hat{\beta}_1 = -.557 \quad \text{and} \quad \hat{\beta}_0 = 6.00$$

The least-squares line is

$$\hat{y} = \hat{\beta}_0 + \hat{\beta}_1 x = 6.00 - .557x.$$

b. The graph of the least-squares line and the six data points is shown below.

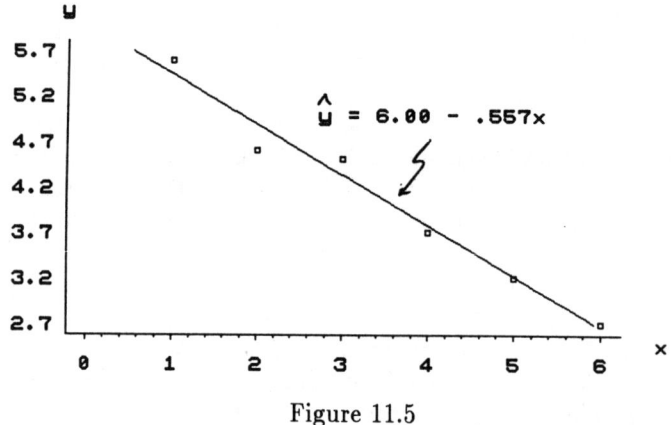

Figure 11.5

c. When $x = 3.5$, the value for y can be predicted using the least-squares line as

$$\hat{y} = 6.000 - .557143(3.5) = 4.05$$

11.10

a-b. From the EXECUSTAT printout,

$$\hat{\beta}_1 = -30.00$$

$$\hat{\beta}_0 = 201.433$$

The least-squares line is

$$\hat{y} = \hat{\beta}_0 + \hat{\beta}_1 x = 201.433 - 30x.$$

The graph of the least-squares line and the six data points is shown below.

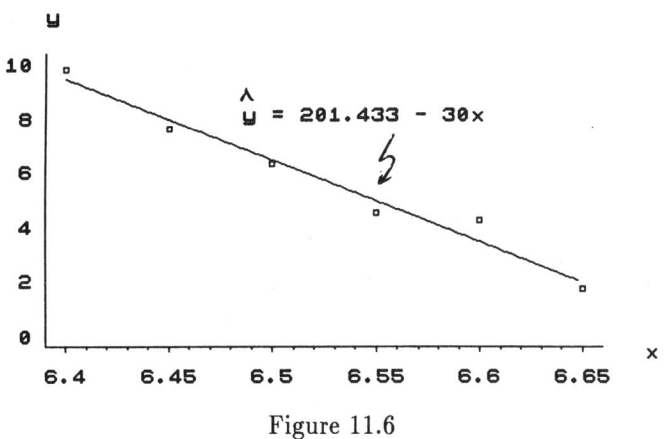

Figure 11.6

c. When the unit price is 6.60, and $\hat{y} = 201.433 - 30(6.6) = 3.433$.

11.11

a-c. From the MINITAB printout,

$$\hat{\beta}_1 = 3.1522$$

$$\hat{\beta}_0 = -2.000$$

and the least-squares line is

$$\hat{y} = \hat{\beta}_0 + \hat{\beta}_1 x = -2.000 + 3.1522x.$$

The graph of the least-squares line and the eight data points is shown below.

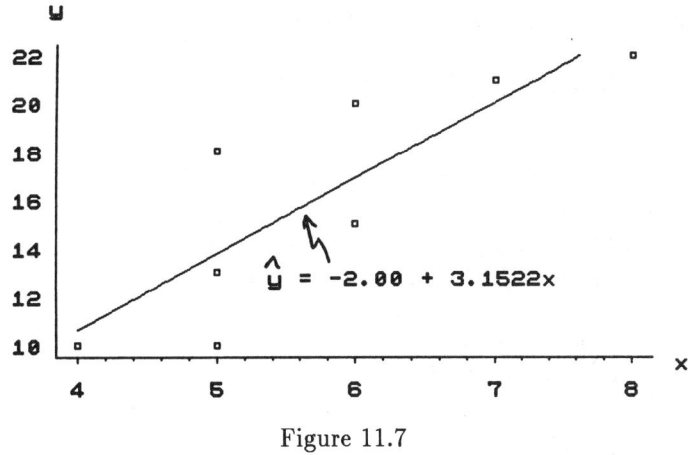

Figure 11.7

d. When $x = 6$, $\hat{y} = -2.000 + 3.1522(6) = 16.91$.

11.12

a-b. The EXECUSTAT plot of the data is shown in Figure 11.8. Provident Federal (x = 518, y = 68) may be an outlier. It does not fit the linear pattern very well.

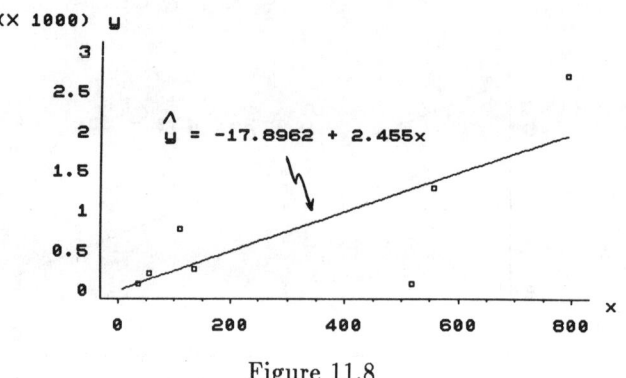

Figure 11.8

c. From the EXECUSTAT printout,

$$\hat{\beta}_1 = 2.45502$$

$$\hat{\beta}_0 = -17.8962$$

and the least-squares line is

$$\hat{y} = \hat{\beta}_0 + \hat{\beta}_1 x = -17.896 + 2.455x$$

d. When $x = 200$, $\hat{y} = -17.896 + 2.455(200) = 473.1$.

11.13

The hypothesis to be tested is

$$H_0: \beta_1 = 0 \qquad H_a: \beta_1 \neq 0$$

and the test statistic is a Student's t, calculated as

$$t = \frac{\hat{\beta}_1 - \beta_1}{s/\sqrt{S_{xx}}} = \frac{1.2 - 0}{.23094} = 5.20$$

given in the line marked "Slope" and the column marked "t value" in the printout. The critical value of t is based on $(n - 2) = 3$ degrees of freedom, and the rejection region for $\alpha = .05$ is $|t| > t_{.025} = 3.182$. Since the observed value of t falls in the rejection region, we reject H_0 and conclude that $\beta_1 \neq 0$. That is, x is useful in the prediction of y.

11.14

A $100(1 - \alpha)\%$ confidence interval for β_1 is given as

$$\hat{\beta}_1 \pm t_{\alpha/2} \times (\text{std error of } \hat{\beta}_1)$$

For this exercise, the 90% confidence interval is

$$\hat{\beta}_1 \pm t_{.05} \times (\text{std error of } \hat{\beta}_1) = 1.2 \pm 2.353(.23094) = 1.2 \pm .543$$

or $.657 < \beta_1 < 1.743$. Intervals constructed in this manner will enclose β_1 90% of the time in repeated sampling. Hence, we are fairly confident that this particular interval encloses β_1.

11.15

This is similar to Exercise 11.13. The hypothesis to be tested is

$$H_0: \beta_1 = 0 \qquad H_a: \beta_1 \neq 0$$

and the test statistic is

$$t = \frac{\hat{\beta}_1 - \beta_1}{s/\sqrt{S_{xx}}} = \frac{-.55714}{.04518} = -12.33$$

given in the printout. The critical value of t is based on $(n - 2) = 4$ degrees of freedom and the rejection region for $\alpha = .05$ is $|t| > t_{.025} = 2.776$. Since the observed value of t falls in the rejection region, we reject H_0 and conclude that $\beta_1 \neq 0$. That is, x is useful in the prediction of y.

11.16

The 90% confidence interval is

$$\hat{\beta}_1 \pm t_{.05} \times \text{(std error of } \hat{\beta}_1) = -.557 \pm 2.132(.04518) = -.557 \pm .096$$

or $-.653 < \beta_1 < -.461$. Intervals constructed in this manner will enclose β_1 95% of the time in repeated sampling. Hence, we are fairly confident that this particular interval encloses β_1.

11.17

The hypothesis to be tested is

$$H_0: \beta_1 = 0 \qquad H_a: \beta_1 \neq 0$$

and the test statistic found in the printout is

$$t = \frac{\hat{\beta}_1 - \beta_1}{s/\sqrt{S_{xx}}} = \frac{-30}{2.49952} = -12.00$$

The critical value of t is based on $(n - 2) = 4$ degrees of freedom and the rejection region for $\alpha = .05$ is $|t| > t_{.025} = 2.776$. Since the observed value of t falls in the rejection region, we reject H_0 and conclude that $\beta_1 \neq 0$. That is, x is useful in the prediction of y.

11.18

The hypothesis to be tested is

$$H_0: \beta_1 = 0 \qquad H_a: \beta_1 \neq 0$$

and the test statistic is

$$t = \frac{\hat{\beta}_1 - \beta_1}{s/\sqrt{S_{xx}}} = \frac{3.1522}{.8393} = 3.76$$

The critical value of t is based on $(n - 2) = 6$ degrees of freedom and the rejection region for $\alpha = .05$ is $|t| > t_{.025} = 2.447$. Since the observed value of t falls in the rejection region, we reject H_0 and conclude that $\beta_1 \neq 0$. That is, x is useful in the prediction of y.

11.19

Refer to Exercise 11.18. The 90% confidence interval is

$$\hat{\beta}_1 \pm t_{.05} \frac{s}{\sqrt{S_{xx}}} = 3.1522 \pm 1.943(.8393) = 3.15 \pm 1.63$$

or $1.52 < \beta_1 < 4.78$. Intervals constructed in this manner will enclose β_1 90% of the time in repeated sampling. Hence, we are fairly confident that this particular interval encloses β_1.

11.20

a. The hypothesis to be tested is

$$H_0: \beta_1 = 0 \qquad H_a: \beta_1 \neq 0$$

and the test statistic is

$$t = \frac{\hat{\beta}_1 - \beta_1}{s/\sqrt{S_{xx}}} = \frac{2.45502}{0.911933} = 2.69$$

The critical value of t is based on $(n - 2) = 5$ degrees of freedom and the rejection region for $\alpha = .05$ is $|t| > t_{.025} = 2.571$. Since the observed value of t falls in the rejection region, we reject H_0 and conclude that $\beta_1 \neq 0$. That is, the amount of total assets contributes information for the prediction of net income for the S & L's.

b. The assumptions for the linear probabilistic model are discussed in Sections 11.2 and 11.8. They appear to be met for these data. However, extrapolation to savings and loans institutions outside of Riverside and San Bernardino Counties would probably not yield accurate predictions.

11.21

a. From the MINITAB printout, $\hat{\beta}_1 = 1.5$ and $\hat{\beta}_0 = 4.3$, so that the least-squares line is

$$\hat{y} = \hat{\beta}_0 + \hat{\beta}_1 x = 4.3 + 1.5x.$$

b. The graph of the least-squares line and the fifteen data points is shown below.

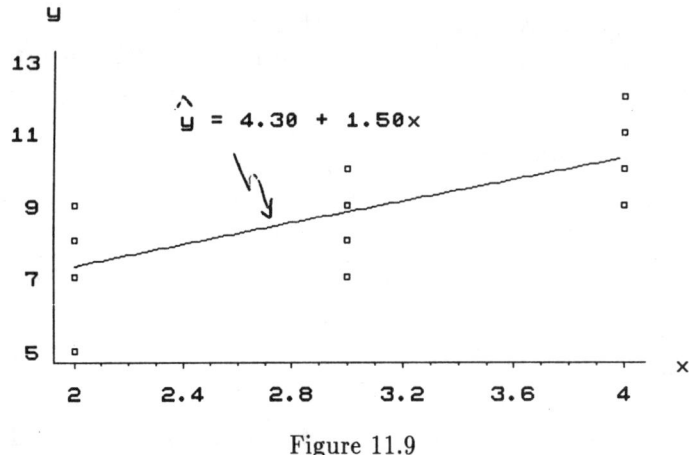

Figure 11.9

c. From the MINITAB printout, $s = 1.237$, so that

$$s^2 = 1.53$$

d. The hypothesis to be tested is

$$H_0: \beta_1 = 0 \qquad H_a: \beta_1 \neq 0$$

and the test statistic is

$$t = \frac{\hat{\beta}_1 - \beta_1}{s/\sqrt{S_{xx}}} = \frac{1.5}{.3913} = 3.83$$

The critical value of t is based on $(n - 2) = 13$ degrees of freedom and the rejection region for $\alpha = .05$ is $|t| > t_{.025} = 2.160$. Since the observed value of t falls in the rejection region, H_0 is rejected. We conclude that x and y are linearly related.

e. From Table 4, notice that $t = 3.83$ is larger than the largest tabulated value with $n - 2 = 13$ degrees of freedom ($t_{.005} = 3.012$). Hence, the observed level of significance for this two-tailed test is

$$\text{p-value} = 2P[t > 3.83] < 2(.005) = .01.$$

That is, H_0 could be rejected for any value of α greater than .01.

11.22

a. From the MINITAB printout, $\hat{\beta}_1 = .4736$ and $\hat{\beta}_0 = 63.86$, so that the least squares line is

$$\hat{y} = \hat{\beta}_0 + \hat{\beta}_1 x = 63.86 + .4736x$$

The graph of the least-squares line and the ten data points is shown below.

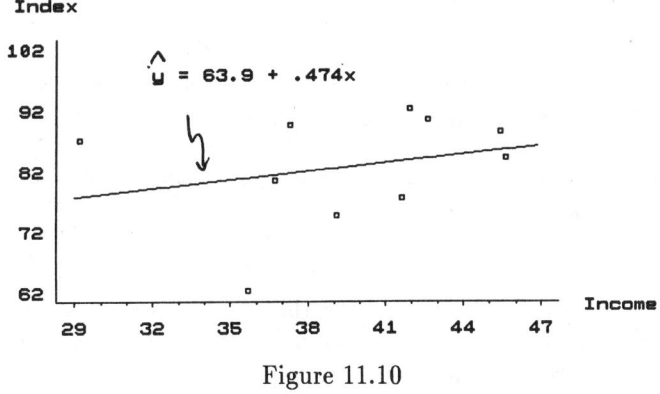

Figure 11.10

In order to test for a linear relationship between x and y, the hypothesis to be tested is

$$H_0: \beta_1 = 0 \qquad H_a: \beta_1 \neq 0$$

and the test statistic is

$$t = \frac{\hat{\beta}_1 - \beta_1}{s/\sqrt{S_{xx}}} = 0.75$$

with p-value $= .475$. Since the observed level of significance is so large, H_0 is not rejected. We cannot conclude that x and y are linearly related.

b. From the MINITAB printout, $\hat{\beta}_1 = -.3773$ and $\hat{\beta}_0 = 116.15$, so that the least-squares line is

$$\hat{y} = \hat{\beta}_0 + \hat{\beta}_1 x = 116.15 - .3773x$$

The graph of the least-squares line and the ten data points is shown below.

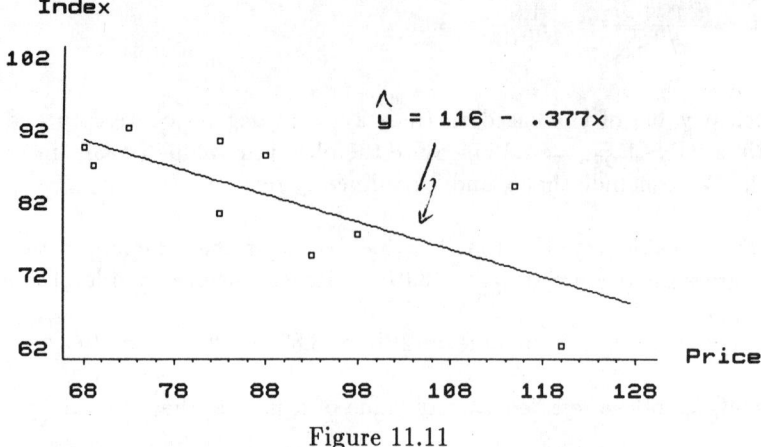

Figure 11.11

In order to test for a linear relationship between x and y, the hypothesis to be tested is

$$H_0: \beta_1 = 0 \qquad H_a: \beta_1 \neq 0$$

and the test statistic is

$$t = \frac{\hat{\beta}_1 - \beta_1}{s/\sqrt{S_{xx}}} = -3.05$$

with p-value = .016. Since the observed level of significance is so small (i.e., smaller than .05), H_0 is rejected. We can conclude that x and y are linearly related.

c. From the results of parts a and b, x = median home price appears to be the better predictor of the housing index.

11.23

a. In order to obtain an estimate for the expected value of y for a given value of x (or for a particular value of y), it would seem reasonable to use the prediction equation, $\hat{y} = \hat{\beta}_0 + \hat{\beta}_1 x$. Notice that x_p represents the given value of x for which we are estimating E(y). The point estimator for E(y) when x = 1 is

$$\hat{y} = 3 + 1.2(1) = 4.2$$

and the 90% confidence interval is

$$\hat{y} \pm t_{.05} s_{\hat{y}}$$

or 3.25865 to 5.14135 as shown on the computer printout. That is,

$$3.259 < E(y) < 5.141$$

b. It is necessary to find a 90% prediction interval for y when x = 1. The interval used in predicting a particular value of y is

$$\hat{y} \pm t_{.05} s_{(y-\hat{y})} \quad \text{or} \quad \hat{y} \pm t_{.05}\sqrt{s^2 + s_{\hat{y}}^2}$$

Using the information given in the column marked "90% Prediction Limits," the interval is

$$2.24042 \text{ to } 6.15958 \quad \text{or} \quad 2.240 < y < 6.160$$

We are 90% confident that the true value of y when x = 1 is in the above interval. Note that the above interval is much wider than the interval calculated for the expected value of y. The variability of predicting a particular value of y is greater than the variability of predicting the population mean for a particular value of x.

11.24

Refer to Exercise 11.9, in which $\hat{y} = 6.0 - .557143x$. When x = 2, the estimate of E(y) is

$$\hat{y} = 6 - (.557143)(2) = 4.886$$

and the 90% confidence interval for E(y) is

$$\hat{y} \pm t_{.05} s_{\hat{y}}$$

$$4.886 \pm 2.132(.102685)$$

$$4.886 \pm .219 \quad \text{or} \quad 4.667 < E(y) < 5.105$$

11.25

a. From the computer printout in Exercise 11.15, $s_y = .1890$.

b. $s_{(y-\hat{y})} = \sqrt{s_y^2 + s_{\hat{y}}^2} = \sqrt{(.1890)^2 + (.102685)^2} = .215093$

c. When x = 2, the prediction of y is

$$\hat{y} = 6 - (.557143)(2) = 4.886$$

and the 95% prediction interval for y is

$$\hat{y} \pm t_{.025} s_{(y-\hat{y})}$$

$$4.886 \pm 2.776(.215093)$$

$$4.886 \pm .597 \quad \text{or} \quad 4.289 < y < 5.483$$

11.26

The MINITAB commands shown on the output include the subcommand PREDICT 7. Hence, the last line of the printout gives the value of $\hat{y}$ when x = 7, along with the 95% confidence interval for E(y) and the 95% prediction interval for y when x = 7. For this exercise, the 95% confidence interval is

$$16.51 < E(y) < 23.62$$

11.27

a. The EXECUSTAT printout provides the predicted values, $\hat{y}$, for the values x = 200, x = 400, and x = 500. For x = 200, the 90% confidence interval for E(y) is read from the printout as

$$-81.5812 < E(y) < 1027.8$$

b. From the printout, with x = 500, the 90% prediction interval is

$$-279.839 < y < 2699.07$$

11.28

a-b. From the computer printout,

$$\hat{\beta}_1 = -28.765$$

$$\hat{\beta}_0 = 1071.42$$

and the least-squares line, $\hat{y} = \hat{\beta}_0 + \hat{\beta}_1 x = 1071.42 - 28.765x$, is plotted in Figure 11.12.

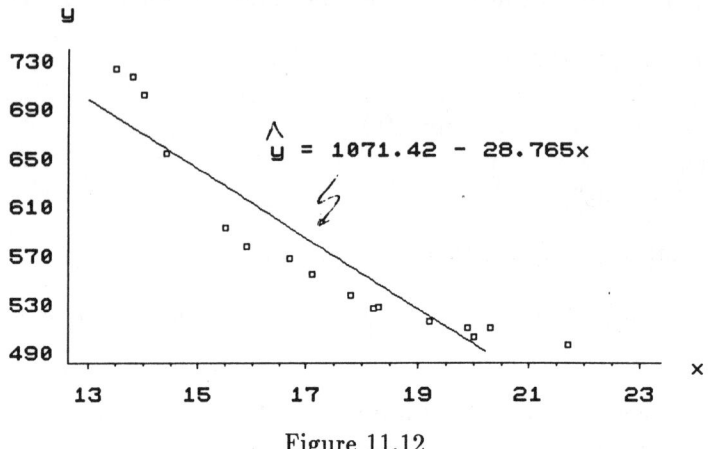

Figure 11.12

c. To test $H_0: \beta_1 = 0$, $H_a: \beta_1 \neq 0$, the test statistic is

$$t = \frac{\hat{\beta}_1 - \beta_1}{s/\sqrt{S_{xx}}} = -9.54$$

The rejection region for $\alpha = .05$ is $|t| > t_{.025} = 2.145$ and we reject H_0. There is sufficient evidence to indicate that the independent variable x does help in predicting values of the dependent variable y.

d. From Table 4, notice that $t = -9.54$ is larger than the largest tabulated value with $n - 2 = 14$ degrees of freedom ($t_{.005} = 2.977$). Hence, the observed level of significance for this two-tailed test is

$$2 P[t > 9.54] < 2(.005) = .01$$

That is, H_0 could be rejected for any value of α greater than .01. The p-value is given in the printout as 0.0000, which confirms the above results.

e. When $x = 16$, $\hat{y} = 611.183$ (last line of the printout) and the 99% confidence interval is

$$586.045 < E(y) < 636.321$$

f. From the last line of the printout, the 99% prediction interval for y when $x = 16$ is

$$518.09 < y < 704.276$$

11.29

a. The mean increase in y for a one-unit change in x is given by β_1, and the 90% confidence interval is

$$\hat{\beta}_1 \pm t_{.05} \times (\text{std error of } \hat{\beta}_1) = 27.406 \pm 1.812(1.828) = 27.406 \pm 3.312$$

or $24.094 < \beta_1 < 30.718$. Intervals constructed in this manner will enclose β_1 90% of the time in repeated sampling. Hence, we are fairly confident that this particular interval encloses β_1.

b. From the last line of the printout, the 95% confidence interval for E(y) when x = 2000 is

$$94676 < E(y) < 97360$$

c. For each house in the sample, the price per square foot is calculated as $z_i = y_i/x_i$, and the results are shown below.

53.90	47.11	52.44	54.10	49.57	43.87
48.25	54.04	53.86	50.14	51.78	45.89

The average cost per square foot is

$$\bar{z} = \frac{\sum_{i=1}^{n} z_i}{n} = \frac{604.963}{12} = 50.41$$

This is not the same as

$$\hat{\beta}_1 = 27.406$$

and should not be, since they are calculated in totally different ways.

d. From the last line of the printout, the 95% prediction interval for y when x = 2000 is

$$91{,}803 < y < 100{,}233$$

11.30

a. From the MINITAB printout, $\hat{\beta}_1 = 2049.5$ and $\hat{\beta}_0 = 648$. The least squares line is then $\hat{y} = 648 + 2049.5x$.

b. The hypothesis of interest is

$$H_0: \beta_1 = 0, \quad H_a: \beta_1 \neq 0,$$

and the test statistic is

$$t = \frac{\hat{\beta}_1 - \beta_1}{s/\sqrt{S_{xx}}} = 18.29$$

The rejection region for $\alpha = .10$ with $n - 2 = 4$ d.f. is $|t| > t_{.05} = 2.132$ and we reject H_0. There is sufficient evidence to indicate that the independent variable x does help in predicting values of the dependent variable y.

c. The data points are shown in Figure 11.13. Note the strong linear trend.

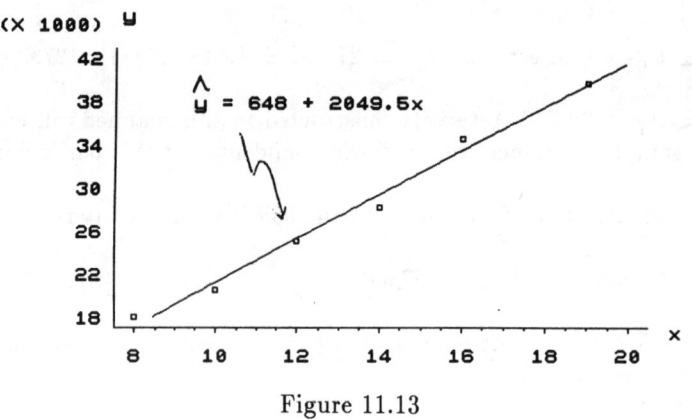

Figure 11.13

d. When x = 18, the estimate of E(y) is

$$\hat{y} = 648 + 2049.5(18) = 37,539$$

and the 95% confidence interval is given in the last line of printout to be

$$35,650 < E(y) < 39,428$$

e. When x = 18, the estimate of y is $\hat{y} = 37,539$ and the 95% prediction interval is

$$34,163 < y < 40,916$$

11.31

The significance of the algebraic sign and the magnitude of r, the coefficient of correlation, will be discussed in Exercise 11.32. Note, however, that r^2 provides a meaningful measure of the strength of the linear relationship between two variables, y and x. It is the ratio of the reduction in the sum of squares of deviations obtained using the model, $y = \beta_0 + \beta_1 x + \epsilon$, to the sum of squares of deviations that would be obtained if the variable x were ignored. That is, r^2 measures the amount of variation that can be attributed to the variable x.

11.32

If the value of r is positive, then the least-squares line slopes upward to the right. Similarly, if the value of r is negative, the line slopes downward to the right. The coefficient of correlation r will be 0 only when $\hat{\beta}_1$ is 0 (see Section 11.7 of the text). Moreover, the least-squares equation when $\hat{\beta}_1 = 0$ is given by $\hat{y} = \hat{\beta}_0$. The variable x has no effect on the value of y and there is no linear correlation between x and y. Finally, r will equal ±1 only when SSE = 0; that is, all points fall exactly on the fitted line.

11.33

Refer to Exercise 11.32. In the first instance, r = +1; in the second, r = −1.

11.34

a. Refer to Figure 11.14. The sample correlation coefficient will be positive.

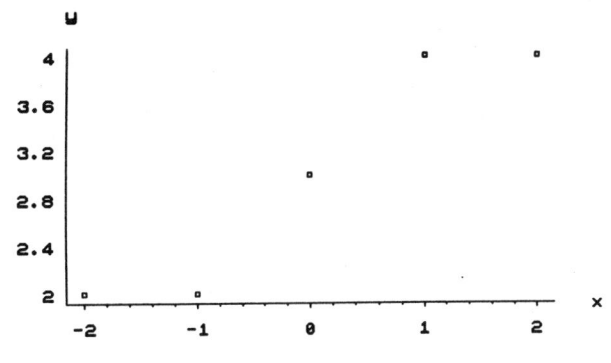

Figure 11.14

b. Calculate

$$r = \frac{S_{xy}}{\sqrt{S_{xx}S_{yy}}} = \frac{6}{\sqrt{40}} = .9487$$

and $r^2 = (.9487)^2 = .9000$. The total sum of squares of deviations was reduced by approximately 90% using the least-squares equation instead of $\overline{y}$ as a predictor of y.

11.35

a. Refer to Figure 11.15. The sample correlation coefficient will be negative.

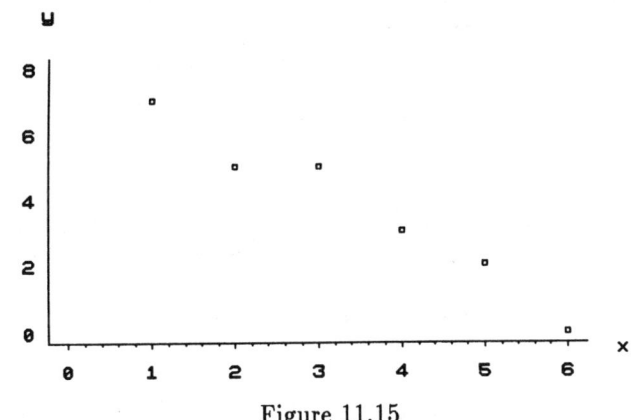

Figure 11.15

b. Calculate

$$r = \frac{S_{xy}}{\sqrt{S_{xx}S_{yy}}} = \frac{-23}{\sqrt{17.5(31.3333)}} = -.9822$$

c. We first calculate the coefficient of determination:

$$r^2 = (-.9822)^2 = .9647.$$

This value implies that the sum of squares of deviations is reduced by 96.47% using the linear model $\hat{y} = \hat{\beta}_0 + \hat{\beta}_1 x$ instead of $\overline{y}$ to predict values of y.

11.36

The data from Exercise 11.35 are reused here, except that the y observations are reordered. The only calculation that has changed from the previous exercise is $S_{xy} = 23$.

a. Refer to Figure 11.16. The sample correlation coefficient will be positive.

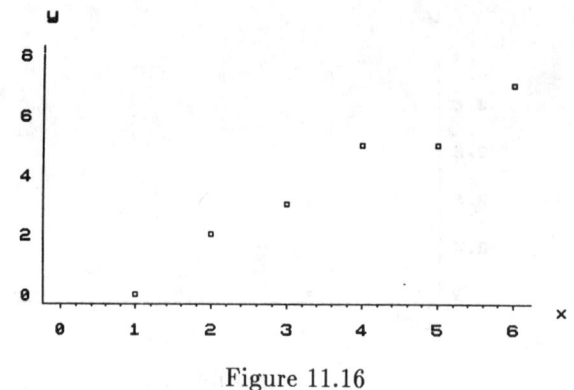

Figure 11.16

b. Calculate

$$r = \frac{S_{xy}}{\sqrt{S_{xx}S_{yy}}} = \frac{23}{\sqrt{17.5(31.3333)}} = .9822$$

Now, since the value of r is near 1, a strong positive linear association between the two variables is implied. Note that this value of r is the negative of the value calculated for r in Exercise 11.35.

c. We first calculate $r^2 = (.9822)^2 = .9647$. As in Exercise 11.35, the sum of squares of deviations can be reduced by 96.47% using the linear model to predict values of the dependent variable y. Since the values of r^2 calculated in these two exercises are identical, we can conclude that reversing the values of y had no effect on the strength of the linear relationship between the two variables.

11.37

a. Calculate $r = \dfrac{S_{xy}}{\sqrt{S_{xx}S_{yy}}} = \dfrac{-26.25}{\sqrt{17.5(40.46833)}} = -.9864$

There is a high negative correlation between x and y.

b. $r^2 = (-.9864)^2 = .9730$. The sum of squares of deviations can be reduced by 97.3% using the linear model to predict y.

11.38

a. Calculate $r = \dfrac{S_{xy}}{\sqrt{S_{xx}S_{yy}}} = \dfrac{36.25}{\sqrt{11.5(162.875)}} = .8376$

There is a reasonably high positive correlation between x and y.

b. $r^2 = (.8376)^2 = .7016$. The sum of squares of deviations can be reduced by 70.2% using the linear model rather than $\bar{y}$ to predict y.

c. The computer printout in Exercise 11.11 shows "R-sq = 70.2%," which agrees with the value calculated in part b. Notice that the printout reports r^2 as a percent, while r^2 is actually calculated to be a proportion.

11.39

a. From the printout, $\hat{\beta}_1 = 34.5833$ and $\hat{\beta}_0 = 307.917$. The least-squares line is

$$\hat{y} = \hat{\beta}_0 + \hat{\beta}_1 x = 307.917 + 34.583x.$$

b. From the printout, $r^2 = .7636$, so that $r = \sqrt{.7636} = .874$. Since r^2 is close to 1, a strong linear relationship between the two variables is implied. Since the sign of r is positive, we know that there is a positive association between x and y.

c. The 90% confidence interval for E(y) when x = 6 is shown in the last line of the printout to be

$$488.4 < E(y) < 542.433$$

d. Similar to part c. The 90% prediction interval for y when x = 6 is

$$418.007 < y < 612.826$$

e. Since $r^2 = .7636$, the total sum of squares of deviations was reduced by approximately 76.36% using the least-squares equation instead of $\bar{y}$ as a predictor of y.

f. The above conclusions are applicable only to the interval where the measurements are taken, from x = 3 to x = 9. In order to state that the relationship between x and y is linear between x = 1 and x = 30, we must take data from this wider range and analyze the new data.

11.40

a. As the number of years of schooling increases, both male and female salaries should increase. Therefore, a high female salary should imply a high male salary. The correlation should be positive. The plot of the data in Figure 11.17 confirms this relationship.

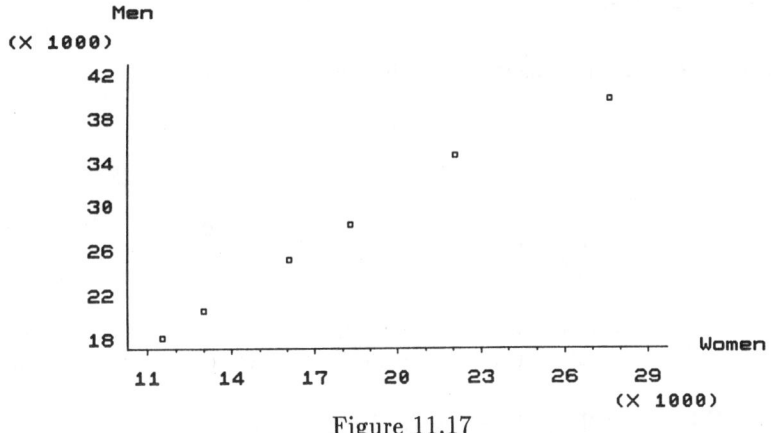

Figure 11.17

b. The hypothesis of interest is

$$H_0: \rho = 0 \qquad H_a: \rho > 0$$

and the test statistic is

$$t = \frac{r\sqrt{n-2}}{\sqrt{1-r^2}} = \frac{.995\sqrt{4}}{\sqrt{1-(.995)^2}} = 19.92$$

The rejection region with $\alpha = .05$ and $n - 2 = 4$ d.f. is $t > t_{.05} = 2.132$. Hence, there is a positive correlation between x and y.

c. The high positive correlation between men's and women's salaries indicates that as men's salaries increase, so do women's salaries. However, it is clear from examining the data that, for the same number of years of schooling, women's salaries are always below the men's salaries. It would be interesting to know if there is a significant difference in the y intercepts of the lines relating median salary to years of schooling for men and women, or if there is a difference in the slopes of the two lines (men's salary increasing faster than women's).

11.41

If $x = 0$, $y = 2$. If $x = 1$, $y = 5$. If $x = 2$, $y = 8$. The graph is shown below.

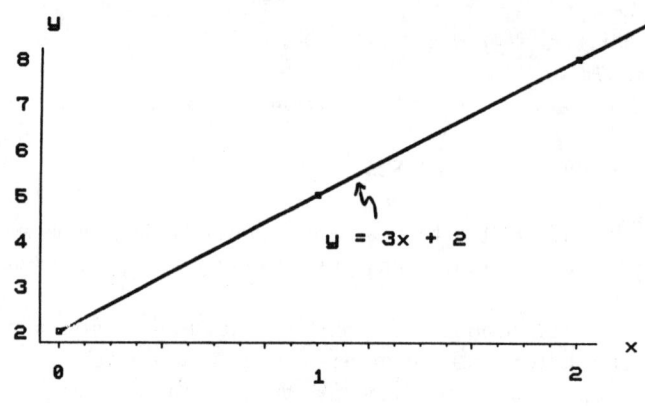

Figure 11.18

11.42

The equation $2x + 3y + 6 = 0$ must be rewritten in the form

$$y = \beta_0 + \beta_1 x$$

so that the slope and y intercept can be determined. Then,

$$2x + 3y + 6 = 0$$

$$3y = -6 - 2x$$

$$y = -2 - \frac{2}{3}x$$

The slope is $\beta_1 = -2/3$ and the y intercept is $\beta_0 = -2$. The graph of the line corresponding to $y = -2 - (2/3)x$ is obtained by locating the points corresponding to various values of x.

When $x = 0$, $y = -2 - (2/3)(0) = -2$

When $x = 1$, $y = -2 - (2/3)(1) = -2\frac{2}{3}$

The graph is shown below.

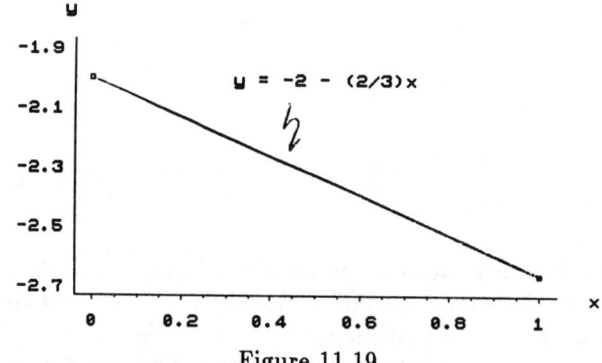

Figure 11.19

11.43

Rewriting the equation, we obtain

$$2x - 3y - 5 = 0$$

$$3y = -5 + 2x$$

$$y = -\frac{5}{3} + \frac{2}{3}x$$

The slope of the line is $\beta_1 = 2/3$ and the y intercept is $\beta_0 = -5/3$. The line is graphed as follows.

$$\text{When } x = 0, \ y = -5/3 + (2/3)(0) = -5/3$$

$$\text{When } x = 1, \ y = -5/3 + (2/3)(1) = -1$$

The graph is shown below.

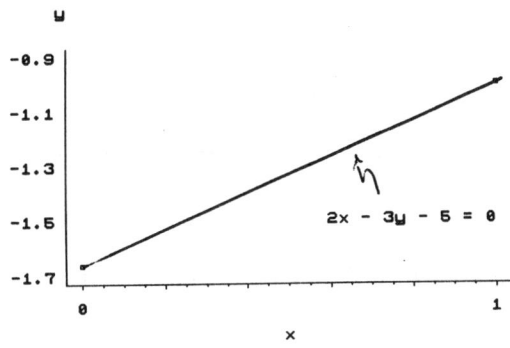

Figure 11.20

11.44

Rewriting the equation,

$$x/y = 1/2$$

$$x = (1/2)y$$

$$y = 2x$$

The slope of the line is $\beta_1 = 2$ and the y intercept is $\beta_0 = 0$. The line is graphed as follows.

$$\text{When } x = 0, \ y = 2(0) = 0$$

$$\text{When } x = 1, \ y = 2(1) = 2$$

The graph is shown below.

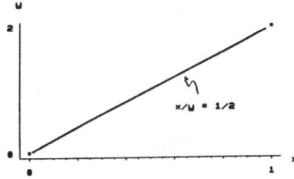

Figure 11.21

11.45

The student may use a computer program, or perform the hand calculations given below:

a. Preliminary calculations:

$$\Sigma x_i = 0 \qquad \Sigma y_i = 5 \qquad \Sigma x_i y_i = 7$$

$$\Sigma x_i^2 = 10 \qquad \Sigma y_i^2 = 11 \qquad n = 5$$

$$S_{xy} = \Sigma x_i y_i - \frac{(\Sigma x_i)(\Sigma y_i)}{n} = 7 - 0 = 7$$

$$S_{xx} = \Sigma x_i^2 - \frac{(\Sigma x_i)^2}{n} = 10 - 0 = 10$$

$$S_{yy} = \Sigma y_i^2 - \frac{(\Sigma y_i)^2}{n} = 11 - 5 = 6$$

Then

$$\hat{\beta}_1 = \frac{7}{10} = .7$$

$$\hat{\beta}_0 = \bar{y} - \hat{\beta}_1 \bar{x} = 1 - (.7)(0) = 1.0$$

The least-squares line is $\hat{y} = \hat{\beta}_0 + \hat{\beta}_1 x = 1 + .7x$.

b. The least-squares line is graphed as follows:

When $x = 0$, $\hat{y} = 1 + .7(0) = 1$

When $x = 1$, $\hat{y} = 1 + .7(1) = 1.7$

The graph is shown below.

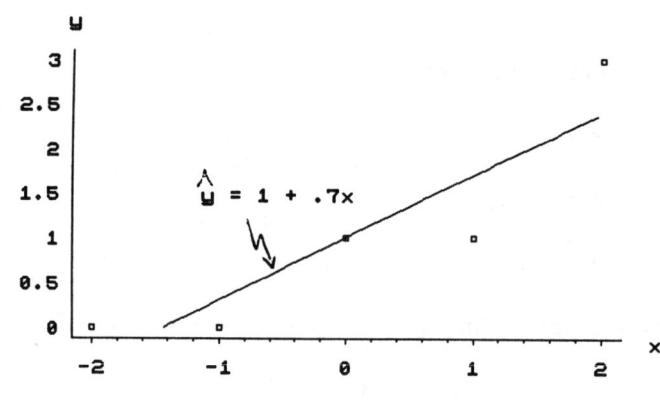

Figure 11.22

c. Calculate $SSE = S_{yy} - \hat{\beta}_1 S_{xy} = 6 - .7(7) = 1.1$
and

$$s^2 = \frac{SSE}{(n-2)} = \frac{1.1}{3} = .3667$$

To test $H_0: \beta_1 = 0$, $H_a: \beta_1 \neq 0$, the test statistic is

$$t = \frac{\hat{\beta}_1 - \beta_1}{s/\sqrt{S_{xx}}} = \frac{.7}{\sqrt{\frac{.3667}{10}}} = 3.656$$

The rejection region for $\alpha = .05$ is $|t| > t_{.025} = 3.182$ and we reject H_0. There is sufficient evidence to indicate that the independent variable x does help in predicting values of the dependent variable y.

d. The 90% confidence interval for β_1 is

$$\hat{\beta}_1 \pm t_{.05}\sqrt{\frac{s^2}{S_{xx}}} = .7 \pm 2.353\sqrt{\frac{.3667}{10}} = .7 \pm .451$$

or $.249 < \beta_1 < 1.151$.

e. When $x = 1$, the estimate of E(y) is $\hat{y} = 1 + .7(1) = 1.7$ and the 90% confidence interval is

$$\hat{y} \pm t_{.05}\sqrt{s^2\left[\frac{1}{n} + \frac{(x_p - \bar{x})^2}{S_{xx}}\right]}$$

$$1.7 \pm 2.353\sqrt{.3667\left[\frac{1}{5} + \frac{(1-0)^2}{10}\right]}$$

$$1.7 \pm .780 \quad\quad \text{or} \quad\quad .920 < E(y) < 2.480$$

f. Calculate
$$r = \frac{S_{xy}}{\sqrt{S_{xx}S_{yy}}} = \frac{7}{\sqrt{10(6)}} = .904$$

Because $r^2 = (.904)^2 = .817$, we know that the sum of squares of deviations was reduced by 81.7% by using $\hat{y}$ rather than $\bar{y}$ as a predictor for y. That is, 81.7% of the total variation in y can be explained by the variable x. There is a high positive correlation between x and y.

11.46

The student may use a computer program or perform the hand calculations shown below.
Preliminary calculations:

$\Sigma x_i = 0$ $\quad\quad\quad$ $\Sigma y_i = 18.5$ $\quad\quad\quad$ $\Sigma x_i y_i = 15.5$

$\Sigma x_i^2 = 28$ $\quad\quad\quad$ $\Sigma y_i^2 = 60.75$ $\quad\quad\quad$ $n = 7$

$S_{xy} = \Sigma x_i y_i - \frac{(\Sigma x_i)(\Sigma y_i)}{n} = 15.5 - 0 = 15.5$

$S_{xx} = \Sigma x_i^2 - \frac{(\Sigma x_i)^2}{n} = 28 - 0 = 28$

$S_{yy} = \Sigma y_i^2 - \frac{(\Sigma y_i)^2}{n} = 60.75 - 48.893 = 11.857$

a. $\hat{\beta}_1 = \frac{S_{xy}}{S_{xx}} = \frac{15.5}{28} = .554$

$\hat{\beta}_0 = \bar{y} - \hat{\beta}_1 \bar{x} = 2.643 - .554(0) = 2.643$

and the least-squares line is $\hat{y} = \hat{\beta}_0 + \hat{\beta}_1 x = 2.643 + .554x$.

b. The graph and the observed y values are shown in Figure 11.23.

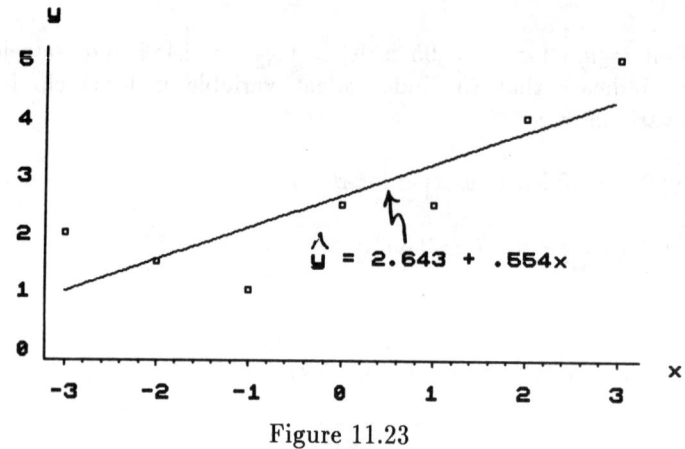

Figure 11.23

c. Calculate

$$SSE = S_{yy} - \hat{\beta}_1 S_{xy} = 11.857 - .554(15.5) = 3.277$$

and

$$s^2 = \frac{SSE}{(n-2)} = \frac{3.277}{5} = .6554$$

To test H_0: $\beta_1 = 0$, H_a: $\beta_1 \neq 0$, the test statistic is

$$t = \frac{\hat{\beta}_1 - \beta_1}{s/\sqrt{S_{xx}}} = \frac{.554}{\sqrt{\frac{.6554}{28}}} = 3.618$$

The rejection region for $\alpha = .05$ is $|t| > t_{.025} = 2.571$ and we reject H_0. There is sufficient evidence to indicate that the independent variable x does help in predicting values of the dependent variable y.

d. The 95% confidence interval for β_1 is

$$\hat{\beta}_1 \pm t_{.025} \sqrt{\frac{s^2}{S_{xx}}} = .554 \pm 2.571 \sqrt{\frac{.6554}{28}} = .554 \pm .393$$

or $.161 < \beta_1 < .947$.

e. When $x = -1$, $\hat{y} = 2.643 + .554(-1) = 2.089$ and the 95% confidence interval is

$$\hat{y} \pm t_{.025} \sqrt{s^2 \left[\frac{1}{n} + \frac{(x_p - \bar{x})^2}{S_{xx}}\right]}$$

$$2.089 \pm 2.571 \sqrt{.6554 \left[\frac{1}{7} + \frac{(-1-0)^2}{28}\right]}$$

$$2.089 \pm .880 \qquad \text{or} \qquad 1.209 < E(y) < 2.969$$

f. The predictor for y when $x = 2$ is $\hat{y} = 2.643 + 2(.554) = 3.750$ and the 90% prediction interval is

$$\hat{y} \pm t_{.05} \sqrt{s^2 \left[1 + \frac{1}{n} + \frac{(x_p - \bar{x})^2}{S_{xx}}\right]}$$

$$3.750 \pm 2.015 \sqrt{.6554 \left[1 + \frac{1}{7} + \frac{(2-0)^2}{28}\right]}$$

$$3.750 \pm 1.850 \quad\quad \text{or} \quad\quad 1.900 < y < 5.600$$

g. Calculate
$$r = \frac{S_{xy}}{\sqrt{S_{xx}S_{yy}}} = \frac{15.5}{\sqrt{28(11.857)}} = .851$$

Again, there is a high positive correlation between x and y.

h. Because $r^2 = (.851)^2 = .7236$, we know that the sum of squares of deviations was reduced by 72.36% by using $\hat{y}$ rather than $\bar{y}$ as a predictor for y. That is, 72.36% of the total variation in y can be explained by the variable x.

11.47

a. From the printout, $\hat{\beta}_1 = 1.4$, $\hat{\beta}_0 = -.333$ and the least-squares line is $\hat{y} = \hat{\beta}_0 + \hat{\beta}_1 x = -.333 + 1.4x$.

b. The graph and the observed y values are shown in Figure 11.24.

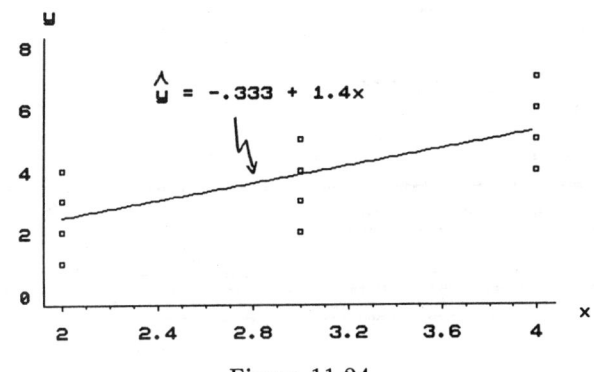

Figure 11.24

c. From the printout, $r = \sqrt{.549} = .741$. The hypothesis of interest is

$$H_0: \rho = 0 \quad\quad H_a: \rho \neq 0$$

and the test statistic is

$$t = \frac{r\sqrt{n-2}}{\sqrt{1-r^2}} = \frac{.741\sqrt{13}}{\sqrt{1-(.741)^2}} = 3.979 \quad \text{(note the slight rounding error from the printout)}$$

The rejection region with $\alpha = .05$ and $n - 2 = 13$ d.f. is $|t| > t_{.025} = 2.160$. Hence, there is a significant correlation between x and y.

11.48

A MINITAB printout is shown below. Answers will vary from student to student, but students should notice a relatively strong positive linear relationship (and correlation) between x and y. The least-squares regression line is

$$\hat{y} = 0.0667 + .51667x$$

```
The regression equation is
C1 = 0.067 + 0.517 C2

Predictor        Coef        Stdev      t-ratio         p
Constant       0.0667       0.3935        0.17       0.870
C2             0.51667      0.09107       5.67       0.000

s = 0.4461      R-sq = 82.1%      R-sq(adj) = 79.6%

Analysis of Variance

SOURCE         DF         SS          MS         F         p
Regression      1       6.4067      6.4067     32.19     0.000
Error           7       1.3933      0.1990
Total           8       7.8000
```

11.49

An EXECUSTAT printout is shown below.

```
              Simple Regression Analysis for 11-49
Linear model: CORN = -95420 + 51.878*YEAR
                      Table of Estimates

                                Standard         t           P
                 Estimate         Error        Value       Value

Intercept        -95420         207516         -0.46       0.6566
Slope             51.878        104.451         0.50       0.6313

R-squared =  2.67%
Correlation coeff. = 0.163
Standard error of estimation = 1245.47
Durbin-Watson statistic = 1.5442
Mean absolute error = 837.732
Sample size (n) = 11

                    Table of Predicted Values
                                      99.00%              99.00%
                         Predicted  Prediction Limits  Confidence Limits
Row      YEAR             CORN     Lower     Upper    Lower     Upper
 1       1993            7972.85   3239.34   12706.4  5518.64   10427.1
```

a. From the printout, $\hat{\beta}_1 = 51.878$, $\hat{\beta}_0 = -95420$ and the least-squares line is

$$\hat{y} = \hat{\beta}_0 + \hat{\beta}_1 x = -95420 + 51.878x$$

b. The graph and the observed y values are shown in Figure 11.25.

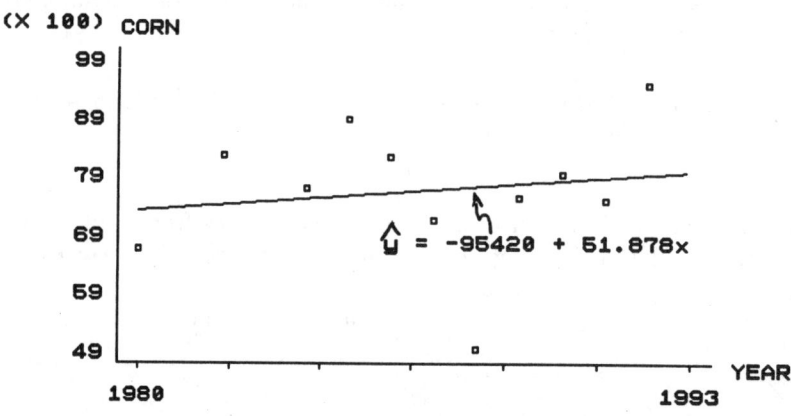

Figure 11.25

c. The hypothesis to be tested is

$$H_0: \beta_1 = 0 \qquad H_a: \beta_1 \neq 0$$

and the test statistic is a Student's t, found on the printout to be

$$t = \frac{\hat{\beta}_1 - \beta_1}{s/\sqrt{S_{xx}}} = \frac{51.878}{104.451} = .50$$

The critical value of t is based on $(n-2) = 9$ degrees of freedom and the rejection region for $\alpha = .05$ is $|t| > t_{.025} = 2.262$. Since the observed value of t does not fall in the rejection region, we do not reject H_0. That is, we cannot conclude that x is useful in the prediction of y.

d. The 90% confidence interval for β_1 is

$$\hat{\beta}_1 \pm t_{.05} \times (\text{standard error of } \hat{\beta}_1) \Rightarrow 51.878 \pm 1.833(104.451)$$

$$51.878 \pm 191.459$$

or $-139.581 < \beta_1 < 243.337$. Intervals constructed in this manner enclose the true value of β_1 90% of the time in repeated sampling. Hence, we are fairly certain that this particular interval encloses β_1.

e. The expected increase in a one-year time period is β_1, which is estimated in part d to fall between -139.581 and 243.337.

f. The predictor for y when $x = 1993$ is $\hat{y} = 7972.85$ (found in the last line of the printout) and the 99% prediction interval is given as

$$3239.34 < y < 12,706.4$$

If all of the least-squares assumptions are met, intervals constructed in this manner should enclose the predicted value of y 99% of the time in repeated sampling. However, in this case, there are two problems. First, we are trying to predict outside the range of the observed values of x. This extrapolation will effect the accuracy of the prediction. Second, since the test for the significance of the linear regression ($H_0: \beta_1 = 0$) in part c was not rejected, we cannot put much faith in the predictions made using the linear regression model.

11.50

This is similar to Exercise 11.49. The EXECUSTAT printout is shown below.

Simple Regression Analysis for 11-49

Linear model: CORN = -95420 + 51.878*YEAR

Table of Estimates

	Estimate	Standard Error	t Value	P Value
Intercept	-95420	207516	-0.46	0.6566
Slope	51.878	104.451	0.50	0.6313

R-squared = 2.67%
Correlation coeff. = 0.163
Standard error of estimation = 1245.47
Durbin-Watson statistic = 1.5442
Mean absolute error = 837.732
Sample size (n) = 11

Table of Predicted Values

Row	YEAR	Predicted CORN	99.00% Prediction Limits Lower	Upper	99.00% Confidence Limits Lower	Upper
1	1993	7972.85	3239.34	12706.4	5518.64	10427.1

a. From the printout, $\hat{\beta}_1 = -27.9744$, $\hat{\beta}_0 = 57885.4$ and the least-squares line is

$$\hat{y} = \hat{\beta}_0 + \hat{\beta}_1 x = 57885.4 - 27.9744x$$

b. The graph and the observed y values are shown in Figure 11.26.

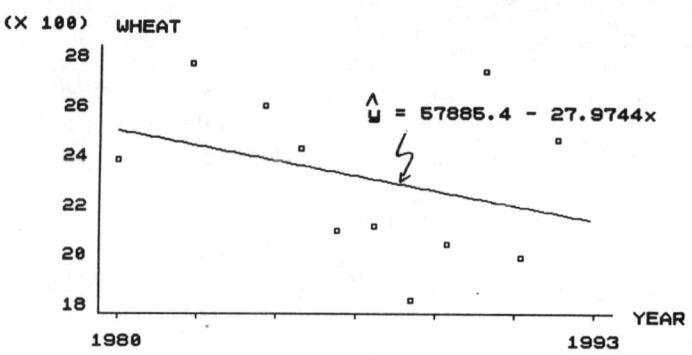

Figure 11.26

c. The hypothesis to be tested is

$$H_0: \beta_1 = 0 \qquad H_a: \beta_1 \neq 0$$

and the test statistic is a Student's t, found on the printout to be

$$t = \frac{\hat{\beta}_1 - \beta_1}{s/\sqrt{S_{xx}}} = \frac{-27.9744}{26.7932} = -1.04$$

The critical value of t is based on $(n - 2) = 9$ degrees of freedom and the rejection region for $\alpha = .05$ is $|t| > t_{.025} = 2.262$. Since the observed value of t does not fall in the rejection region, we do not reject H_0. That is, we cannot conclude that x is useful in the prediction of y.

d. The 90% confidence interval for β_1 is

$$\hat{\beta}_1 \pm t_{.05} \times (\text{standard error of } \hat{\beta}_1) \Rightarrow -27.9744 \pm 1.833(26.7932)$$

$$-27.9744 \pm 49.112$$

or $-77.0864 < \beta_1 < 21.1376$. Intervals constructed in this manner enclose the true value of β_1 90% of the time in repeated sampling. Hence, we are fairly certain that this particular interval encloses β_1.

e. The expected increase in a one-year time period is β_1, which is estimated in part d to fall between -77.0864 and 21.1376.

f. The predictor for y when x = 1993 is $\hat{y} = 2132.54$ (found in the last line of the printout) and the 99% prediction interval is given as

$$918.326 < y < 3346.76$$

11.51

Refer to Exercises 11.49 and 11.50 with x = corn and y = wheat. Using an EXECUSTAT command, the following printout is generated, with r = .5616.

Correlation Analysis for 11-49

	CORN	WHEAT
CORN		0.5616 (11) 0.0722
WHEAT	0.5616 (11) 0.0722	

The table shows estimated product-moment correlation (sample size), and two-tailed P value.

The hypothesis of interest is

$$H_0: \rho = 0 \qquad H_a: \rho > 0$$

and the test statistic is

$$t = \frac{r\sqrt{n-2}}{\sqrt{1-r^2}} = \frac{.5616\sqrt{9}}{\sqrt{1-(.5616)^2}} = 2.036$$

The rejection region with $\alpha = .05$ and $n - 2 = 9$ d.f. is $t > t_{.05} = 1.833$. Hence, there is a significant positive correlation between x and y.

11.52

a-b. From the printout, $\hat{\beta}_1 = 1.31176$, $\hat{\beta}_0 = -4.58706$, and the least-squares line is

$$\hat{y} = \hat{\beta}_0 + \hat{\beta}_1 x = -4.58706 + 1.31176x.$$ The plot is omitted.

c. The hypothesis to be tested is

$$H_0: \beta_1 = 0 \qquad H_a: \beta_1 \neq 0$$

and the test statistic is a Student's t, found on the printout:

$$t = \frac{\hat{\beta}_1 - \beta_1}{s/\sqrt{S_{xx}}} = \frac{1.31176}{0.3140} = 4.18$$

The critical value of t is based on $(n - 2) = 10$ degrees of freedom and the rejection region for $= .05$ is $|t| > t_{.025} = 2.228$. Since the observed value of t falls in the rejection region, we reject H_0. That is, we can conclude that x is useful in the prediction of y.

d. From the printout, $r^2 = .6357$. That is, 63.57% of the variation in y can be accounted for by the independent variable x. The sum of squares of deviations has been reduced by 63.57% by using $\hat{y}$ rather than $\bar{y}$ to predict y.

e. The estimate of E(y) when x = 80 is $\hat{y} = -4.58706 + 1.31176(80) = 100.354$ and the 90% confidence interval is given in the printout:

$$91.494 < E(y) < 109.213$$

f. The predictor for y when x = 90 is $\hat{y} = -4.58706 + 1.31176(90) = 113.471$ and the 90% prediction interval is

$$82.285 < y < 144.658$$

11.53

A MINITAB printout is shown below. Answers will vary from student to student, but students should notice a positive linear relationship (and correlation) between x and y. This correlation is not extremely strong ($r^2 = .55$). The least-squares regression line is

$$\hat{y} = 2.20 + .575x$$

```
MTB > Regress 'y' 1 'x'.

The regression equation is
y = 2.20 + 0.575 x

Predictor      Coef      Stdev     t-ratio        p
Constant      2.200      3.121        0.70    0.501
x            0.5750     0.1839        3.13    0.014

s = 3.290      R-sq = 55.0%       R-sq(adj) = 49.4%

Analysis of Variance

SOURCE       DF        SS         MS         F        p
Regression    1     105.80     105.80      9.77    0.014
Error         8      86.60      10.82
Total         9     192.40
```

11.54

a. The hypothesis of interest is

$$H_0: \beta_1 = 0, \quad H_a: \beta_1 \neq 0,$$

and the test statistic is

$$t = \frac{\hat{\beta}_1 - \beta_1}{s/\sqrt{S_{xx}}} = \frac{15.8}{\sqrt{\frac{41.2}{4.95}}} = 5.48$$

The rejection region for $\alpha = .05$ with $n - 2 = 8$ d.f. is $|t| > t_{.025} = 2.306$ and we reject H_0. There is sufficient evidence to indicate that the independent variable x does help in predicting values of the dependent variable y.

b. When $x = 1$, the estimate of $E(y)$ is

$$\hat{y} = 22.4 + 15.8(1) = 38.2$$

and the 90% confidence interval is

$$\hat{y} \pm t_{.05}\sqrt{s^2\left[\frac{1}{n} + \frac{(x_p - \bar{x})^2}{S_{xx}}\right]}$$

$$38.2 \pm 1.86\sqrt{41.2\left[\frac{1}{10} + \frac{(1 - 1.5)^2}{4.95}\right]}$$

$$38.2 \pm 4.63 \qquad \text{or} \qquad 33.57 < E(y) < 42.83$$

Intervals constructed in this manner enclose $E(y)$ 90% of the time in repeated sampling. Hence, we are fairly certain that this particular interval encloses $E(y)$.

11.55

When x = 2, $\hat{y} = 22.4 + 2(15.8) = 54.0$ and the 90% prediction interval is

$$\hat{y} \pm t_{.05}\sqrt{s^2\left[1 + \frac{1}{n} + \frac{(x_p - \bar{x})^2}{S_{xx}}\right]}$$

$$54.0 \pm 1.86\sqrt{41.2\left[1 + \frac{1}{10} + \frac{(2 - 1.5)^2}{4.95}\right]}$$

$$54.0 \pm 12.806 \qquad \text{or} \qquad 41.194 < y < 66.806$$

11.56

Since $\hat{\beta}_1 = \frac{S_{xy}}{S_{xx}}$, we can calculate S_{xy} as

$$S_{xy} = \hat{\beta}_1 S_{xx} = 15.8(4.95) = 78.21$$

Similarly, SSE = $S_{yy} - \hat{\beta}_1 S_{xy}$ and we can calculate S_{yy} as

$$S_{yy} = \text{SSE} + \hat{\beta}_1 S_{xy} = 329.6 + 15.8(78.21) = 1565.318$$

Then

$$r^2 = \frac{S_{xy}^2}{S_{xx}S_{yy}} = \frac{(78.21)^2}{4.95(1565.318)} = .789$$

The sum of squares of deviations can be reduced by 78.9% using the linear model rather than $\bar{y}$ to predict y. The fit is reasonably good.

11.57

a. From the printout, the least-squares line is $\hat{y} = .141786 + .0840476x$.

b. The graph is omitted.

c. Since $r^2 = .815$, 81.5% of the total variation can be accounted for by x.

d. To test $H_0: \beta_1 = 0$, $H_a: \beta_1 \neq 0$, the test statistic is t = 5.13 with p-value = .0021. Hence, x is useful in predicting y.

e. A 95% confidence interval for β_1 is

$$.084 \pm 2.447(.016368)$$

$$.084 \pm .040 \qquad \text{or} \qquad .044 < \beta_1 < .124$$

11.58

A MINITAB printout is shown below. Answers will vary from student to student, but students should notice a very strong positive linear relationship (and correlation) between x and y. Either "Buyer's Bid" or "Seller's Asking Price" will be very useful in predicting "Closing Bid".

```
MTB > regress c3 1 c2

The regression equation is
C3 = 181 + 0.986 C2                                          MTB > Correlation C1-C3.

Predictor       Coef       Stdev      t-ratio      p                    C1        C2
Constant      181.06       28.93       6.26      0.000       C2       0.985
C2           0.98609      0.03981     24.77      0.000       C3       0.991     0.994

s = 48.76         R-sq = 98.7%      R-sq(adj) = 98.6%

Analysis of Variance

SOURCE          DF          SS          MS          F          p
Regression       1       1458791      1458791     613.52     0.000
Error            8         19022        2378
Total            9       1477813
```

MTB > regress c3 1 c1

The regression equation is
C3 = - 52.6 + 0.875 C1

```
Predictor       Coef       Stdev      t-ratio      p
Constant      -52.63       44.73      -1.18      0.273
C1           0.87513      0.04243     20.62      0.000

s = 58.40         R-sq = 98.2%      R-sq(adj) = 97.9%

Analysis of Variance

SOURCE          DF          SS          MS          F          p
Regression       1       1450531      1450531     425.36     0.000
Error            8         27281        3410
Total            9       1477813
```

11.59

A MINITAB printout is shown below. Answers will vary from student to student, but students should notice a mildly negative linear relationship (and correlation) between x and y. There is insufficient evidence (t = −1.36 with p-value = .231) to indicate that x will be useful in predicting y.

MTB > Regress C1 1 C2.

The regression equation is
C1 = 1076 - 0.616 C2

```
Predictor       Coef       Stdev      t-ratio      p
Constant      1076.3      416.9       2.58      0.049
C2           -0.6159     0.4521      -1.36      0.231

s = 479.6         R-sq = 27.1%      R-sq(adj) = 12.5%

Analysis of Variance

SOURCE          DF          SS          MS          F          p
Regression       1        426841      426841      1.86      0.231
Error            5       1150119      230024
Total            6       1576960
```

MTB > corr c1 c2

Correlation of C1 and C2 = -0.520

CHAPTER 12
Multiple Regression Analysis

12.1

a. When $x_2 = 2$, $E(y) = 3 + x_1 - 2(2) = x_1 - 1$.
When $x_2 = 1$, $E(y) = 3 + x_1 - 2(1) = x_1 + 1$.
When $x_2 = 0$, $E(y) = 3 + x_1 - 2(0) = x_1 + 3$.

These three straight lines are graphed in Figure 12.1.

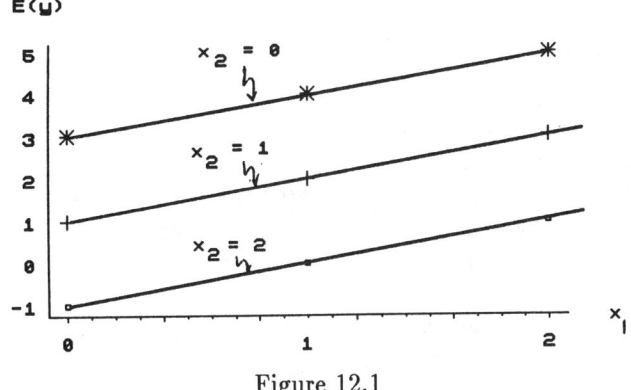

Figure 12.1

b. Notice that the lines are parallel (they have the same slope).

12.2

This is similar to Exercise 12.1.

a. When $x_1 = 0$, $E(y) = 3 + 0 - 2x_2 = 3 - 2x_2$.
When $x_1 = 1$, $E(y) = 3 + 1 - 2x_2 = 4 - 2x_2$.
When $x_1 = 2$, $E(y) = 3 + 2 - 2x_2 = 5 - 2x_2$.

These three straight lines are graphed in Figure 12.2.

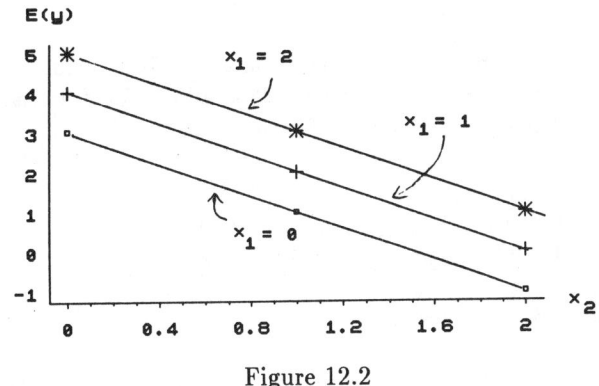

Figure 12.2

b. The lines are parallel (they have the same slope).

c. The first-order model, $E(y) = \beta_0 + \beta_1 x_1 + \beta_2 x_2$, will be a line in three-dimensional space. This line will always have the same slope, but will move along the x_1 or x_2 axis, depending on the values of x_1 and x_2 (Figure 12.3).

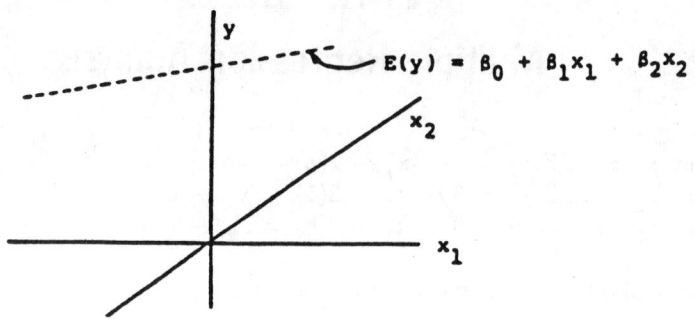

Figure 12.3

12.3

This is similar to Exercise 12.1.

a. When $x_2 = 0$, $E(y) = 3 + x_1 - 2(0) + x_1(0) = x_1 + 3$.
When $x_2 = 2$, $E(y) = 3 + x_1 - 2(2) + x_1(2) = 3x_1 - 1$.
When $x_2 = -2$, $E(y) = 3 + x_1 - 2(-2) + x_1(-2) = 7 - x_1$.

These three straight lines are shown in Figure 12.4.

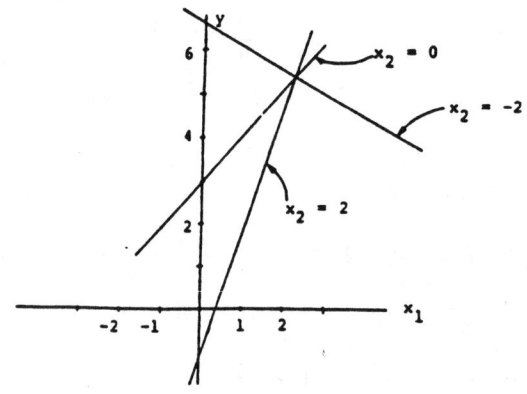

Figure 12.4

b-c. Notice that the lines are no longer parallel. The inclusion of the term $x_1 x_2$ allows the slope of the line to change as x_2 changes. This allows for "interaction," which is a difference in the relationship between x_1 and y for different values of x_2.

12.4

a. If $R^2 = .94$, the total sum of squares of deviations of the y values about their mean has been reduced by 94% by using the linear model to predict y.

b. The hypothesis to be tested is

$$H_0: \beta_1 = \beta_2 = \beta_3 = 0 \qquad H_a: \text{at least one } \beta_i \text{ differs from } 0$$

and the test statistic (given in Example 12.1, "assessing the utility of the model") is

$$F = \frac{R^2/k}{(1 - R^2)/[n - (k + 1)]} = \frac{.94/3}{(1 - .94)/(15 - 4)} = 57.44$$

The rejection region, which is found in the upper tail of the F distribution with $\nu_1 = 3$ and $\nu_2 = 11$ degrees of freedom, is $F > 3.59$ and H_0 is rejected. There is evidence that the model contributes information for the prediction of y.

12.5

The hypothesis of interest is

$$H_0: \beta_i = 0 \qquad H_a: \beta_i \neq 0 \qquad \text{for } i = 1, 2, 3$$

and the test statistic is $t = \dfrac{\hat{\beta}_i - \beta_i}{s_{\hat{\beta}_i}}$.

For $i = 1$, $t = \dfrac{1.29}{.42} = 3.071$. For $i = 2$, $t = \dfrac{2.72}{.65} = 4.185$.

For $i = 3$, $t = \dfrac{.41}{.17} = 2.412$.

The rejection region, with $n - (k + 1) = 11$ degrees of freedom, is $|t| > t_{.025} = 2.201$ and all three hypotheses are rejected. All of the three independent variables contribute to the prediction of y, in the presence of the other two variables.

12.6

a. From the information given in the exercise, the prediction equation is

$$\hat{y} = 1.04 + 1.29x_1 + 2.72x_2 + .41x_3$$

b. When $x_2 = 1$ and $x_3 = 0$, $\hat{y} = 3.76 + 1.29x_1$. When $x_2 = 1$ and $x_3 = .5$, $\hat{y} = (1.04 + 2.72 + .205) + 1.29x_1$ or $\hat{y} = 3.965 + 1.29x_1$. The two lines are graphed together in Figure 12.5. Notice that the two lines are parallel.

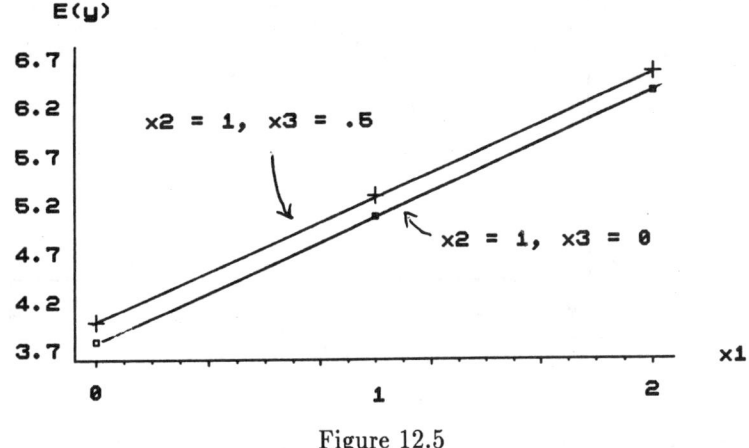

Figure 12.5

c. β_1 represents the change in y for a one unit change in x_1 <u>when x_2 and x_3 are held constant</u>.

12.7

The 90% confidence interval for β_1 is given as

$$\hat{\beta}_1 \pm t_{.05} s_{\hat{\beta}_1} = 1.29 \pm 1.796(.42) = 1.29 \pm .754$$

or $.536 < \beta_1 < 2.044$. Intervals constructed using this procedure will enclose β_1 90% of the time. Hence, we are fairly certain that this particular interval encloses β_1.

12.8
a. The model is quadratic.

b. Since $R^2 = .762$, the sum of squares of deviations is reduced by 76.2% using the quadratic model rather than $\bar{y}$ to predict y.

c. The hypothesis to be tested is

$$H_0: \beta_1 = \beta_2 = 0 \qquad H_a: \text{at least one } \beta_i \text{ differs from 0}$$

and the test statistic is

$$F = \frac{R^2/k}{(1-R^2)/[n-(k+1)]} = \frac{.762/2}{(1-.762)/(20-3)} = 27.214$$

The rejection region, which is found in the upper tail of the F distribution with $\nu_1 = 2$ and $\nu_2 = 17$ degrees of freedom, is $F > 4.45$ and H_0 is rejected. There is evidence that the model contributes information for the prediction of y.

12.9
a. The prediction equation is

$$\hat{y} = 1.21 + 7.60x - .94x^2$$

b. The graph is shown in Figure 12.6.

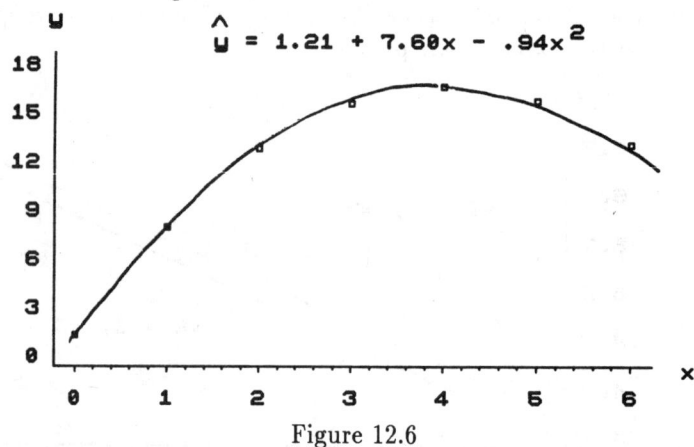

Figure 12.6

12.10
a. Refer to Exercise 12.9. When $x = 0$, the estimate of E(y) is $\hat{y} = 1.21 + 7.60(0) - .94(0^2) = 1.21$.

b. Since $E(y) = \beta_0 + \beta_1 x + \beta_2 x^2$, when $x = 0$, $E(y) = \beta_0$. A test of E(y given $x = 0$) = 0 is equivalent to a test of

$$H_0: \beta_0 = 0, \qquad H_a: \beta_0 \neq 0$$

The individual t test is

$$t = \frac{\hat{\beta}_0}{s_{\hat{\beta}_0}} = \frac{1.21}{.62} = 1.952$$

and the rejection region is $|t| > t_{.05} = 1.74$ with $n - (k + 1) = 17$ d.f. Hence, H_0 is rejected. The mean value of y differs from 0 when $x = 0$.

c. A 90% confidence interval for β_0 is

$$\hat{\beta}_0 \pm t_{.05} s_{\hat{\beta}_0} = 1.21 \pm 1.74(.62) = 1.21 \pm 1.079$$

or $.131 < \beta_0 < 2.289$. Intervals constructed using this procedure will enclose β_0 90% of the time. Hence, we are fairly certain that this particular interval encloses β_0.

12.11

a. If E(y) and x can be modeled by a straight line, then $\beta_2 = 0$.

b. The hypothesis of interest is

$$H_0: \beta_2 = 0, \quad H_a: \beta_2 \neq 0$$

The individual t test is

$$t = \frac{\hat{\beta}_2}{s_{\hat{\beta}_2}} = \frac{-.94}{.33} = -2.848$$

and the rejection region for $\alpha = .05$ is $|t| > t_{.025} = 2.11$ with $n - (k + 1) = 17$ d.f. Hence, H_0 is rejected. There is evidence to indicate curvature.

12.12

a. Rate of increase is measured by the slope of a line tangent to the curve; this line is given by an equation obtained as dy/dx, the derivative of y with respect to x. In particular,

$$\frac{dy}{dx} = \frac{d}{dx}(\beta_0 + \beta_1 x + \beta_2 x^2) = \beta_1 + 2\beta_2 x$$

which has slope $2\beta_2$. If β_2 is negative, then the rate of increase is decreasing. Hence, the hypothesis of interest is

$$H_0: \beta_2 = 0, \quad H_a: \beta_2 < 0$$

b. The individual t test is

$$t = \frac{\hat{\beta}_2}{s_{\hat{\beta}_2}} = \frac{-.94}{.33} = -2.848$$

and the rejection region for $\alpha = .05$ is $t < -t_{.05} = -1.74$ with $n - (k + 1) = 17$ d.f. Hence, H_0 is rejected. There is evidence to indicate a decreasing rate of increase.

12.13

a. The publisher expects y to increase as x increases to a point and then to decrease as x increases further. Hence, the parabola should be cupped downward. The student can discover by inspecting several parabolas (see Example 12.2) that those that are cupped downward have **negative** coefficients for x^2. Hence, we would expect β_2 to be **negative**. From the printout, $\hat{\beta}_2 = -0.8198$.

b. Refer to the printout given in the exercise. SSE is found in the column labeled "SUM OF SQUARES" and the row labeled "ERROR" to be

$$SSE = 1.05985748 \quad \text{and} \quad s^2 = \frac{SSE}{3} = 0.35328583$$

c. The degrees of freedom for SSE and s^2 are 3.

d-e. The hypothesis of interest is

$$H_0: \beta_1 = \beta_2 = 0 \qquad H_a: \text{at least one } \beta_i \text{ differs from } 0$$

and the test statistic is

$$F = 332.53$$

found in the column labeled "F VALUE" and row labeled "MODEL" in the printout. Since the observed level of significance for F is .0003, H_0 can be rejected for any value of α greater than .0003. For $\alpha = .05$, H_0 is rejected and we conclude that the model contributes information to the prediction of y.

f-g. The hypothesis of interest is

$$H_0: \beta_2 = 0 \qquad H_a: \beta_2 \neq 0$$

and the test statistic is $\quad t = \dfrac{\hat{\beta}_2 - \beta_2}{s_{\hat{\beta}_2}} = -4.49$

found in the column labeled "T FOR H_0: PARAMETER = 0" and the row labeled "X*X". Since the observed significance level is p-value = .0206, H_0 can be rejected for any $\alpha \geq .0206$. Hence, for $\alpha = .05$, H_0 is rejected. We conclude that there is curvature in the relationship.

h. The prediction equation is $\hat{y} = -44.192 + 16.334x - 0.820x^2$ and is graphed in Figure 12.7.

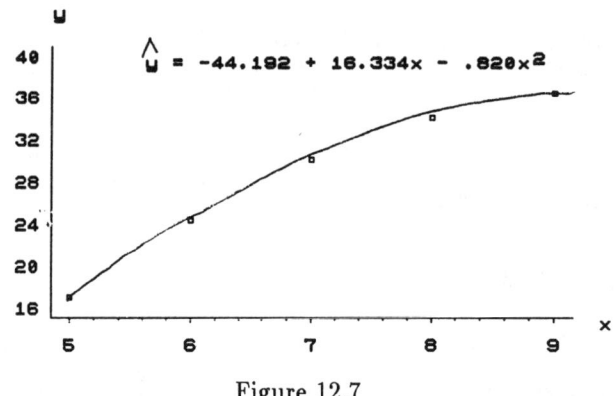

Figure 12.7

i. R^2 is found in the column labeled "R SQUARE" and the row labeled "MODEL" to be $R^2 = .9955$. Hence, 99.55% of the total variation is accounted for by using x and x^2 in the model.

j. Substituting $x = 6.5$ into the prediction equation, the estimate of y is

$$\hat{y} = -44.19249551 + 16.33386317(6.5) - 0.81976920(6.5)^2 = 27.3424$$

This value can be found at the bottom of the printout.

12.14

Refer to Exercise 12.13.

a. From the printout, SSE = 1.05985748, SSR = 234.95514252 and S_{yy} = 236.015. Then by inspection,

$$S_{yy} = SSR + SSE.$$

b. $R^2 = 1 - \dfrac{SSE}{S_{yy}} = 1 - \dfrac{1.05985748}{236.015} = .9955$

which agrees with the printout.

c. Calculate $R^2(\text{adj}) = \left(1 - \dfrac{MSE}{s_y^2}\right)100\% = \left(1 - \dfrac{.35328583}{236.015/5}\right)100\% = 99.25\%$

The value of $R^2(\text{adj})$ can be used to compare two or more regression models using different numbers of independent predictor variables.

12.15

a. Refer to the printout to find SSE = 2,238,508 and s^2 = MSE = 559,627.

b. The prediction equation is $\hat{y} = -383 + 2.418x_1 + 69.1x_2$.

c. R^2 is found in the section labeled "R-sq" to be $R^2 = .597$. Hence, 59.7% of the total variation is accounted for by using x and x^2 in the model.

d. $R^2(\text{adj}) = \left(1 - \dfrac{MSE}{s_y^2}\right)100\% = \left(1 - \dfrac{559,627}{5,550,659/6}\right)100\% = 39.5\%$

Since the adjusted value of R^2 is smaller for this model than for the linear model with one variable in Exercise 11.12, we would conclude that x_2 is not contributing much information to the model.

e. The hypothesis of interest is

$H_0: \beta_1 = \beta_2 = 0$ H_a: at least one β_i differs from 0

and the test statistic is

$F = 2.96$

found in the column labeled "F" and row labeled "REGRESSION" in the printout. Since the observed level of significance for F is .163, H_0 can be rejected for any value of α greater than .163. For $\alpha = .10$, H_0 is not rejected and we cannot conclude that the model contributes information for the prediction of y.

f. With such a large observed level of significance, $H_0: \beta_2 = 0$ cannot be rejected. Therefore, we cannot conclude that x_2 contributes information for the prediction of y.

g. The fit of the model is not good.

12.16

a. $E(y) = \beta_0 + \beta_1 x_1 + \beta_2 x_2$

where x_1 = average seller's asking price and x_2 = average buyer's bid.

The assumptions required are given in Section 12.2 of the text.

b. The prediction equation is given in the printout as

$$\hat{y} = 81.6 + .356 \, x_1 + .591 \, x_2$$

c. From the printout, $R^2 = .992$, so that 99.2% of the variation in y can be explained by x_1 and x_2.

d. The hypothesis of interest is

$$H_0: \beta_1 = \beta_2 = 0 \qquad H_a: \text{at least one } \beta_i \text{ differs from } 0$$

and the test statistic is $F = 426.37$. The rejection region with 2 and $10 - 3 = 7$ d.f. is 4.74 and H_0 is rejected. The model contributes information for the prediction of y.

e. The p-value is given in the printout as p-value = .000.

f. The individual linear regressions used in Exercise 11.58 were almost as effective ($R^2 = .98$) as the multiple linear regression analysis used in this exercise. The high correlation between x_1 and x_2 ($r = .9854$) means that once one of these two variables is included in the model, the other will provide little additional information.

12.17

a. $E(y) = \beta_0 + \beta_1 x_1 + \beta_2 x_2$

where x_1 = median home price and x_2 = median family income.

The assumptions required are given in Section 12.2 of the text.

b. The prediction equation is given in the printout as

$$\hat{y} = 83.9 - .471 \, x_1 + 1.03 \, x_2$$

c. From the printout, $R^2 = .814$, so that 81.4% of the variation in y can be explained by x_1 and x_2.

d. The hypothesis of interest is

$$H_0: \beta_1 = \beta_2 = 0 \qquad H_a: \text{at least one } \beta_i \text{ differs from } 0$$

and the test statistic is $F = 15.30$. The rejection region with 2 and $10 - 3 = 7$ d.f. is 4.74 and H_0 is rejected. The model contributes information for the prediction of y.

e. The p-value is given in the printout as p-value = .003.

f. The individual linear regressions used in Exercise 11.22 were not as effective ($r^2 = .066$ and $r^2 = .538$) as the multiple linear regression analysis used in this exercise. The two variables taken together provide a much better method for prediction, with $R^2 = .814$.

12.18

a. Quantitative **b.** Quantitative

c. Qualitative **d.** Quantitative **e.** Qualitative

12.19

a. The variable x_2 must be the quantitative variable, since it appears as a quadratic term in the model. Qualitative variables appear only with exponent 1, although they may appear as the coefficient of another quantitative variable with exponent 2 or greater.

b. When $x_1 = 0$, $\hat{y} = 12.6 + 3.9x_2^2$, while when $x_1 = 1$,

$$\hat{y} = 12.6 + .54(1) - 1.2x_2 + 3.9x_2^2$$

$$= 13.14 - 1.2x_2 + 3.9x_2^2$$

c. The graph in Figure 12.8 shows the two parabolas.

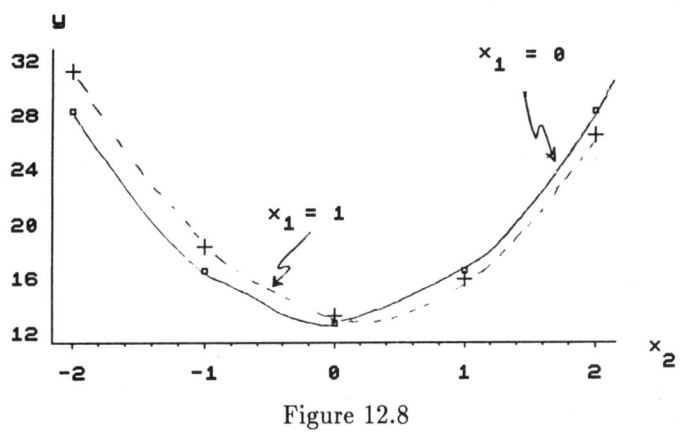

Figure 12.8

12.20

a. The prediction equation is $\hat{y} = -5125 + 1763.9x_1 + 9533.3x_2$.

b. The hypothesis of interest is

$$H_0: \beta_1 = \beta_2 = 0 \qquad H_a: \text{at least one } \beta_i \text{ differs from } 0$$

and the test statistic is $F = 170.74$. The rejection region with 2 and $12 - 3 = 9$ d.f. is 3.86 (for $\alpha = .05$) and H_0 is rejected. The model contributes information for the prediction of y.

c. Both independent variables are significant, with p-values = .000.

d. When $x_1 = 18$ and $x_2 = 0$, $\hat{y} = -5125 + 1763.9(18) = 26{,}625.2$, which differs slightly from the prediction on the printout, due to rounding error in the coefficients of the prediction equation.

e. When $x_1 = 18$ and $x_2 = 1$, $\hat{y} = -5125 + 1763.9(18) + 9533.3(1) = 36{,}158.5$, which differs slightly from the prediction on the printout, due to rounding error in the coefficients of the prediction equation.

12.21

a. The predictor variable x_1 = type of meat (beef or chicken) is a qualitative variable, while the predictor variable x_2 = year $-$ 1969 is a quantitative variable. An interaction term, x_1x_2, is included to allow for a different relationship between consumption and year, depending on which type of meat is being measured.

b. From the printout, $R^2 = .954$, so that 95.4% of the total variation in the experiment is explained by the model. This is very high, indicating that the model is quite effective in predicting y.

c. The hypothesis of interest is

$$H_0: \beta_1 = \beta_2 = \beta_3 = 0 \qquad H_a: \text{at least one } \beta_i \text{ differs from } 0$$

and the test statistic is

$$F = 69.83$$

with p-value $= .000$. Since the p-value is so small, H_0 is rejected. The model contributes significant information for the prediction of y.

d. From the last line of the printout, the 95% confidence interval for $E(y)$ is

$$48.44 < E(y) < 64.13$$

while the 95% prediction interval for y is

$$44.13 < y < 68.45$$

12.22

a. From the information given in the exercise, $R^2 = .68$, with 44 d.f. To determine the significance of the regression, the hypothesis of interest is

$$H_0: \beta_1 = \beta_2 = \beta_3 = \beta_4 = 0 \qquad H_a: \text{at least one } \beta_i \text{ differs from } 0$$

and the test statistic is

$$F = \frac{R^2/k}{(1-R^2)/[n-(k+1)]} = \frac{.68/4}{(1-.68)/44} = 23.375$$

The p-value with 4 and 44 d.f. is less than .005, and H_0 is rejected. The regression is significant.

b-c. The asterisks in the printout indicate that K/L, AGR, and SEV are significant predictor variables, while PY is not. It is possible that the last variable, PY, could be dropped from the model.

12.23

a. When $x_2 = 10$, the model becomes

$$E(y) = \beta_0 + \beta_1 x_1 + \beta_2 x_1^2 + 10\beta_3 + 10\beta_4 x_1 + 10\beta_5 x_1^2$$

$$= (\beta_0 + 10\beta_3) + (\beta_1 + 10\beta_4)x_1 + (\beta_2 + 10\beta_5)x_1^2$$

which is the equation for a parabola (since the coefficients of x_1 and x_1^2 are simply constants).

b. If consumption ($E(y)$) drops as the temperature increases to a point and then increases as temperature gets very high (and air conditioning is used), the parabola should be cupped upward. Hence, the coefficient of x_1^2 will be positive.

c. When $x_1 = 50$,

$$E(y) = (\beta_0 + 50\beta_1 + 2500\beta_2) + (\beta_3 + 50\beta_4 + 2500\beta_5)x_2$$

which is the equation of a straight line.

d. As price increases, consumption should decrease, and the slope of the line in part c, represented by the coefficient of x_2, should be negative.

e. The interaction terms, $\beta_4 x_1 x_2$ and $\beta_5 x_1^2 x_2$, allow for different parabolas, depending on the value of x_2.

12.24

a-b. Refer to the printout given in this exercise. SSE = 152.17748 and s^2 = 8.4543 with 18 degrees of freedom from the row labeled "ERROR." Note that

$$s^2 = \frac{SSE}{18} = \frac{152.17748}{18} = 8.4543$$

c. The hypothesis of interest is

$$H_0: \beta_1 = \beta_2 = \beta_3 = \beta_4 = \beta_5 = 0 \qquad H_a: \text{at least one } \beta_i \text{ differs from 0}$$

and the test statistic is

$$F = 31.85$$

with 5 and 18 degrees of freedom. The rejection region is one-tailed and the critical value for α = .10 is F > 2.20. Hence, H_0 is rejected and we conclude that the model contributes information for the prediction of y.

d. The critical values of F with 5 and 18 d.f. for a = .10, .05, .025, .01 and .005 are found in Table 6 and are shown in the following table.

a	F_a
.10	2.20
.05	2.77
.025	3.38
.01	4.25
.005	4.96

Since F = 31.85 exceeds $F_{.005}$ = 4.96, the observed level of significance is

p-value < .005.

Notice that the printout reports p-value = .0001.

e. Refer to the printout in the row labeled "CORRECTED TOTAL." This gives total SS as S_{yy} = 1498.625. Then

$$R^2 = 1 - \frac{SSE}{S_{yy}} = 1 - \frac{152.17748}{1498.625} = .898$$

which is given in the column marked "R-SQUARE." That is, 89.8% of the total variation is accounted for by using the model rather than $\bar{y}$ to predict y.

f. Substituting $x_1 = 60$ and $x_2 = 8$ into the prediction equation gives the estimate $\hat{y}$ = 41.262.

12.25

a. When $x_2 = 8$,
$$\hat{y} = 325.606445 - 11.38256x_1 + .113497x_1^2 - 21.69921(8) \\ + .873029(8)x_1 - .008869(8)x_1^2$$

$$= 152.01277 - 4.39832x_1 + .042541x_1^2$$

When $x_2 = 10$,
$$\hat{y} = 325.606445 - 11.38256x_1 + .113497x_1^2 - 21.69921(10) \\ + .873029(10)x_1 - .008869(10)x_1^2$$

$$= 108.61435 - 2.652264x_1 + .0248024x_1^2$$

The graphs are parabolas and are shown in Figure 12.9. Note the differences in the curves.

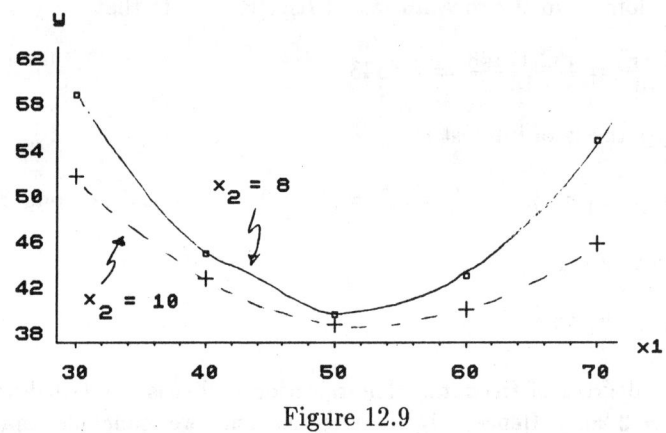

Figure 12.9

b. The hypothesis of interest is

$$H_0: \beta_3 = \beta_4 = \beta_5 = 0 \qquad H_a: \text{at least one } \beta_i \text{ differs from 0 for } i = 3, 4, 5$$

i-ii. From the printout in Exercise 12.24, $SSE_2 = 152.1774845$ with 18 degrees of freedom, while the printout in this exercise gives $SSE_1 = 465.13432996$ with 21 degrees of freedom. Hence, the degrees of freedom associated with $SSE_1 - SSE_2 = 313.1709$ is $21 - 18 = 3$.

iii. The test statistic is

$$F = \frac{(SSE_1 - SSE_2)/3}{SSE_2/18} = \frac{313.1709/3}{8.4543047} = 12.35$$

iv-v. The rejection region is $F > F_{.05} = 3.16$ (with 3 and 18 d.f.) and H_0 is rejected. There is evidence that x_2 is important in predicting usage.

12.26

From Table 12.6, $\hat{\beta}_1 = 969.0$, $s_{\hat{\beta}_1} = 63.67050315$. Then

$$t = \frac{969.0}{63.6705} = 15.218978 \quad \text{or } t = 15.22$$

as given in the column labeled "T FOR H_0: PARAMETER = 0"

12.27

a. In Table 4 of Appendix II, we find $t_{.025} = 2.306$ with 8 d.f.

b. Then $F_{1,8} = 5.32$, and $t^2 = (2.306)^2 = 5.32$.

12.28
$t^2 = F$ only when $\nu_1 = 1$.

12.29
a. When the observation y occurs in the men's department, then $x_2 = 0$ and $x_3 = 0$. Hence, the model becomes

$$E(y) = \beta_0 + \beta_1 x_1 + \beta_2(0) + \beta_3(0) + \beta_4 x_1(0) + \beta_5 x_1(0) = \beta_0 + \beta_1 x_1$$

b. When the observation y occurs in the children's department, then $x_2 = 1$ and $x_3 = 0$. Hence, the model becomes

$$E(y) = \beta_0 + \beta_1 x_1 + \beta_2(1) + \beta_3(0) + \beta_4 x_1(1) + \beta_5 x_1(0)$$
$$= (\beta_0 + \beta_2) + (\beta_1 + \beta_4) x_1$$

c. When the observation y occurs in the women's department, then $x_2 = 0$ and $x_3 = 1$. Hence, the model becomes

$$E(y) = \beta_0 + \beta_1 x_1 + \beta_3 + \beta_5 x_1 = (\beta_0 + \beta_3) + (\beta_1 + \beta_5) x_1$$

d. Referring to part b and part a, the difference in intercepts is $(\beta_0 + \beta_2) - \beta_0 = \beta_2$.

e. Referring to part c and part a, the difference in slopes is $(\beta_1 + \beta_5) - \beta_1 = \beta_5$.

f. The slopes of the lines corresponding to men's, children's, and women's departments are β_1, $\beta_1 + \beta_4$, and $\beta_1 + \beta_5$, respectively. Hence, if the slopes are all equal, we must have $\beta_4 = 0$ and $\beta_5 = 0$. The hypothesis to be tested is

$$H_0: \beta_4 = \beta_5 = 0 \qquad H_a: \text{at least one } \beta_i \text{ differs from 0 for } i = 4, 5$$

12.30
A computer program was run for this data, and the results are shown in the following printout.

```
RSQ = .98397093
SEY = .36802778
                COEFF       S.E.        F
INTERCEPT      4.10000     0.38599    112.83 ***
VAR 1          1.04000     0.11638     79.86 ***
VAR 2          3.53000     0.54587     41.82 ***
VAR 3          4.76000     0.54587     76.04 ***
VAR 4         -0.43000     0.16459      6.83 *
VAR 5         -0.08000     0.16459      0.24 NS

ANALYSIS OF VARIANCE

SOURCE
REGRESSION     5    74.83033    14.96607    110.50 ***
RESIDUAL       9     1.21900     0.13544
TOTAL         14    76.04933
```

a-b. Note that SSE = 1.2190, $s^2 = 0.13544$ with 9 degrees of freedom (from the row marked "RESIDUAL").

c-d. From the row marked "TOTAL," total SS is $S_{yy} = 76.04933$ and

$$R^2 = 1 - \frac{SSE}{S_{yy}} = 1 - \frac{1.2190}{76.04933} = .98397.$$

e-f. The parameter estimates are found in the column marked "COEFF" and the prediction equation is

$$\hat{y} = 4.10 + 1.04x_1 + 3.53x_2 + 4.76x_3 - 0.43x_1x_2 - 0.08x_1x_3$$

From parts a, b, and c, the coefficients can be combined to give the three lines that are graphed in Figure 12.10.

Men: $\hat{y} = 4.10 + 1.04x_1$
Children: $\hat{y} = 7.63 + 0.61x_1$
Women: $\hat{y} = 8.86 + 0.96x_1$

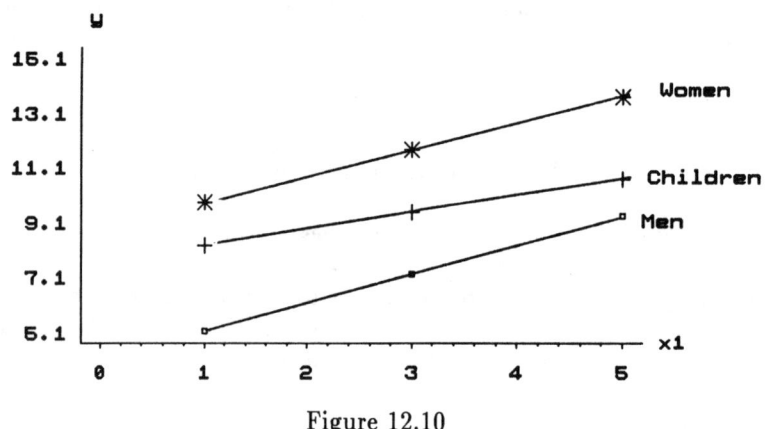

Figure 12.10

g. The hypothesis of interest is

$$H_0: \beta_4 = 0 \qquad H_a: \beta_4 \neq 0$$

and the test statistic is $t = \dfrac{\hat{\beta}_4 - \beta_4}{s_{\hat{\beta}_4}} = \dfrac{-0.43}{.16459} = -2.613.$

The rejection region with $\alpha = .05$ and 9 degrees of freedom is $|t| > t_{.025} = 2.262$ and H_0 is rejected. There is a difference in the slopes. Note that if the student chooses to use the approximation $s_{\hat{\beta}_4} = .165$, he will obtain $t = -2.606$.

h. From part g, the 95% confidence interval is

$$\hat{\beta}_4 \pm t_{.025}\, s_{\hat{\beta}_4} = -0.43 \pm 2.262(.16459) = -0.43 \pm .372$$

or $-0.802 < \beta_4 < -0.058$.

i. The hypothesis of interest is

$$H_0: \beta_5 = 0 \qquad H_a: \beta_5 \neq 0$$

and the test statistic is $\quad t = \dfrac{\hat{\beta}_5 - \beta_5}{s_{\hat{\beta}_5}} = \dfrac{-0.08}{.16459} = -.486$.

The rejection region with $\alpha = .05$ and 9 degrees of freedom is $|t| > t_{.025} = 2.262$ and H_0 is not rejected. There is no difference in the slopes.

12.31

If both variables are quantitative, the response surface is a straight line in three dimensions. See Figure 12.11. The line is $E(y) = \beta_0 + \beta_1 x_1 + \beta_2 x_2$.

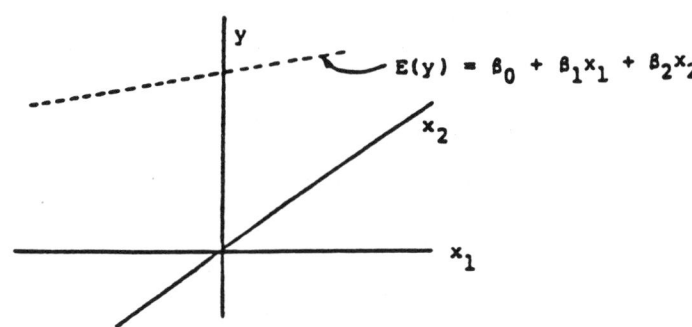

Figure 12.11

12.32

$E(y)$ follows a pattern traced by a parabola. Hence,

$$E(y) = \beta_0 + \beta_1 x + \beta_2 x^2$$

12.33

Since the model is first order, no interaction terms are used. The three variables must be independent.

$$E(y) = \beta_0 + \beta_1 x_1 + \beta_2 x_2 + \beta_3 x_3$$

12.34

For three levels of a single qualitative variable, two dummy variables are used.

$$E(y) = \beta_0 + \beta_1 x_1 + \beta_2 x_2$$

where
$\quad x_1 = 1$ if level 2
$\qquad\quad\;\; 0$ otherwise

$\quad x_2 = 1$ if level 3
$\qquad\quad\;\; 0$ otherwise

12.35

A complete second order model includes quadratic and interaction terms.

$$E(y) = \beta_0 + \beta_1 x_1 + \beta_2 x_2 + \beta_3 x_1^2 + \beta_4 x_2^2 + \beta_5 x_1 x_2$$

12.36

This is similar to Exercise 12.35 with three variables.

$E(y) = \beta_0 + \beta_1 x_1 + \beta_2 x_2 + \beta_3 x_3 + \beta_4 x_1^2 + \beta_5 x_2^2 + \beta_6 x_3^2 + \beta_7 x_1 x_2 + \beta_8 x_1 x_3 + \beta_9 x_2 x_3$

12.37

The first variable is represented by two dummy variables; the second variable is represented by one dummy variable. There is no interaction.

$$E(y) = \beta_0 + \beta_1 x_1 + \beta_2 x_2 + \beta_3 x_3$$

where
$x_1 = 1$ if level 2 of variable 1
 0 otherwise

$x_3 = 1$ if level 2 of variable 2
 0 otherwise

$x_2 = 1$ if level 3 of variable 1
 0 otherwise

12.38

Refer to Exercise 12.37. With x_1, x_2 and x_3 as defined there, the model is

$$E(y) = \beta_0 + \beta_1 x_1 + \beta_2 x_2 + \beta_3 x_3 + \beta_4 x_1 x_3 + \beta_5 x_2 x_3$$

12.39

a-c. The least-squares prediction equation, given in the printout as

$$\hat{y} = .9555 - .25272x + .0254924x^2$$

is plotted along with the ten data points in Figure 12.12.

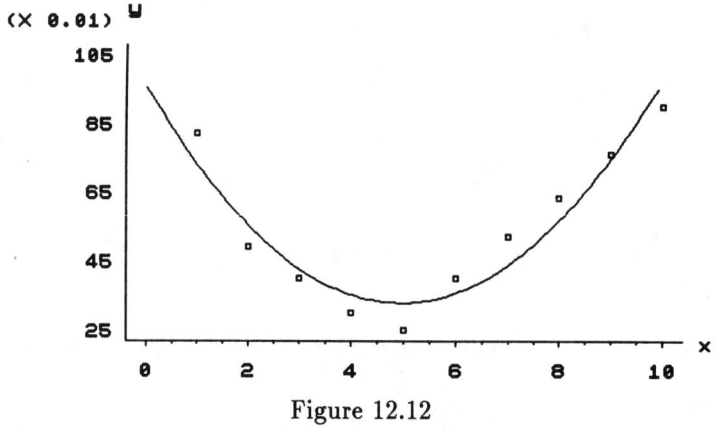

Figure 12.12

d-e. The test statistic is $t = 7.62$ with p-value $= .0001$. Hence, x^2 is useful for the prediction of y.

f. $R^2 = .9076$. Hence, 91% of the variation is explained by the model.

g. The number of data points increases the degrees of freedom, hence inflating the value of

$$F = \frac{R^2/k}{(1 - R^2)/[n - (k + 1)]}$$

and causing rejection of $H_0: \beta_i = 0$ for all i.

CHAPTER 13
Time Series and Index Numbers

13.1

The simple index for each month, using January 1991 as the base month, is given as

$$I_t = \frac{y_t}{y_0}(100) = \frac{\text{production in month t}}{\text{production in January, 1991}}(100)$$

For example, for February 1991,

$$I_1 = \frac{y_1}{y_0}(100) = \frac{6.7}{7.6}(100) = 88.158$$

The indices are shown below.

Month	t	y_t	I_t
January	0	7.6	100.000
February	1	6.7	88.158
March	2	7.3	96.053
April	3	7.1	93.421
May	4	7.1	93.421
June	5	7.0	92.105
July	6	7.3	96.053
August	7	7.4	97.368
September	8	7.5	98.684
October	9	7.7	101.316
November	10	7.5	98.684
December	11	7.3	96.053

13.2

This is similar to Exercise 13.1. Using y_t = number of failures in year t and $y_0 = 11432$, the indices are given below.

Year	t	y_t	I_t
1980	80	11742	102.71
1981	81	16794	146.90
1982	82	24908	217.88
1983	83	31534	275.84
1984	84	52078	455.55
1985	85	57252	500.80
1986	86	61601	538.85
1987	87	61384	536.95
1988	88	57099	499.47
1989	89	50361	440.53
1990	90	60432	528.62

For example, $I_{83} = \frac{y_{83}}{y_0}(100) = \frac{31534}{11432}(100) = 275.84$.

13.3

This is similar to Exercise 13.2.

a. Using y_t = price of plywood grade #1 and $y_0 = 9.3$,

$$I_2 = \frac{y_2}{y_0}(100) = \frac{18.1}{9.3}(100) = 194.62$$

$$I_3 = \frac{y_3}{y_0}(100) = \frac{17.6}{9.3}(100) = 189.25$$

$$I_4 = \frac{y_4}{y_0}(100) = \frac{17.4}{9.3}(100) = 187.10$$

b. Using y_t = price of plywood grade #2 and $y_0 = 11.1$,

$$I_2 = \frac{y_2}{y_0}(100) = \frac{22.4}{11.1}(100) = 201.80$$

$$I_3 = \frac{y_3}{y_0}(100) = \frac{20.6}{11.1}(100) = 185.59$$

$$I_4 = \frac{y_4}{y_0}(100) = \frac{19.3}{11.1}(100) = 173.87$$

c. Using y_t = price of plywood grade #3 and $y_0 = 7.2$,

$$I_2 = \frac{y_2}{y_0}(100) = \frac{18.8}{7.2}(100) = 261.11$$

$$I_3 = \frac{y_3}{y_0}(100) = \frac{15.9}{7.2}(100) = 220.83$$

$$I_4 = \frac{y_4}{y_0}(100) = \frac{15.7}{7.2}(100) = 218.06$$

13.4

Define P_1 = price of plywood grade #1, P_2 = price of plywood grade #2, and P_3 = price of plywood grade #3. Then $y = P_1 + P_2 + P_3$. For $t = 0$ (year 1),

$$y_0 = 9.3 + 11.1 + 7.2 = 27.6.$$

The three simple composite index numbers are calculated as

$$I_t = \frac{y_t}{y_0}(100) \quad \text{where } y_0 = 27.6 \text{ and } y_t = P_{1t} + P_{2t} + P_{3t}$$

That is,

$$I_2 = \frac{y_2}{y_0}(100) = \frac{59.3}{27.6}(100) = 214.86$$

$$I_3 = \frac{y_3}{y_0}(100) = \frac{54.1}{27.6}(100) = 196.01$$

$$I_4 = \frac{y_4}{y_0}(100) = \frac{52.4}{27.6}(100) = 189.86$$

13.5

Define P_{1t}, P_{2t}, and P_{3t} as in Exercise 13.4. For the Laspeyres index, define w_1, w_2, and w_3 as the production during the base year (year 1). That is,

$$w_1 = 56.1 \qquad w_2 = 33.7 \qquad w_3 = 88.6$$

Then

$$y_0 = \Sigma w_i P_{i0} = 9.3(56.1) + 11.1(33.7) + 7.2(88.6) = 1533.72$$

$$y_2 = \Sigma w_i P_{i2} = 18.1(56.1) + 22.4(33.7) + 18.8(88.6) = 3435.97$$

$$y_3 = \Sigma w_i P_{i3} = 17.6(56.1) + 20.6(33.7) + 15.9(88.6) = 3090.32$$

$$y_4 = \Sigma w_i P_{i4} = 17.4(56.1) + 19.3(33.7) + 15.7(88.6) = 3017.57$$

The three Laspeyres indexes are

$$I_2 = \frac{y_2}{y_0}(100) = \frac{3435.97}{1533.72}(100) = 224.03$$

$$I_3 = \frac{y_3}{y_0}(100) = \frac{3090.32}{1533.72}(100) = 201.49$$

$$I_4 = \frac{y_4}{y_0}(100) = \frac{3017.57}{1533.72}(100) = 196.75$$

13.6

For the Paasche index, the weights w_1, w_2, and w_3 for year t are the production during year t.

(1) For year 2, $w_1 = 112.7$, $w_2 = 69.5$, and $w_3 = 204.1$, so that

$$y_0 = \Sigma w_i P_{i0} = 9.3(112.7) + 11.1(69.5) + 7.2(204.1) = 3289.08$$

$$y_2 = \Sigma w_i P_{i2} = 18.1(112.7) + 22.4(69.5) + 18.8(204.1) = 7433.75$$

and

$$I_2 = \frac{y_2}{y_0}(100) = \frac{7433.75}{3289.08}(100) = 226.01$$

(2) For year 3, $w_1 = 38.6$, $w_2 = 40.3$, and $w_3 = 79.9$, so that

$$y_0 = \Sigma w_i P_{i0} = 9.3(38.6) + 11.1(40.3) + 7.2(79.9) = 1381.59$$

$$y_3 = \Sigma w_i P_{i3} = 18.1(38.6) + 22.4(40.3) + 18.8(79.9) = 2779.95$$

and

$$I_3 = \frac{y_3}{y_0}(100) = \frac{2779.95}{1381.59}(100) = 201.21$$

(3) For year 4, $w_1 = 30.9$, $w_2 = 37.2$, and $w_3 = 65$, so that

$$y_0 = \Sigma w_i P_{i0} = 9.3(30.9) + 11.1(37.2) + 7.2(65) = 1168.29$$

$$y_4 = \Sigma w_i P_{i4} = 18.1(30.9) + 22.4(37.2) + 18.8(65) = 2276.12$$

and

$$I_4 = \frac{y_4}{y_0}(100) = \frac{2276.12}{1168.29}(100) = 194.82$$

13.7

This is similar to Exercise 13.4. Define

$$P_1 = \text{price of pig iron and } P_2 = \text{price of steel}$$

Then $y = P_1 + P_2$, and, for $t = 0$ (year 1967),

$$y_0 = 87.0 + 127.2 = 214.2$$

The simple composite index numbers are calculated as

$$I_t = \frac{y_t}{y_0}(100) \quad \text{where } y_0 = 214.2 \text{ and } y_t = P_{1t} + P_{2t}$$

That is,

$$I_{87} = \frac{y_{87}}{y_0}(100) = \frac{48.4 + 89.2}{214.2}(100) = 64.2$$

$$I_{88} = \frac{y_{88}}{y_0}(100) = \frac{157.6}{214.2}(100) = 73.58$$

$$I_{89} = \frac{y_{89}}{y_0}(100) = \frac{153.8}{214.2}(100) = 71.80$$

$$I_{90} = \frac{y_{90}}{y_0}(100) = \frac{152.9}{214.2}(100) = 71.38$$

$$I_{91} = \frac{y_{91}}{y_0}(100) = \frac{135.8}{214.2}(100) = 63.40$$

13.8

a. Define

P_1 = price for Dow Chemical P_2 = price for Du Pont P_3 = price for Monsanto

Then $y = P_1 + P_2 + P_3$, and $y_0 = 23.0 + 19.5 + 24.0 = 66.5$

The simple composite index numbers are calculated as

$$I_t = \frac{y_t}{y_0}(100) \quad \text{where } y_0 = 66.5 \text{ and } y_t = P_{1t} + P_{2t} + P_{3t}$$

That is,

$$I_{90} = \frac{y_{90}}{y_0}(100) = \frac{142.8}{66.5}(100) = 214.74$$

$$I_{94} = \frac{y_{94}}{y_0}(100) = \frac{61.5 + 54.1 + 77.2}{66.5}(100) = 289.92$$

b. The Laspeyres index uses the index year weights,

$$w_1 = 285 \quad w_2 = 722 \quad w_3 = 154$$

Then $y_0 = \Sigma w_i P_{i0} = 285(23.0) + 722(19.5) + 154(24.0) = 24{,}330$

$y_{90} = \Sigma w_i P_{i,90} = 285(56.4) + 722(36.9) + 154(49.5) = 50{,}338.8$

$y_{94} = \Sigma w_i P_{i,94} = 285(61.5) + 722(54.1) + 154(77.2) = 68{,}476.5$

The two Laspeyres indexes are

$$I_{90} = \frac{y_{90}}{y_0}(100) = \frac{50{,}338.8}{24{,}330}(100) = 206.90$$

$$I_{94} = \frac{y_{94}}{y_0}(100) = \frac{68{,}476.5}{24{,}330}(100) = 281.45$$

c. The Paasche index uses the index year weights. Hence,

$$I_{90} = \frac{270(56.4) + 670(36.9) + 126(49.5)}{270(23.0) + 670(19.5) + 126(24.0)}(100) = 207.13$$

$$I_{94} = \frac{283(61.5) + 680(54.1) + 115(77.2)}{283(23.0) + 680(19.5) + 115(24.0)}(100) = 279.95$$

13.9

a. The time series, with the twelve time periods plotted along the horizontal axis and the production (in millions of short tons) plotted along the vertical axis, is shown in Figure 13.1.

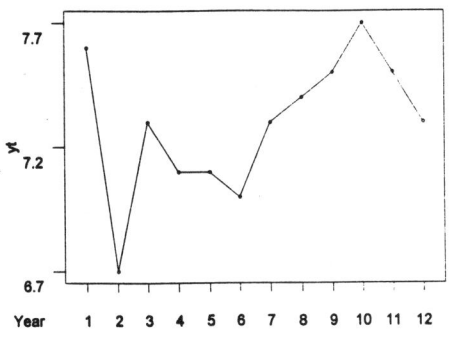

Figure 13.1

b-d. The first two-point moving average is available at a time between the first and second observations and all eleven values are calculated using the formula

$$M_{t.5} = \frac{y_t + y_{t+1}}{2}$$

Similarly, the first three-point moving average is available at the second observation and the ten values are calculated as

$$M_t = \frac{y_{t-1} + y_t + y_{t+1}}{3}$$

The two moving averages and the original series are shown in Table 13.1, and are plotted along with the original series in Figure 13.2. Note that the secular trend is more obvious in the graph of the three-point moving average.

Table 13.1

t	y_t	2-Point Moving Average	3-Point Moving Average
1	7.6		—
		7.15	
2	6.7		7.20
		7.00	
3	7.3		7.03
		7.20	
4	7.1		7.17
		7.10	
5	7.1		7.07
		7.05	
6	7.0		7.13
		7.15	
7	7.3		7.23
		7.35	
8	7.4		7.40
		7.45	
9	7.5		7.53
		7.60	
10	7.7		7.57
		7.60	
11	7.5		7.50
		7.40	
12	7.3		—

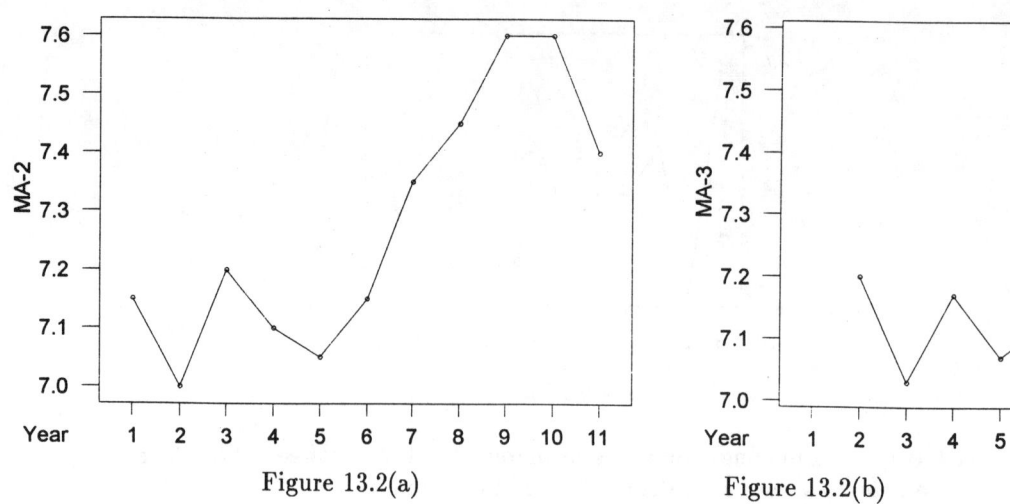

Figure 13.2(a) Figure 13.2(b)

13.10

This is similar to Exercise 13.9. The three- and four-point moving averages are shown in Table 13.2. Notice that the four-point moving averages fall between the two middle values used to compute the numerator of M_t and hence, must be centered by averaging each two adjacent four-point moving averages.

Table 13.2

t	y_t	3-Point Moving Average	4-Point Moving Average	4-pt Centered Moving Average
1	8377	*		*
2	7568	8190.00	8258.50	*
3	8625	8219.00	7944.00	8101.25
4	8464	8069.33	8217.50	8080.75
5	7119	8081.67	8374.25	8295.88
6	8662	8344.33	8655.25	8514.75
7	9252	9167.33	8716.25	8685.75
8	9588	8734.33	8288.50	8502.37
9	7363	7967.33	8098.50	8193.50
10	6951	7602.00	7944.25	8021.38
11	8492	8138.00	8394.50	8169.38
12	8971	8875.67	8714.25	8554.38
13	9164	8788.33	8236.50	8475.37
14	8230	7991.67	7546.00	7891.25
15	6581	7006.67	6694.50	7120.25
16	6209	6182.67	6335.25	6514.88
17	5758	6253.33	6678.00	6506.62
18	6793	6834.33	7177.00	6927.50
19	7952	7650.00	7791.25	7484.13
20	8205	8124.00	7863.25	7827.25
21	8215	7833.67	7760.00	7811.63
22	7081	7611.67	7478.25	7619.13
23	7539	7232.67	7149.00	7313.62
24	7078	7171.67	6913.00	7031.00
25	6898	6704.33		*
26	6137	*		*

Figure 13.3 shows the original time series (a) with the three-point (b) and centered four-point (c) moving averages superimposed. Note the smoothing effect of the averaging.

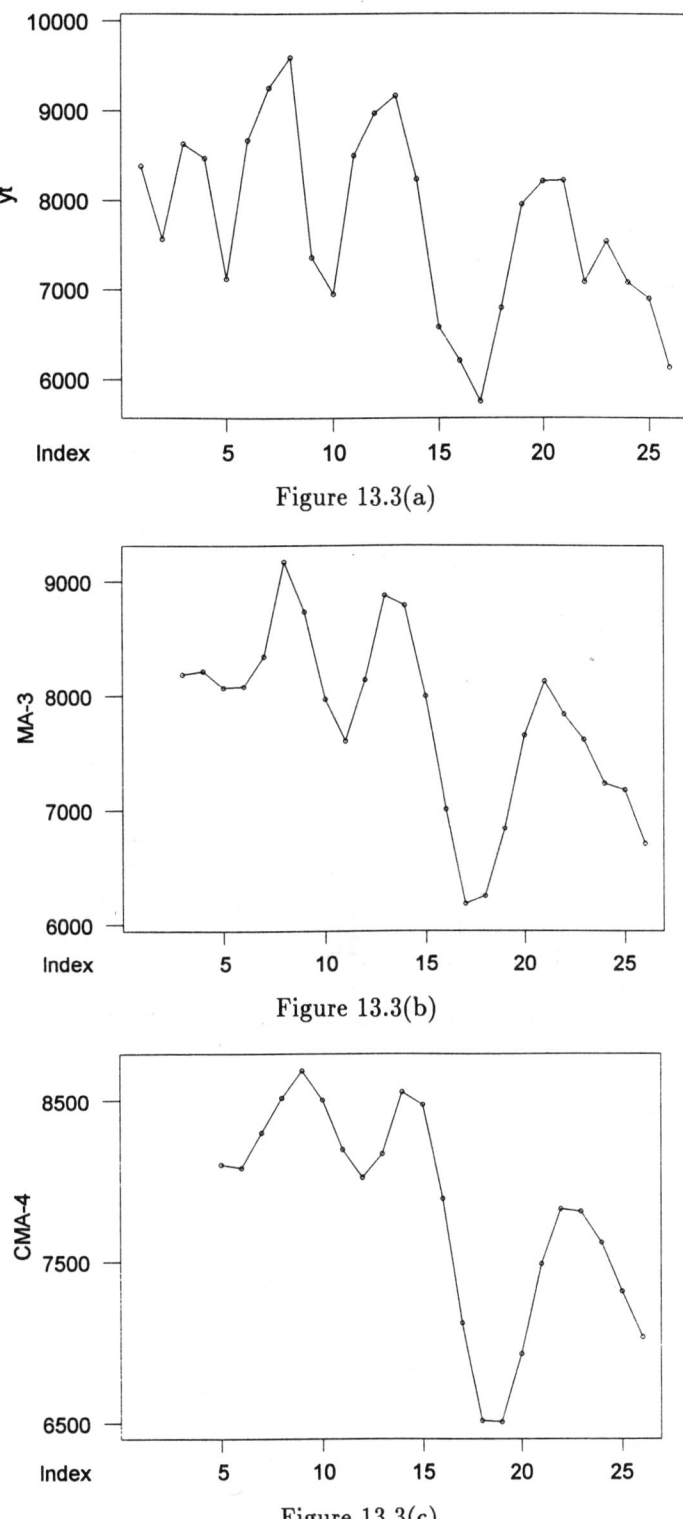

Figure 13.3(a)

Figure 13.3(b)

Figure 13.3(c)

13.11

This is similar to Exercise 13.10, except that y_t is now annual imported sales. Table 13.3 shows the three-point, four-point, and four-point centered moving averages along with the original series. Figures 13.4 (a)-(c) show the original time series, the three-point moving average, and the centered four-point moving average.

Table 13.3

t	y_t	3-Point Moving Average	4-Point Moving Average	4-pt Centered Moving Average
1	658	*	*	*
2	779	822.33	*	*
3	1030	975.33	896.00	974.12
4	1117	1143.33	1052.25	1150.63
5	1283	1322.00	1249.00	1322.88
6	1566	1490.00	1396.75	1477.38
7	1621	1649.67	1558.00	1574.13
8	1762	1598.33	1590.25	1592.87
9	1412	1587.00	1595.50	1580.62
10	1587	1500.33	1565.75	1604.87
11	1502	1721.33	1644.00	1717.50
12	2075	1859.00	1791.00	1883.75
13	2000	2134.67	1976.50	2088.50
14	2329	2242.33	2200.50	2231.87
15	2398	2351.00	2263.25	2290.88
16	2326	2315.00	2318.50	2325.63
17	2221	2311.00	2332.75	2338.25
18	2386	2349.67	2343.75	2407.25
19	2442	2554.00	2470.75	2597.50
20	2834	2837.00	2724.25	2825.63
21	3235	3088.67	2927.00	3009.12
22	3197	3177.00	3091.25	3090.12
23	3099	3040.33	3089.00	3009.75
24	2825	2841.67	2930.50	2812.25
25	2601	2559.00	2694.00	*
26	2251	*	*	*

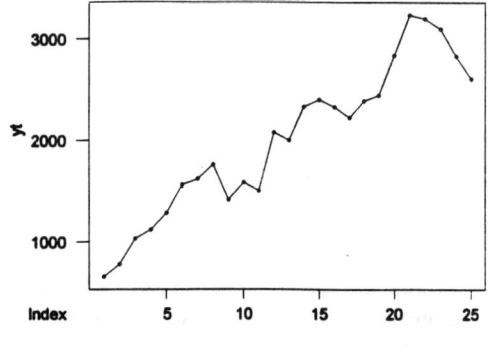

Figure 13.4(a)

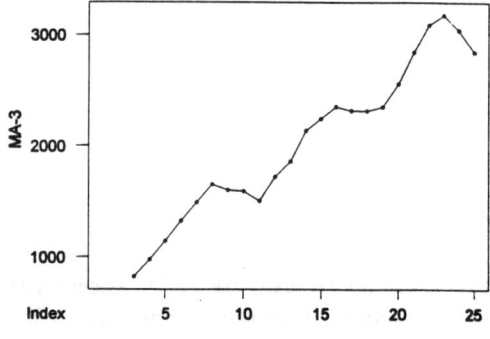

Figure 13.4(b)

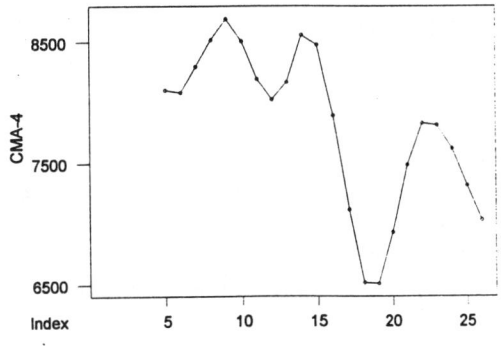

Figure 13.4(c)

13.12

This is similar to previous exercises. The first three-point moving average is available at t = 2. Table 13.4 shows the calculations, while Figure 13.5 show a graph of the original time series and the three-point moving average.

Table 13.4

t	y_t	3-Point Moving Average	t	y_t	3-Point Moving Average
1	86.0	—	19	53.0	55.6333
2	72.0	81.2667	20	57.9	58.6000
3	85.8	75.1333	21	64.9	60.3333
4	67.6	71.4333	22	58.2	64.1667
5	60.9	63.9667	23	69.4	65.2000
6	63.4	65.8000	24	68.0	67.7000
7	73.1	69.2667	25	65.7	65.7667
8	71.3	71.0667	26	63.6	68.9333
9	68.8	69.3333	27	77.5	70.6667
10	67.9	68.4667	28	70.9	70.1000
11	68.7	70.1667	29	61.9	66.8000
12	73.9	67.2333	30	67.6	64.1000
13	59.1	62.9000	31	62.8	66.0000
14	55.7	56.8333	32	67.6	63.6667
15	55.7	56.2667	33	60.6	64.1333
16	57.4	55.0333	34	64.2	63.4667
17	52.0	55.1333	35	65.6	65.1333
18	56.0	53.6667	36	65.6	—

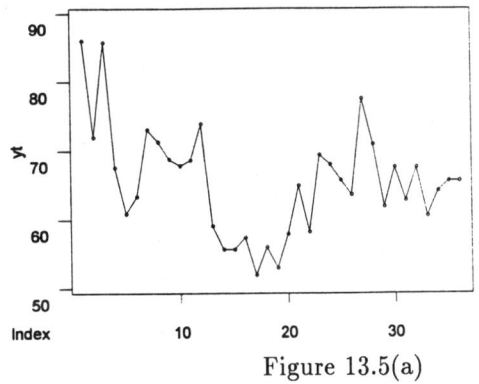

Figure 13.5(a)

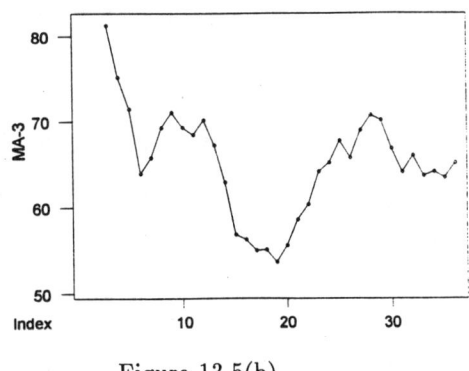

Figure 13.5(b)

13.13

The exponential smoothing method is employed. Let $E_1 = y_1$ at the first period. Then, for each succeeding time period t, the smoothed value E_t is found by calculating

$$E_t = \omega y_t + (1 - \omega)E_{t-1}$$

The calculations for $\omega = .2$, $\omega = .5$, and $\omega = .8$ are found in Table 13.5 and the smoothed series along with the original series are plotted in Figure 13.6. Notice that, although the values presented in the table are rounded to three decimals, the more accurate values of E_t are used in the computation of E_{t+1}. The author has used calculator or computer accuracy in all computations.

Table 13.5

t	y_t	$E_t(\omega = .2)$	$E_t(\omega = .5)$	$E_t(\omega = .8)$
1	7.6	7.600	7.600	7.600
2	6.7	7.420	7.510	7.528
3	7.3	7.396	7.453	7.468
4	7.1	7.337	7.395	7.410
5	7.1	7.289	7.342	7.356
6	7.0	7.232	7.287	7.301
7	7.3	7.245	7.266	7.273
8	7.4	7.276	7.271	7.271
9	7.5	7.321	7.296	7.291
10	7.7	7.397	7.346	7.335
11	7.5	7.417	7.382	7.373
12	7.3	7.394	7.388	7.385

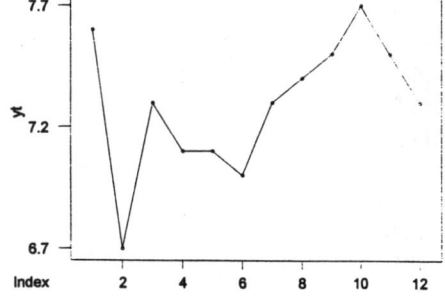
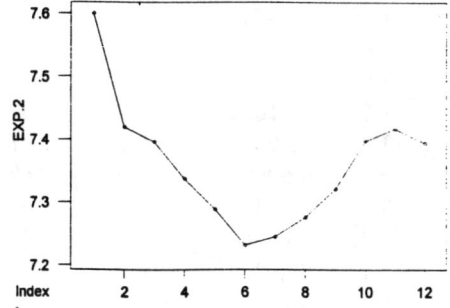

Figure 13.6(a)

Figure 13.6(b)

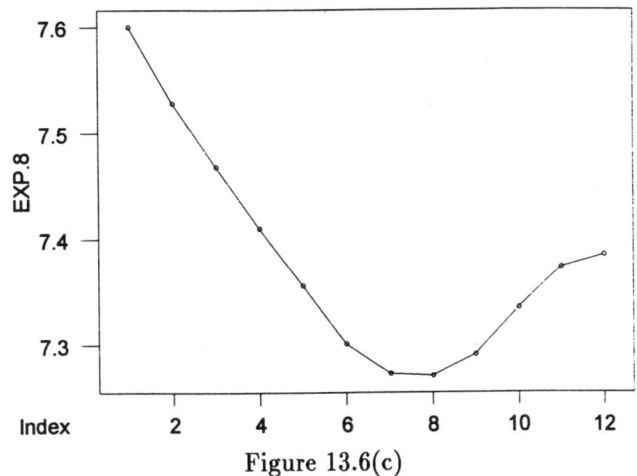

Figure 13.6(c)

13.14

This is similar to Exercise 13.13. The computations are shown in Table 13.6. The graph of the original and smoothed series is shown in Figure 13.7.

Table 13.6

t	y_t	$E_t(\omega = .4)$	t	y_t	$E_t(\omega = .4)$
1	8377	8377.000	13	9164	8723.660
2	7568	8053.400	14	8230	8526.196
3	8625	8282.040	15	6581	7748.118
4	8464	8354.824	16	6209	7132.471
5	7119	7860.494	17	5758	6582.682
6	8662	8181.097	18	6793	6666.809
7	9252	8609.458	19	7952	7180.886
8	9588	9000.875	20	8205	7590.531
9	7363	8345.725	21	8215	7840.319
10	6951	7787.835	22	7081	7536.591
11	8492	8069.501	23	7539	7537.555
12	8971	8430.101	24	7078	7353.733
			25	6898	7171.440
			26	6137	6757.664

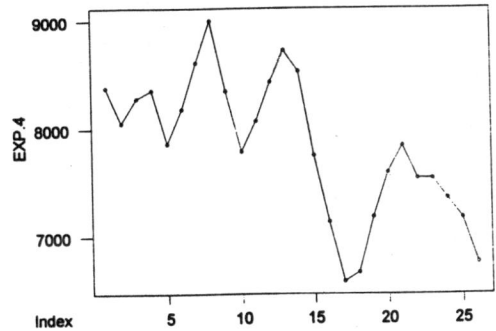 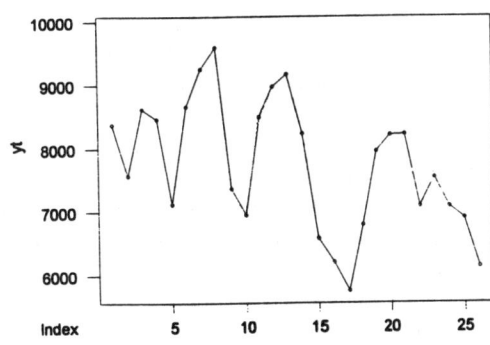

Figure 13.7

13.15

The computations are shown in Table 13.7 with $E_t = .3y_t + .7E_{t-1}$. The graph of the original and smoothed series is shown in Figure 13.8.

Table 13.7

t	y_t	$E_t(\omega = .3)$	t	y_t	$E_t(\omega = .3)$
1	86.0	86.000	19	53.0	55.943
2	72.0	81.800	20	57.9	56.530
3	85.8	83.000	21	64.9	59.041
4	67.6	78.380	22	58.2	58.789
5	60.9	73.136	23	69.4	61.972
6	63.4	70.215	24	68.0	63.780
7	73.1	71.081	25	65.7	64.356
8	71.3	71.146	26	63.6	64.129
9	68.8	70.443	27	77.5	68.141
10	67.9	69.680	28	70.9	68.968
11	68.7	69.386	29	61.9	66.848
12	73.9	70.740	30	67.6	67.074
13	59.1	67.248	31	62.8	65.791
14	55.7	63.784	32	67.6	66.334
15	55.7	61.359	33	60.6	64.614
16	57.4	60.171	34	64.2	64.490
17	52.0	57.720	35	65.6	64.823
18	56.0	57.204	36	65.6	65.056

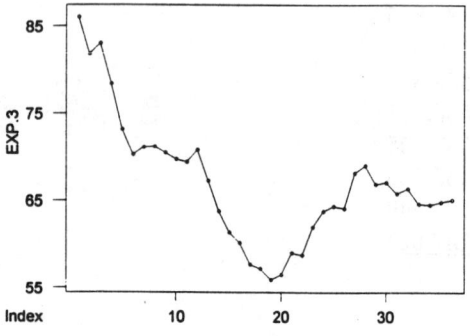 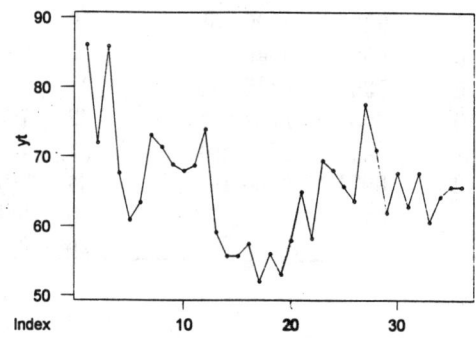

Figure 13.8

The prices show a gradual upward trend, perhaps quadratic, and also show some sort of cyclical or seasonal variation, perhaps on a twelve-month cycle.

13.16

The computations for the three- and five-point moving averages are shown in Table 13.8, while Figure 13.9 shows the two smoothed series superimposed on the original series. The five-point smoothing appears better than the three-point smoothing.

Table 13.8

t	y_t	3-Point Moving Average	5-Point Moving Average	t	y_t	3-Point Moving Average	5-Point Moving Average
1	248.0	—	—	19	266.1	252.700	242.66
2	210.5	226.533	—	20	255.2	252.033	242.84
3	221.1	213.500	224.48	21	234.8	237.100	239.74
4	208.9	221.300	224.52	22	221.3	225.800	235.34
5	233.9	230.333	236.72	23	221.3	228.900	233.46
6	248.2	251.200	246.04	24	244.1	237.067	231.44
7	271.5	262.467	251.04	25	245.8	238.200	234.10
8	267.7	257.700	248.90	26	224.7	235.033	232.10
9	233.9	241.600	243.50	27	234.6	223.533	227.76
10	223.2	226.100	235.92	28	211.3	222.767	228.52
11	221.2	226.000	231.18	29	222.4	227.767	240.04
12	233.6	232.933	227.96	30	249.6	251.433	248.94
13	244.0	231.800	228.26	31	282.3	270.333	253.98
14	217.8	228.833	226.18	32	279.1	265.967	254.22
15	224.7	217.767	223.54	33	236.5	246.400	249.46
16	210.8	218.633	222.10	34	223.6	228.633	—
17	220.4	222.667	231.76	35	225.8	—	—
18	236.8	241.100	237.86				

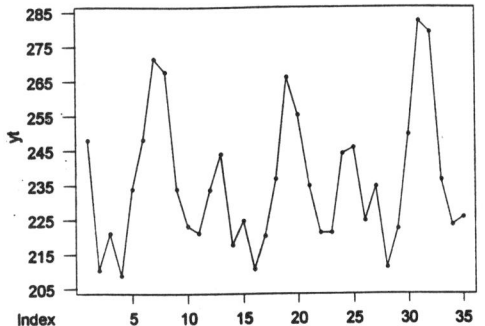
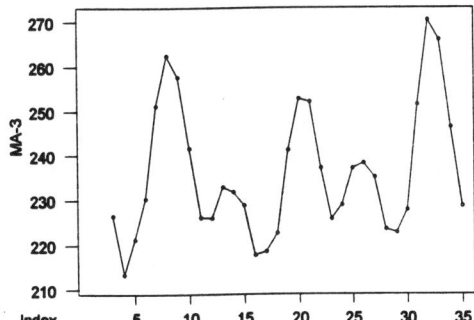

Figure 13.9(a)

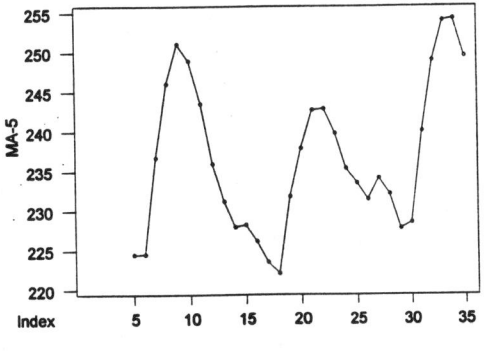

Figure 13.9(b)

13.17

The calculations are given in Table 13.9, and Figure 13.10 shows the original and smoothed series.

Table 13.9

t	y_t	$E_t(\omega = .3)$	t	y_t	$E_t(\omega = .3)$
1	248.0	248.000	19	266.1	238.689
2	210.5	236.750	20	255.2	243.643
3	221.1	232.055	21	234.8	240.990
4	208.9	225.109	22	221.3	235.083
5	233.9	227.746	23	221.3	230.948
6	248.2	233.882	24	244.1	234.894
7	271.5	245.168	25	245.8	238.166
8	267.7	251.927	26	224.7	234.126
9	233.9	246.519	27	234.6	234.268
10	223.2	239.523	28	211.3	227.378
11	221.2	234.026	29	222.4	225.884
12	233.6	233.898	30	249.6	232.999
13	244.0	236.929	31	282.3	247.789
14	217.8	231.190	32	279.1	257.183
15	224.7	229.243	33	236.5	250.978
16	210.8	223.710	34	223.6	242.764
17	220.4	222.717	35	225.8	237.675
18	236.8	226.942			

There is a strong seasonal component as well as a slight secular (upward) trend over the years.

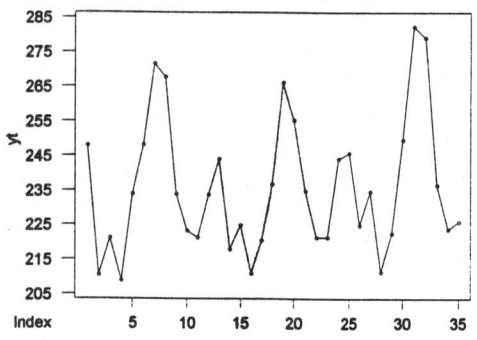 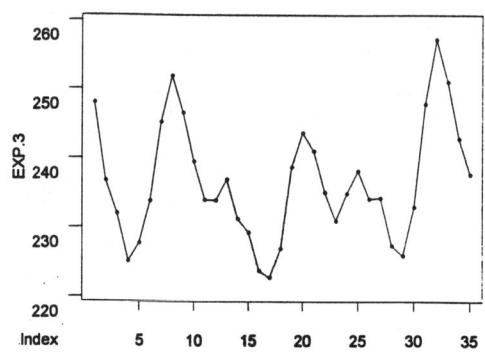

Figure 13.10

13.18

The twelve-point moving averages fall between the observed monthly values of y_t, starting with $t = 6.5$. Hence, a second smoothing is performed so that the centered moving average begins at the time $t = 7$. Six points are lost at the beginning and end of the time series. The twelve-point moving average and the centered moving averages are shown in Tables 13.10(a) and (b).

Table 13.10(a)

t	12-Point Moving Average	t	12-Point Moving Average	t	12-Point Moving Average
6.5	231.525	21.5	234.142	36.5	233.800
7.5	232.000	22.5	233.983	37.5	233.350
8.5	231.467	23.5	234.925	38.5	232.308
9.5	231.408	24.5	234.867	39.5	232.383
10.5	231.667	25.5	235.308	40.5	232.225
11.5	231.900	26.5	235.267	41.5	232.233
12.5	233.025	27.5	234.950	42.5	233.108
13.5	233.817	28.5	234.817	43.5	233.258
14.5	234.642	29.5	235.450	44.5	233.833
15.5	235.542	30.5	235.142	45.5	234.658
16.5	236.017	31.5	234.808	46.5	234.700
17.5	235.567	32.5	235.417	47.5	234.867
18.5	233.792	33.5	235.717	48.5	235.933
19.5	234.708	34.5	235.875	49.5	237.283
20.5	234.525	35.5	234.750	50.5	239.275
				51.5	239.417
				52.5	239.608
				53.5	239.983

Table 13.10(b)

t	12-Point Centered Moving Average	t	12-Point Centered Moving Average	t	12-Point Centered Moving Average
7	231.762	22	234.062	37	233.575
8	231.733	23	234.454	38	232.829
9	231.437	24	234.896	39	232.346
10	231.538	25	235.087	40	232.304
11	231.783	26	235.288	41	232.229
12	232.462	27	235.108	42	232.671
13	233.421	28	234.883	43	233.183
14	234.229	29	235.133	44	233.546
15	235.092	30	235.296	45	234.246
16	235.779	31	234.975	46	234.679
17	235.792	32	235.113	47	234.783
18	234.679	33	235.567	48	235.400
19	234.250	34	235.796	49	236.608
20	234.617	35	235.312	50	238.279
21	234.333	36	234.275	51	239.346
				52	239.513
				53	239.796

13.19

This is similar to Exercise 13.18. Tables 13.11(a) and (b) show the four-point moving averages and the centered four-point moving averages, respectively.

Table 13.11(a)

t	4-Point Moving Average	t	4-Point Moving Average
2.5	.2950	10.5	.3650
3.5	.3000	11.5	.3750
4.5	.3075	12.5	.3850
5.5	.3175	13.5	.4000
6.5	.3250	14.5	.4125
7.5	.3350	15.5	.4250
8.5	.3450	16.5	.4425
9.5	.3525	17.5	.4625
		18.5	.4750

Table 13.11(b)

t	4-Point Moving Average	t	4-Point Moving Average
3	.29750	11	.37000
4	.30375	12	.38000
5	.31250	13	.39250
6	.32125	14	.40625
7	.33000	15	.41875
8	.34000	16	.43375
9	.34875	17	.45250
10	.35875	18	.46875

13.20

a. The linear trend model is given as

$$\hat{T}_t = 836.6 + 87.702t$$

This linear model fits the data fairly well, since $R^2 = 83.9\%$ (or $R^2(adj) = 83.2\%$).

b. The series is detrended by dividing each observed value of y_t by the estimated value of the trend line at time t, $\hat{T}_t$. The values of y_t are given in the printout in the column labeled "sales", while the values of $\hat{T}_t$ are found in the column labeled "Fits1." For example, for $t = 1$,

$$\frac{y_1}{\hat{T}_1} = \frac{658.0}{924.34} = .7119$$

The detrended series is shown in Table 13.12, and is plotted in Figure 13.11. There is a pronounced cyclic effect.

Table 13.12

t	$y_t/\hat{T}_t$	t	$y_t/\hat{T}_t$	t	$y_t/\hat{T}_t$
1	0.7119	10	0.9261	19	0.9756
2	0.7697	11	0.8338	20	1.0939
3	0.9366	12	1.0984	21	1.2078
4	0.9407	13	1.0118	22	1.1558
5	1.0062	14	1.1281	23	1.0859
6	1.1491	15	1.1142	24	0.9604
7	1.1175	16	1.0385	25	0.8586
8	1.1455	17	0.9542	26	0.7222
9	0.8684	18	0.9879		

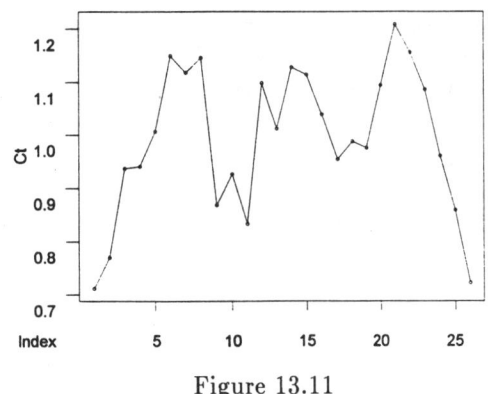

Figure 13.11

c. In order to forecast y_t for $t = 27$ (1992), we must approximate the cyclic effect, C_t, at time $t = 27$. Since the cycle seems to repeat with approximately an eight-year period, we use Figure 13.11 to <u>approximate</u> the value

$$\hat{C}_{27} \approx 1.1$$

[Note that each student will arrive at a different approximation for $\hat{C}_{27}$.]

Then the estimated trend component at time $t = 27$ is

$$\hat{T}_{27} = 836.6 + 87.702(27) = 3204.554$$

and the estimated value of y_{27} is

$$\hat{y}_{27} = \hat{T}_{27} \cdot \hat{C}_{27} = 3204.554(1.1) = 3525.01$$

13.21

This is similar to Exercise 13.20.

a. The time series is plotted in Figure 13.12. Notice the cubic effect.

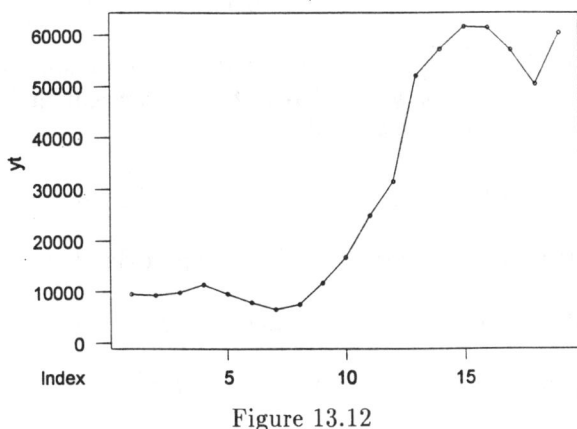

Figure 13.12

b. The cubic trend model is given as

$$\hat{T}_t = 27{,}449 - 11{,}887t + 1655.2t^2 - 49.70t^3$$

This cubic model fits the data fairly well, since $R^2 = 92.3\%$ (or $R^2(\text{adj}) = 90.8\%$).

c. The series is detrended by dividing each observed value of y_t by the estimated value of the trend line at time t, $\hat{T}_t$. The values of y_t are given in the printout in the column labeled "y", while the values of $\hat{T}_t$ are found in the column labeled "Fits1". For example, for $t = 1$,

$$\frac{y_1}{\hat{T}_1} = \frac{9566}{17{,}167.6} = .5572$$

The detrended series is shown in Table 13.13, and is plotted in Figure 13.13. The cyclic effect is very pronounced.

Table 13.13

t	$y_t/\hat{T}_t$	t	$y_t/\hat{T}_t$	t	$y_t/\hat{T}_t$
1	0.5572	7	0.7977	13	1.1984
2	0.9441	8	0.5891	14	1.1667
3	1.8557	9	0.6414	15	1.1444
4	3.5686	10	0.6865	16	1.0691
5	3.0262	11	0.8082	17	0.9589
6	1.5904	12	0.8460	18	0.8405
				19	1.0378

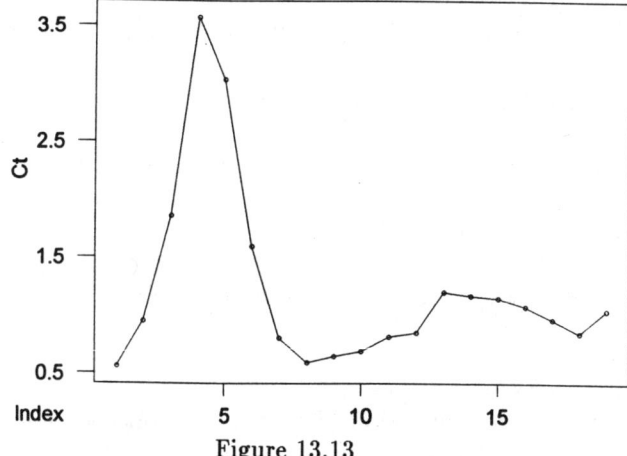

Figure 13.13

d. In order to forecast y_t for $t = 20$ (1991), we must approximate the cyclic effect, C_t, at time $t = 20$. Since the cycle seems to repeat with approximately an eight to nine year period, we use Figure 13.13 to <u>approximate</u> the value

$$\hat{C}_{20} \approx 1.05$$

[Note that each student will arrive at a different approximation for $\hat{C}_{20}$.]

Then the estimated trend component at time $t = 20$ is

$$\hat{T}_{20} = 27{,}449 - 11{,}887(20) + 1655.2(20^2) - 49.70(20^3) = 54{,}189$$

and the estimated value of y_{20} is

$$\hat{y}_{20} = \hat{T}_{20} \cdot \hat{C}_{20} = (54{,}189)(1.05) = 56{,}898.45$$

This estimate is highly volatile and hence not very reliable, since its value changes very quickly, depending on the choice of the cyclical estimate, $\hat{C}_{20}$.

13.22

The twelve-point centered moving averages for this time series (t = 1 to t = 59) were calculated in Exercise 13.18 for the time periods t = 7 to t = 53. Using these moving averages, we calculate the specific seasonals, $s_t = y_t/M_t$ for times t = 7 to t = 53. For example, for t = 7,

$$s_7 = \frac{y_7}{M_7} = \frac{256.7}{231.762} = 1.10760$$

These specific seasonals are calculated using the data given in Table 13.14 and are shown in a different format in Table 13.13.

Table 13.13

Month	Jan.	Feb.	March	April	May	June
1989	—	—	—	—	—	—
1990	1.01533	0.90809	0.96005	0.89406	0.94405	1.06060
1991	1.05493	0.89465	0.94042	0.88938	0.99476	1.05484
1992	1.04463	0.93545	0.96709	0.90743	0.94906	1.01775
1993	1.03885	0.94301	0.98017	0.88221	0.92745	—
$\bar{s}_i$	1.03844	0.92030	0.96193	0.89327	0.95383	1.04440
$\hat{S}_i$	1.039	0.921	0.963	0.894	0.955	1.045

Month	July	Aug.	Sept.	Oct.	Nov.	Dec.
1989	1.10760	1.11464	0.98040	0.94628	0.94485	1.11244
1990	1.13639	1.14314	1.01437	0.96043	0.91105	1.01023
1991	1.15544	1.13860	0.99292	0.94658	0.94003	0.99712
1992	1.14116	1.09272	1.00236	0.94299	0.94257	1.03696
1993	—	—	—			
$\bar{s}_i$	1.13515	1.12228	0.99751	0.94907	0.93463	1.03919
$\hat{S}_i$	1.136	1.123	0.998	0.950	0.935	1.040

Table 13.14

t	y_t	M_t	s_t	$\dfrac{y_t}{\hat{S}_t}$	t	y_t	M_t	s_t	$\dfrac{y_t}{\hat{S}_t}$
1	231.3			222.618	31	271.5	234.975	1.15544	238.996
2	219.1			237.894	32	267.7	235.113	1.13860	238.379
3	226.4			235.099	33	233.9	235.567	0.99292	234.369
4	207.7			232.327	34	223.2	235.796	0.94658	234.947
5	219.8			230.157	35	221.2	235.312	0.94003	236.578
6	235.4			225.263	36	233.6	234.275	0.99712	224.615
7	256.7	231.762	1.10760	225.968	37	244.0	233.575	1.04463	234.841
8	258.3	231.733	1.11464	230.009	38	217.8	232.829	0.93545	236.482
9	226.9	231.437	0.98040	227.355	39	224.7	232.346	0.96709	233.333
10	219.1	231.538	0.94628	230.632	40	210.8	232.304	0.90743	235.794
11	219.0	231.783	0.94485	234.225	41	220.4	232.229	0.94906	230.785
12	258.6	232.462	1.11244	248.654	42	236.8	232.671	1.01775	226.603
13	237.0	233.421	1.01533	228.104	43	266.1	233.183	1.14116	234.243
14	212.7	234.229	0.90809	230.945	44	255.2	233.546	1.09272	227.248
15	225.7	235.092	0.96005	234.372	45	234.8	234.246	1.00236	235.271
16	210.8	235.779	0.89406	235.794	46	221.3	234.679	0.94299	232.947
17	222.6	235.792	0.94405	233.089	47	221.3	234.783	0.94257	236.684
18	248.9	234.679	1.06060	238.182	48	244.1	235.400	1.03696	234.712
19	266.2	234.250	1.13639	234.331	49	245.8	236.608	1.03885	236.574
20	268.2	234.617	1.14314	238.825	50	224.7	238.279	0.94301	243.974
21	237.7	234.333	1.01437	238.176	51	234.6	239.346	0.98017	243.614
22	224.8	234.062	0.96043	236.632	52	211.3	239.513	0.88221	236.353
23	213.6	234.454	0.91105	228.449	53	222.4	239.796	0.92475	232.880
24	237.3	234.896	1.01023	228.173	54	249.6			238.852
25	248.0	235.087	1.05493	238.691	55	282.3			248.504
26	210.5	235.288	0.89465	228.556	56	279.1			248.531
27	221.1	235.108	0.94042	229.595	57	236.5			236.974
28	208.9	234.883	0.88938	233.669	58	223.6			235.368
29	233.9	235.133	0.99476	244.921	59	225.8			241.497
30	248.2	235.296	1.05484	237.512					

Once the specific seasonals have been calculated, they are averaged over each of the twelve months in the cycle to obtain $\bar{s}_i$, for $i = 1, 2, \ldots, 12$ (see Table 13.13). These average indexes are then normalized by calculating

$$S = \bar{s}_1 + \bar{s}_2 + \cdots + \bar{s}_{12}$$

and we find the seasonal indexes as

$$\hat{S}_i = \bar{s}_i \left(\frac{12}{S}\right)$$

These calculations are shown in Table 13.13. Finally, the original series is deseasonalized by using the seasonal indexes to find

$$\frac{y_t}{\hat{S}_t}$$

with $\hat{S}_t$ the seasonal index for the month corresponding to time t. The deseasonalized series is shown in Table 13.14.

13.23

a. Using the deseasonalized series from Exercise 13.22, a MINITAB program is used to fit a linear model to the data. The printout is shown below. The regression is significant, but notice the small value of $R^2 = .164$. The model does not fit very well; there is much unexplained variation.

```
The regression equation is
Deseason = 231 + 0.136 t

Predictor         Coef       Stdev      t-ratio       p
Constant       230.640       1.404       164.32   0.000
t              0.13582     0.04069         3.34   0.001

s = 5.322        R-sq = 16.4%       R-sq(adj) = 14.9%

Analysis of Variance

SOURCE         DF         SS         MS         F       p
Regression      1     315.65     315.65     11.14   0.001
Error          57    1614.66      28.33
Total          58    1930.31

Unusual Observations
Obs.        t     Deseason       Fit    Stdev.Fit    Residual
  12     12.0      248.654   232.270        1.008      16.384
  36     36.0      224.615   235.530        0.735     -10.914
  55     55.0      248.504   238.110        1.231      10.393
```

t	FITS2	t	FITS2	t	FITS2	t	FITS2
1	230.776	19	233.221	34	235.258	48	237.159
2	230.912	20	233.356	35	235.394	49	237.295
3	231.047	21	233.492	36	235.530	50	237.431
4	231.183	22	233.628	37	235.665	51	237.567
5	231.319	23	233.764	38	235.801	52	237.703
6	231.455	24	233.900	39	235.937	53	237.839
7	231.591	25	234.036	40	236.073	54	237.974
8	231.727	26	234.171	41	236.209	55	238.110
9	231.862	27	234.307	42	236.345	56	238.246
10	231.998	28	234.443	43	236.480	57	238.382
11	232.134	29	234.579	44	236.616	58	238.518
12	232.270	30	234.715	45	236.752	59	238.654
13	232.406	31	234.850	46	236.888		
14	232.541	32	234.986	47	237.024		
15	232.677	33	235.122				
16	232.813						
17	232.949						
18	233.085						

The series is detrended by obtaining the fit $\hat{T}_t$ for each value of t and then finding $y_t/\hat{T}_t$, where y_t are the values of the deseasonalized series. The calculations are shown in Table 13.15 and the deseasonalized and detrended series is shown in Figure 13.14.

Table 13.15

t	$y_t/\hat{T}_t$	t	$y_t/\hat{T}_t$	t	$y_t/\hat{T}_t$
1	0.96465	20	1.02343	39	0.98896
2	1.03024	21	1.02006	40	0.99882
3	1.01753	22	1.01286	41	0.97704
4	1.00495	23	0.97726	42	0.95878
5	0.99498	24	0.97552	43	0.99054
6	0.97325	25	1.01989	44	0.96041
7	0.97572	26	0.97602	45	0.99374
8	0.99259	27	0.97989	46	0.98337
9	0.98056	28	0.99670	47	0.99857
10	0.99411	29	1.04409	48	0.98968
11	1.00901	30	1.01192	49	0.99696
12	1.07054	31	1.01765	50	1.02756
13	0.98149	32	1.01444	51	1.02545
14	0.99313	33	0.99680	52	0.99432
15	1.00728	34	0.99868	53	0.97915
16	1.01280	35	1.00503	54	1.00369
17	1.00060	36	0.95366	55	1.04365
18	1.02187	37	0.99650	56	1.04317
19	1.00476	38	1.00289	57	0.99409
				58	0.98680
				59	1.01192

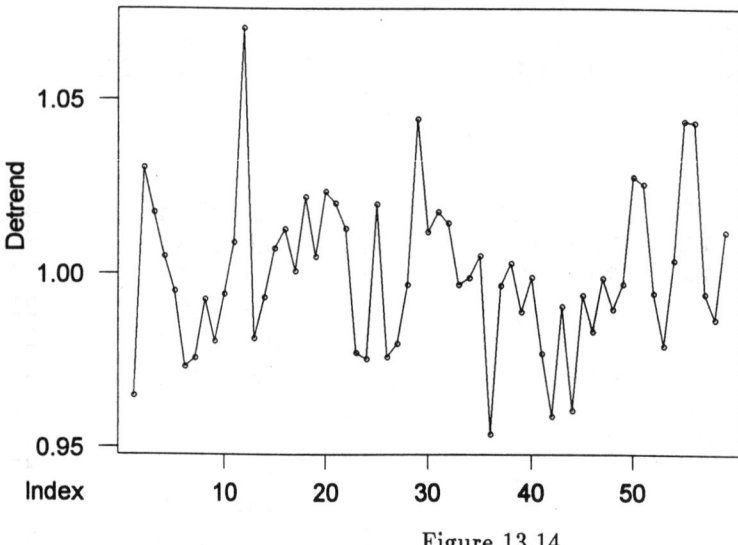

Figure 13.14

The deseasonalized and detrended series has no pronounced cyclic effects.

b. Because there is no pronounced cyclic effect, we can estimate

$$\hat{y}_{60} = \hat{T}_{60} \cdot \hat{S}_{12} = [230.640 + 0.13582(60)](1.04) = 248.34$$

13.24

The four-point centered moving averages for this time series (t = 1 to t = 20) were calculated in Exercise 13.19 for the time periods t = 3 to t = 18. Using these moving averages, we calculate

296

the specific seasonals, $s_t = y_t/M_t$ for times $t = 3$ to $t = 18$. For example, for $t = 3$,

$$s_3 = \frac{y_3}{M_3} = \frac{.36}{.29750} = 1.21008$$

These specific seasonals are calculated using the data given in Table 13.17 and are shown in a different format in Table 13.16.

Table 13.16

Quarter	Mar. 31	June 30	Sept. 30	Dec. 31
1991	—	—	1.21008	0.88889
1992	0.80000	1.08949	1.21212	0.88235
1993	0.83154	1.08711	1.16216	0.92105
1994	0.84076	1.05846	1.17015	0.92219
1995	0.83978	1.06667	—	—
$\bar{s}_i$	0.82802	1.07543	1.18863	0.90362
$\hat{S}_i$	0.82891	1.07659	1.18991	0.90459

Table 13.17

t	y_t	M_t	s_t	$\frac{y_t}{\hat{S}_t}$	t	y_t	M_t	s_t	$\frac{y_t}{\hat{S}_t}$
1	.23			.277	11	.43	.37000	1.16216	.361
2	.32			.297	12	.35	.38000	0.92105	.387
3	.36	.29750	1.21008	.303	13	.33	.39250	0.84076	.398
4	.27	.30375	0.88889	.298	14	.43	.40625	1.05846	.399
5	.25	.31250	0.80000	.302	15	.49	.41875	1.17015	.412
6	.35	.32125	1.08949	.325	16	.40	.43375	0.92219	.442
7	.40	.33000	1.21212	.336	17	.38	.45250	0.83978	.458
8	.30	.34000	0.88235	.332	18	.50	.46875	1.06667	.464
9	.29	.34875	0.83154	.350	19	.57			.479
10	.39	.35875	1.08711	.362	20	.45			.497

Once the specific seasonals have been calculated, they are averaged over each of the four quarters in the cycle to obtain $\bar{s}_i$, for $i = 1, 2, ..., 4$ (see Table 13.16). These average indexes are then normalized by calculating

$$S = \bar{s}_1 + \bar{s}_2 + \cdots + \bar{s}_4$$

and we find the **seasonal indexes** as

$$\hat{S}_i = \bar{s}_i(\tfrac{4}{S})$$

These calculations are shown in Table 13.16. Finally, the original series is deseasonalized by using the seasonal indexes to find

$$\frac{y_t}{\hat{S}_t}$$

with $\hat{S}_t$ the seasonal index for the quarter corresponding to time t. The deseasonalized series is shown in Table 13.17.

13.25

a. Using the deseasonalized series from Exercise 13.24, a MINITAB program is used to fit a linear model to the data. The printout is shown below. The regression is highly significant.

```
The regression equation is
Deseason = 0.256 + 0.0113 t

Predictor       Coef       Stdev     t-ratio       p
Constant     0.255732    0.005293      48.32    0.000
t            0.0112708   0.0004418     25.51    0.000

s = 0.01139     R-sq = 97.3%     R-sq(adj) = 97.2%

Analysis of Variance

SOURCE       DF        SS          MS         F         p
Regression    1     0.084475    0.084475    650.78    0.000
Error        18     0.002337    0.000130
Total        19     0.086812
```

t	FITS1
1	0.267002
2	0.278273
3	0.289544
4	0.300815
5	0.312085
6	0.323356
7	0.334627
8	0.345898
9	0.357168
10	0.368439
11	0.379710
12	0.390981
13	0.402252
14	0.413522
15	0.424793
16	0.436064
17	0.447335
18	0.458605
19	0.469876
20	0.481147

The series is detrended by obtaining the fit $\hat{T}_t$ for each value of t and then finding $y_t/\hat{T}_t$, where y_t are the values of the seasonalized series. The calculations are shown in Table 13.18 and the deseasonalized and detrended series is shown in Figure 13.15.

Table 13.18

t	$y_t/\hat{T}_t$	t	$y_t/\hat{T}_t$	t	$y_t/\hat{T}_t$
1	1.03921	8	0.95879	15	0.96940
2	1.06814	9	0.97953	16	1.01405
3	1.04490	10	0.98321	17	1.02481
4	0.99223	11	0.95170	18	1.01270
5	0.96640	12	0.98960	19	1.01948
6	1.00539	13	0.98971	20	1.03391
7	1.00458	14	0.96587		

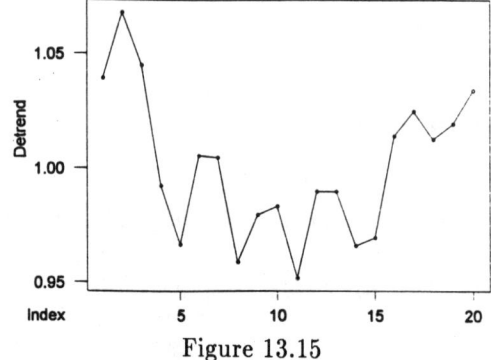

Figure 13.15

The deseasonalized and detrended series has no pronounced cyclic effect.

b. Because there is no cyclic effect, we can estimate

$$\hat{y}_{21} = \hat{T}_{21} \cdot \hat{S}_1 = [.255732 + 0.0112708(21)](.82891) = .408$$

13.26
a. The time series, with the twelve time periods plotted along the horizontal axis and the investment (in billions of dollars) plotted along the vertical axis, is shown in Figure 13.16. Notice the seasonal trend that is present.

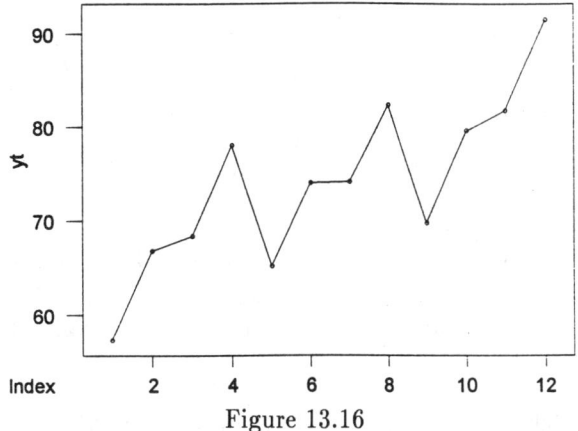

Figure 13.16

b. The model is given as

$$E(y_t) = \beta_0 + \beta_1 t + \beta_2 x_1 + \beta_3 x_2 + \beta_4 x_3$$

where $x_1 = 1$ if Quarter 1 $x_2 = 1$ if Quarter 2 $x_3 = 1$ if Quarter 3
       0 if not   0 if not   0 if not

c. In order to form a complete first-order autoregressive model, add the residual component,

$$z_t = \phi z_{t-1} + \epsilon_t$$

so that the entire model is

$$y_t = \beta_0 + \beta_1 t + \beta_2 x_1 + \beta_3 x_2 + \beta_4 x_3 + z_t$$

with x_1, x_2, and x_3 defined in part a and z_t defined above.

d. Referring to the fourth section of the printout in the column labeled "B VALUE," the estimates of the β_i are found and

$$\hat{y}_t = 70.3000 + 1.7045t - 14.6078x_1 - 6.9717x_2 - 7.4449x_3 + \hat{z}_t$$

where $\hat{z}_t = .5586\hat{z}_{t-1}$. Note that $\hat{\phi} = .5586$ is found in the second section of the printout as the negative of the value shown in the column labeled "COEFFICIENT" for LAG 1.

e. The last section of the printout provides the values $\hat{y}_t$ for the twelve quarters. The plotted values are shown below.

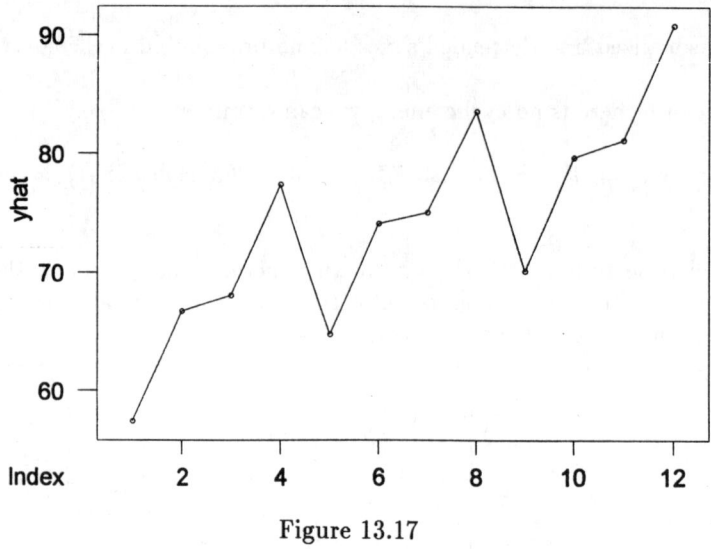

Figure 13.17

13.27

a. Refer to Exercise 13.26. For year 4, the first quarter represents the time period $t = 13$. Also, for Quarter 1, $x_1 = 1$, while $x_2 = x_3 = 0$. Finally, $\hat{z}_{13} = .5586\hat{z}_{12}$, where $\hat{z}_{12}$ is the residual at time $t = 12$ found in the last section of the computer printout to be $\hat{z}_{12} = .6750$. Hence, the forecast is

$$\hat{y}_{13} = 70.3000 + 1.7045(13) - 14.6078 + .5586(.6750) = 78.2278$$

b. The third section of the printout shows MSE $= .6968278$.

c. Using parts a and b, the approximate 95% prediction interval is

$$\hat{y}_t \pm 1.96\sqrt{MSE} \Rightarrow 78.2278 \pm 1.6361$$

13.28

Refer to the computer printout. For year 18, $t = 18$ and the forecast equation is given in the last section of the printout (ESTIMATES OF THE AUTOREGRESSIVE PARAMETERS: COEFFICIENTS and YULE-WALKER ESTIMATES: B VALUE) as

$$\hat{y}_t = 8.4424 - .0973t + \hat{z}_t$$

where $\hat{z}_t = .6827\hat{z}_{t-1} - .4601\hat{z}_{t-2}$. For $t = 18$,

$$\hat{z}_{18} = .6827\hat{z}_{17} - .4601\hat{z}_{16} = .6827(.4034) - .4601(.2847) = .1444$$

and $\hat{y}_{18} = 8.4424 - .0973(18) + .1444 = 6.8354$.

13.29

From Table 13.16, MSE $= .91595$ and the approximate 95% prediction equation for y_{18} in Exercise 13.28 is

$$\hat{y}_t \pm 1.96\sqrt{MSE} \Rightarrow 6.8354 \pm 1.96\sqrt{.91595} \Rightarrow 6.8354 \pm 1.8758$$

13.30

a-b. The data along with a three-point moving average is shown in Table 13.19, while the time series and the three-point moving averages are plotted in Figure 13.18.

Table 13.19

t	y_t	3-Point Moving Average	t	y_t	3-Point Moving Average
1	10.11	—	24	7.21	7.0967
2	10.00	10.0367	25	6.50	6.7367
3	10.00	10.0000	26	6.50	6.5000
4	10.00	10.0000	27	6.50	6.5000
5	10.00	10.0000	28	6.50	6.5000
6	10.00	10.0000	29	6.50	6.5000
7	10.00	10.0000	30	6.50	6.3400
8	10.00	10.0000	31	6.02	6.1733
9	10.00	10.0000	32	6.00	6.0067
10	10.00	10.0000	33	6.00	6.0000
11	10.00	10.0000	34	6.00	6.0000
12	10.00	9.8400	35	6.00	6.0000
13	9.52	9.5233	36	6.00	6.0000
14	9.05	9.1900	37	6.00	6.0000
15	9.00	9.0167	38	6.00	6.0000
16	9.00	8.8333	39	6.00	6.0000
17	8.50	8.6667	40	6.00	6.0000
18	8.50	8.5000	41	6.00	6.0000
19	8.50	8.5000	42	6.00	6.0000
20	8.50	8.4000	43	6.00	6.0000
21	8.20	8.2333	44	6.00	6.0000
22	8.00	7.9267	45	6.00	6.0000
23	7.58	7.5967	46	6.00	—

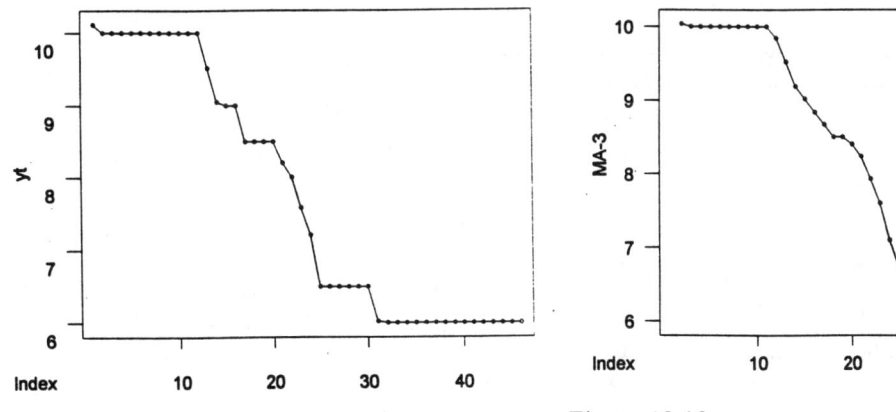

Figure 13.18

c. From the table in this exercise, in the section labeled "YULE WALKER ESTIMATES," we find $\hat{\beta}_0 = 11.1427$, $\hat{\beta}_1 = -.2037$, $\hat{\beta}_2 = .00182$, and $\hat{\phi} = .33766$. The fitted model is then

$$\hat{y}_t = 11.1427 - .2037t + .00182t^2 + .33766\hat{z}_{t-1}$$

d. From the computer printout, the predicted values are labeled "YHAT" as the final entries in the table. The plotted estimates and the observed values are shown in Figure 13.19.

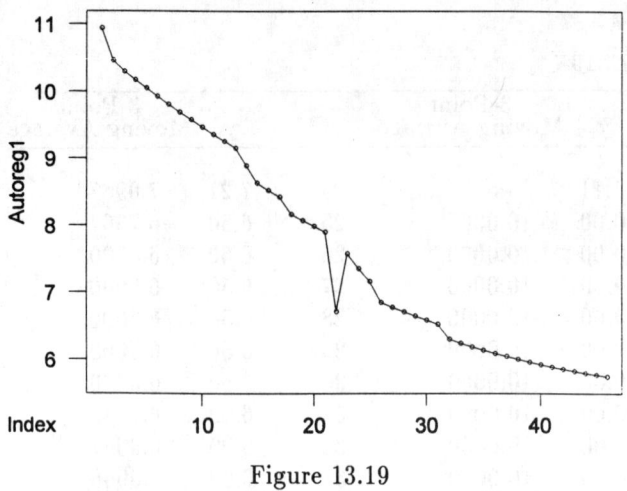

Figure 13.19

13.31

a. For November 1993, $t = 47$ and

$$\hat{y}_{47} = 11.1427 - .2037(47) + .00182(47^2) + .33766(.25750) = 5.676$$

b-c. From the printout, MSE = .332954 and the approximate 95% prediction interval is

$$\hat{y}_t \pm 1.96 \sqrt{MSE} \Rightarrow 5.676 \pm 1.96\sqrt{.332954} \Rightarrow 5.676 \pm 1.131$$

13.32

This is similar to previous exercises. The relevant data are shown in the table below, along with the sum, $P_1 + P_2$ = residential price + industrial price. The simple composite index is given in column 5.

Table 13.20

t	P_1	P_2	$P_1 + P_2$	I_t
1980	3.68	2.56	6.24	69.03
1981	4.29	3.14	7.43	82.19
1982	5.17	3.87	9.04	100.00
1983	6.06	4.18	10.24	113.27
1984	6.12	4.22	10.34	114.38
1985	6.12	3.95	10.07	111.39
1986	5.83	3.23	9.06	100.22
1987	5.54	2.94	8.48	93.81
1988	5.47	2.95	8.42	93.14
1989	5.64	2.96	8.60	95.13
1990	5.80	2.93	8.73	96.57
1991	5.82	2.69	8.51	94.14
1992	5.89	2.84	8.73	96.57

13.33

Refer to Exercise 13.32. For the Laspeyres index, the weights for the base year 1982 are

$$w_1 = 4770 \quad w_2 = 6794$$

and

$$y_0 = 4770(5.17) + 6794(3.87) = 50{,}953.68$$

The Laspeyres indexes are calculated below, with P_1 and P_2 as given in Table 13.20 above.

$$I_{88} = \frac{4770(5.47) + 6794(2.95)}{50{,}953.68}(100) = 90.54$$

$$I_{89} = \frac{4770(5.64) + 6794(2.96)}{50{,}953.68}(100) = 92.27$$

$$I_{90} = \frac{4770(5.80) + 6794(2.93)}{50{,}953.68}(100) = 93.36$$

$$I_{91} = \frac{4770(5.82) + 6794(2.69)}{50{,}953.68}(100) = 90.35$$

$$I_{92} = \frac{4770(5.89) + 6794(2.84)}{50{,}953.68}(100) = 93.01$$

The plot is omitted.

13.34

Refer to Exercise 13.32. The weights for the Paasche index are index-year weights, and the prices are P_1 and P_2, given in Table 13.20 in the solution to Exercise 13.32. Then

$$I_{88} = \frac{4692(5.47) + 2204(2.95)}{4692(5.17) + 2204(3.87)}(100) = 98.11$$

$$I_{89} = \frac{4798(5.64) + 1962(2.96)}{4798(5.17) + 1962(3.87)}(100) = 101.45$$

$$I_{90} = \frac{4471(5.80) + 1890(2.93)}{4471(5.17) + 1890(3.87)}(100) = 103.42$$

$$I_{91} = \frac{4550(5.82) + 1742(2.69)}{4550(5.17) + 1742(3.87)}(100) = 102.98$$

$$I_{92} = \frac{4678(5.89) + 1721(2.84)}{4678(5.17) + 1721(3.87)}(100) = 105.17$$

The plot is omitted.

13.35

a. The graph of average price vs. total consumption is shown in Figure 13.20. Notice that as the consumption decreases, the price decreases. However, the relationship does not appear to be very strong.

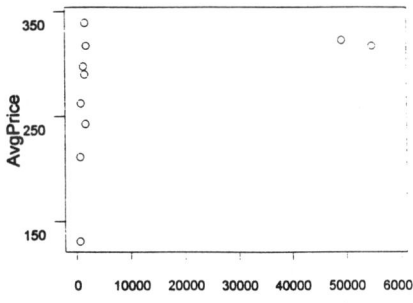

Figure 13.20

b-c. The quarterly prices, along with the four-point centered moving average and the exponentially smoothed series with $\omega = .3$, are shown in Table 13.21. The plot is shown in Figure 13.21.

Table 13.21

t	y_t	4-pt Centered Moving Average	$E_t(\omega = .3)$	t	y_t	4-pt Centered Moving Average	$E_t(\omega = .3)$
1	338.91	—	338.910	23	349.29	333.755	333.058
2	300.23	—	327.306	24	293.45	325.828	321.175
3	299.05	312.663	318.829	25	330.00	309.880	323.823
4	331.69	311.011	322.687	26	308.86	299.241	319.334
5	300.45	316.434	316.016	27	265.13	293.852	303.073
6	325.48	317.249	318.855	28	292.50	285.600	299.901
7	317.18	317.539	318.353	29	287.84	281.894	296.283
8	320.08	317.226	318.871	30	285.00	267.420	292.898
9	314.38	316.150	317.524	31	259.34	236.485	282.831
10	309.05	317.742	314.982	32	182.50	198.644	252.731
11	325.00	310.615	317.987	33	150.36	157.820	222.020
12	325.00	298.750	320.091	34	119.75	133.605	191.339
13	252.44	276.741	299.796	35	98.00	133.262	163.337
14	276.07	245.659	292.678	36	150.12	147.311	159.372
15	181.91	228.874	259.448	37	180.00	171.342	165.561
16	219.43	227.992	247.442	38	202.50	192.202	176.642
17	223.73	250.194	240.329	39	207.50	202.812	185.900
18	297.73	285.042	257.549	40	207.50	206.100	192.380
19	337.86	316.181	281.642	41	207.50	203.137	196.916
20	342.27	338.784	299.831	42	201.30	196.262	198.231
21	350.00	347.031	314.881	43	185.00	—	194.262
22	352.28	342.358	326.101	44	175.00	—	188.483

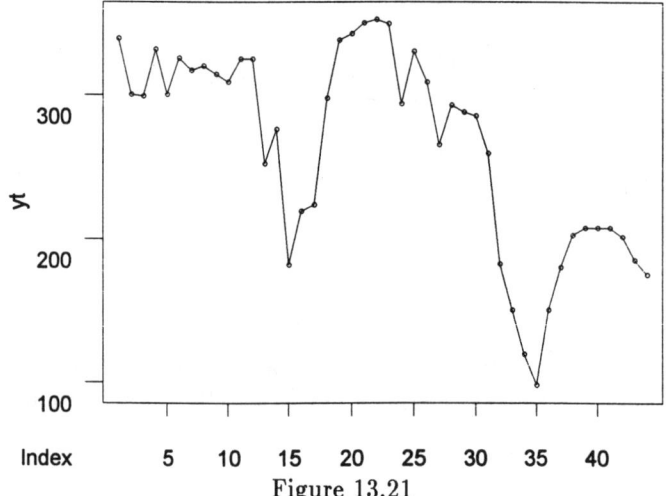

Figure 13.21

13.36

a. Figure 13.22 shows the secular trend in the time series.

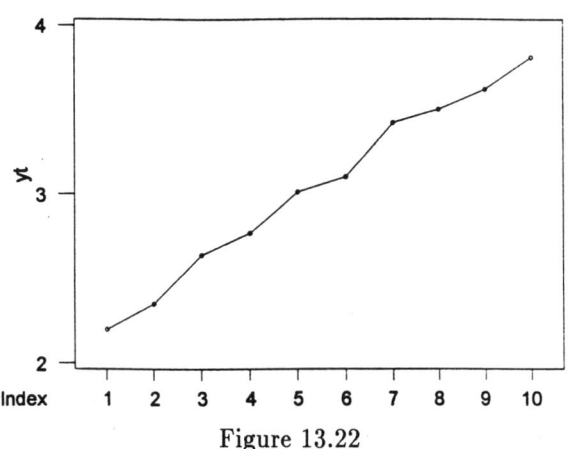

Figure 13.22

b. The secular trend is represented by

$$E(y_t) = \beta_0 + \beta_1 t$$

c. The complete first-order autoregressive model is

$$E(y_t) = \beta_0 + \beta_1 t + z_t \quad \text{where} \quad z_t = \phi z_{t-1}$$

d. From the printout given in the exercise, the prediction equation is

$$\hat{y}_t = 2.05173936 + 0.180217t + \hat{z}_t$$

where

$$\hat{z}_t = -0.14021658\hat{z}_{t-1}$$

13.37

a. For $t = 11$ in Exercise 13.36,

$$\hat{y}_{11} = 2.05173936 + 0.180217(11) - .14021658(-.05144) = 4.0413$$

b. The approximate 95% prediction interval is

$$\hat{y}_t \pm 1.96\sqrt{\text{MSE}} \Rightarrow 4.0413 \pm 1.96\sqrt{.003878} \Rightarrow 4.0413 \pm .1221$$

13.38

a. From the computer printout,

$$\hat{y}_t = 11.02788 - 0.19534t + 0.001703t^2 + \hat{z}_t$$

where

$$\hat{z}_t = 0.26093\hat{z}_{t-1} + .22725\hat{z}_{t-2}$$

b. The two plots are superimposed in Figure 13.23.

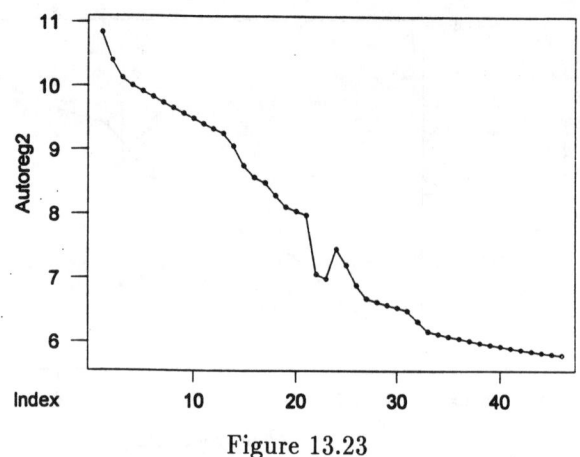

Figure 13.23

c. Since MSE = .31838 compared to MSE = .332954 for the first order model, there is a slight improvement in the fit.

13.39

a. The twelve-point centered moving averages for this time series (t = 1 to t = 46) are calculated in Table 13.23 for the time periods t = 7 to t = 40. Using these moving averages, we calculate the specific seasonals, $s_t = y_t/M_t$ for times t = 7 to t = 40. For example, for t = 7,

$$s_7 = \frac{y_7}{M_7} = \frac{10.00}{9.98458} = 1.00154$$

These specific seasonals are calculated using the data given in Table 13.23 and are shown in a different format in Table 13.22.

Table 13.22

Month	Jan.	Feb.	March	April	May	June
1990	—	—	—	—	—	—
1991	1.01259	0.97557	0.98477	1.00213	0.96628	0.99072
1992	0.90962	0.93682	0.96403	0.98972	1.01272	1.03140
1993	0.99986	1.00000	1.00000	1.00000	—	—
$\bar{s}_i$	0.97402	0.97080	0.98293	0.99728	0.98950	1.01106
$\hat{S}_i$	0.97578	0.97255	0.98471	0.99908	0.99129	1.01289

Month	July	Aug.	Sept.	Oct.	Nov.	Dec.
1990	1.00154	1.00802	1.01635	1.02503	1.03609	1.04969
1991	1.01949	1.04868	1.03863	1.04076	1.01078	0.98329
1992	0.96616	0.96944	0.97601	0.98267	0.98942	0.99626
1993	—	—	—	—		
$\bar{s}_i$	0.99573	1.00871	1.01033	1.01615	1.01210	1.00975
$\hat{S}_i$	0.99753	1.01053	1.01215	1.01799	1.01393	1.01157

Table 13.23

t	y_t	M_t	s_t	$\dfrac{y_t}{\hat{S}_t}$	t	y_t	M_t	s_t	$\dfrac{y_t}{\hat{S}_t}$
1	10.11	—	—	10.3609	24	7.21	7.33250	0.98329	7.1275
2	10.00	—	—	10.2822	25	6.50	7.14583	0.90962	6.6613
3	10.00	—	—	10.1553	26	6.50	6.93833	0.93682	6.6835
4	10.00	—	—	10.0092	27	6.50	6.74250	0.96403	6.6009
5	10.00	—	—	10.0879	28	6.50	6.56750	0.98972	6.5060
6	10.00	—	—	9.8727	29	6.50	6.41833	1.01272	6.5571
7	10.00	9.98458	1.00154	10.0248	30	6.50	6.30208	1.03140	6.4173
8	10.00	9.92042	1.00802	9.8958	31	6.02	6.23083	0.96616	6.0349
9	10.00	9.83917	1.01635	9.8800	32	6.00	6.18917	0.96944	5.9375
10	10.00	9.75583	1.02503	9.8233	33	6.00	6.14750	0.97601	5.9280
11	10.00	9.65167	1.03609	9.8626	34	6.00	6.10583	0.98267	5.8940
12	10.00	9.52667	1.04969	9.8856	35	6.00	6.06417	0.98942	5.9176
13	9.52	9.40167	1.01259	9.7563	36	6.00	6.02250	0.99626	5.9314
14	9.05	9.27667	0.97557	9.3054	37	6.00	6.00083	0.99986	6.1489
15	9.00	9.13917	0.98477	9.1397	38	6.00	6.00000	1.00000	6.1693
16	9.00	8.98083	1.00213	9.0083	39	6.00	6.00000	1.00000	6.0932
17	8.50	8.79667	0.96628	8.5747	40	6.00	6.00000	1.00000	6.0055
18	8.50	8.57958	0.99072	8.3918	41	6.00	—	—	6.0527
19	8.50	8.33750	1.01949	8.5210	42	6.00	—	—	5.9236
20	8.50	8.10542	1.04868	8.4114	43	6.00	—	—	6.0149
21	8.20	7.89500	1.03863	8.1016	44	6.00	—	—	5.9375
22	8.00	7.68667	1.04076	7.8586	45	6.00	—	—	5.9280
23	7.58	7.49917	1.01078	7.4759	46	6.00	—	—	5.8940

Once the specific seasonals have been calculated, they are averaged over each of the twelve months in the cycle to obtain $\bar{s}_i$, for $i = 1, 2, \ldots, 12$ (see Table 13.22). These average indexes are then normalized by calculating

$$S = \bar{s}_1 + \bar{s}_2 + \cdots + \bar{s}_{12}$$

and we find the seasonal indexes as

$$\hat{S}_i = \bar{s}_i \left(\frac{12}{S}\right)$$

These calculations are shown in Table 13.22. Finally, the original series is deseasonalized by using the seasonal indexes to find

$$\frac{y_t}{\hat{S}_t}$$

with $\hat{S}_t$ the seasonal index for the month corresponding to time t. The deseasonalized series is shown in Table 13.23.

13.40
a. Using the deseasonalized series from Exercise 13.39, a MINITAB program is used to fit a cubic model to the data. The printout is shown below. The regression is highly significant.

The regression equation is
Deseason = 10.2 + 0.0622 t +0.000193 t-cu - 0.0120 t-sq

Predictor	Coef	Stdev	t-ratio	p
Constant	10.1644	0.1529	66.48	0.000
t	0.06215	0.02787	2.23	0.031
t-cu	0.00019290	0.00001917	10.06	0.000
t-sq	-0.012029	0.001370	-8.78	0.000

s = 0.2384   R-sq = 98.2%   R-sq(adj) = 98.1%

Analysis of Variance

SOURCE	DF	SS	MS	F	p
Regression	3	130.190	43.397	763.62	0.000
Error	42	2.387	0.057		
Total	45	132.577			

SOURCE	DF	SEQ SS
t	1	121.622
t-cu	1	4.186
t-sq	1	4.382

ROW	FITS2
1	10.2147
2	10.2421
3	10.2478
4	10.2329
5	10.1985
6	10.1459
7	10.0762
8	9.9905
9	9.8900
10	9.7759
11	9.6493
12	9.5113
13	9.3632
14	9.2061
15	9.0411
16	8.8694
17	8.6922
18	8.5106
19	8.3258
20	8.1389
21	7.9511
22	7.7635
23	7.5774
24	7.3938
25	7.2139
26	7.0389
27	6.8700
28	6.7082
29	6.5548
30	6.4109
31	6.2776
32	6.1562
33	6.0477
34	5.9534
35	5.8744
36	5.8118
37	5.7669
38	5.7407
39	5.7344
40	5.7492
41	5.7862
42	5.8467
43	5.9317
44	6.0424
45	6.1799
46	6.3455

The series is detrended by obtaining the fit $\hat{T}_t$ for each value of t and then finding $y_t/\hat{T}_t$, where y_t are the values of the deseasonalized series. The calculations are shown in Table 13.24 and the deseasonalized and detrended series is shown in Figure 13.24.

Table 13.24

t	$y_t/\hat{T}_t$	t	$y_t/\hat{T}_t$	t	$y_t/\hat{T}_t$
1	1.01432	17	0.98648	32	0.96447
2	1.00392	18	0.98604	33	0.98020
3	0.99097	19	1.02345	34	0.99001
4	0.97814	20	1.03348	35	1.00735
5	0.98915	21	1.01892	36	1.02057
6	0.97307	22	1.01225	37	1.06625
7	0.99490	23	0.98660	38	1.07468
8	0.99052	24	0.96399	39	1.06257
9	0.99898	25	0.92340	40	1.04459
10	1.00485	26	0.94950	41	1.04606
11	1.02211	27	0.96084	42	1.01317
12	1.03935	28	0.96986	43	1.01403
13	1.04198	29	1.00036	44	0.98264
14	1.01079	30	1.00100	45	0.95923
15	1.01091	31	0.96134	46	0.92884
16	1.01566				

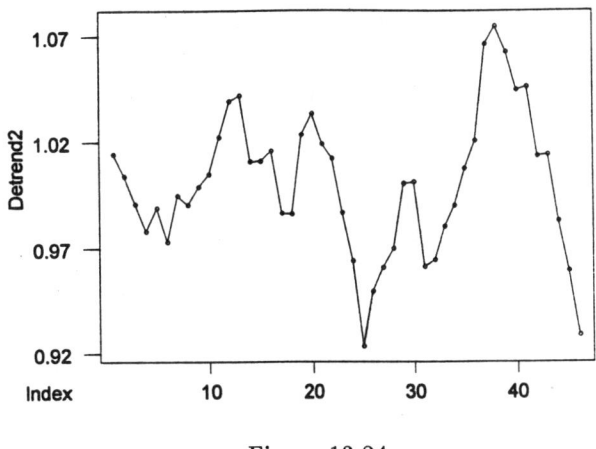

Figure 13.24

The deseasonalized and detrended series does not appear to have a cyclic effect.

b. Since there is no cyclic effect, we can predict y_{47} as

$$\hat{y}_{47} = \hat{T}_{47} \cdot \hat{S}_{11} = [10.1644 + 0.06215(47) + .0001929(47^2) - .012029(47^3)](1.01393)$$

$$= 6.541$$

If the student chooses to predict using the cyclic effect, he should guess the value of C_{47} and multiply the above prediction by that value. Because of the highly variable results, mostly based on educated guesses by the experimenter, no measure of goodness can be given.

CHAPTER 14
Sampling Methods

14.1

Remove two cards from the deck. Assign each of the fifty cards remaining to represent a population sampling unit. Thoroughly shuffle the deck, and draw five cards. These five sampling units represent the sample.

14.2

Each student will obtain a different answer. Two-digit random numbers should be chosen, only including 00-21 in the sample (or some equivalent system of choice).

14.3

Answers will vary; five-digit random numbers will be used.

14.4

Answers will vary; three-digit random numbers will be used.

14.5

Referring to Table 13 in Appendix II, we will select twenty numbers. First choose a starting point and consider the first four digits in each number. If the four digits are a number greater than 7000, discard it. Continue until twenty numbers have been chosen. The customers have already been numbered from 0001 to 7000. One possible selection for the sample size $n = 20$ is obtained by starting in line 1, column 1 of Table 13 as follows:

1048	2891	5108	4866
2236	6355	0236	5416
2413	0942	0101	3263
4216	1036	5216	2933
3757	0711	0705	0248

14.6

The student should use the same procedure as in Exercise 14.5 and refer to Table 13 in Appendix II.

14.7

No. The readers who do not respond to the questionnaire are probably those who are satisfied with the status quo, or have no strong feelings about a given subject. Those who respond are those who have strong and possible radical opinions about issues.

14.8

If all of the town citizenry is likely to pass this corner, a sample obtained by selecting every tenth person is probably a fairly random sample. However, since this is an unlikely assumption, it is unlikely that the sample will be random.

14.9

No. Only homemakers, retired persons, people who work nights, or others who work at home will respond to the survey. All workers who are out of the house from 9 A.M. to 5 P.M. will be systematically excluded.

14.10

No. The sample was not random. Only those employees with a particularly strong opinion on the subject will return the questionnaire.

14.11

When simple random sampling is used, the estimate of μ is the sample mean $\bar{x}$. The variance of $\bar{x}$ depends on the population and sample sizes.

a. When N is large, $\sigma_{\bar{x}} = \sigma/\sqrt{n}$, and the 95% confidence interval is approximately

$$\bar{x} \pm 1.96 \frac{s}{\sqrt{n}}$$

$$84.1 \pm 1.96\sqrt{\frac{122.44}{50}}$$

$$84.1 \pm 3.067 \qquad \text{or} \qquad 81.033 < \mu < 87.167$$

b. When N is small, a finite population correction factor must be used, and the 95% confidence interval is approximately

$$\bar{x} \pm 1.96 \frac{s}{\sqrt{n}}\sqrt{\frac{N-n}{N}}$$

$$84.1 \pm 1.96\sqrt{\frac{122.44}{50}}\sqrt{\frac{100-50}{100}}$$

$$84.1 \pm 2.169 \qquad \text{or} \qquad 81.931 < \mu < 86.269$$

14.12

The estimate of the population total, τ, is

$$\hat{\tau} = N\bar{x} = 100(84.1) = 8410$$

and the approximate 95% confidence interval is

$$\hat{\tau} \pm 1.96 \frac{Ns}{\sqrt{n}}\sqrt{\frac{N-n}{N}}$$

$$8410 \pm 1.96(100)\sqrt{\frac{122.44}{50}\left(\frac{100-50}{100}\right)}$$

$$8410 \pm 216.879 \qquad \text{or} \qquad 8193.121 < \tau < 8626.879$$

14.13

When $n = 100$, $N = 400$, and $x = 34$, the 95% confidence interval for p is approximately

$$\hat{p} \pm 1.96\sqrt{\frac{\hat{p}\hat{q}}{n-1}\left(\frac{N-n}{N}\right)}$$

$$.34 \pm 1.96\sqrt{\frac{.34(.66)}{99}\left(\frac{300}{400}\right)}$$

$$.34 \pm .081 \qquad \text{or} \qquad .249 < p < .431$$

14.14

The finite population correction factor, given in the text as

$$\sqrt{\frac{N-n}{N}} = \sqrt{1 - \frac{n}{N}}$$

is shown below for various values of n/N. As n/N increases (approaches 1), the finite correction factor approaches 0. Hence, as the sample size approaches the population size, the variance of $\bar{x}$ becomes quite small and the interval estimates become quite narrow.

n/N	.001	.1	.3	.5	.7
FPC	.999	.949	.837	.707	.548

14.15

The 95% confidence interval for μ is approximately

$$\bar{x} \pm 1.96 \frac{s}{\sqrt{n}}\sqrt{\frac{N-n}{N}}$$

$$1248 \pm 1.96\frac{175}{\sqrt{50}}\sqrt{\frac{421-50}{421}}$$

$$1248 \pm 45.536 \quad \text{or} \quad 1202.464 < \mu < 1293.536$$

14.16

The approximate 95% confidence interval for τ is

$$N\bar{x} \pm 1.96 \frac{Ns}{\sqrt{n}}\sqrt{\frac{N-n}{N}}$$

$$421(1248) \pm 1.96\sqrt{\frac{421(175)}{50}\left(\frac{421-50}{421}\right)}$$

$$525{,}408 \pm 19{,}170.662$$

14.17

With $N = 8746$, $n = 500$, and $x = 29$, calculate $\hat{p} = 29/500 = .058$, and the approximate 95% confidence interval is

$$\hat{p} \pm 1.96\sqrt{\frac{\hat{p}\hat{q}}{n-1}\left(\frac{N-n}{N}\right)}$$

$$.058 \pm 1.96\sqrt{\frac{.058(.942)}{499}\left(\frac{8746-500}{8746}\right)}$$

$$.058 \pm .020$$

14.18

The 95% confidence interval for μ is approximately

$$\bar{x} \pm 1.96 \frac{s}{\sqrt{n}}\sqrt{\frac{N-n}{N}}$$

$$6751 \pm 1.96\frac{1463}{\sqrt{46}}\sqrt{\frac{9706-1000}{9706}}$$

$$6751 \pm 400.415$$

Notice that the FPC, $\sqrt{(9706-1000)/9706}$, is used since the total number of the bank's customers wishing to borrow money is unknown.

14.19

With $N = 9706$, $n = 1000$, and $x = 46$, the approximate 95% confidence interval is

$$\hat{p} \pm 1.96\sqrt{\frac{\hat{p}\hat{q}}{n-1}\left(\frac{N-n}{N}\right)}$$

$$.046 \pm 1.96\sqrt{\frac{.046(.954)}{999}\left(\frac{8706}{9706}\right)}$$

$$.046 \pm .012$$

14.20

When stratified sampling is used, estimates of μ are obtained within each stratum and then a weighted average of the estimates is calculated. Formulas for estimating population means and proportions are given in Section 14.4. The stratum sizes, N_i, sample sizes, n_i, and sample observations, x_i, are given. The table below shows the sample means and variances for each stratum.

Stratum	1	2	3	4
$\bar{x}_i$	6.25	8.50	6.75	5.75
s_i^2	1.5833	1.6667	.9167	.9167

Then the estimate of μ is

$$\bar{x}_{st} = \frac{1}{N}\sum N_i\bar{x}_i = \frac{1}{150}[40(6.25) + 30(8.50) + 30(6.75) + 50(5.75)] = 6.633$$

and the 95% confidence interval is

$$\bar{x}_{st} \pm 1.96\sqrt{\frac{1}{N^2}\sum N_i^2\left(\frac{N_i-n_i}{N_i}\right)\frac{s_i^2}{n_i}}$$

$$6.633 \pm 1.96\sqrt{\frac{1}{150^2}\left\{\frac{40(36)s_1^2}{4} + \frac{30(26)s_2^2}{4} + \frac{30(26)s_3^2}{4} + \frac{50(46)s_4^2}{4}\right\}}$$

$$6.633 \pm .523 \qquad \text{or} \qquad 6.110 < \mu < 7.156$$

14.21

Refer to Exercise 14.20. The estimate of τ is

$$\hat{\tau} = N\bar{x}_{st} = 150(6.63333) = 995$$

and the approximate 95% confidence interval is

$$995 \pm 150(.52280266) \qquad \text{(from Exercise 14.20)}$$

$$995 \pm 78.42 \qquad \text{or} \qquad 916.58 < \tau < 1073.42$$

14.22

The estimate of μ is

$$\bar{x}_{st} = \frac{1}{N}\sum N_i\bar{x}_i = \frac{1}{7000}[1000(421) + 3000(502) + 2000(325) + 1000(280)] = 408.143$$

and the 95% confidence interval is approximately

$$\bar{x}_{st} \pm 1.96 \sqrt{\frac{1}{N^2}\sum N_i^2\left(\frac{N_i - n_i}{N_i}\right)\frac{s_i^2}{n_i}}$$

408.143 ± 1.96 (times)

$$\sqrt{\frac{1}{7000^2}\left\{(1000)^2(.8)\frac{2410}{200} + (3000)^2(.9333)\frac{2938}{200} + (2000)^2(.9)\frac{2047}{200} + (1000)^2(.8)\frac{2214}{200}\right\}}$$

408.143 ± 3.743

14.23

Refer to Exercise 14.22. The estimate of τ is

$$\hat{\tau} = N\bar{x}_{st} = 7000\left(\frac{2{,}857{,}000}{7000}\right) = 2{,}857{,}000$$

and the approximate 95% confidence interval, using the calculations from Exercise 14.22, is

$$2{,}857{,}000 \pm 1.96\sqrt{178{,}738{,}000}$$

$$2{,}857{,}000 \pm 26{,}203.815$$

14.24

In the case of population proportions, a weighted average of stratum estimates is again used. To estimate p, use

$$\hat{p}_{st} = \frac{1}{N}\sum N_i\hat{p}_i = \frac{1}{4500}[1000(.3) + 1200(.25) + 800(.29) + 1500(.34)] = .298$$

and the 95% confidence interval is approximately

$$\hat{p}_{st} \pm 1.96 \sqrt{\frac{1}{N^2}\sum N_i^2\left(\frac{N_i - n_i}{N_i}\right)\frac{\hat{p}_i\hat{q}_i}{n_i - 1}}$$

$$.298 \pm 1.96 \sqrt{\frac{1}{4500^2}\left[1000(900)\frac{.3(.7)}{99} + 1200(1000)\frac{.25(.75)}{99} + 800(700)\frac{.29(.71)}{99} + 1500(1400)\frac{.34(.66)}{99}\right]}$$

$$.298 \pm 1.96 \sqrt{.0005103}$$

.298 ± .0443   or   .2537 < p < .3423

14.25

The estimate of p is

$$\hat{p}_{st} = \frac{1}{N}\sum N_i\hat{p}_i = \frac{1}{900}[400(.62) + 200(.74) + 300(.55)] = .7929$$

and the 95% confidence interval is approximately

$$\hat{p}_{st} \pm 1.96 \sqrt{\frac{1}{N^2}\sum N_i^2\left(\frac{N_i - n_i}{N_i}\right)\frac{\hat{p}_i\hat{q}_i}{n_i - 1}}$$

$$.7929 \pm 1.96 \sqrt{\frac{1}{900^2}\left[400(300)\frac{.62(.38)}{99} + 200(100)\frac{.74(.26)}{99} + 300(200)\frac{.55(.45)}{99}\right]}$$

.7929 ± .0474

14.26

This is similar to Exercise 14.20. The table shows the sample means and variances for each stratum.

	Stratum I	Stratum II
N_i	110	168
n_i	20	30
$\bar{x}_i$	12,000	14,000
s_i^2	5,263,157.9	4,482,758.6

Then $\bar{x}_{st} = \frac{1}{N}\sum N_i \bar{x}_i = \frac{1}{278}[110(12,000) + 168(14,000)] = 13,208.633$

and the 95% confidence interval is approximately

$$\bar{x}_{st} \pm 1.96 \sqrt{\frac{1}{N^2}\sum N_i^2 \left(\frac{N_i - n_i}{N_i}\right)\frac{s_i^2}{n_i}}$$

$$13,208.633 \pm 1.96 \sqrt{\frac{1000^2}{278^2}\left\{110(90)\frac{5.2631579}{20} + 168(138)\frac{4.4827586}{30}\right\}}$$

$$13,208.633 \pm 549.2741$$

14.27

Refer to Exercise 14.26.

$N\bar{x}_{st} \pm N(549.2741)$

$278(13,208.633) \pm 152,698.2$

$3,672,000 \pm 152,698.2$

14.28

This is similar to Exercise 14.24. Calculate $\hat{p}_i$ for each stratum.

Stratum	1	2	3
N_i	249	432	316
n_i	50	80	60
$\hat{p}_i$	.28	.425	.4833

To estimate p, use

$$\hat{p}_{st} = \frac{1}{N}\sum N_i \hat{p}_i = \frac{1}{997}[249(.28) + 432(.425) + 316(.4833)] = .4073$$

and the 95% confidence interval is approximately

$$\hat{p}_{st} \pm 1.96 \sqrt{\frac{1}{N^2}\sum N_i^2 \left(\frac{N_i - n_i}{N_i}\right)\frac{\hat{p}_i \hat{q}_i}{n_i - 1}}$$

$$.4073 \pm 1.96 \sqrt{\frac{1}{997^2}\left[249(199)\frac{.28(.72)}{49} + 432(352)\frac{.425(.575)}{79} + 316(256)\frac{.4833(.5167)}{59}\right]}$$

$$.4073 \pm .0627 \quad \text{or} \quad .3446 < p < .4700$$

14.29

This is similar to previous exercises. The estimate of μ is

$$\bar{x}_{st} = \frac{1}{N}\sum N_i \bar{x}_i = \frac{1}{1300}[425(287) + 316(389) + 559(316)] = 324.2638$$

and the 95% confidence interval is approximately

$$\bar{x}_{st} \pm 1.96 \sqrt{\frac{1}{N^2}\sum N_i^2\left(\frac{N_i - n_i}{N_i}\right)\frac{s_i^2}{n_i}}$$

$$324.2638 \pm 1.96 \sqrt{\frac{1}{1300^2}\left\{425(375)\frac{41,116}{50} + 316(266)\frac{35,488}{50} + 559(509)\frac{59,106}{50}\right\}}$$

$$324.2638 \pm 34.613$$

14.30

From Exercise 14.29, $\hat{\tau} = N\bar{x}_{st} = 421{,}543$ and the approximate 95% confidence interval is

$$421{,}543 \pm 1300(34.613)$$

$$421{,}543 \pm 44{,}997$$

14.31

From Exercises 14.29 and 14.30,

$$\hat{\tau}_1 = 425(287) = 121{,}975 \qquad \hat{\tau}_2 = 316(389) = 122{,}924$$

$$\hat{\tau}_1 - \hat{\tau}_2 = -949$$

and the approximate 95% confidence interval is

$$-949 \pm 1.96\sqrt{425^2\left(\frac{41{,}116}{50}\right)\left(\frac{375}{425}\right) + 316^2\left(\frac{35{,}488}{50}\right)\left(\frac{266}{316}\right)}$$

$$-949 \pm 27{,}067.65$$

14.32

The estimate of μ for a cluster sample pools the combined information for all the clusters. Formulas for estimating means and proportions are given in Section 14.5. For this exercise, $\Sigma x_i = 4944$ and $\Sigma m_i = 200(10) = 2000$. Hence,

$$\bar{x} = \frac{\Sigma x_i}{\Sigma m_i} = \frac{4944}{2000} = 2.472$$

and the approximate 95% confidence interval is

$$\bar{x} \pm 1.96\sqrt{\left(\frac{N-n}{Nn\bar{M}^2}\right)\frac{\Sigma(x_i - \bar{x}m_i)^2}{n-1}}$$

$$2.472 \pm 1.96 \sqrt{\left(\frac{19{,}800}{20{,}000(200)10^2}\right)\left(\frac{483}{199}\right)}$$

$$2.472 \pm .02148 \quad \text{or} \quad 2.45052 < \mu < 2.49348$$

14.33

Refer to Exercise 14.32, with $\hat{\tau} = M\bar{x} = 200{,}000\left(\frac{4944}{2000}\right) = 494{,}400$. The approximate 95% confidence interval is

$$494{,}400 \pm 1.96(200{,}000)\sqrt{\left(\frac{19{,}800}{20{,}000(200)10^2}\right)\left(\frac{483}{199}\right)}$$

$$494{,}400 \pm 4296.71$$

14.34

This is similar to Exercise 14.32, with

$$n = 10 \qquad \Sigma m_i = 31 \qquad \Sigma x_i = 190$$

$$N = 2000 \qquad \Sigma m_i^2 = 117 \qquad \Sigma x_i^2 = 4650$$

$$\bar{M} = 3.2 \qquad \Sigma m_i x_i = 732$$

Then $\Sigma(x_i - \bar{x}m_i)^2 = 4650 - 2\left(\frac{190}{31}\right)(732) + \left(\frac{190}{31}\right)^2(117) = 72.206035$

and $\quad \bar{x} = \frac{\Sigma x_i}{\Sigma m_i} = \frac{190}{31} = 6.129$

The approximate 95% confidence interval for μ is

$$\bar{x} \pm 1.96 \sqrt{\left(\frac{N-n}{Nn\bar{M}^2}\right)\frac{\Sigma(x_i - \bar{x}m_i)^2}{n-1}}$$

$$6.129 \pm 1.96 \sqrt{\left(\frac{1990}{2000(10)(3.2)^2}\right)\left(\frac{72.206035}{9}\right)}$$

$$6.129 \pm .5472 \quad \text{or} \quad 5.5818 < \mu < 6.6762$$

14.35

Refer to Exercise 14.34, with $\hat{\tau} = M\bar{x} = 6400\left(\frac{190}{31}\right) = 39{,}225.8$. The approximate 95% confidence interval is

$$39{,}225.8 \pm 1.96(6400)\sqrt{\left(\frac{1990}{2000(10)(3.2)^2}\right)\left(\frac{72.206035}{9}\right)}$$

$$39{,}225.8 \pm 3502.38$$

14.36

Let m_i be the number of elements in cluster i, and a_i be the number of successes in cluster i. The following information is necessary.

$$n = 10 \qquad \Sigma m_i = 105 \qquad \Sigma a_i = 53$$

$$N = 2000 \qquad \Sigma m_i^2 = 1199 \qquad \Sigma a_i^2 = 319$$

$$\bar{M} = 10.7 \qquad \Sigma a_i m_i = 609$$

Then $\Sigma(a_i - \hat{p}m_i)^2 = 319 - 2\left(\frac{53}{105}\right)(609) + \left(\frac{53}{105}\right)^2(1199) = 9.686712$

and $\quad \hat{p} = \frac{\Sigma a_i}{\Sigma m_i} = \frac{53}{105} = .5048$

The approximate 95% confidence interval for p is

$$\hat{p} \pm 1.96\sqrt{\left(\frac{N-n}{Nn\bar{M}^2}\right)\frac{\Sigma(a_i - \hat{p}m_i)^2}{n-1}}$$

$$.5048 \pm 1.96\sqrt{\left(\frac{1990}{2000(10)(10.7)^2}\right)\left(\frac{9.686712}{9}\right)}$$

$.5048 \pm .0599 \quad$ or $\quad .4449 < p < .5647$

14.37

This is similar to Exercise 14.34. The following information is necessary:

$n = 25 \qquad \Sigma m_i = 151 \qquad \Sigma x_i = 2658(1000)$

$N = 415 \qquad \Sigma m_i^2 = 1047 \qquad \Sigma x_i^2 = 328,156(1000)^2$

$\bar{m} = 6.04 \qquad \Sigma m_i x_i = 16,806(1000)$

Then $\Sigma(x_i - \bar{x}m_i)^2 = (1000)^2[328,156 - 2\left(\frac{2658}{151}\right)(16,806) + \left(\frac{2658}{151}\right)^2(1047)]$

$\qquad = 60,912.116(1000)^2$

and $\quad \bar{x} = \frac{\Sigma x_i}{\Sigma m_i} = \frac{2658(1000)}{151} = 17,602.649$

The approximate 95% confidence interval for μ is

$$\bar{x} \pm 1.96\sqrt{\left(\frac{N-n}{Nn\bar{M}^2}\right)\frac{\Sigma(x_i - \bar{x}m_i)^2}{n-1}}$$

$$17,602.649 \pm 1.96(1000)\sqrt{\left(\frac{390}{415(25)(6.04)^2}\right)\left(\frac{60,912.116}{24}\right)}$$

14.38

$17,602.649 \pm 3169.5938 \quad$ or $\quad 14,433.055 < \mu < 20,772.243$

With $M = 2562$, the estimate of τ is $\hat{\tau} = M\bar{x} = 2562(17,602.649) = 45,097,986.75$, and the approximate 95% confidence interval is

$$M\bar{x} \pm 1.96\sqrt{N^2\left(\frac{N-n}{Nn}\right)\frac{\Sigma(x_i - \bar{x}m_i)^2}{n-1}}$$

$$45,097,986.75 \pm 1.96(1000)\sqrt{415^2\left(\frac{390}{415(25)}\right)\left(\frac{60,912.116}{24}\right)}$$

$45,097,986.75 \pm 7,944,903.77 \quad$ or $\quad 37,153,082.98 < \tau < 53,042,890.52$

14.39

This is similar to Exercise 14.36 with

m_i = number of employees at plant i
a_i = number favoring new policy at plant i

The following information is necessary:

$$n = 15 \qquad \Sigma m_i = 911 \qquad \Sigma a_i = 646$$

$$N = 87 \qquad \Sigma m_i^2 = 58{,}075 \qquad \Sigma a_i^2 = 29{,}104$$

$$\bar{m} = 911/15 = 60.733 \qquad \Sigma a_i m_i = 40{,}730$$

Then $\Sigma(a_i - \hat{p}m_i)^2 = 29{,}104 - 2\left(\frac{646}{911}\right)(40{,}730) + \left(\frac{646}{911}\right)^2(58{,}075) = 542.1585$

and $\hat{p} = \frac{\Sigma a_i}{\Sigma m_i} = \frac{646}{911} = .70911$

The approximate 95% confidence interval for p is

$$\hat{p} \pm 1.96 \sqrt{\left(\frac{N-n}{Nn\bar{M}^2}\right)\frac{\Sigma(a_i - \hat{p}m_i)^2}{n-1}}$$

$$.70911 \pm 1.96 \sqrt{\left(\frac{72}{87(15)(60.733)^2}\right)\left(\frac{542.1585}{14}\right)}$$

$$.70911 \pm .0472$$

14.40

Answers will vary from student to student. Two-digit numbers from 01 to 49 should be selected, and n = 5 will be chosen for inclusion in the sample.

14.41

Answers will vary from student to student. Basically, six-digit random numbers will be selected n = 1000 times to determine which homeowners will be chosen for inclusion. If the student chooses to select random numbers from 000001 to 474159, approximately half of all random numbers will have to be eliminated. A more efficient method would be to assign two six-digit random numbers to each homeowner and to choose this homeowner if either of the two random numbers are selected. For example, homeowner #1 would be assigned numbers 000001 and 474160. Homeowner #2 would be assigned 000002 and 474161. Homeowner #4323 would be assigned 004323 and 474159 + 004323 = 478482. In this way, 2(474159) = 948318 six-digit numbers will be used, and only 5% of numbers chosen will need to be eliminated.

14.42

This is similar to previous exercises with N = 6300. Calculate

$$\bar{x}_{st} = \frac{1}{N}\Sigma N_i \bar{x}_i = \frac{1}{6300}[2400(12.1) + 3000(13.4) + 900(9.7)] = 12.37619$$

and the 95% confidence interval is approximately

$$\bar{x}_{st} \pm 1.96 \sqrt{\frac{1}{N^2}\Sigma N_i^2 \left(\frac{N_i - n_i}{N_i}\right)\frac{s_i^2}{n_i}}$$

$$12.37619 \pm 1.96 \sqrt{\frac{1}{6300^2}\left\{2400(2300)\frac{3.22}{100} + 3000(2880)\frac{2.45}{120} + 900(850)\frac{1.07}{50}\right\}}$$

$$12.376 \pm .189 \quad \text{or} \quad 12.187 < \mu < 12.565$$

14.43

From Exercise 14.42, $\hat{\tau} = N\bar{x}_{st} = 6300\left(\frac{77{,}970}{6300}\right) = 77{,}970$ and the approximate 95% confidence interval is

$$77{,}970 \pm 6300(1.96)\sqrt{\frac{1}{6300^2}\left\{2400(2300)\frac{3.22}{100} + 3000(2880)\frac{2.45}{120} + 900(850)\frac{1.07}{50}\right\}}$$

$$77{,}970 \pm 1193.05$$

14.44

This is similar to previous exercises, with $N = 2700$. Calculate

$$\hat{p}_{st} = \frac{1}{N}\sum N_i\hat{p}_i = \frac{1}{2700}[900(.43) + 1200(.52) + 600(.38)] = .4589$$

and the 95% confidence interval is approximately

$$\hat{p}_{st} \pm 1.96\sqrt{\frac{1}{N^2}\sum N_i^2\left(\frac{N_i - n_i}{N_i}\right)\frac{\hat{p}_i\hat{q}_i}{n_i - 1}}$$

$$.4589 \pm 1.96\sqrt{\frac{1}{2700^2}\left[900(800)\frac{.43(.57)}{99} + 1200(1100)\frac{.52(.48)}{99} + 600(500)\frac{.38(.62)}{99}\right]}$$

$$.4589 \pm .0554 \quad \text{or} \quad .4035 < p < .5143$$

14.45

a. This is similar to previous exercises, with two strata (male and female), $N_1 = 8435$, $N_2 = 6453$, and $N = 14{,}888$. Calculate

$$\bar{x}_{st} = \frac{1}{N}\sum N_i\bar{x}_i = \frac{1}{14{,}888}[8435(864) + 6453(717)] = 800.28486$$

and the 95% confidence interval is approximately

$$\bar{x}_{st} \pm 1.96\sqrt{\frac{1}{N^2}\sum N_i^2\left(\frac{N_i - n_i}{N_i}\right)\frac{s_i^2}{n_i}}$$

$$800.285 \pm 1.96\sqrt{\frac{1}{14888^2}\left\{8435(8335)\frac{31{,}000}{100} + 6453(6353)\frac{22{,}000}{100}\right\}}$$

$$800.285 \pm 23.110 \quad \text{or} \quad 777.175 < \mu < 823.395$$

b.
$$\bar{x}_2 \pm 1.96\sqrt{\frac{s_2^2}{n_2}\left(\frac{N_2 - n_2}{N_2}\right)}$$

$$717 \pm 1.96\sqrt{\frac{22{,}000}{100}\left(\frac{6453 - 100}{6453}\right)}$$

$$717 \pm 28.845 \quad \text{or} \quad 688.155 < \mu_2 < 745.845$$

14.46

Refer to Exercise 14.45. The estimate of τ is $\hat{\tau} = N\bar{x}_{st} = 14{,}888(800.28486) = 11{,}914{,}614$ and the approximate 95% confidence interval is

$$11{,}914{,}614 \pm 1.96\sqrt{8435(8335)\tfrac{31{,}000}{100} + 6453(6353)\tfrac{22{,}000}{100}}$$

$$11{,}194{,}614 \pm 344{,}056.07$$

14.47

a. Refer to Exercise 14.45, with $N_1 = 8435$, $N_2 = 6453$, $N = 14{,}888$, $\hat{p}_1 = .61$, and $\hat{p}_2 = .38$. Calculate

$$\hat{p}_{st} = \tfrac{1}{N}\sum N_i\hat{p}_i = \tfrac{1}{14{,}888}[8435(.61) + 6453(.38)] = .5103$$

and the 95% confidence interval is approximately

$$\hat{p}_{st} \pm 1.96\sqrt{\tfrac{1}{N^2}\sum N_i^2\left(\tfrac{N_i - n_i}{N_i}\right)\tfrac{\hat{p}_i\hat{q}_i}{n_i - 1}}$$

$$.5103 \pm 1.96\sqrt{\tfrac{1}{14888^2}\left[8435(8335)\tfrac{.61(.39)}{99} + 6453(6353)\tfrac{.38(.62)}{99}\right]}$$

$$.510 \pm .068 \quad \text{or} \quad .442 < p < .578$$

b. For males, the 95% confidence interval for p is approximately

$$\hat{p}_1 \pm 1.96\sqrt{\tfrac{\hat{p}_1\hat{q}_1}{n_1 - 1}\left(\tfrac{N_1 - n_1}{N_1}\right)}$$

$$.61 \pm 1.96\sqrt{\tfrac{.61(.39)}{99}\left(\tfrac{8435 - 100}{8435}\right)}$$

$$.61 \pm .096 \quad \text{or} \quad .514 < p_1 < .706$$

14.48

Cluster sampling has been used, and the following information is necessary:

$$n = 12 \quad \quad \sum m_i = 60 \quad \quad \sum x_i = 121$$

$$N = 5000 \quad \quad \sum m_i^2 = 322 \quad \quad \sum x_i^2 = 1401$$

$$\bar{M} = 4.9 \quad \quad \sum m_i x_i = 662$$

Then $\sum(x_i - \bar{x}m_i)^2 = 1401 - 2\left(\tfrac{121}{60}\right)(662) + \left(\tfrac{121}{60}\right)^2(322) = 40.489445$

and $\bar{x} = \tfrac{\sum x_i}{\sum m_i} = \tfrac{121}{60} = 2.0167$

The approximate 95% confidence interval for μ is

$$\bar{x} \pm 1.96 \sqrt{\left(\frac{N-n}{Nn\bar{M}^2}\right)\frac{\Sigma(x_i - \bar{x}m_i)^2}{n-1}}$$

$$2.0167 \pm 1.96 \sqrt{\left(\frac{4988}{5000(12)(4.9)^2}\right)\left(\frac{40.489445}{11}\right)}$$

$$2.0167 \pm .2213 \quad \text{or} \quad 1.7954 < \mu < 2.2380$$

14.49

With $M = N\bar{M} = 24{,}500$, the estimate of τ is $\hat{\tau} = M\bar{x} = 24{,}500(\frac{121}{60}) = 49{,}408.33$ and the approximate 95% confidence interval is

$$49{,}408.33 \pm (24{,}500)(1.96) \sqrt{\left(\frac{4988}{5000(12)(4.9)^2}\right)\left(\frac{40.489445}{11}\right)}$$

$$49{,}408.33 \pm 5421.11$$

14.50

This is similar to previous exercises. The following information is necessary:

$n = 12$ $\quad \Sigma m_i = 188 \quad$ $\Sigma a_i = 57$

$N = 5000$ $\quad \Sigma m_i^2 = 3066 \quad$ $\Sigma a_i^2 = 297$

$\bar{M} = 15.2$ $\quad \Sigma a_i m_i = 920$

Then $\Sigma(a_i - \hat{p}m_i)^2 = 297 - 2\left(\frac{57}{188}\right)(920) + \left(\frac{57}{188}\right)^2(3066) = 20.969952$

and $\quad \hat{p} = \frac{\Sigma a_i}{\Sigma m_i} = \frac{57}{188} = .303$

The approximate 95% confidence interval for p is

$$\hat{p} \pm 1.96 \sqrt{\left(\frac{N-n}{Nn\bar{M}^2}\right)\frac{\Sigma(a_i - \hat{p}m_i)^2}{n-1}}$$

$$.303 \pm 1.96 \sqrt{\left(\frac{4988}{5000(12)(15.2)^2}\right)\left(\frac{20.969952}{11}\right)}$$

$$.303 \pm .0513 \quad \text{or} \quad .2517 < p < .3543$$

14.51

This is similar to previous exercises, with $n = 5$, $\frac{N-n}{N} \approx 1$, and $m_i = 12$ for all i, so that $\bar{m} = \bar{M} = 12$. For each cluster, calculate x_i, the total of all observations.

Cluster	m_i	x_i
1	12	192.0
2	12	192.1
3	12	192.0
4	12	192.5
5	12	191.7

Then $n = 5$ $\Sigma m_i = 60$ $\Sigma x_i = 960.3$

$\Sigma m_i^2 = 720$ $\Sigma x_i^2 = 184{,}435.55$

$\bar{m} = 60/5 = 12$ $\Sigma m_i x_i = 11{,}523.6$

Then $\Sigma(x_i - \bar{x}m_i)^2 = 184{,}435.55 - 2\left(\frac{960.3}{60}\right)(11{,}523.6) + \left(\frac{960.3}{60}\right)^2(720) = .332$

and $\bar{x} = \frac{\Sigma x_i}{\Sigma m_i} = \frac{960.3}{60} = 16.005$

The approximate 95% confidence interval for μ is

$$\bar{x} \pm 1.96 \sqrt{\left(\frac{N-n}{Nn\bar{M}^2}\right)\frac{\Sigma(x_i - \bar{x}m_i)^2}{n-1}}$$

$$16.005 \pm 1.96 \sqrt{\left(\frac{1}{5(12)^2}\right)\left(\frac{.332}{4}\right)}$$

$16.005 \pm .021$ or $15.984 < \mu < 16.026$

14.52

For this exercise, each cab is considered a cluster, with $m_i = 4$ for each car. Then

$n = 25$ $\Sigma m_i = 25(4) = 100$ $\Sigma a_i = 40$

$N = 175$ $\Sigma m_i^2 = 25(4)^2 = 400$ $\Sigma a_i^2 = 102$

$\bar{M} = \bar{m} = 4$ $\Sigma a_i m_i = 4\Sigma a_i = 160$

Calculate $\Sigma(a_i - \hat{p}m_i)^2 = 102 - 2\left(\frac{40}{100}\right)(160) + \left(\frac{40}{100}\right)^2(400) = 38$

and $\hat{p} = \frac{\Sigma a_i}{\Sigma m_i} = \frac{40}{100} = .4$

The approximate 95% confidence interval for p is

$$\hat{p} \pm 1.96 \sqrt{\left(\frac{N-n}{Nn\bar{M}^2}\right)\frac{\Sigma(a_i - \hat{p}m_i)^2}{n-1}}$$

$$.4 \pm 1.96 \sqrt{\left(\frac{150}{175(25)(4)^2}\right)\left(\frac{38}{24}\right)}$$

$.4 \pm .1142$ or $.2858 < p < .5142$

CHAPTER 15
The Chi-Square Goodness-of-Fit Test

15.1
See Section 15.1 of the text.

15.2
Use Table 5, Appendix II, with χ_α^2 and the appropriate degrees of freedom.

a. $\chi_{.05}^2 = 7.81473$ **b.** $\chi_{.01}^2 = 20.0902$

c. $\chi_{.10}^2 = 22.3072$ **d.** $\chi_{.10}^2 = 17.2750$

15.3
For a test of specified cell probabilities, the degrees of freedom are $k - 1$. Use Table 5, Appendix II:

a. d.f. = 6; $\chi_{.10}^2 = 10.6446$; reject H_0 if $X^2 > 10.6446$

b. d.f. = 9; $\chi_{.01}^2 = 21.6660$; reject H_0 if $X^2 > 21.6660$

c. d.f. = 13; $\chi_{.05}^2 = 22.3621$; reject H_0 if $X^2 > 22.3621$

d. d.f. = 2; $\chi_{.05}^2 = 5.99147$; reject H_0 if $X^2 > 5.99147$

15.4
Three hundred responses were each classified into one of five categories. The objective is to determine whether or not one category is preferred over another. The test statistic to test

$$H_0: p_1 = p_2 = p_3 = p_4 = p_5 = \tfrac{1}{5}$$

will be

$$X^2 = \sum_{i=1}^{k} \frac{[n_i - E(n_i)]^2}{E(n_i)}$$

which, when n is large, possesses an approximate chi-square distribution in repeated sampling. The values of n_i are the actual counts observed in the experiment, and

$$E(n_i) = np_i = 300(1/5) = 60.$$

a-b. To obtain the rejection region for this test, the degrees of freedom associated with X^2 must be determined. The number of degrees of freedom is equal to the number of cells, k, less one degree of freedom for each linearly independent restriction placed on $n_1, n_2, \ldots, n_k$. For this exercise, k = 5 and 1 degree of freedom is lost because of the restriction that

$$\sum_{i=1}^{k} n_i = n$$

Hence, X^2 has $k - 1 = 4$ degrees of freedom and the appropriate upper-tailed rejection region is $X^2 > \chi_{.05}^2 = 9.4877$.

c. The alternative hypothesis states only that H_0 is false. That is, at least one of the p_i, i = 1, 2, ..., 5 is not equal to 1/5. This can be alternatively stated as

$$H_a: p_i \neq p_j \text{ for some pair } i \neq j$$

d. A table of observed and expected cell counts follows:

Category	1	2	3	4	5
n_i	47	63	74	51	65
$E(n_i)$	60	60	60	60	60

Then
$$X^2 = \frac{(47-60)^2}{60} + \frac{(63-60)^2}{60} + \frac{(74-60)^2}{60} + \frac{(51-60)^2}{60} + \frac{(65-60)^2}{60}$$
$$= \frac{480}{60} = 8.00$$

Since the observed value of X^2 does not fall in the rejection region, we cannot conclude that there is a difference in the preference for the five categories.

e. From Table 5, with 4 degrees of freedom, the observed value $X^2 = 8.00$ lies between $\chi^2_{.10}$ and $\chi^2_{.05}$. Hence, $.05 < $ p-value $< .10$.

15.5

This is similar to Exercise 15.4.

a-b. The degrees of freedom are $k - 1 = 2$ and the rejection region with $\alpha = .05$ is

$$X^2 > \chi^2_{.05} = 5.99147$$

c. The alternative hypothesis states that at least one of the p_i does not equal the other two. That is,
$$H_a: p_i \neq p_j \text{ for some pair } i \neq j$$

d. Since $p_1 = p_2 = p_3$ and $\Sigma p_i = 1$ under H_0, we must have $p_1 = p_2 = p_3 = 1/3$. Hence,
$$E(n_i) = np_i = 300(1/3) = 100 \text{ for } i = 1, 2, 3$$
Then
$$X^2 = \frac{(130-100)^2}{100} + \frac{(98-100)^2}{100} + \frac{(72-100)^2}{100} = 16.88$$

Since the observed value of X^2 falls in the rejection region, we conclude that there is a difference in the preference for the three categories.

e. From Table 5, with 2 degrees of freedom, the observed value $X^2 = 16.88$ exceeds $\chi^2_{.005}$. Hence, p-value $< .005$.

15.6

One thousand cars were each classified according to the lane that they occupied (1 through 4). If no lane is preferred over another, the probability that a car will be driven in lane i, i = 1, 2, 3, 4, is 1/4. The null hypothesis is then

$$H_0: p_1 = p_2 = p_3 = p_4 = \tfrac{1}{4}$$

and the test statistic is

$$X^2 = \sum_{i=1}^{4} \frac{[n_i - E(n_i)]^2}{E(n_i)}$$

with $E(n_i) = np_i = 1000(1/4) = 250$ for $i = 1, 2, 3, 4$. A table of observed and expected cell counts follows:

Lane	1	2	3	4
n_i	294	276	238	192
$E(n_i)$	250	250	250	250

Then
$$X^2 = \frac{(294-250)^2}{250} + \frac{(276-250)^2}{250} + \frac{(238-250)^2}{250} + \frac{(192-250)^2}{250}$$
$$= \frac{6120}{250} = 24.48$$

The rejection region with $k - 1 = 3$ d.f. is $X^2 > \chi^2_{.05} = 7.81$. Since the observed value of X^2 falls in the rejection region, we reject H_0. There is difference in preference for the four lanes.

15.7

The null hypothesis to be tested is

$$H_0: p_1 = p_2 = p_3 = \tfrac{1}{3}$$

and the test statistic is

$$X^2 = \sum_{i=1}^{3} \frac{[n_i - E(n_i)]^2}{E(n_i)}$$

with $E(n_i) = np_i = 200(1/3) = 66.67$ for $i = 1, 2, 3$. A table of observed and expected cell counts follows:

Entrance	1	2	3
n_i	83	61	56
$E(n_i)$	66.67	66.67	66.67

Then
$$X^2 = \frac{(83-66.67)^2}{66.67} + \frac{(61-66.67)^2}{66.67} + \frac{(56-66.67)^2}{66.67} = 6.190$$

The rejection region with $k - 1 = 2$ d.f. is $X^2 > \chi^2_{.05} = 5.99$. Since the observed value of X^2 falls in the rejection region, we reject H_0. There is difference in preference for the three doors.

A 90% confidence interval for p_1 is given as

$$\frac{x_1}{n} \pm z_{.05}\sqrt{\frac{\hat{p}_1\hat{q}_1}{n}} = \frac{83}{200} \pm 1.645\sqrt{\frac{.415(.585)}{200}} = .415 \pm .057$$

or $.358 < p_1 < .472$.

15.8

This is similar to previous exercises. The null hypothesis is

$$H_0: p_1 = .02;\ p_2 = .16;\ \ldots;\ p_7 = .11$$

against the alternative that at least one proportion is not as specified.

A table of observed and expected cell counts follows:

n_i	27	193	234	322	568	482	174
$E(n_i)$	40	320	260	380	400	380	220

Then
$$X^2 = \frac{(27-40)^2}{40} + \frac{(193-320)^2}{320} + \cdots + \frac{(174-220)^2}{220}$$
$$= 173.638$$

The rejection region with $k - 1 = 6$ d.f. is $X^2 > \chi^2_{.05} = 12.59$. Since the observed value of X^2 falls in the rejection region, we reject H_0. The distribution of incomes within this city differs from the national distribution.

15.9

a. Let p be the percentage of the population of consumers having a particular opinion (*very annoying, somewhat annoying, not annoying*) about advertisements on videos. If n is the size of the sample, the margin of error for estimation of p is approximately

$$1.96\sqrt{\frac{pq}{n}} = 1.96\sqrt{\frac{(.5)(.5)}{n}}$$

and the sample size necessary to achieve a bound of .031 is

$$\sqrt{n} = \frac{1.96(.5)}{.031} = 31.613 \quad \Rightarrow \quad n = 999.38 \quad \text{or } n = 1000$$

b. The null hypothesis to be tested is

$$H_0: p_1 = p_2 = p_3 = \tfrac{1}{3}$$

and the test statistic is

$$X^2 = \sum_{i=1}^{3} \frac{[n_i - E(n_i)]^2}{E(n_i)}$$

with $E(n_i) = np_i = 1000(1/3) = 333.33$ for $i = 1, 2, 3$. A table of observed and expected cell counts follows:

Opinion	1	2	3
n_i	363	312	325
$E(n_i)$	333.33	333.33	333.33

Then
$$X^2 = \frac{(363-333.33)^2}{333.33} + \frac{(312-333.33)^2}{333.33} + \frac{(325-333.33)^2}{333.33} = 4.214$$

The rejection region with $k - 1 = 2$ d.f. is $X^2 > \chi^2_{.01} = 9.21$. Since the observed value of X^2 does not fall in the rejection region, we do not reject H_0. There is insufficient evidence to indicate a difference in the proportions falling in the three categories.

15.10

The null hypothesis to be tested is

$$H_0: p_1 = .12; \ p_2 = .17; \ p_3 = .41; \ p_4 = .30$$

and the test statistic is

$$X^2 = \sum_{i=1}^{4} \frac{[n_i - E(n_i)]^2}{E(n_i)}$$

with $E(n_i) = np_i$ for i = 1, 2, 3, 4. A table of observed and expected cell counts follows:

Category	1	2	3	4
n_i	30	38	75	57
$E(n_i)$	24	34	82	60

Then

$$X^2 = \frac{(30 - 24)^2}{24} + \frac{(38 - 34)^2}{34} + \frac{(75 - 82)^2}{82} + \frac{(57 - 60)^2}{60} = 2.718$$

The rejection region with k − 1 = 3 d.f. is $X^2 > \chi^2_{.01} = 11.3449$. Since the observed value of X^2 does not fall in the rejection region, we do not reject H_0. There is insufficient evidence to indicate that the survey percentages are inaccurate.

15.11

a. The experiment is analyzed as a 3 x 4 contingency table. Hence, the expected cell counts must be obtained for each of the cells. Since values for the cell probabilities are not specified by the null hypothesis, they must be estimated, and the appropriate estimator is

$$\hat{E}(n_{ij}) = \frac{R_i C_j}{n}$$

where R_i is the total for row i and C_j is the total for column j (see Section 10.3). The contingency table, including column and row totals and the estimated expected cell counts, follows.

Rows	Columns 1	2	3	4	Total
1	120 (67.68)	70 (66.79)	55 (67.97)	16 (58.56)	261
2	79 (84.27)	108 (83.17)	95 (84.64)	43 (72.91)	325
3	31 (78.05)	49 (77.03)	81 (78.39)	140 (67.53)	301
Total	230	227	231	199	887

The test statistic is

$$X^2 = \frac{(120 - 67.68)^2}{67.68} + \cdots + \frac{(140 - 67.53)^2}{67.53} = 211.71$$

using the two-decimal accuracy given above. The degrees of freedom are

$$(r - 1)(c - 1) = (3 - 1)(4 - 1) = 6$$

b. This is similar to part a. The estimated expected cell counts are calculated as

$$\hat{E}(n_{ij}) = \frac{R_i C_j}{n}$$

and are shown in parentheses in the table below.

Rows	Columns 1	2	3	Total
1	35 (37.84)	16 (26.37)	84 (70.80)	135
2	120 (117.16)	92 (81.63)	206 (219.20)	418
Total	155	108	290	553

The test statistic is calculated (using calculator accuracy rather than the two-decimal accuracy given in part a as

$$X^2 = \frac{(35 - 37.84)^2}{37.84} + \cdots + \frac{(206 - 219.20)^2}{219.20} = 8.93$$

The degrees of freedom are $(r - 1)(c - 1) = (2 - 1)(3 - 1) = 2$.

15.12

Refer to Section 15.3 of the text. For a 3 x 5 contingency table with $r = 3$ and $c = 5$, there are $(r - 1)(c - 1) = (2)(4) = 8$ degrees of freedom.

15.13

a. Since $r = 2$ and $c = 3$, the total degrees of freedom are $(r - 1)(c - 1) = (1)(2) = 2$.

b. The experiment is analyzed as a 2 x 3 contingency table. The contingency table, including column and row totals and the estimated expected cell counts, follows.

Rows	Columns 1	2	3	Total
1	37 (42.23)	34 (37.31)	93 (84.46)	164
2	66 (60.77)	57 (53.69)	113 (121.54)	236
Total	103	91	206	400

The estimated expected cell counts were calculated as:

$$\hat{E}(n_{11}) = \frac{R_1 C_1}{n} = \frac{164(103)}{400} = 42.23$$

$$\hat{E}(n_{12}) = \frac{R_1 C_2}{n} = \frac{164(91)}{400} = 37.31 \quad \text{and so on. Then}$$

$$X^2 = \frac{(37 - 42.23)^2}{42.23} + \frac{(34 - 37.31)^2}{37.31} + \cdots + \frac{(113 - 121.54)^2}{121.54} = 3.059$$

c. With $\alpha = .10$, a one-tailed rejection region is found, using Table 5, to be $X^2 > \chi^2_{.10} = 4.605$.

d-e. The observed value of $X^2 = 3.059$ does not fall in the rejection region. Hence, H_0 is not rejected. There is no reason to expect a dependence between rows and columns. In fact, $X^2 = 3.059$ has an observed significance level of p-value $> .10$.

15.14

The experiment is analyzed as a 2 x 3 contingency table. The contingency table, including column and row totals and the estimated expected cell counts, follows.

Rows	Columns 1	2	3	Total
1	69 (74.74)	126 (114.40)	16 (21.86)	211
2	78 (72.26)	99 (110.60)	27 (21.14)	204
Total	147	225	43	415

The estimated expected cell counts were calculated as:

$$\hat{E}(n_{11}) = \frac{R_1 C_1}{n} = \frac{211(147)}{415} = 74.74$$

$$\hat{E}(n_{12}) = \frac{R_1 C_2}{n} = \frac{211(225)}{415} = 114.40 \quad \text{and so on. Then}$$

$$X^2 = \frac{(69 - 74.74)^2}{74.74} + \frac{(126 - 114.40)^2}{114.40} + \cdots + \frac{(27 - 21.14)^2}{21.14} = 6.49$$

With $\alpha = .05$, a one-tailed rejection region is found, using Table 5, to be $X^2 > \chi^2_{.05} = 5.99$. The observed value of $X^2 = 6.49$ falls in the rejection region. Hence, H_0 is rejected. There is difference in the percentage favoring carpet from one area to another.

b. Let p_1 and p_2 denote the percentage favoring carpet in area 1 and area 2, respectively. Then

$$\hat{p}_1 = \frac{x_1}{n_1} = \frac{69}{147} = .469 \quad \text{and} \quad \hat{p}_2 = \frac{x_2}{n_2} = \frac{126}{225} = .56$$

A 95% confidence interval for $p_1 - p_2$ is obtained as in Chapter 7.

$$(\hat{p}_1 - \hat{p}_2) \pm z_{.025} \sqrt{\frac{\hat{p}_1 \hat{q}_1}{n_1} + \frac{\hat{p}_2 \hat{q}_2}{n_2}}$$

$$(.469 - .56) \pm 1.96 \sqrt{\frac{(.469)(.531)}{147} + \frac{(.56)(.44)}{225}}$$

$$-.091 \pm .104 \quad \text{or} \quad -.195 < (p_1 - p_2) < .013$$

15.15

This is similar to previous exercises. The contingency table analysis for the counts given in this exercise was done using a MINITAB computer analysis, and the printout is shown below.

```
Expected counts are printed below observed counts

                1        2        3        4     Total
        1      11       42       61       24      138
              8.83    36.43    58.51    34.22

        2      19       77      140       96      332
             21.25    87.65   140.77    82.34

        3       2       13       11        4       30
              1.92     7.92    12.72     7.44

Total          32      132      212      124      500

ChiSq =  0.532 +  0.851 +  0.106 +  3.054 +
         0.238 +  1.294 +  0.004 +  2.268 +
         0.003 +  3.258 +  0.233 +  1.591 = 13.431
df = 6
1 cells with expected counts less than 5.0
```

The observed value of the test statistic is $X^2 = 13.431$ and the rejection region with $\alpha = .05$ is $X^2 > 12.5916$. Hence, H_0 is rejected. There is a difference in advertising believability depending on the educational attainment of the viewer.

15.16

a. A 2×2 contingency table is shown below with estimated expected cell counts in parentheses.

	Antilock	No Antilock	Total
Accident	3 (6.45)	12 (8.55)	15
No Accident	40 (36.55)	45 (48.45)	85
Total	43	57	100

The test statistic is

$$X^2 = \frac{(3 - 6.45)^2}{6.45} + \cdots + \frac{(45 - 48.45)^2}{48.45} = 3.809$$

With $\alpha = .05$, a one-tailed rejection region is found, using Table 5, to be $X^2 > \chi^2_{.05} = 3.84$. The null hypothesis is not rejected. We cannot conclude that the proportion of cars that have accidents depends on whether or not the car has antilock brakes.

b. From Table 5, the observed value of the test statistic falls between $\chi^2_{.05} = 3.84$ and $\chi^2_{.10} = 2.7055$, so that
$$.05 < \text{p-value} < .10$$

15.17

a. H_0: $p_1 - p_2 = 0$ where p_1 = proportion of accidents for cars with antilock brakes and p_2 = proportion of accidents for cars without antilock brakes

b. H_a: $p_1 - p_2 < 0$

c. Calculate $\hat{p}_1 = 3/43 = .06976744$, $\hat{p}_2 = 12/57 = .21052632$, and $\hat{p} = \frac{x_1 + x_2}{n_1 + n_2} = \frac{15}{100} = .15$.

The test statistic is then

$$z = \frac{\hat{p}_1 - \hat{p}_2}{\sqrt{\hat{p}\hat{q}\left(\frac{1}{n_1} + \frac{1}{n_2}\right)}} = \frac{.06976744 - .21052632}{\sqrt{.15(.85)(1/43 + 1/52)}} = -1.91$$

The p-value for this one-tailed test is

$$\text{p-value} = P[z < -1.91] = .5 - .4719 = .0281$$

and H_0 is rejected for any value of α greater than or equal to .0281. There is a difference in the two proportions. (*Note:* The two-directional alternative of the chi-square test in Exercise 15.16 prevented rejection of H_0 for $\alpha = .05$ in that exercise.)

15.18

This is similar to previous exercises. The contingency table analysis for the counts given in this exercise was done using a MINITAB computer analysis, and the printout is shown in the exercise.

The observed value of the test statistic is $X^2 = 17.597$ and the rejection region with $\alpha = .05$ is $X^2 > 15.5073$. Hence, H_0 is rejected. There is a dependence between one's opinion on the president's economic policy and one's household income.

15.19

A 2×2 contingency table is shown below with estimated expected cell counts in parentheses.

	Men	Women	Total
Read	66 (53.40)	23 (35.60)	89
Do Not Read	84 (96.60)	77 (64.40)	161
Total	150	100	250

The test statistic is

$$X^2 = \frac{(66 - 53.40)^2}{53.40} + \cdots + \frac{(77 - 64.40)^2}{64.40} = 11.541$$

The observed value, $X^2 = 11.541$ is greater than $\chi^2_{.005} = 7.87944$ in Table 5, so that p-value $< .005$. Since the p-value is so small, the null hypothesis is rejected. We can conclude that there is a difference in the proportion of men and women who use office PCs and read computer literature.

15.20

A 2×2 contingency table is shown below with estimated expected cell counts in parentheses.

	Make Effort	Don't Make Effort	Total
1991	433 (405.5)	567 (594.5)	1000
1993	378 (405.5)	622 (594.5)	1000
Total	811	1189	2000

The test statistic is

$$X^2 = \frac{(433 - 405.5)^2}{405.5} + \cdots + \frac{(622 - 594.5)^2}{594.5} = 6.274$$

The observed value, $X^2 = 6.274$ with 1 d.f., falls between $\chi^2_{.01} = 6.635$ and $\chi^2_{.025} = 5.0239$ in Table 5, so that

$$.01 < \text{p-value} < .025$$

15.21

This is similar to previous exercises, except that the number of observations per row was selected prior to the experiment. The test procedure is identical to that used for an r x c contingency table. The contingency table, including column and row totals and the estimated expected cell counts, follows.

Popn	Category 1	2	3	Total
1	108 (102.33)	52 (47.33)	40 (50.33)	200
2	87 (102.33)	51 (47.33)	62 (50.33)	200
3	112 (102.33)	39 (47.33)	49 (50.33)	200
Total	307	142	151	600

a. The test statistic is

$$X^2 = \frac{(108 - 102.33)^2}{102.33} + \frac{(52 - 47.33)^2}{47.33} + \cdots + \frac{(49 - 50.33)^2}{50.33} = 10.597$$

using calculator accuracy.

b. With $(r-1)(c-1) = 4$ d.f. and $\alpha = .10$, the rejection region is $X^2 > 7.779$.

c. The null hypothesis is rejected. There is evidence to indicate that the proportions depend upon the population from which they were drawn.

d. Since the observed value, $X^2 = 10.597$, falls between $\chi^2_{.05}$ and $\chi^2_{.025}$,

$$.025 < \text{p-value} < .05$$

15.22

The test procedure is identical to that used for an r x c contingency table. The contingency table, including column and row totals and the estimated expected cell counts, follows.

	Population 1	Population 2	Population 3	Total
Number of Successes	24 (25.33)	19 (25.33)	33 (25.33)	76
Number of Failures	76 (74.67)	81 (74.67)	67 (74.67)	224
Total	100	100	100	300

a. The test statistic is

$$X^2 = \frac{(24-25.33)^2}{25.33} + \frac{(19-25.33)^2}{25.33} + \cdots + \frac{(67-74.67)^2}{74.67} = 5.322$$

b. With $(r-1)(c-1) = 2$ d.f. and $\alpha = .05$, the rejection region is $X^2 > 5.99$.

c. The null hypothesis is not rejected. There is no evidence to indicate that the proportions depend upon the population from which they were drawn.

d. Since the observed value, $X^2 = 5.322$, falls between $\chi^2_{.05}$ and $\chi^2_{.10}$,

$$.05 < \text{p-value} < .10$$

15.23

The contingency table, including column and row totals and the estimated expected cell counts, follows.

Result	Manager A	Manager B	Manager C	Total
Profit	63 (63)	71 (63)	55 (63)	189
No Profit	37 (37)	29 (37)	45 (37)	111
Total	100	100	100	300

The test statistic is

$$X^2 = \frac{(63-63)^2}{63} + \frac{(71-63)^2}{63} + \cdots + \frac{(45-37)^2}{37} = 5.491$$

The rejection region with 2 d.f. and $\alpha = .05$ is $X^2 > 5.99$ and H_0 is not rejected. There is not enough information to conclude with $\alpha = .05$ that the proportion of successful purchases will differ among the managers. Notice that if the experimenter takes $\alpha = .10$ with rejection region $X^2 > 4.61$, the null hypothesis can be rejected.

15.24

The objective of the experiment is to determine whether the number of defective and non-defective buttons is affected by the particular machine on which they are produced. Hence, buttons have been classified according to machine and whether they were defective or nondefective. A sample of 400 buttons was selected from each machine.

The complete contingency table, with column totals fixed at 400, is shown below.

Result	Machine 1	Machine 2	Machine 3	Total
Defective	16 (16.33)	24 (16.33)	9 (16.33)	49
Nondefective	384 (383.67)	376 (383.67)	391 (383.67)	1151
Total	400	400	400	1200

The test statistic is

$$X^2 = \frac{(16-16.33)^2}{16.33} + \frac{(24-16.33)^2}{16.33} + \cdots + \frac{(391-383.67)^2}{383.67} = 7.193$$

The rejection region with 2 d.f. and $\alpha = .05$ is $X^2 > 5.99$ and H_0 is rejected. There is a difference in the fraction defective from machine to machine.

15.25

a. A 2×2 contingency table is shown below with estimated expected cell counts in parentheses.

	Dual Earner	Single Earner	Total
Agree	217 (195.75)	44 (65.25)	261
Disagree	158 (179.25)	81 (59.75)	239
Total	375	125	500

The test statistic is

$$X^2 = \frac{(217-195.75)^2}{195.75} + \cdots + \frac{(81-59.75)^2}{59.75} = 19.304$$

With $\alpha = .01$, a one-tailed rejection region is found, using Table 5, to be $X^2 > \chi^2_{.01} = 6.63$ and H_0 is rejected. There is a difference in the two proportions.

b. The hypothesis to be tested is

$$H_0: p_1 - p_2 = 0 \qquad H_a: p_1 - p_2 \neq 0$$

Calculate

$$\hat{p}_1 = \frac{217}{375} = .57867, \quad \hat{p}_2 = \frac{44}{125} = .352, \text{ and } \hat{p} = \frac{x_1 + x_2}{n_1 + n_2} = \frac{217 + 44}{375 + 125} = .522$$

The test statistic is then

$$z = \frac{\hat{p}_1 - \hat{p}_2}{\sqrt{\hat{p}\hat{q}\left(\frac{1}{n_1} + \frac{1}{n_2}\right)}} = \frac{.57867 - .352}{\sqrt{.522(.478)(1/375 + 1/125)}} = 4.39$$

The rejection region with $\alpha = .01$ is $|z| > 2.58$ and H_0 is rejected. There is evidence of a difference in the two proportions.

15.26

a-b. A 3×2 contingency table is shown in the printout with estimated expected cell counts below the observed cell counts.

The test statistic is

$$X^2 = \frac{(58 - 52)^2}{52} + \cdots + \frac{(1 - 1)^2}{1} = 2.917$$

From Table 5, the observed value $X^2 = 2.917$ is less than $\chi^2_{.10} = 4.605$ with 2 d.f. Therefore,

p-value $> .10$

and the null hypothesis is not rejected. There is no evidence to indicate a difference of opinion between American and Japanese workers.

c. Yes. The estimated expected cell counts in row 3 are less than five.

15.27

a. The data are displayed in a 2×2 contingency table shown below.

	Male	Female	Total
Smoker	62 (58)	54 (58)	116
Nonsmoker	138 (142)	146 (142)	284
Total	200	200	400

The test statistic is

$$X^2 = \frac{(62 - 58)^2}{58} + \cdots + \frac{(146 - 142)^2}{142} = .777$$

With $\alpha = .10$, a one-tailed rejection region is found, using Table 5, to be $X^2 > \chi^2_{.10} = 2.706$ and H_0 is not rejected. There is no significant difference in the two proportions.

b. The hypothesis to be tested is

$H_0: p_1 - p_2 = 0 \qquad H_a: p_1 - p_2 > 0$

which cannot be tested using the X^2 test statistic. You can use the z test from Section 8.8.

15.28

The hypothesis to be tested is

$H_0: p_1 = p_2 = p_3 = p_4 = p_5 = p_6 = \frac{1}{6}$

against the alternative that at least one proportion is not as specified. A table of observed and expected cell counts follows:

	1	2	3	4	5	6
n_i	89	113	98	104	117	79
$E(n_i)$	100	100	100	100	100	100

Then
$$X^2 = \frac{(89-100)^2}{100} + \frac{(113-100)^2}{100} + \cdots + \frac{(79-100)^2}{100}$$
$$= 10.4$$

The rejection region with $k - 1 = 5$ d.f. is $X^2 > \chi^2_{.05} = 11.07$. Since the observed value of X^2 does not fall in the rejection region, we do not reject H_0. There is insufficient evidence to suggest that the die is unbalanced.

15.29

a. The random variable of interest is $\hat{p}$, the fraction of dice on which a 6 is observed. Hence, a binomial situation is implied, where p is the probability of observing a 6 and q is the probability that a 6 is not observed. Proceeding as in Chapter 8, the hypothesis to be tested is:
$$H_0: p = \tfrac{1}{6} \qquad H_a: p < \tfrac{1}{6}$$
and the test statistic is
$$z = \frac{\hat{p} - p_0}{\sqrt{\frac{p_0 q_0}{n}}} = \frac{\frac{79}{600} - \frac{1}{6}}{\sqrt{\frac{(1/6)(5/6)}{600}}} = -2.300$$

The rejection region with $\alpha = .05$ is $z < -1.645$ and the null hypothesis is rejected.

b. The procedure that was used in making the decision to test the above hypothesis was statistically invalid. That is, the experimenter examined the observed data and, seeing that the number of dice falling in cell 6 was small, decided to test this particular hypothesis. In the first place, it is unlikely that the experimenter would have selected cell #6 had he not looked at the data. Second, if the decision to test the hypothesis that $p = 1/6$ had been made before observing the data, the experimenter would most likely have chosen the alternative hypothesis that $p \neq 1/6$, thus creating a two-tailed test of hypothesis. Instead of using a rejection region with $\alpha/2 = .025$ in each tail, he has used a rejection region with $\alpha = .05$ in one tail. If the same rejection region (i.e., $|z| > 1.645$) was used correctly as a two-tailed test, the experimenter's probability of error is actually .10 instead of .05.

15.30

The hypothesis to be tested is
$$H_0: p_1 = .2; \; p_2 = .05; \; p_3 = .03 \; ; \; p_4 = .72$$

against the alternative that at least one proportion is not as specified. A table of observed and expected cell counts follows:

n_i	45	21	8	224
$E(n_i)$	59.6	14.9	8.94	214.56

Then
$$X^2 = \frac{(45 - 59.6)^2}{59.6} + \frac{(21 - 14.9)^2}{14.9} + \cdots + \frac{(224 - 214.56)^2}{214.56}$$
$$= 6.588$$

The rejection region with $k - 1 = 3$ d.f. is $X^2 > \chi^2_{.10} = 6.251$. Since the observed value of X^2 falls in the rejection region, H_0 is rejected. The data disagree with the company's contention.

15.31

The data is analyzed as a 2×3 contingency table with the observed and estimated expected cell counts shown in the printout.

The test statistic is
$$X^2 = \frac{(67 - 61.43)^2}{61.43} + \frac{(26 - 28.04)^2}{28.04} + \cdots + \frac{(46 - 42.47)^2}{42.47} = 1.886$$

The rejection region with 2 d.f. and $\alpha = .05$ is $X^2 > 5.99$ and H_0 is not rejected. There is insufficient evidence to indicate that the frequency of fatal accidents is dependent on the plant department.

15.32

The data is analyzed as a 2×4 contingency table with the observed and estimated expected cell counts shown below.

Number of Radios	Income Brackets				Total
	1	2	3	4	
1	126 (107.725)	362 (347.5)	129 (146.645)	78 (93.13)	695
2	29 (47.275)	138 (152.5)	82 (64.355)	56 (40.87)	305
Total	155	500	211	134	1000

The test statistic is
$$X^2 = \frac{(126 - 107.725)^2}{107.725} + \frac{(362 - 347.5)^2}{347.5} + \cdots + \frac{(56 - 40.87)^2}{40.87} = 27.17$$

The rejection region with 3 d.f. and $\alpha = .10$ is $X^2 > 6.25$ and H_0 is rejected. There is evidence to indicate that the number of radios per household is dependent on family income.

15.33

The hypothesis to be tested is

$$H_0: p_1 = .30; \; p_2 = .45; \; p_3 = .25$$

against the alternative that at least one proportion is not as specified. A table of observed and expected cell counts follows:

n_i	21	39	40
$E(n_i)$	30	45	25

Then
$$X^2 = \frac{(21-30)^2}{30} + \frac{(39-45)^2}{45} + \frac{(40-25)^2}{25} = 12.5$$

The rejection region with $k - 1 = 2$ d.f. is $X^2 > \chi^2_{.10} = 4.61$ and H_0 is rejected. The data disagree with the company's claim.

15.34

The contingency table analysis for the counts given in this exercise was done using a MINITAB computer analysis, and the printout is shown below.

The observed value of the test statistic is $X^2 = 680.397$ and the rejection region with $\alpha = .05$ and 4 d.f. is $X^2 > 9.49$. Hence, H_0 is rejected. There is a difference in the survival rates among these three industries.

```
Expected counts are printed below observed counts

                1          2          3      Total
     1       3217        354        144       3715
            2609.40     819.47     286.13

     2       1393        726        235       2354
            1653.44     519.25     181.31

     3       5914       2225        775       8914
            6261.16    1966.28     686.56

  Total     10524       3305       1154      14983

ChiSq =141.479 +264.391 + 70.602 +
        41.023 + 82.319 + 15.901 +
        19.249 + 34.042 + 11.392 = 680.397
df = 4
```

15.35

The MINITAB printout is shown below.

```
Expected counts are printed below observed counts

                1          2          3      Total
     1         55         35         12        102
              47.23      41.76      13.01

     2         12         19          6         37
              17.13      15.15       4.72

     3          2          7          1         10
               4.63       4.09       1.28

  Total       69         61         19        149

ChiSq =  1.277 + 1.094 + 0.078 +
         1.538 + 0.980 + 0.348 +
         1.495 + 2.063 + 0.059 = 8.932
df = 4
4 cells with expected counts less than 5.0
```

The observed value of the test statistic is $X^2 = 8.932$ and the rejection region with $\alpha = .05$ is $X^2 > 9.49$. Hence, H_0 is not rejected. However, notice the warning about the cells with expected cell counts less than 5. Any conclusions should be drawn with care.

15.36

The null hypothesis to be tested is

$$H_0: p_1 = .18; \; p_2 = .11; \; \ldots ; p_9 = .03$$

against the alternative that at least one proportion is not as specified.

A table of observed and expected cell counts follows:

n_i	40	24	15	12	3	26	35	42	3
$E(n_i)$	36	22	14	10	6	24	32	50	6

Then

$$X^2 = \frac{(40-36)^2}{36} + \frac{(24-22)^2}{22} + \cdots + \frac{(3-6)^2}{6} = 5.826$$

The rejection region with $k - 1 = 8$ d.f. is $X^2 > \chi^2_{.01} = 20.09$, and H_0 is not rejected. The figures are not significantly different from those reported in the table.

15.37

The data is analysed as a 2×4 contingency table with the observed and estimated expected cell counts shown below.

Computer Literacy	Educational Level				Total
	1	2	3	4	
User	2 (5.60)	13 (20.16)	30 (24.27)	11 (5.97)	56
Nonuser	13 (9.40)	41 (33.84)	35 (40.73)	5 (10.03)	94
Total	15	54	65	16	150

The test statistic is

$$X^2 = \frac{(2-5.60)^2}{5.60} + \frac{(13-20.16)^2}{20.16} + \cdots + \frac{(5-10.03)^2}{10.03} = 16.66 \text{ (computer accuracy)}$$

The rejection region with 3 d.f. and $\alpha = .05$ is $X^2 > 7.81$ and H_0 is rejected. There is evidence of a difference in computer literacy for the four educational levels.

15.38

The data is analyzed as a 5×3 contingency table with the observed and estimated expected cell counts shown below.

	Type of Repair			
Make	1	2	3	Total
A	17 (14.395)	19 (18.695)	7 (9.908)	43
B	14 (10.043)	7 (13.043)	9 (6.913)	30
C	6 (13.056)	21 (16.956)	12 (8.987)	39
D	33 (32.139)	44 (41.739)	19 (22.121)	96
E	7 (7.365)	9 (9.565)	6 (5.069)	22
Total	195	89	62	346

The test statistic is

$$X^2 = \frac{(17-14.395)^2}{14.395} + \frac{(19-18.695)^2}{18.695} + \cdots + \frac{(6-5.069)^2}{5.069} = 12.915$$

The rejection region with 8 d.f. and $\alpha = .05$ is $X^2 > 15.5$, and H_0 is not rejected. There is insufficient evidence to indicate a dependence between "make" and "type of repair". Notice that even if $\alpha = .10$ with rejection region $X^2 > 13.36$, the null hypothesis is not rejected.

15.39

a. A 2×2 contingency table is shown below with estimated expected cell counts in parentheses.

	Participative	Nonparticipative	Total
Approve	73 (62)	51 (62)	124
Do not Approve	27 (38)	49 (38)	76
Total	100	100	200

The test statistic is

$$X^2 = \frac{(73-62)^2}{62} + \cdots + \frac{(49-38)^2}{38} = 10.272$$

With $\alpha = .05$, a one-tailed rejection region is found, using Table 5, to be $X^2 > \chi^2_{.05} = 3.84$ and H_0 is rejected. Approval or disapproval depends on whether workers participate in decision making.

b. The hypothesis to be tested is

$$H_0: p_1 - p_2 = 0 \qquad H_a: p_1 - p_2 > 0$$

where p_1 is the probability that a participative worker approves of the firm's decisions and p_2 is the probability that a nonparticipative worker approves of the firm's decisions.

Calculate

$$\hat{p}_1 = \frac{73}{100} = .73, \quad \hat{p}_2 = \frac{51}{10} = .51, \text{ and } \hat{p} = \frac{x_1 + x_2}{n_1 + n_2} = \frac{124}{200} = .62$$

The test statistic is then

$$z = \frac{\hat{p}_1 - \hat{p}_2}{\sqrt{\hat{p}\hat{q}\left(\frac{1}{n_1} + \frac{1}{n_2}\right)}} = \frac{.73 - .51}{\sqrt{.62(.38)(1/100 + 1/100)}} = 3.205$$

The rejection region with $\alpha = .05$ is $z > 1.645$ and H_0 is rejected. Workers with participative decision making tend to approve more of the firm's decisions.

CHAPTER 16
Nonparametric Statistics

16.1

a. If a paired difference experiment has been used and the sign test is one-tailed (H_a: p > .5), then the experimenter would like to show that one population of measurements lies above the other population. An exact practical statement of the alternative hypothesis would depend on the experimental situation.

b. It is necessary that α (the probability of rejecting the null hypothesis when it is true) take values less than $\alpha = .15$. Assuming the null hypothesis to be true, the two populations are identical and consequently, p = P[A exceeds B for a given pair of observations] is 1/2. The binomial probability was discussed in Chapter 4. In particular, it was noted that the distribution of the random variable x is symmetrical about the mean np when p = 1/2. For example, with n = 25, P[x = 0] = P[x = 25]. Similarly, P[x = 1] = P[x = 24], and so on. Hence, the lower-tailed probabilities tabulated in Table 1, Appendix II will be identical to their upper-tailed equivalent probabilities. The values of α available for this upper-tailed test and the corresponding rejection regions are shown below.

Rejection Region	α
x $\geq$ 20	.002
x $\geq$ 19	.007
x $\geq$ 18	.022
x $\geq$ 17	.054
x $\geq$ 16	.115

16.2

a. The experimenter wants to prove that the populations differ in location.

b. The probabilities given in Exercise 16.1 b are doubled for a two-tailed test. Hence, fewer rejection regions are available for $\alpha \leq .15$.

Rejection Region	α
x $\leq$ 5; x $\geq$ 20	.004
x $\leq$ 6; x $\geq$ 19	.014
x $\leq$ 7; x $\geq$ 18	.044
x $\leq$ 8; x $\geq$ 17	.108

16.3

a. See solution to Exercise 16.1 a.

b. The rejection regions and levels of α are given in the table for the n = 15 and a one-tailed test.

n = 15
x $\geq$ 13 $\alpha = .004$
x $\geq$ 12 $\alpha = .018$
x $\geq$ 11 $\alpha = .059$

16.4

a. See solution to Exercise 16.2 a.

b. For the two-tailed test, the rejection regions with $\alpha < .15$ are shown below.

n = 15
$x \leq 2, x \geq 13$ $\alpha = .008$
$x \leq 3, x \geq 12$ $\alpha = .036$
$x \leq 4, x \geq 11$ $\alpha = .118$

16.5

Refer to Exercise 16.2. From the table given there, the appropriate rejection region for $\alpha \approx .05$ is $x \leq 7$ and $x \geq 18$. For large samples, H_0 is rejected for $|z| > 1.96$ where

$$z = \frac{x - .5n}{\sqrt{.25n}} = \frac{x - 12.5}{2.5}$$

The critical values of x are then

$$\frac{x - 12.5}{2.5} = 1.96 \qquad \text{or } x = 17.4$$

$$\frac{x - 12.5}{2.5} = -1.96 \qquad \text{or } x = 7.6$$

The rejection regions are almost equivalent.

16.6

a. With $\alpha \approx .05$, the rejection region is chosen as $x \geq 17$ (see Exercise 16.1) with $\alpha = .054$. Since the observed value of x is $x = 17$, H_0 is rejected and we conclude that $p > .5$. If we insist that α does not exceed .05, H_0 cannot be rejected.

b. Since the observed value of x is $x = 17$, the p-value for this one-tailed test is

$$P[x \geq 17] = .054$$

16.7

a-b. The hypothesis to be tested is

$$H_0: p = .5 \qquad H_a: p > .5$$

where $p = P[x_A > x_B]$ for a given pair.

c. With $n = 9$, we need to find a rejection region with $\alpha \approx .05$, consisting of large values of x. Three possible rejection regions are found using Table 1 in Appendix II, with $n = 9$ and $p = .5$.

Rejection Region	α
$x = 9$	.002
$x \geq 8$	.020
$x \geq 7$	.090

The experimenter should choose either $x \geq 7$ or $x \geq 8$ as the rejection region, depending on the value of α he prefers. In any case, since we observe $x = 8$, H_0 is rejected, and we conclude $p > .5$.

d. The p-value is $P[x \geq 8] = .020$.

16.8

a-b. The hypothesis to be tested is

$$H_0: p = .5 \qquad H_a: p \neq .5 \text{ where } p = P[x_A > x_B] \text{ for a given pair.}$$

c. With n = 9, we need to find a rejection region with $\alpha \approx .05$, consisting of large and small values of x. Three possible rejection regions are found using Table 1 in Appendix II, with n = 9 and p = .5.

Rejection Region	α
x = 9; x = 0	.004
x ≥ 8; x ≤ 1	.040
x ≥ 7; x ≤ 2	.180

The experimenter should choose $x \geq 8$ or $x \leq 1$ as the rejection region, with $\alpha = .04$. Since we observe x = 8, H_0 is rejected, and we conclude $p \neq .5$.

d. The p-value is $P[x \geq 8 \text{ or } x \leq 1] = .040$.

16.9

a. The hypothesis to be tested is

$$H_0: p = .5 \qquad H_a: p \neq .5$$

where $p = P[x_A > x_B]$ for a given pair.

b. For large samples, H_0 is rejected for $|z| > 1.96$ where

$$z = \frac{x - .5n}{\sqrt{.25n}}$$

c. For x = 10 and n = 30,

$$z = \frac{10 - 15}{\sqrt{7.5}} = -1.826$$

and H_0 is not rejected. There is insufficient evidence to indicate that $p \neq .5$.

16.10

a. The hypothesis to be tested is

$$H_0: p = .5 \qquad H_a: p > .5$$

where $p = P[x_A > x_B]$ for a given pair.

b. For large samples, H_0 is rejected for $z > 1.645$ where

$$z = \frac{x - .5n}{\sqrt{.25n}}$$

c. For x = 18 and n = 35,

$$z = \frac{18 - 35(.5)}{\sqrt{.25(35)}} = .169$$

and H_0 is not rejected. There is insufficient evidence to indicate that $p > .5$.

d. The p-value is $P[z > .169] = .5 - .0675 = .4325$

16.11

a. If assessors A and B are equal in their property assessments, then p, the probability that A's assessment exceeds B's assessment for a given property, should equal 1/2. If one of the assessors tends to be more conservative than the other, then either $p > 1/2$ or $p < 1/2$. Hence, we can test the equivalence of the two assessors by testing the hypothesis

$$H_0: p = 1/2 \quad \text{versus} \quad H_a: p \neq 1/2$$

using the test statistic x, the number of times that assessor A exceeds assessor B for a particular property assessment. To find a two-tailed rejection region with α close to .05, use Table 1 with $n = 8$ and $p = .5$. For the rejection region $\{x = 0, x = 8\}$ the value of α is $.004 + .004 = .008$, while for the rejection region $\{x = 0, 1, 7, 8\}$ the value of α is $.035 + .035 = .070$, which is closer to .05.

Hence, using the rejection region $\{x \leq 1 \text{ or } x \geq 7\}$, the null hypothesis is not rejected, since x = number of properties for which A exceeds B = 6. The p-value for this two-tailed test is

$$\text{p-value} = 2\, P[x \geq 6] = 2(1 - .855) = .290$$

b. The t statistic used in Exercise 8.43 allows the experimenter to reject H_0, while the sign test fails to reject H_0. This is because the sign test uses less information and makes fewer assumptions than does the t test. If all normality assumptions are met, the t test is the more powerful test and can reject when the sign test cannot.

16.12

Define $p = P[\text{gourmet A's rating exceeds gourmet B's rating for a given meal}]$ and $x =$ number of meals for which gourmet A exceeds B. The hypothesis to be tested is

$$H_0: p = 1/2 \quad \text{versus} \quad H_a: p \neq 1/2$$

using the sign test with x as the test statistic. Notice that for this exercise $n = 17$, since a tie rating was given to meals 7, 14, and 20. The observed value of the test statistic is $x = 8$.

Various two-tailed rejection regions are tried in order to find a region with $\alpha \approx .05$. These are shown in the following table.

Rejection Region	α
$x \leq 2;\ x \geq 15$	$2(C_0^{17} + C_1^{17} + C_2^{17})(.5)^{17} = .00235$
$x \leq 4;\ x \geq 13$	$2(C_0^{17} + C_1^{17} + C_2^{17} + C_3^{17} + C_4^{17})(.5)^{17} = .04904$

We choose to reject H_0 if $x \leq 4$ or $x \geq 13$ with $\alpha = .049$. Since $x = 8$, H_0 is not rejected. There is insufficient evidence to indicate a difference between the two gourmets.

16.13

a. This is similar to previous exercises. The hypothesis to be tested is

$$H_0: p = 1/2 \quad \text{versus} \quad H_a: p \neq 1/2$$

where $p = P[\text{customer prefers style A}]$ and $x =$ number of customers preferring style A. For this exercise, x = 6. The rejection region (two-tailed) will be $x \leq 5$ or $x \geq 15$ with

$$\alpha = P[x \leq 5 \text{ when } p = 1/2] + P[x \geq 15 \text{ when } p = 1/2] = .042$$

Since the observed value of the test statistic does not fall in the rejction region, H_0 is not rejected. There is insufficient evidence to indicate a difference in preference for the two styles.

b.   With B = .10 and $p \approx .5$, the inequality to be solved is

$$1.96\sqrt{\tfrac{pq}{n}} \leq B$$

$$1.96\sqrt{\tfrac{.5(.5)}{n}} \leq .10 \quad \Rightarrow \quad \sqrt{n} \geq 9.8 \quad \Rightarrow \quad n \geq 96.04 \quad \text{or } n = 97$$

16.14

This is similar to previous exercises. Define

x = number of days B exceeds A in number of defectives

p = probability that B exceeds A on a given day

n = 10

If both lines produce the same fraction of defectives, one would expect p to be 1/2; however, if B produces more defectives than A, one would expect p to be greater than 1/2. Thus, a one-tailed test is implied, with a rejection region defined for large values of x:

$$H_0: p = 1/2 \quad \text{versus} \quad H_a: p > 1/2$$

The next problem is to determine an appropriate rejection region for the test. In order to do this, the probability of a type I error, α, must be evaluated for various rejection regions. Since α measures the risk of incorrectly rejecting H_0 when true, we might select the value of α close to .05. As α is decreased, the probability of type II error, β, increases (for a fixed sample size). Hence, we must be satisfied with a .05 probability of type I error, in order that β will not be unduly large. Choosing a rejection region of x = 10 implies

$$\alpha = P[x = 10 \text{ when } p = 1/2] = 1 - P[x \leq 9 \text{ when } p = 1/2] = 1 - .999 = .001$$

using the binomial tables. This value of α is smaller than our target, $\alpha = .05$. Choosing a rejection region of x = 9 or x = 10 implies

$$\alpha = 1 - P[x \leq 8 \text{ when } p = 1/2] = 1 - .989 = .011$$

which is still too small. Choosing x = 8, 9, or 10 yields

$$\alpha = 1 - P[x \leq 7 \text{ when } p = 1/2] = 1 - .945 = .055$$

which is close to the desired value of .05. Thus, using x = 8, 9, or 10 as the rejection region, and noting that the value of the test statistic, x = 8, falls in the rejection region, we reject H_0 and conclude that B produces more defectives than A. The probability of error is $\alpha = .05$.

16.15

a.   This is similar to previous exercises. Define

x = number of objects for which A exceeds B

p = probability that A exceeds B for a given object

n = 7

The hypothesis of interest is

$$H_0: p = 1/2 \quad \text{versus} \quad H_a: p \neq 1/2$$

and the null hypothesis will be rejected for very small or very large values of x. Using Table 1 in Appendix II for n = 7, we can find a rejection region for which $\alpha \approx .10$.

Rejection Region	α
x = 0; x = 7	2(.008) = .016
x ≤ 1; x ≥ 6	2(.062) = .124

The second rejection region is closest to $\alpha = .10$; hence, we choose to reject H_0 if $x \leq 1$ or $x \geq 6$ with $\alpha = .124$.

b. Using this rejection region, the value of β for p = .9 is

$$\beta = P[\text{accept } H_0 \text{ when } H_a \text{ true}] = P[2 \leq x \leq 5 \text{ when } p = .9]$$

$$= .150 - .000 = .150$$

c. If the observed value of x is x = 5, H_0 is not rejected. There is not enough evidence to conclude that the two appraisers are different in their level of appraisal.

16.16

With n = 50 − 2 = 48 and x = 38, the hypothesis of interest is

$$H_0: p = .5 \qquad H_a: p > .5$$

where p = P[treated window preferred to untreated window]. The test statistic is

$$z = \frac{x - .5(48)}{\sqrt{.25(48)}} = \frac{38 - 24}{\sqrt{12}} = 4.04$$

and the rejection region is z > 2.33 with $\alpha = .01$. The null hypothesis is rejected and we conclude that the treated glass was preferred to untreated.

16.17

a. If H_a is true and population 1 lies to the right of population 2, then sample 1 will contain the higher ranks, T_A will be large, and U_A will be small. Hence, U_A will be the test statistic, with a lower-tailed rejection region.

b-c. From Table 14, Appendix II, index $n_1 = 6$ and $n_2 = 8$. For $\alpha \leq .10$, the appropriate rejection region is $U \leq 13$ with $\alpha = .0906$.

16.18

a. If the alternative is two-tailed, then one of the two samples (we don't know which one) will contain high ranks and produce a small value of U. Hence, we choose the smaller of U_A and U_B as the test statistic.

b-c. From Table 14, Appendix II, we choose a rejection region with $\alpha/2 \leq .05$. The appropriate rejection region is $U \leq 10$ with $\alpha/2 = .0406$ or $\alpha = .0812$.

16.19

a. Since H_a implies distribution B to the right of distribution A, sample B will have high ranks, T_B will be large, and U_B will be small. Hence, U_B is used as the test statistic.

b-c. From Table 14, with $n_1 = 4$, $n_2 = 5$, the appropriate rejection region is $U \leq 4$ with $\alpha = .0952$.

16.20

a. See solution to Exercise 16.18 a.

b. From Table 14, choose a rejection region with $\alpha/2 \leq .05$. The appropriate rejection region is $U \leq 2$ with $\alpha/2 = .0317$ or $\alpha = .0634$.

16.21

If H_a is true and population A lies to the right of population B, then T_A will be large and U_A will be small. Hence, the test statistic will be U_A and the large sample approximation can be used. Calculate

$$U_A = n_1 n_2 + \frac{n_1(n_1 + 1)}{2} - T_A = 12(14) + \frac{12(13)}{2} - 193 = 53$$

$$E(U) = \frac{n_1 n_2}{2} = \frac{12(14)}{2} = 84$$

$$\sigma_U^2 = \frac{n_1 n_2 (n_1 + n_2 + 1)}{12} = \frac{12(14)(27)}{12} = 378$$

The test statistic is

$$z = \frac{U - E(U)}{\sigma_U} = \frac{53 - 84}{\sqrt{378}} = -1.59$$

The rejection region with $\alpha = .05$ is $z < -1.645$ and H_0 is not rejected. There is insufficient evidence to indicate a difference in the two population distributions.

16.22

This is similar to Exercise 16.21. The large sample approximation to the Mann-Whitney U test is used. Calculate

$$U_B = n_1 n_2 + \frac{n_2(n_2 + 1)}{2} - T_B = 15(15) + \frac{15(16)}{2} - 251 = 94$$

$$E(U) = \frac{n_1 n_2}{2} = \frac{15(15)}{2} = 112.5$$

$$\sigma_U^2 = \frac{n_1 n_2 (n_1 + n_2 + 1)}{12} = \frac{15(15)(31)}{12} = 581.25$$

The test statistic is

$$z = \frac{U - E(U)}{\sigma_U} = \frac{94 - 112.5}{\sqrt{581.25}} = -.77$$

The rejection region with $\alpha = .10$ is $|z| > 1.645$ and H_0 is not rejected. There is insufficient evidence to indicate a difference in the two population distributions.

16.23

It is necessary to test the hypothesis that the two populations are identical using the Mann-Whitney U test. The test statistic is U, the number of observations in sample A that precede each observation in sample B. The data, with corresponding ranks, are shown in the following table.

A	B
15 (14.5)	17 (17.5)
11 (10)	6 (2.5)
20 (20)	15 (14.5)
14 (13)	10 (8.5)
9 (7)	6 (2.5)
12 (11)	8 (5.5)
5 (1)	10 (8.5)
17 (17.5)	16 (16)
13 (12)	8 (5.5)
18 (19)	7 (4)
$T_A = 125$	$T_B = 85$
$n_1 = 10$	$n_2 = 10$

Calculate

$$U_A = n_1 n_2 + \frac{n_1(n_1 + 1)}{2} - T_A = 10(10) + \frac{10(11)}{2} - 125 = 30$$

$$U_B = n_1 n_2 + \frac{n_2(n_2 + 1)}{2} - T_B = 10(10) + \frac{10(11)}{2} - 85 = 70$$

From Table 14, the rejection region will be chosen so that $\alpha \approx .05$. For a two-tailed test with $n_1 = n_2 = 10$, choosing the lower portion of the rejection region as $U \leq 24$ gives $\alpha/2 = .0262$ or $\alpha = .0524$. Choosing the smaller of U_A and U_B as the test statistic ($U = 30$), the null hypothesis is not rejected. There is insufficient evidence to suggest a difference in the two distributions.

16.24

This is similar to Exercise 16.23. The data, with corresponding ranks, are shown in the following table.

A	B
32 (4)	41 (13)
25 (1)	39 (10.5)
40 (12)	36 (8)
31 (3)	47 (17)
35 (7)	45 (16)
29 (2)	34 (6)
37 (9)	48 (18)
39 (10.5)	44 (15)
	43 (14)
	33 (5)
$T_A = 48.5$	$T_B = 122.5$

Calculate

$$U_A = n_1 n_2 + \frac{n_1(n_1 + 1)}{2} - T_A = 8(10) + \frac{8(9)}{2} - 48.5 = 67.5$$

$$U_B = n_1 n_2 + \frac{n_2(n_2 + 1)}{2} - T_B = 8(10) + \frac{10(11)}{2} - 122.5 = 12.5$$

The test statistic is $U = 12.5$ and the rejection region with $n_1 = 8$ and $n_2 = 10$, found in Table 14, is $U \leq 21$ with $\alpha = 2(.0506) = .1012$. The null hypothesis is rejected. We conclude that the life in months of service before failure of the picture tube differs for the picture tubes manufactured by each firm.

16.25

This is similar to previous exercises. The data, with corresponding ranks, are shown in the following table.

Standard	Enhanced
1.56 (17.5)	1.59 (19)
1.41 (13.5)	1.68 (20)
1.48 (15)	1.17 (5)
1.37 (10)	0.94 (1)
1.39 (12)	1.56 (17.5)
1.20 (6)	0.96 (2)
1.38 (11)	1.09 (3)
1.54 (16)	1.26 (8)
1.41 (13.5)	1.23 (7)
1.16 (4)	1.30 (9)

a. With $n_1 = 10$, $n_2 = 10$, calculate $T_A = 118.5$ and $T_B = 91.5$. Then

$$U_A = n_1 n_2 + \frac{n_1(n_1 + 1)}{2} - T_A = 10(10) + \frac{10(11)}{2} - 118.5 = 36.5$$

$$U_B = n_1 n_2 + \frac{n_2(n_2 + 1)}{2} - T_B = 10(10) + \frac{10(11)}{2} - 91.5 = 63.5$$

and the test statistic is $U_A = 36.5$. From Table 14, we need a two-tailed rejection region with $\alpha/2 \approx .05$ in the lower tail. This region is $U \leq 28$ with $\alpha/2 = .0526$ or $\alpha = .1052$. Since the observed value $U_A = 36.5$ does not fall in the rejection region, H_0 is not rejected. There is no evidence of a difference.

b. The parametric test was unable to reject H_0; the tests achieve the same result.

16.26

This is similar to previous exercises. The data, with corresponding ranks, are shown in the following table.

A	B
7 (1)	19 (13)
12 (4)	24 (18)
29 (20)	14 (6)
8 (2)	17 (9.5)
15 (7.5)	25 (19)
11 (3)	21 (15)
17 (9.5)	13 (5)
15 (7.5)	23 (17)
22 (16)	18 (11.5)
20 (14)	18 (11.5)
$T_A = 84.5$	$T_B = 125.5$
$n_1 = 10$	$n_2 = 10$

a. The random variable x = number of defective bolts has a hypergeometric distribution, which could be approximated as a binomial distribution. It does not have a normal distribution.

b. Calculate

$$U_A = n_1 n_2 + \frac{n_1(n_1 + 1)}{2} - T_A = 10(10) + \frac{10(11)}{2} - 84.5 = 70.5$$
$$U_B = n_1 n_2 + \frac{n_2(n_2 + 1)}{2} - T_B = 10(10) + \frac{10(11)}{2} - 125.5 = 29.5$$

The test statistic is $U = 29.5$ and the rejection region with $n_1 = 10$ and $n_2 = 10$, found in Table 14, is $U \leq 24$ with $\alpha = 2(.0262) = .0524$. The null hypothesis is not rejected. We cannot conclude that there is a difference in the proportions shipped by the two suppliers.

16.27

This is similar to previous exercises. The data and their ranks are shown below.

A	B
13 (4)	16 (8.5)
18 (12)	10 (1)
17 (10.5)	12 (3)
20 (14.5)	15 (6.5)
16 (8.5)	19 (13)
20 (14.5)	17 (10.5)
14 (5)	11 (2)
15 (6.5)	
$T_A = 75.5$	$T_B = 44.5$
$n_1 = 8$	$n_2 = 7$

Calculate

$$U_A = n_1 n_2 + \frac{n_1(n_1 + 1)}{2} - T_A = 8(7) + \frac{8(9)}{2} - 75.5 = 16.5$$
$$U_B = n_1 n_2 + \frac{n_2(n_2 + 1)}{2} - T_B = 8(7) + \frac{7(8)}{2} - 44.5 = 39.5$$

The test statistic is $U = 16.5$ and the rejection region, found in Table 14, is $U \leq 13$ with $\alpha = 2(.0469) = .0938$. The null hypothesis is not rejected. We cannot conclude that there is a difference in the scores for the two brokers.

16.28

a. H_0: population distributions A and B are identical
H_a: the distributions differ in location

b. Since Table 15, Appendix II gives critical values for rejection in the lower tail of the distribution, we use the smaller of T^+ and T^- as the test statistic.

c. From Table 15 with $n = 30$, $\alpha = .05$, and a two-tailed test, the rejection region is $T \leq 137$.

d. Since $T^+ = 249$, we can calculate
$$T^- = \frac{n(n+1)}{2} - T^+ = \frac{30(31)}{2} - 249 = 216$$

The test statistic is the smaller of T^+ and T^- or $T = 216$ and H_0 is not rejected. There is no evidence of a difference between the two distributions.

16.29

a. H_0: population distributions A and B are identical
H_a: population distribution A is shifted to the right of distribution B

b. If H_a is true, most of the differences, $x_A - x_B$, should be large and positive. Hence, T^+ would tend to be large and T^- small. Therefore, T^- will be the test statistic.

c. From Table 15 with $n = 30$, $\alpha = .05$, and a one-tailed test, the rejection region is $T \leq 152$.

d. Since $T^+ = 249$, we can calculate
$$T^- = \frac{n(n+1)}{2} - T^+ = \frac{30(31)}{2} - 249 = 216$$

The test statistic is $T = 216$ and H_0 is not rejected. There is no evidence of a difference between the two distributions.

16.30

Since $n > 25$, the large sample approximation to the signed-rank test can be used to test the hypothesis given in Exercise 16.28 a. Calculate
$$E(T) = \frac{n(n+1)}{4} = \frac{30(31)}{4} = 232.5$$
$$\sigma_T^2 = \frac{n(n+1)(2n+1)}{24} = \frac{30(31)(61)}{24} = 2363.75$$

The test statistic is
$$z = \frac{T - E(T)}{\sigma_T} = \frac{216 - 232.5}{\sqrt{2363.75}} = -.34$$

The two-tailed rejection region with $\alpha = .05$ is $|z| > 1.96$ and H_0 is not rejected. The results agree with Exercise 16.28 d.

16.31

Refer to Exercise 16.30. For a one-tailed test, the test statistic is again $z = -.34$, but the rejection region is $z < -1.645$. The null hypothesis is not rejected and the results agree with Exercise 16.29 d.

16.32

a. The hypothesis to be tested is

H_0: population distributions A and B are identical
H_a: the distributions differ in location

and the test statistic is T, the rank sum of the positive (or negative) differences. The differences, along with their ranks (according to absolute magnitude) are shown in the following table.

d_i	1.2	1.6	2.9	4.1	−0.4	1.8	0.8	−0.1
Rank $|d_i|$	4	5	7	8	2	6	3	1

The rank sum for positive differences is $T^+ = 33$ and the rank sum for negative differences is $T^- = 3$ with $n = 8$. Indexing $n = 8$ and $\alpha = .05$ in Table 15, the lower portion of the two-tailed rejection region is $T \leq 4$ and H_0 is rejected. There is a difference in the two population locations.

b. Notice that the signed-rank test allows rejection of the null hypothesis, as did the t test. It is slightly more powerful than the sign test in this situation and still does not rely on the assumption of normality.

16.33

a. This is similar to Exercise 16.32. The Wilcoxon signed-rank test is used, and the differences, along with their ranks (according to absolute magnitude) are shown in the following table.

d_i	−4	2	−2	−5	−3	0	1	1	−6
Rank $|d_i|$	6	3.5	3.5	7	5	—	1.5	1.5	8

The sixth pair is tied and is hence eliminated from consideration. Pairs 7 and 8, 2 and 3 are tied and receive an average rank. Then $T^+ = 6.5$ and $T^- = 29.5$ with $n = 8$. Indexing $n = 8$ and $\alpha = .05$ in Table 15, the lower portion of the two-tailed rejection region is $T \leq 4$ and H_0 is not rejected. There is insufficient evidence to detect a difference in the two machines.

b. If a machine continually breaks down, it will eventually be fixed, and the breakdown rate for the following month will decrease.

16.34

This is similar to previous exercises. The hypothesis to be tested is

H_0: population distributions 1 and 2 are identical
H_a: the distributions differ in location

and the test statistic is T, the rank sum of the positive (or negative) differences. The differences, along with their ranks (according to absolute magnitude) are shown in the following table.

d_i	8	−8	−14	3	10	4	6	6	−2	6
Rank$|d_i|$	7.5	7.5	10	2	9	3	5	5	1	5

The rank sum for positive differences is $T^+ = 36.5$ and the rank sum for negative differences is $T^- = 18.5$ with $n = 10$. Indexing $n = 10$ and $\alpha = .10$ in Table 15, the lower portion of the two-tailed rejection region is $T \leq 11$ and H_0 is not rejected. There is insufficient evidence to suggest that one counselor tends to rate higher than the other.

16.35

a. This is similar to previous exercises. The hypothesis to be tested is

H_0: population distributions 1 and 2 are identical
H_a: the distributions differ in location

and the test statistic is T, the rank sum of the positive (or negative) differences. The differences, along with their ranks (according to absolute magnitude) are shown in the following table.

d_i	2	−1	−1	1	2	1	0	2	0	0	3	0
Rank $|d_i|$	6	2.5	2.5	2.5	6	2.5	—	6	—	—	8	—

The rank sum for positive differences is $T^+ = 31$ and the rank sum for negative differences is $T^- = 5$ with $n = 8$. Indexing $n = 8$ and $\alpha = .05$ in Table 15, the lower portion of the two-tailed rejection region is $T \leq 4$ and H_0 is not rejected. There is insufficient evidence to indicate a difference in location.

b. The random variable $x =$ number of accidents has a Poisson, not a normal distribution.

Hence, if the means are different, so will be the variance. (For a Poisson random variable, $\sigma = \sqrt{\mu}$.) Hence, the assumptions of normality and equal variances are violated.

16.36

a. This is similar to previous exercises. The differences, along with their ranks (according to absolute magnitude), are shown in the following table.

d_i	1	1	−3	−2	−1	−5	−4	0
Rank $|d_i|$	2	2	5	4	2	7	6	—

The rank sum for positive differences is $T^+ = 4$ and the rank sum for negative differences is $T^- = 24$ with n = 7. Indexing n = 7 and $\alpha = .05$ in Table 15, the lower portion of the two-tailed rejection region is $T \leq 2$ and H_0 is not rejected. There is no evidence of a significant difference in the two population locations.

b. From Table 15, the observed value T = 4 is the critical value for a two-tailed test if $\alpha = .10$. Hence, p-value $\approx .10$.

16.37

This is similar to previous exercises. The differences, along with their ranks (according to absolute magnitude), are shown in the following table.

d_i	.54	−.94	−.23	.30	.28	.76
Rank $|d_i|$	4	6	1	3	2	5

a. The rank sum for positive differences is $T^+ = 14$ and the rank sum for negative differences is $T^- = 7$ with n = 6. Indexing n = 6 and $\alpha = .05$ in Table 15, the one-tailed rejection region is $T \leq 2$ and H_0 is not rejected. There is no evidence of a significant difference in the two population locations.

b. From Table 15, all that we can say is that p-value > .05. The conclusion agrees with the results of Exercise 8.44.

16.38

The Kruskal-Wallis H test provides a nonparametric analog to the analysis of variance F test for a completely randomized design presented in Chapter 9. The data are jointly ranked from smallest to largest, with ties treated as in the Mann-Whitney U test. The data with corresponding ranks in parentheses are shown below.

Treatment		
1	2	3
26 (9.5)	27 (11.5)	25 (8)
29 (15.5)	31 (19)	24 (6)
23 (4)	30 (17.5)	27 (11.5)
24 (6)	28 (13.5)	22 (3)
28 (13.5)	29 (15.5)	24 (6)
26 (9.5)	32 (20)	20 (1)
	30 (17.5)	21 (2)
	33 (21)	
$T_1 = 58$	$T_2 = 135.5$	$T_3 = 37.5$
$n_1 = 6$	$n_2 = 8$	$n_3 = 7$

The test statistic, based on the rank sums, is

$$H = \frac{12}{n(n+1)} \sum \frac{T_i^2}{n_i} - 3(n+1)$$

$$= \frac{12}{21(22)}\left[\frac{(58)^2}{6} + \frac{(135.5)^2}{8} + \frac{(37.5)^2}{7}\right] - 3(22) = 13.39$$

The rejection region with $\alpha = .05$ and $k - 1 = 2$ d.f. is based on the chi-square distribution, or $H > \chi^2_{.05} = 5.99$. The null hypothesis is rejected and we conclude that there is a difference in location among the three treatments.

16.39

This is similar to Exercise 16.38. The data with corresponding ranks in parentheses are shown below.

	Treatment		
1	2	3	4
124 (9)	147 (20)	141 (17)	117 (4.5)
167 (26)	121 (7)	144 (18.5)	128 (10.5)
135 (14)	136 (15)	139 (16)	102 (1)
160 (24)	114 (3)	162 (25)	119 (6)
159 (23)	129 (12)	155 (22)	128 (10.5)
144 (18.5)	117 (4.5)	150 (21)	123 (8)
133 (13)	109 (2)		
$T_1 = 127.5$	$T_2 = 63.5$	$T_3 = 119.5$	$T_4 = 40.5$
$n_1 = 7$	$n_2 = 7$	$n_3 = 6$	$n_4 = 6$

The test statistic, based on the rank sums, is

$$H = \frac{12}{n(n+1)} \sum \frac{T_i^2}{n_i} - 3(n+1)$$

$$= \frac{12}{26(27)}\left[\frac{(127.5)^2}{7} + \frac{(63.5)^2}{7} + \frac{(119.5)^2}{6} + \frac{(40.5)^2}{6}\right] - 3(27) = 13.90$$

The rejection region with $\alpha = .05$ and $k - 1 = 3$ d.f. is $H > \chi^2_{.05} = 7.81$. The null hypothesis is rejected and we conclude that there is a difference in location among the four treatments.

16.40

This is similar to Exercise 16.38. The data with corresponding ranks in parentheses are shown below.

	Location		
1	2	3	4
1.59 (5.5)	1.58 (3.5)	1.54 (1)	1.69 (13)
1.63 (9.5)	1.61 (7.5)	1.59 (5.5)	1.70 (14)
1.65 (12)	1.64 (11)	1.55 (2)	
1.61 (7.5)	1.63 (9.5)	1.58 (3.5)	
$T_1 = 34.5$	$T_2 = 31.5$	$T_3 = 12$	$T_4 = 27$
$n_1 = 4$	$n_2 = 4$	$n_3 = 4$	$n_4 = 2$

a. The test statistic, based on the rank sums, is

$$H = \frac{12}{n(n+1)} \sum \frac{T_i^2}{n_i} - 3(n+1)$$

$$= \frac{12}{14(15)} \left[\frac{(34.5)^2}{4} + \frac{(31.5)^2}{4} + \frac{(12)^2}{4} + \frac{(27)^2}{2} \right] - 3(15) = 9.064$$

The rejection region with $\alpha = .05$ and $k - 1 = 3$ d.f. is $H > \chi^2_{.05} = 7.81$. The null hypothesis is rejected and we conclude that there is a difference among the four locations.

b. From Table 5, with d.f. = 3, $.025 <$ p-value $< .05$.

c. From the printout, Exercise 9.14, we find $p = .0009$.

d. The observed level of significance is smaller for the more powerful parametric test. However, the results of the test are the same, and the restrictive parametric assumption of normality was not used.

16.41

Similar to Exercise 16.38. The data with corresponding ranks in parentheses are shown below.

	Program	
A	B	C
59 (6)	52 (1)	58 (4.5)
64 (9.5)	58 (4.5)	65 (11)
57 (3)	54 (2)	71 (12)
62 (7)		63 (8)
		64 (9.5)
$T_1 = 25.5$	$T_2 = 7.5$	$T_3 = 45$
$n_1 = 4$	$n_2 = 3$	$n_3 = 5$

The test statistic, based on the rank sums, is

$$H = \frac{12}{n(n+1)} \sum \frac{T_i^2}{n_i} - 3(n+1)$$

$$= \frac{12}{12(13)} \left[\frac{(25.5)^2}{4} + \frac{(7.5)^2}{3} + \frac{(45)^2}{5} \right] - 3(13) = 6.101$$

The rejection region with $\alpha = .05$ and $k - 1 = 2$ d.f. is $H > \chi^2_{.05} = 5.99$. The null hypothesis is rejected and we conclude that there is a difference in location among the three programs.

b. From Table 5, with d.f. = 2, $.025 <$ p-value $< .05$.

c. From the printout, Exercise 9.15, we find $p = .0251$.

d. The test results are the same.

16.42

Similar to previous exercises. The data with corresponding ranks in parentheses are shown below.

	Type of Music	
1	2	3
94 (15)	84 (7.5)	81 (5)
87 (10)	89 (11)	76 (2)
90 (12.5)	82 (6)	73 (1)
86 (9)	90 (12.5)	79 (4)
91 (14)	78 (3)	84 (7.5)
$T_1 = 60.5$	$T_2 = 40$	$T_3 = 19.5$
$n_1 = 5$	$n_2 = 5$	$n_3 = 5$

The test statistic, based on the rank sums, is

$$H = \frac{12}{n(n+1)} \sum \frac{T_i^2}{n_i} - 3(n+1)$$

$$= \frac{12}{15(16)}\left[\frac{(60.5)^2}{5} + \frac{(40)^2}{5} + \frac{(19.5)^2}{5}\right] - 3(16) = 8.405$$

The rejection region with $\alpha = .10$ and $k - 1 = 2$ d.f. is $H > \chi^2_{.10} = 4.605$. The null hypothesis is rejected and we conclude that there is a difference in location among the music types.

16.43

a. In using the Friedman F_r test, data are ranked **within a block** from 1 to k. The treatment rank sums are then calculated as usual. The data and their corresponding ranks are shown below.

Block	Treatment		
	1	2	3
1	3.2 (3)	3.1 (2)	2.4 (1)
2	2.8 (2)	3.0 (3)	1.7 (1)
3	4.5 (2)	5.0 (3)	3.9 (1)
4	2.5 (1)	2.7 (3)	2.6 (2)
5	3.7 (2)	4.1 (3)	3.5 (1)
6	2.4 (2.5)	2.4 (2.5)	2.0 (1)
	$T_1 = 12.5$	$T_2 = 16.5$	$T_3 = 7$

The test statistic is

$$F_r = \frac{12}{bk(k+1)} \sum T_i^2 - 3b(k+1)$$

$$= \frac{12}{6(3)(4)}\left[(12.5)^2 + (16.5)^2 + 7^2\right] - 3(6)(4)$$

$$= 7.58$$

and the rejection region is $F_r > \chi^2_{.05} = 5.99$. Hence, H_0 is rejected and we conclude that there is a difference among the three treatments.

b. The observed value, $F_r = 7.58$ falls between $\chi^2_{.025}$ and $\chi^2_{.01}$. Hence, $.01 < $ p-value $ < .025$.

c-e. The analysis of variance is performed as in Chapter 9. The ANOVA table is shown below.

Source	df	SS	MS	F
Treatments	2	1.56	0.78	10.833
Blocks	5	10.965	2.193	30.458
Error	10	0.72	0.072	
Total	17	13.245		

The analysis of variance F test for treatments is F = 10.833 and the approximate p-value with 2 and 10 d.f. is p-value < .005.

f. In both instances, H_0 is rejected. However, the p-value for the parametric test is smaller. Hence, if all of the parametric assumptions are met, the parametric test will be more powerful than its nonparametric analog.

16.44

This is similar to Exercise 16.43. The ranks of the data are shown below.

	Treatment			
Block	1	2	3	4
1	4	1	2	3
2	4	1.5	1.5	3
3	4	1	3	2
4	4	1	2	3
5	4	1	2.5	2.5
6	4	1	2	3
7	4	1	3	2
8	4	1	2	3

a. The test statistic is

$$F_r = \frac{12}{8(4)(5)}\left[32^2 + (8.5)^2 + 18^2 + (21.5)^2\right] - 3(8)(5) = 21.19$$

and the rejection region is $F_r > \chi^2_{.05} = 7.81$. Hence, H_0 is rejected and we conclude that there is a difference among the four treatments.

b. Since $F_r = 21.19$ exceeds $\chi^2_{.005}$, we have p-value < .005.

c-e. The analysis of variance is performed as in Chapter 9. The ANOVA table is shown below.

Source	df	SS	MS	F
Treatments	3	198.34375	66.114583	75.43
Blocks	7	220.46875	31.495536	
Error	21	18.40625	.876488	
Total	31	437.21875		

The analysis of variance F test for treatments is F = 75.43 and the approximate p-value with 3 and 21 d.f. is p-value < .005. The result is identical to the parametric result.

16.45

This is similar to previous exercises. The ranks within each block are shown below.

	Management Level		
Company	1	2	3
1	3	1	2
2	3	1	2
3	2	3	1
4	3	2	1
5	3	1.5	1.5
6	3	2	1
7	3	2	1
8	3	1.5	1.5
9	3	1	2
10	3	2	1
T_i	29	17	14

The test statistic is

$$F_r = \frac{12}{(30)(4)}\left[29^2 + 17^2 + 14^2\right] - 3(10)(4) = 12.6$$

and the rejection region with $\alpha = .10$ is $F_r > \chi^2_{.10} = 4.605$. Hence, H_0 is rejected and we conclude that there is a difference among the management levels.

16.46

This is similar to previous exercises. The ranks within each block are given below.

	Year						
	1	2	3	4	5	6	T_i
Campbell Soup	3	1	2	2	2	2	12
General Mills	2	3	3	3	3	3	17
Pillsbury	1	2	1	1	1	1	7

a. Since PE levels will differ from year to year, years act as blocks to isolate unwanted variation.

b. The test statistic is

$$F_r = \frac{12}{6(3)(4)}\left[12^2 + 17^2 + 7^2\right] - 3(6)(4) = 8.33$$

and the rejection region with $\alpha = .05$ is $F_r > \chi^2_{.05} = 5.99$. Hence, H_0 is rejected and we conclude that there is a difference among the three companies.

16.47

Table 16, Appendix II gives critical values r_0 such that $P[r_s \geq r_0] = \alpha$. Hence, for an upper-tailed test, the critical value for rejection can be read directly from the table.

a. $r_s \geq .425$ **b.** $r_s \geq .601$

16.48

For a test of negative correlation, r_s must be large and negative. The critical value for a lower-tailed rejection region can be read from the table and a negative sign is attached.

a. $r_s \leq -.497$ **b.** $r_s \leq -.703$

16.49

For a two-tailed test of correlation, the value of α given along the top of the table is doubled to obtain the **actual** value of α for the test.

a. To obtain $\alpha = .05$, index $.025$ and the rejection region is $|r_s| \geq .400$.
b. To obtain $\alpha = .01$, index $.005$ and the rejection region is $|r_s| \geq .526$.

16.50

a. To calculate the Spearman's rank correlation coefficient, the data are ranked separately according to the variables x and y.

Rank x	2	1	4	6	5	3
Rank y	5	6	3	1	2	4

Since there were no tied observations, the simpler formula for r_s is used, and

$$r_s = 1 - \frac{n\Sigma d_i^2}{n(n^2-1)} = 1 - \frac{6[(-3)^2 + (-5)^2 + \cdots + (-1)^2]}{6(35)}$$

$$= 1 - \frac{420}{210} = -1$$

b. To test for correlation with $\alpha = .05$, index $.025$ in Table 16 and the rejection region is $|r_s| \geq .886$. Hence, H_0 is rejected; there is a correlation between x and y.

16.51

Since the two variables are already ranked, with no ties existing, the shortcut formula may be used for the differences:

$$1, \quad -2, \quad -1, \quad -1, \quad -1, \quad -1, \quad 2, \quad 1, \quad 1, \quad 1$$

The test statistic is

$$r_s = 1 - \frac{n\Sigma d_i^2}{n(n^2-1)} = 1 - \frac{6[(1)^2 + (-2)^2 + \cdots + (1)^2]}{10(99)} = .903$$

To test for positive correlation with $\alpha = .05$, index $.05$ in Table 16 and the rejection region is $r_s \geq .564$. Hence, H_0 is rejected; there is a positive correlation between x and y.

16.52

Since the first variable (interview rating) is already in ranked form, we need only rank the second variable (test score). This variable will be ranked from low to high. The ranks (x_i and y_i) are shown in the following table.

Subject	Rank, x	Rank, y
1	8	5
2	5	6
3	10	2.5
4	3	7
5	6	2.5
6	1	9
7	4	10
8	7	4
9	9	1
10	2	8

Calculate $\Sigma x_i y_i = 233$ $\Sigma x_i^2 = 385$ $\Sigma y_i^2 = 384.5$

$n = 10$ $\Sigma x_i = 55$ $\Sigma y_i = 55$

Then
$$S_{xy} = 233 - \frac{55^2}{10} = -69.5$$

$$S_{xx} = 385 - \frac{55^2}{10} = 82.5$$

$$S_{yy} = 384.5 - \frac{55^2}{10} = 82$$

and
$$r_s = \frac{S_{xy}}{\sqrt{S_{xx}S_{yy}}} = \frac{-69.5}{\sqrt{82(82.5)}} = -.845$$

16.53

Refer to Exercise 16.52. To test for correlation with $\alpha = .05$, index .025 in Table 16, and the rejection region is $|r_s| \geq .564$. The null hypothesis is rejected and we conclude that there is a correlation between the two variables.

16.54

a. Define p = P[assembly time for method A exceeds that for method B] and x = number of times the assembly time for method A exceeds that for method B. The hypothesis to be tested is

$$H_0: p = 1/2 \quad \text{versus} \quad H_a: p \neq 1/2$$

using the sign test with x as the test statistic. Notice that for this exercise $n = 9$, and the observed value of the test statistic is $x = 2$.

Various two-tailed rejection regions are tried in order to find a region with $\alpha \approx .05$. These are shown in the following table.

Rejection Region	α
$x = 0; x = 9$	.004
$x \leq 1; x \geq 8$	.040
$x \leq 2; x \geq 7$	.180

We choose to reject H_0 if $x \leq 1$ or $x \geq 8$ with $\alpha = .040$. Since $x = 2$, H_0 is not rejected. There is insufficient evidence to indicate a difference between the two methods.

b. The experiment has been designed in a paired manner, and the paired-difference test is used. The differences are shown below.

| d_i | $-.9$ | -1.1 | 1.5 | -2.6 | -1.8 | -2.9 | -2.5 | 2.5 | -1.4 |

The hypothesis to be tested is

$$H_0: \mu_1 - \mu_2 = 0 \qquad H_a: \mu_1 - \mu_2 \neq 0$$

Calculate
$$\bar{d} = \frac{\Sigma d_i}{n} = \frac{-9.2}{9} = -1.022$$

$$s_d^2 = \frac{\Sigma d_i^2 - \frac{(\Sigma d_i)^2}{n}}{n - 1} = \frac{37.14 - 9.404}{8} = 3.467$$

and the test statistic is

$$t = \frac{\overline{d}}{\sqrt{\frac{s_d^2}{n}}} = \frac{-1.022}{\sqrt{\frac{3.467}{9}}} = -1.646$$

The rejection region with $\alpha = .05$ and 8 d.f. is $|t| > 2.306$ and H_0 is not rejected.

16.55

The differences are ranked according to absolute magnitude and the rank sums for the positive and negative differences are calculated.

d_i	−.9	−1.1	1.5	−2.6	−1.8	−2.9	−2.5	2.5	−1.4
Rank $|d_i|$	1	2	4	8	5	9	6.5	6.5	3

The rank sum for positive differences is $T^+ = 10.5$ and the rank sum for negative differences is $T^- = 34.5$. Indexing $n = 9$ and $\alpha = .05$ in Table 15, the lower portion of the two-tailed rejection region is $T \leq 6$ and H_0 is not rejected. We cannot conclude that there is a difference in the distributions of the responses for the two methods.

16.56

Similar to previous exercises. The data, with corresponding ranks, are shown in the following table.

Process A	Process B
6.1 (1)	9.1 (16)
9.2 (17)	8.2 (8)
8.7 (12)	8.6 (11)
8.9 (13.5)	6.9 (2)
7.6 (5)	7.5 (4)
7.1 (3)	7.9 (7)
9.5 (18)	8.3 (9.5)
8.3 (9.5)	7.8 (6)
9.0 (15)	8.9 (13.5)
$T_A = 94$	$T_B = 77$

To test the null hypothesis that the two population distributions are identical, calculate

$$U_A = n_1 n_2 + \frac{n_1(n_1 + 1)}{2} - T_A = 9(9) + \frac{9(10)}{2} - 94 = 32$$

$$U_B = n_1 n_2 + \frac{n_2(n_2 + 1)}{2} - T_B = 9(9) + \frac{9(10)}{2} - 77 = 49$$

The test statistic is $U = 32$ and the rejection region with $n_1 = 9$ and $n_2 = 9$, found in Table 14, is $U \leq 21$ with $\alpha = 2(.047) = .094$. The null hypothesis is not rejected. We cannot conclude that there is a difference in brightness for the two processes.

b. The hypothesis to be tested is $H_0: \mu_1 - \mu_2 = 0$ versus $H_a: \mu_1 - \mu_2 \neq 0$. Calculate

$$\overline{x}_1 = \frac{\Sigma x_{1j}}{n_1} = \frac{74.4}{9} = 8.2667 \qquad \overline{x}_2 = \frac{\Sigma x_{2j}}{n_2} = \frac{73.2}{9} = 8.1333$$

$$s^2 = \frac{\Sigma x_{1j}^2 - \frac{(\Sigma x_{1j})^2}{n_1} + \Sigma x_{2j}^2 - \frac{(\Sigma x_{2j})^2}{n_2}}{n_1 + n_2 - 2} = \frac{625.06 - \frac{(74.4)^2}{9} + 599.22 - \frac{(73.2)^2}{9}}{16}$$
$$= .8675$$

and the test statistic is

$$t = \frac{\bar{x}_1 - \bar{x}_2}{\sqrt{s^2\left(\frac{1}{n_1} + \frac{1}{n_2}\right)}} = \frac{8.27 - 8.13}{\sqrt{.8675\left(\frac{2}{9}\right)}} = .304$$

The rejection region with $\alpha = .05$ and 16 degrees of freedom is $|t| > 1.746$ and H_0 is not rejected.

16.57

a. The measurements are ordered according to magnitude, and ranked "from the outside in" as described in part (ii). The resulting ranks are

Instrument	Response	Rank
A	1060.21	1
B	1060.24	3
A	1060.27	5
B	1060.28	7
B	1060.30	9
B	1060.32	8
A	1060.34	6
A	1060.36	4
A	1060.40	2

The hypothesis to be tested is

$$H_0: \sigma_A^2 = \sigma_B^2 \quad \text{versus} \quad H_a: \sigma_A^2 > \sigma_B^2$$

and the test statistic is the Mann-Whitney U. If the alternative hypothesis is true (that is, the variance for instrument A is greater than the variance for instrument B), then the measurements for instrument A should be very low and very high in the sequence of measurements. Hence, they will be assigned the lower ranks, and the "sum of ranks" for the A observations will be small. A one-tailed test of hypothesis is required, with α near .05. Calculate

$$U_1 = n_1 n_2 + \frac{n_1(n_1 + 1)}{2} - T_A = 5(4) + \frac{5(6)}{2} - 18 = 17$$
$$U_2 = n_1 n_2 + \frac{n_2(n_2 + 1)}{2} - T_B = 5(4) + \frac{4(5)}{2} - 27 = 3$$

The test statistic is $U = 3$ and the rejection region with $n_1 = 5$ and $n_2 = 4$, found in Table 14, is $U \leq 3$ with $\alpha = .0556$. The test statistic falls in the rejection region and H_0 is rejected.

b. The sample variances, s_1^2 and s_2^2, must be calculated in order to use the F test of Section 8.10.

$$s_1^2 = \frac{\Sigma x_{1j}^2 - \frac{(\Sigma x_{1j})^2}{n_1}}{n_1 - 1} = \frac{.0230}{4} = .00575$$

$$s_2^2 = \frac{\Sigma x_{2j}^2 - \frac{(\Sigma x_{2j})^2}{n_2}}{n_2 - 1} = \frac{.0035}{3} = .00117$$

Then the test statistic is

$$F = \frac{s_1^2}{s_2^2} = \frac{.00575}{.00117} = 4.914$$

The rejection region with 4 and 3 d.f. and $\alpha = .05$ is $F > 9.12$ and the null hypothesis is not rejected.

16.58

a. This is similar to previous exercises. The hypothesis to be tested is

$$H_0: p = 1/2 \quad \text{versus} \quad H_a: p > 1/2$$

using the sign test with x = number of investments for which CPA #1 exceeds CPA #2 as the test statistic. Notice that for this exercise, $n = 6$, and the observed value of the test statistic is $x = 5$. The rejection region, with $\alpha \approx .01$, is $x \geq 6$, since

$$\alpha = P[x \geq 6 \text{ when } p = .5] = 1 - .984 = .016$$

Since the observed value of x does not fall in the rejection region, H_0 is not rejected. There is insufficient evidence to suggest that CPA #1 tends to give higher estimates.

b. Using the Wilcoxon signed-rank test requires ranking the differences, d_i, according to their absolute value, as shown below.

d_i	3100	−5000	3000	5000	3000	1000
Rank $|d_i|$	4	5.5	2.5	5.5	2.5	1

The rank sum for positive differences is $T^+ = 15.5$ and the rank sum for negative differences is $T^- = 5.5$. Indexing $n = 6$ in Table 15, the smallest possible one-sided rejection region is $T \leq 1$ with $\alpha = .025$. The observed value of the test statistic does not fall in the rejection region and H_0 is not rejected, as in part a.

16.59

A completely randomized design has been used and the Kruskal-Wallis H test is the appropriate nonparametric test. The data and their ranks are shown below.

	Region	
South	West	Northeast
1.11 (4)	1.49 (13)	.85 (3)
1.37 (10)	1.45 (12)	1.22 (7)
1.21 (6)	1.97 (14)	.82 (2)
1.28 (8)	2.51 (15)	1.16 (5)
1.31 (9)	1.42 (11)	.80 (1)
$T_1 = 37$	$T_2 = 65$	$T_3 = 18$
$n_1 = 5$	$n_2 = 5$	$n_3 = 5$

The test statistic, based on the rank sums, is

$$H = \frac{12}{n(n+1)} \sum \frac{T_i^2}{n_i} - 3(n+1)$$

$$= \frac{12}{15(16)}\left[\frac{37^2}{5} + \frac{65^2}{5} + \frac{18^2}{5}\right] - 3(16) = 11.18$$

The rejection region with $\alpha = .10$ and $k - 1 = 2$ d.f. is $H > \chi^2_{.10} = 4.605$. The null hypothesis is rejected and we conclude that there is a difference in location among the three regions.

16.60

A randomized block design has been used, and the Friedman F_r test is the appropriate nonparametric test. The ranks within blocks are shown below.

Investment	Firm 1	2	3
1	2	1	3
2	2.5	1	2.5
3	3	1.5	1.5
4	1	2	3
5	1	2	3
6	2	1	3
T_i	11.5	8.5	16

The test statistic is

$$F_r = \frac{12}{(18)(4)}\left[(11.5)^2 + (8.5)^2 + 16^2\right] - 3(6)(4) = 4.75$$

and the rejection region with $\alpha = .05$ is $F_r > \chi^2_{.05} = 5.99$. Hence, H_0 is not rejected and we cannot conclude that there is a difference among the firms.

16.61

The two variables, grader's rating and moisture content, are ranked from low to high. The ranks (x_i and y_i) are shown in the following table.

Leaf	Rank, x	Rank, y
1	10.5	12
2	5.5	7.5
3	7.5	9
4	7.5	6
5	4	4.5
6	9	10
7	2	3
8	5.5	4.5
9	1	1
10	12	11
11	10.5	7.5
12	3	2

Calculate $\Sigma x_i y_i = 636.25$ $\Sigma x_i^2 = 648.5$ $\Sigma y_i^2 = 649$

$n = 12$ $\Sigma x_i = 78$ $\Sigma y_i = 78$

Then

$S_{xy} = 636.25 - 507 = 129.25$

and
$$S_{xx} = 648.5 - 507 = 141.5$$

$$S_{yy} = 649 - 507 = 142$$

$$r_s = \frac{S_{xy}}{\sqrt{S_{xx}S_{yy}}} = \frac{129.25}{\sqrt{141.5(142)}} = .912$$

To test for correlation with $\alpha = .05$, index .025 in Table 16, and the rejection region is $|r_s| \geq .591$. The null hypothesis is rejected and we conclude that there is a correlation between the two variables.

APPENDIX III
Additional Case Studies

Case Study 1: Screening Tests

1. It is given that $P(T|D) = .98$ so that $P(\overline{T}|D) = .02$. Further, $P(\overline{T}|\overline{D}) = .99$ so that $P(T|\overline{D}) = .01$. The disease rate is estimated to be

$$P(D) = \frac{45,000}{242,200,000} = .000185797$$

and $P(\overline{D}) = .999814203$. Use Bayes' Law to find the probabilities of a false positive and a false negative.

$$P[\text{false positive}] = P(\overline{D}|T) = \frac{P(T|\overline{D})P(\overline{D})}{P(T|\overline{D})P(\overline{D}) + P(T|D)P(D)} = \frac{(.01)(.999814203)}{(.01)P(\overline{D}) + (.98)P(D)}$$

$$= .9821$$

$$P[\text{false negative}] = P(D|\overline{T}) = \frac{P(\overline{T}|D)P(D)}{P(\overline{T}|\overline{D})P(\overline{D}) + P(\overline{T}|D)P(D)} = \frac{(.02)(.000185797)}{(.99)P(\overline{D}) + (.02)P(D)}$$

$$= .0000038$$

Note that the false-positive rate is extremely high, causing many people to be falsely diagnosed as having AIDS.

2. The sensitivity and specificity of the HIV test must be increased.

3. $P(T|D) = P[\text{test shows defective}|\text{part is defective}] =$ "sensitivity".
$P(\overline{T}|\overline{D}) = P[\text{test does not show defective part}|\text{part is not defective}] =$ "specificity".

4. Sensitivity and specificity must be determined by sampling the process over a period of time.

Case Study 2: Capital Budgeting Criteria

1. The expected value of the estimated present value of operating profits must be calculated in two stages. Let x be the estimated present value of operating profits. Then E(x) can take three possible values. First, if the project is correctly evaluated,

$$E(x) = 200,000(.04) + 100,000(.48) + 0(.48) = 56,000$$

If the project is overestimated,

$$E(x) = 300,000(.04) + 200,000(.48) + 100,000(.48) = 156,000$$

and if the project is underestimated,

$$E(x) = 100,000(.04) + 0(.48) - 100,000(.48) = -44,000$$

The probability distribution for E(x) is shown below.

E(x)	Probability
56,000	.50
156,000	.25
−44,000	.25

The expected value of the estimated present values is

$$56{,}000(.50) + 156{,}000(.25) - 44{,}000(.25) = 56{,}000$$

Notice that this is exactly the same as the expected value of the actual present values, calculated as

$$200{,}000(.04) + 100{,}000(.48) = 56{,}000$$

2. Since the actual and estimated present values are the same, the decision-maker's estimates are, on the average, the same as the actual. That is, in repeated sampling, the estimated present values will have an average that is equal to the actual present values, which is a desirable attribute.

3. Refer to the table in Case Study 2. The joint probabilities of each outcome in the center of the table can be obtained as the product of the marginal probabilities. For example,

$$P[\$100{,}000, \text{ good}] = P[\text{underestimate, good site}] = (.25)(.04) = .01$$

Further,

$$P[\$100{,}000] = P[\$100{,}000, \text{ good}] + P[\$100{,}000, \text{ medium}] + P[\$100{,}000, \text{ poor}]$$

$$= .01 + .5(.48) + .25(.48) = .37$$

The probabilities of interest are:

$$P[\text{good}|\$100{,}000] = \frac{P[\$100{,}000, \text{ good}]}{P[\$100{,}000]} = \frac{.01}{.37} = .0270$$

$$P[\text{medium}|\$100{,}000] = \frac{P[\$100{,}000, \text{ medium}]}{P[\$100{,}000]} = \frac{.5(.48)}{.37} = .6486$$

$$P[\text{poor}|\$100{,}000] = \frac{P[\$100{,}000, \text{ poor}]}{P[\$100{,}000]} = \frac{.25(.48)}{.37} = .3243$$

Then, when the estimated present value is $100,000, the expected present value is

$$200{,}000(.0270) + 100{,}000(.6486) + 0(.3243) = \$70{,}260$$

If the cost of opening a new store is $90,000, the project will not cover the costs.

Case Study 3: Western Energy Services

1. Using the Western Energy Service program, an investor who pays $14,000 will receive 400 filings in BLM leases. Since the chance that a Western Energy client wins any given lottery is 1/360, the chance that he wins at least one lottery in 400 trials is

$$P[\text{win at least one}] = 1 - P[\text{win none}] = 1 - \left(\frac{359}{360}\right)^{400} = .6713$$

His expected number of lease acquisitions is $E(x) = np$, where n is the number of trials (lease filings) and p is the probability of winning at any trial. Thus,

$$E(x) = 400(\tfrac{1}{360}) = 1.1111$$

2. Without using the Western Energy Service program, lease filing costs are $10 each, allowing 1400 filings for a $14,000 investment. Since the chance of any applicant winning a lottery is 1/552, the probability of winning at least one lottery in 1400 trials is

$$P[\text{win at least one}] = 1 - P[\text{win none}] = 1 - \left(\tfrac{551}{552}\right)^{1400} = .921$$

3. The expected payoff for an investor with Western Energy is approximately

$$\$23{,}715(1.1111)(.6713) - \$14{,}000(.3287) = \$13{,}086.78$$

4. For an individual who does not use the services of Western Energy, we first compute the expected number of winning leases in 1400 trials as

$$E(x) = 1400(1/552) = 2.5362$$

Then the expected payoff is approximately

$$\$12{,}000(2.5362)(.921) - \$14{,}000(.0790) = \$26{,}924.08$$

Case Study 4: West Coast Container Corporation

This is a sample size determination case. There are several approaches that students may take to evaluate the cost estimate. The most commonly used approach will be to use the sample information provided to estimate the variance of the population and then assume a new independent sample for the actual estimate of breakage. The second approach will be to assume that the information already gathered will be included in the total sample needed for the estimate of breakage. If the first approach is used, the student should conclude that the estimate is indeed reasonable. If the second approach is used, the estimate appears to be about $2400 higher than necessary.

Computations on Cost Estimates

$$B = 1; \quad z_{\alpha/2} = z_{.005} = 2.575$$

$$s^2 = \frac{\Sigma x^2 - \frac{(\Sigma x)^2}{n}}{n-1} = \frac{13{,}954 - \frac{(500)^2}{20}}{19} = 76.5263$$

$$s = \sqrt{76.5263} = 8.748$$

$$n = \frac{z_{\alpha/2}^2 \sigma^2}{B^2} = \frac{(2.575)^2 (8.748)^2}{(1)^2} = 507.42 \quad \text{or } n = 508$$

Assuming New Independent Sample   Assuming Present Sample Included

(508)($200) = $101,600   (508 − 20)($200) = $97,600

A $20,000 budget for additional research essentially puts a limit on the sample size, which forces the accuracy levels to be adjusted. This can be done by altering the bounds on the error, or the confidence in the estimate, or both. For the independent sample approach, holding the bound on the error constant and allowing the confidence level to vary, a sample size of 100 produces about a .7458 confidence level. If the confidence level is held constant and the bound is allowed to vary, the bound produced by n = 100 is about 2.25 taco shells. However, since taco shells come in increments of one, it is reasonable to fix the bound at 2 and allow the confidence level to vary. This produces a confidence level of .9774.

If the previously gathered information is included (n = 120), then fixing the bound produces a confidence level of .7988. Fixing the confidence level produces a bound of 2.056. Setting the bound at 2 produces a confidence level of .9876.

Computations on Accuracy

Independent Sample	Present Sample Included
n = 100 B = 1 s = 8.748	n = 120 B = 1 s = 8.748
$n = \dfrac{z_{\alpha/2}^2 \sigma^2}{B^2} \Rightarrow 100 = \dfrac{z_{\alpha/2}^2 (8.748)^2}{1^2}$	$n = \dfrac{z_{\alpha/2}^2 \sigma^2}{B^2} \Rightarrow 120 = \dfrac{z_{\alpha/2}^2 (8.748)^2}{1^2}$
$\Rightarrow z_{\alpha/2} = 1.14$	$\Rightarrow z_{\alpha/2} = 1.25$
or .7458 confidence level from Table 3.	or .7888 confidence level from Table 3.
Confidence level = .99 n = 100 z = 2.575 s = 8.748	*Confidence level = .99* n = 120 z = 2.575 s = 8.748
$n = \dfrac{z_{\alpha/2}^2 \sigma^2}{B^2} \Rightarrow 100 = \dfrac{(2.575)^2(8.748)^2}{B^2}$	$n = \dfrac{z_{\alpha/2}^2 \sigma^2}{B^2} \Rightarrow 120 = \dfrac{(2.575)^2(8.748)^2}{B^2}$
$\Rightarrow B = 2.25$	$\Rightarrow B = 2.056$

If we take B = 2 instead of B = 1:

n = 100 B = 2 s = 8.748	n = 120 B = 2 s = 8.748
$n = \dfrac{z_{\alpha/2}^2 \sigma^2}{B^2} \Rightarrow 100 = \dfrac{z_{\alpha/2}^2 (8.748)^2}{2^2}$	$n = \dfrac{z_{\alpha/2}^2 \sigma^2}{B^2} \Rightarrow 120 = \dfrac{z_{\alpha/2}^2 (8.748)^2}{2^2}$
$\Rightarrow z_{\alpha/2} = 2.286$	$\Rightarrow z_{\alpha/2} = 2.5$
or .9774 confidence level from Table 3.	or .9876 confidence level from Table 3.

A tremendous savings in cost can be had for just a small sacrifice in accuracy of the estimate (1 taco shell). This is a worthwhile trade-off. The question should be raised whether this type of information is worth enough to the sales staff to warrant the additional cost.

Case Study 5: Noodle Soup Brand Confusion

1. Each of the variables is measured for two samples, those confused and those not confused. Hence, we need to test the hypothesis

$$H_0: \mu_1 - \mu_2 = 0 \quad \text{versus} \quad H_a: \mu_1 - \mu_2 \neq 0$$

for each of the six variables and the test statistic is approximately

$$z = \frac{(\bar{x}_1 - \bar{x}_2) - 0}{\sqrt{\frac{s_1^2}{n_1} + \frac{s_2^2}{n_2}}}$$

For example, for the "brand familiarity" variable, the test statistic is

$$z = \frac{5.88 - 6.31}{\sqrt{\frac{(1.49)^2}{30} + \frac{(1.27)^2}{30}}} = -1.20$$

The rejection region with $\alpha = .05$ is $|z| > 1.96$ and H_0 is not rejected. There is no difference in brand familiarity.

Similar tests can be done for the other five variables:

Brand usage: $z = -1.59$

Experience: $z = -1.35$

Involvement: $z = -1.22$

Certainty of judgment: $z = -4.17$

Attitude: $z = -.22$

Only for the "certainty of judgment" category is there a significant difference between the two groups.

2. The results of this study indicate that the product imitation scheme is not making a difference except for those shoppers who are not certain before entering the store as to what they want to buy. The marketing director should aim his marketing techniques at increasing the brand familiarity, usage, and involvement for this segment of the population, perhaps with free samples or other incentives.

Case Study 6: Custom Precast

The experiment is strictly designed in a paired manner; hence, a paired-difference analysis is appropriate. A MINITAB printout below gives the results of the analysis for testing

$$H_0: \mu_1 - \mu_2 = 0 \qquad H_a: \mu_1 - \mu_2 \neq 0$$

MTB> ttest 0 c3

TEST OF MU = 0.000 VS MU N.E. 0.000

	N	MEAN	STDEV	SE MEAN	T	P VALUE
C3	12	−158.333	190.494	54.991	−2.88	0.015

The results are significant with p-value = .015, and we can conclude that there is a difference in the breaking strength with and without the additive.

If the experiment had been *incorrectly analyzed* in an unpaired manner, the following MINITAB printout would result:

MTB> twosample c1 c2;
SUBC> pooled.

TWOSAMPLE T FOR C1 VS C2
```
       N     MEAN    STDEV    SE MEAN
C1    12    5442     502       145
C2    12    5600     619       179
```

95 PERCENT CI FOR MU C1 − MU C2: (−735, 319)
TTEST MU C1 = MU C2 (VS NE): T = −0.69 P = 0.50 DF = 22.0

The results are not significant, since p-value = .50. Notice the difference in the variation for the two tests. In particular, $s_d = 190.494$, while for the unpaired analysis,

$$s^2 \approx \frac{11(502)^2 + 11(619)^2}{22} = 317{,}582.5$$

so that $s \approx 563.5$. The excessive variation from batch to batch is masking the effect of the additive on breaking strength.

Case Study 7: EXXON's Response to the Regulators

This case study involved an analysis of variance for a randomized block design. Within the design, storage tanks constitute blocks, and the four reprocessing procedures constitute treatments. Computations are as follows:

$$CM = \frac{(\Sigma\Sigma x_{ij})^2}{n} = 256{,}292.2013$$

Total SS $= \Sigma\Sigma x_{ij}^2 - CM = 256{,}476.4 - CM = 184.1987$

$$SST = \frac{\Sigma T_j^2}{b} - CM = \frac{705.4^2 + \cdots + 720.5^2}{8} - CM = 39.22625$$

$$SSB = \frac{\Sigma B_i^2}{t} - CM = \frac{363.1^2 + \cdots + 360.9^2}{4} - CM = 79.23875$$

The ANOVA table is

Source	d.f.	SS	MS	F
Treatments	3	39.22625	13.075417	4.18
Blocks	7	79.23875	11.319821	3.62
Error	21	65.73370	3.130176	
Total	31	184.1987		

(1) From Table 6, we find $F_{.05} = 2.49$ with 7 and 21 d.f. while $F_{.05} = 3.07$ with 3 and 21 d.f. Thus, sufficient evidence exists to indicate *both* a difference in octane reading of gasoline from among the different storage tanks and a difference in the ability of the reprocessing procedures to increase octane readings.

(2) Combining treatments 1 and 2 versus treatments 3 and 4, the test of interest is

$$H_0: \mu_{12} - \mu_{34} = 0 \qquad H_a: \mu_{12} - \mu_{34} \neq 0$$

and the t test is

$$t = \frac{\overline{T}_{12} - \overline{T}_{34}}{s\sqrt{\frac{1}{n_{12}} + \frac{1}{n_{34}}}} = \frac{(705.4 + 709.9)/16 - (728 + 720.5)/16}{\sqrt{3.130176\left(\frac{1}{16} + \frac{1}{16}\right)}} = -3.317$$

The rejection region is $|t| > 2.080$ and H_0 is rejected. There is a difference in the two types of processes.

(3) Although reforming provides a higher average octane reading than alkylation, the latter should probably be preferred since it appears to exhibit less variability in results. The primary weakness of this analysis is the small sample size involved. Experimental costs would probably have remained within control by dividing gasoline from each storage tank into several batches of equal volume, and applying the four reprocessing procedures to each batch. This requires a more complicated experimental design called the "split plot" design not discussed in this text.

Case Study 8: Does Statistical Process Control Really Work?

As the power on the coil is increased, the depth of hardness also increases. Hence, if the team noticed a decrease in the hardness depth, they increased the power to adjust the coil. This was done at point A. After this adjustment, the depth of hardness falls within control limits. However, after a bent coil is straightened at point B, the depth of hardness begins to increase and the team decreases the power. [Perhaps the bent coil was the problem at point A also.] Neither adjustment at points D or E succeeds in keeping the depth of hardness at the proper levels, and the first coil is replaced at point F.

The control chart for coil B indicates a stabilized process. From the control chart,

$$3\frac{\hat{\sigma}}{\sqrt{n}} = .93 \qquad \text{so that}$$

$$\hat{\sigma} = \frac{.93\sqrt{5}}{3} = .693, \text{ and } \mu \approx \overline{\overline{x}} = 4.43. \text{ Then the probability that a single part will fall}$$

within specification limits is

$$P[3.5 < x < 10.5] = P\left[\frac{3.5 - 4.43}{.693} < x < \frac{10.5 - 4.43}{.693}\right] = P[-1.34 < z < 8.76]$$

$$= .5 + .4099 = .9099$$

Alternately, calculate
$$C_p = \frac{10.5 - 3.5}{6(.693)} = 1.68$$

The process is capable of producing individual parts within the specification limits.

Coil C again produces a process that is in control. Calculate

$$\hat{\sigma} = \frac{.59\sqrt{5}}{3} = .494$$

and
$$C_p = \frac{7}{6(.494)} = 2.36$$

The process is even more capable than with coil B. Statistical process control has been successful.

Case Study 9: The Wood Stove Industry

The appropriate multivariable regression model is $y = \beta_0 + \beta_1 x_1 + \beta_2 x_2 + \beta_3 x_3 + \epsilon$ where

y = relative particulate matter concentration

x_1 = .25 if air intake setting is 1/4
  = .50 if air intake setting is 1/2
  = .75 if air intake setting is 3/4
  = 1.00 if air intake setting is open

x_2 = -1 if flue size is S
  = 1 if flue size is L

x_3 = flue temperature

The computer analysis is shown below.

```
RSQ = .31454621
SEY = 5.4541055
                COEFF      S.E.        F
INTERCEPT       39.27507   2.76596     501.62***
VAR 1          -48.29490   21.25175    5.16*
VAR 2            0.38435   1.13671     0.11 NS
VAR 3            0.07630   0.04109     3.45 NS
ANALYSIS OF VARIANCE
SOURCE      DF    SS          MS         F
REGRESSION  3     273.01300   91.00433   3.06 NS
RESIDUAL    20    594.94533   29.74727
TOTAL       23    867.95833
```

The regression equation is

$$\hat{y} = 39.27507 - 48.29490 x_1 + 0.38435 x_2 + 0.07630 x_3$$

but the regression is not significant, based on $F = 3.06$. Note the small value of $R^2 = .3146$. Only x_1 = air intake setting appears to be significant. The chosen variables are not very helpful in predicting relative particulate matter concentration.

Case Study 10: Household Energy Consumption and Expenditure

(1) Since the consumption quantities are given as well as the expenditures, these can be used to form a more accurate index of energy consumption costs. We would choose the Laspeyres index, since it is easier to use than either Paasche or Fisher's index.

(2)-(3) Indexes are calculated using 1980 quantities for both "Space Heating" and "Total" categories. The calculations are performed as in Exercise 13.5.

(4) Since the base-year quantities are used throughout, a strong change in the quantity consumed may tend to bias the index as the reference years change. The only category in which this appears to have happened is "Distillate fuel oil and kerosene." This category may be overweighted in the Laspeyres index.

Case Study 11: Clark County Port District

There are many correct solutions for this case study. However, there are a number of points the student should make.

—The population of interest is individuals who will vote, not those who are registered to vote. If there is a means of identifying the voters who will vote, it should be used. Perhaps randomized response sampling would be helpful.

—The sample frame is the list of registered voters available from the elections director. The sampling element is the individual registered voter. The sampling unit may also be the individual voter; however, some sampling designs, such as one incorporating cluster sampling, may call for a sampling unit that is not an element.

—If students select a design based on the variation among legislative districts (see the table), they should discuss the appropriateness of characteristics such as the number of registered voters, economic base, and political party affiliation as bases for stratification or clustering.

Case Study 12: The Gray Flannel Suit Syndrome

1. The given numbers of women falling in each dress category allow us to approximate the number of women falling in each cell of the contingency table. These estimates, along with the MINITAB contingency table analysis, are as shown.

```
Expected counts are printed below observed counts

            C1       C2       C3       C4     Total
    1       12       12        7        1       32
          9.88    10.20     6.34     5.58

    2       56       39       18       16      129
         39.83    41.12    25.54    22.51

    3       20       42       32       26      120
         37.05    38.26    23.76    20.94

    4        4        2        2        9       17
          5.25     5.42     3.37     2.97

Total       92       95       59       52      298

ChiSq =   0.455 + 0.317 + 0.070 + 3.763 +
          6.569 + 0.110 + 2.226 + 1.883 +
          7.844 + 0.367 + 2.859 + 1.223 +
          0.297 + 2.158 + 0.554 + 12.272 = 42.966
df = 9
2 cells with expected counts less than 5.0
```

Notice that two cells have estimated expected cell counts less than 5, and hence, the chi-square approximation may not be too accurate. To avoid this problem, the table can be collapsed by combining rows 3 and 4 into a category called "average, below average, and failing." The chi-square analysis for such a contingency table is shown below.

Expected counts are printed below observed counts

	C1	C2	C3	C4	Total
1	12	12	7	1	32
	9.88	10.20	6.34	5.58	
2	56	39	18	16	129
	39.83	41.12	25.54	22.51	
3	24	44	34	35	137
	42.30	43.67	27.12	23.91	
Total	92	95	59	52	298

ChiSq = 0.455 + 0.317 + 0.070 + 3.763 +
        6.569 + 0.110 + 2.226 + 1.883 +
        7.914 + 0.002 + 1.743 + 5.148 = 30.200

df = 6

In any event, the highly significant results ($X^2 = 42.97$, p-value < .005 or $X^2 = 30.20$, p-value < .005) imply that job performance evaluations depend on women's choice of attire.

2. The statement implies that, if the women in question were otherwise equally qualified for promotion, the probability of promotion for conservatively dressed women, p_1, will be twice the probability of promotion, p_2, for women who dressed frivolously. That is,

$$p_1 = 2p_2$$

Case Study 13: Market Proxies: How Well Do They Perform?

A MINITAB program was used to evaluate r_s for each pair of six rankings (Indexes 1-5 and Mean Return). The ranks were entered into C1-C6 of a MINITAB worksheet, and the command CORRELATION C1-C6 was used to generate the matrix of rank correlations. The results are shown.

```
MTB > CORR C1-C6
          INDEX1    INDEX2    INDEX3    INDEX4    INDEX5
INDEX2    0.951
INDEX3    0.775     0.630
INDEX4    0.867     0.829     0.871
INDEX5    0.898     0.899     0.768     0.922
MEANRTN   0.637     0.782     0.158     0.466     0.626
```

With n = 32, a positive association between any two indexes is indicated if $r_s \geq .305$. Hence, all the estimated correlations are significantly positive except for Index 3 and the mean returns. That is, the five indexes generate rankings that are substantially similar to one another. However, notice that a sharp change in the level of correlation occurs when we change from Index 2 to Index 3. This phenomenon is also indicated by the fact that Index 3 is not correlated with mean returns. Average market performance drops off substantially when measured against Index 3, which corresponds to the introduction of real estate into the market portfolio.

TRANSPARENCY MASTERS

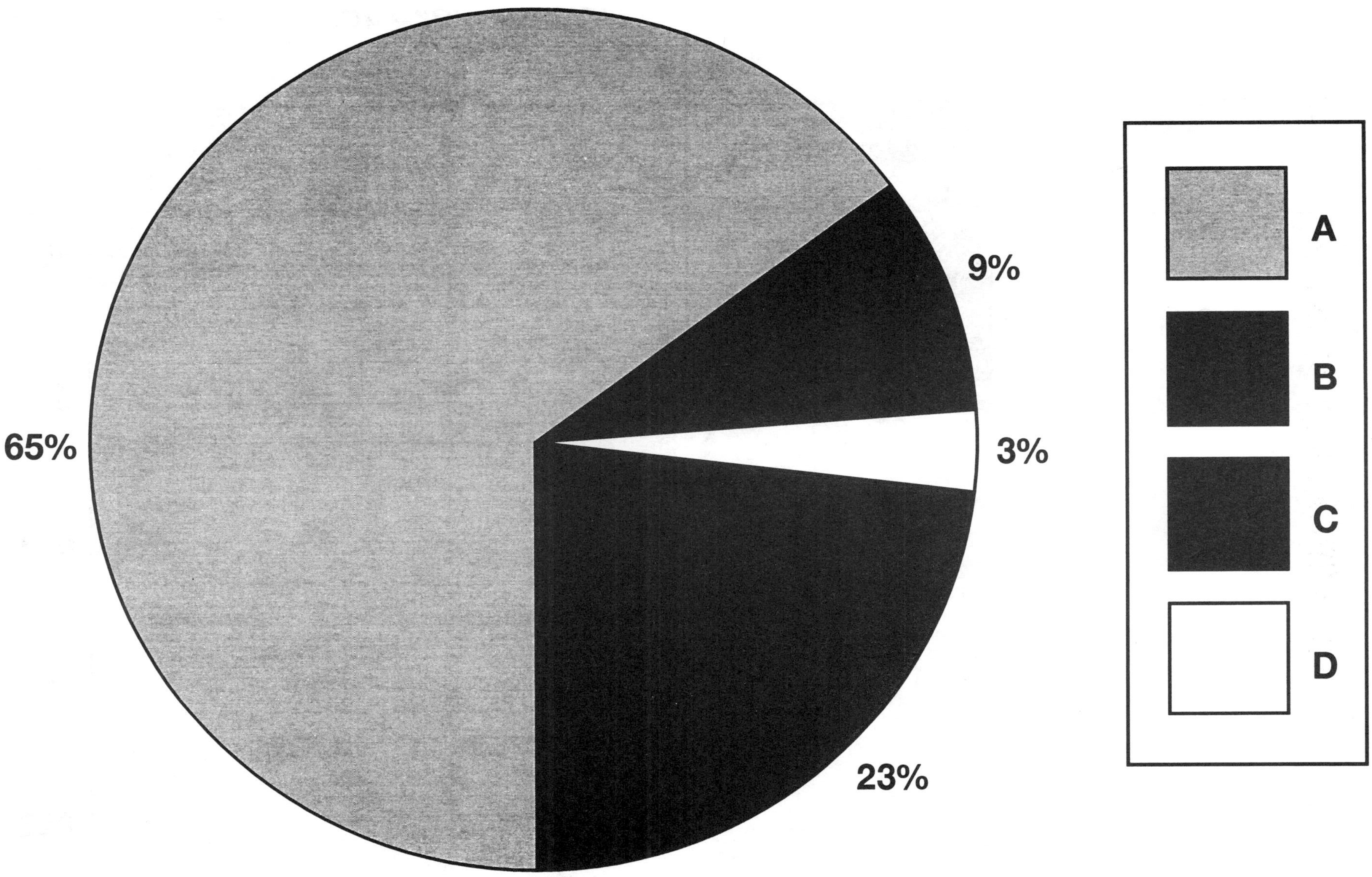

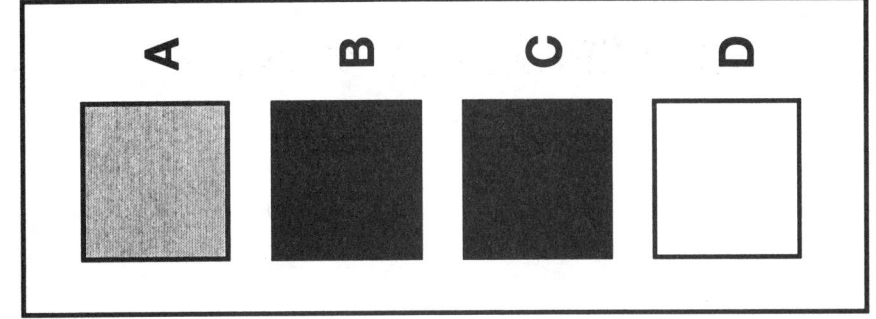

Figure 2.2 Pie chart for Example 2.3.

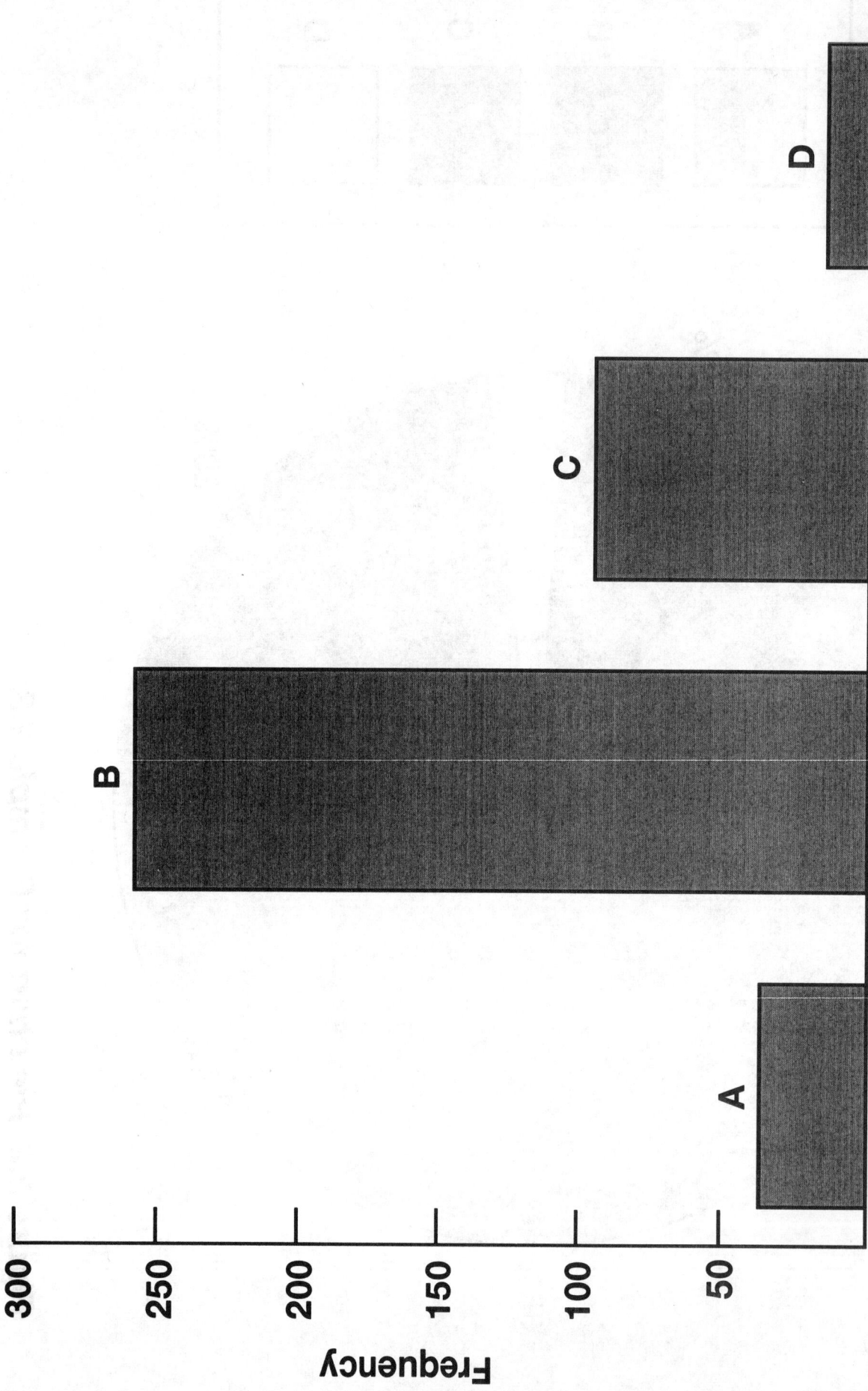

Figure 2.3 Bar graph for Example 2.3.

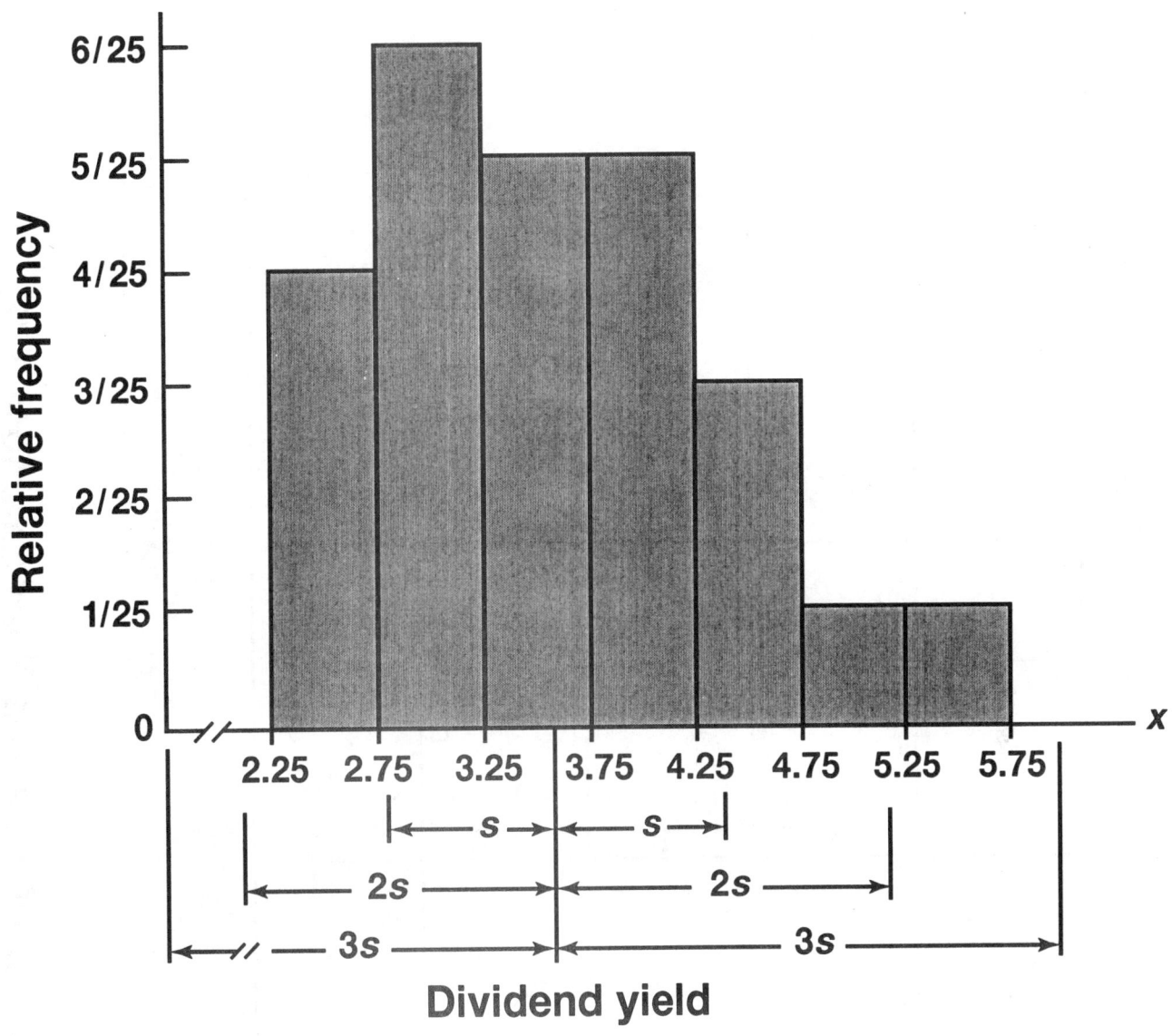

Figure 2.19 Histogram for the dividend yield data with intervals superimposed.

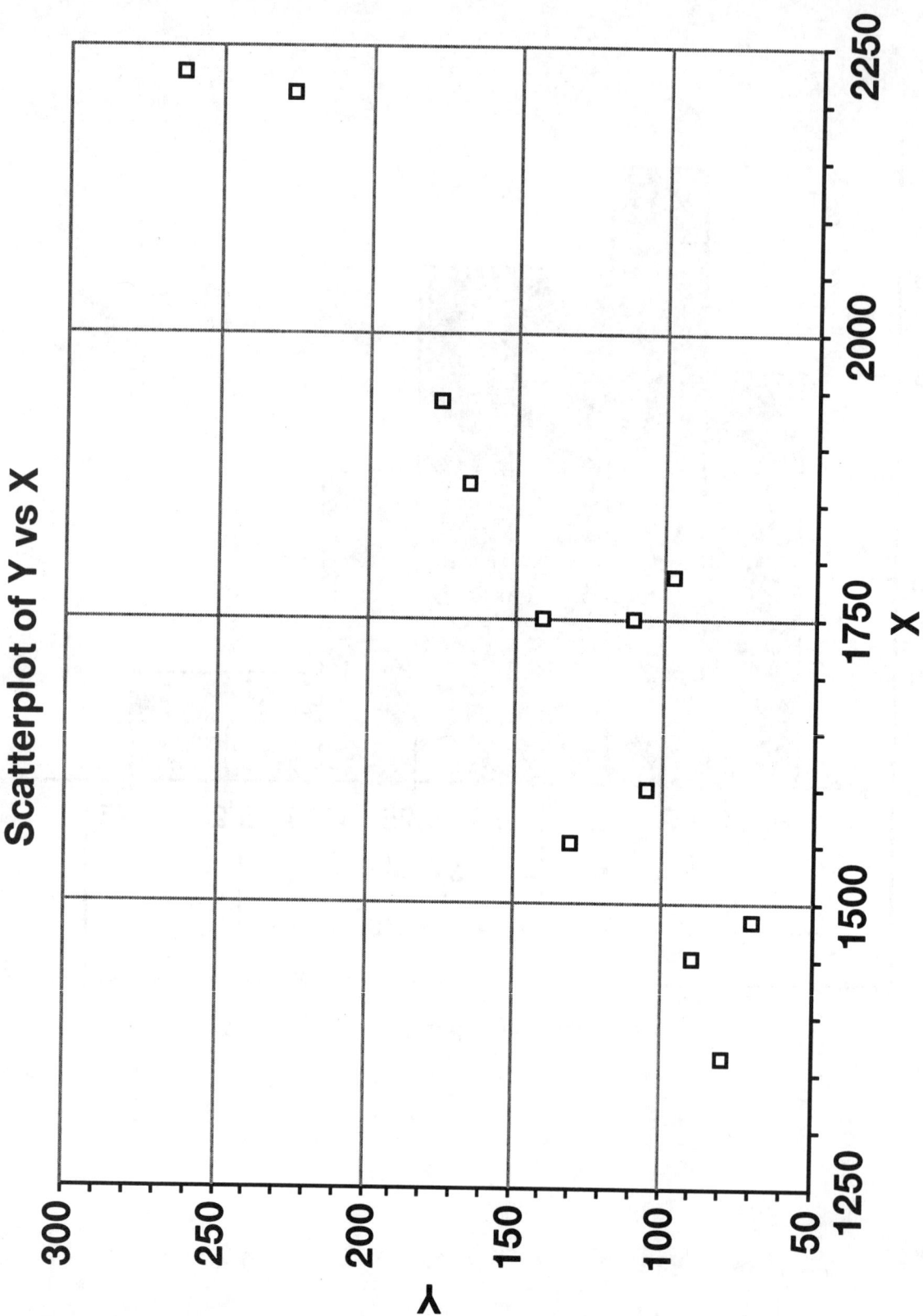

Figure 2.22 Scatterplot of X versus Y for Example 2.17.

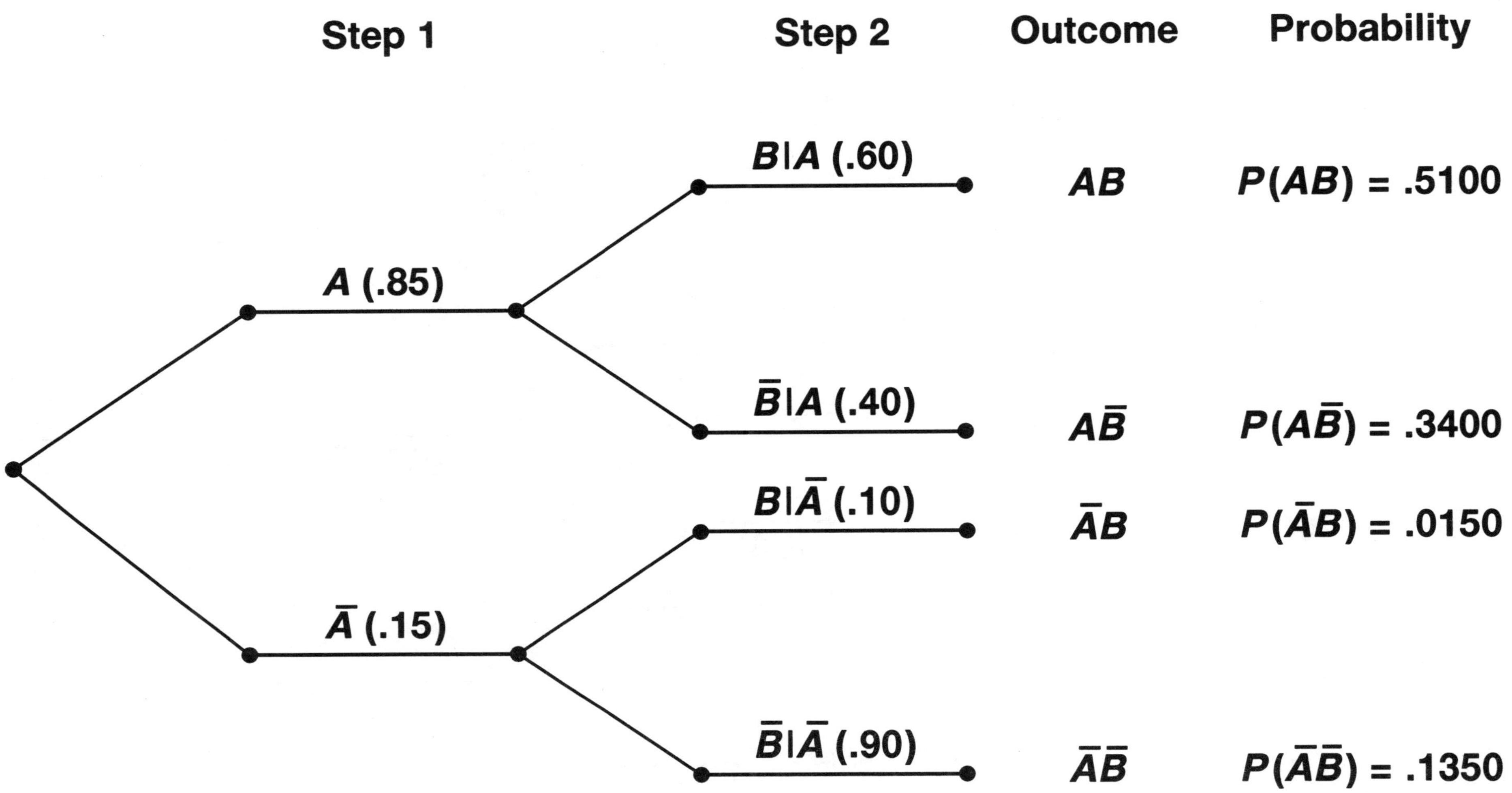

Figure 3.2 Tree diagram for Example 3.10.

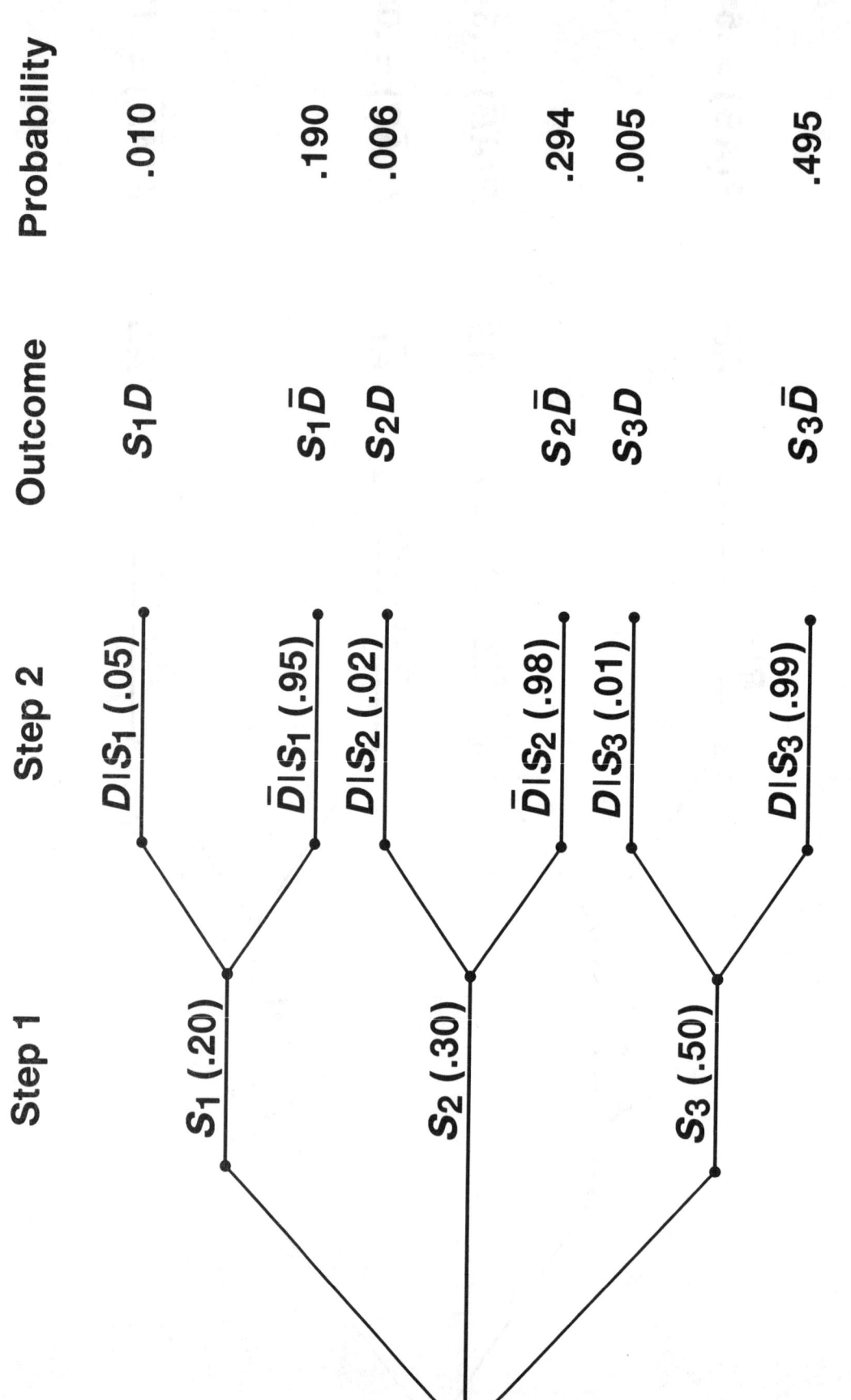

Figure 3.3 Tree diagram for text example.

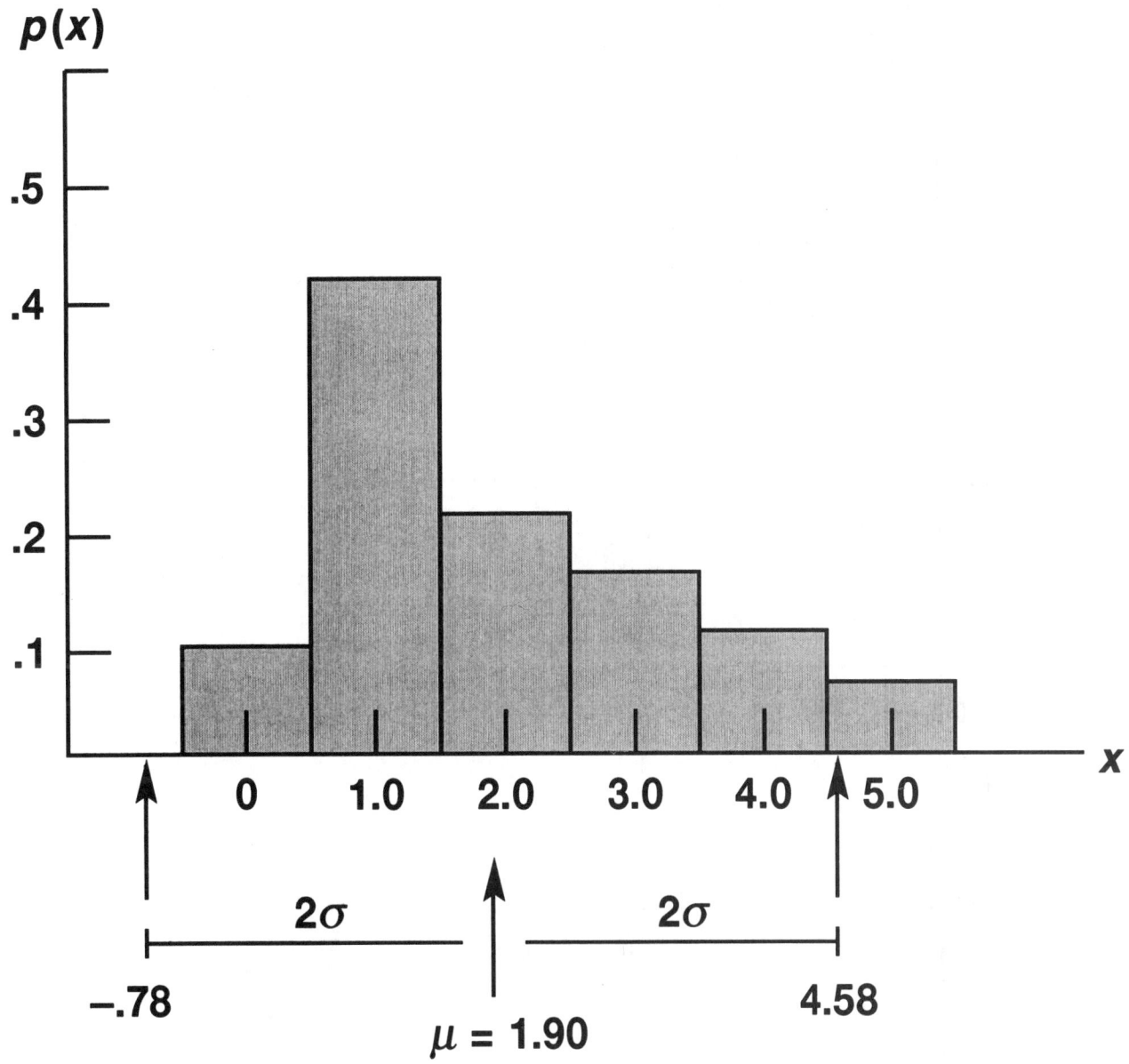

Figure 3.5 The probability histogram for p(x) for Example 3.15.

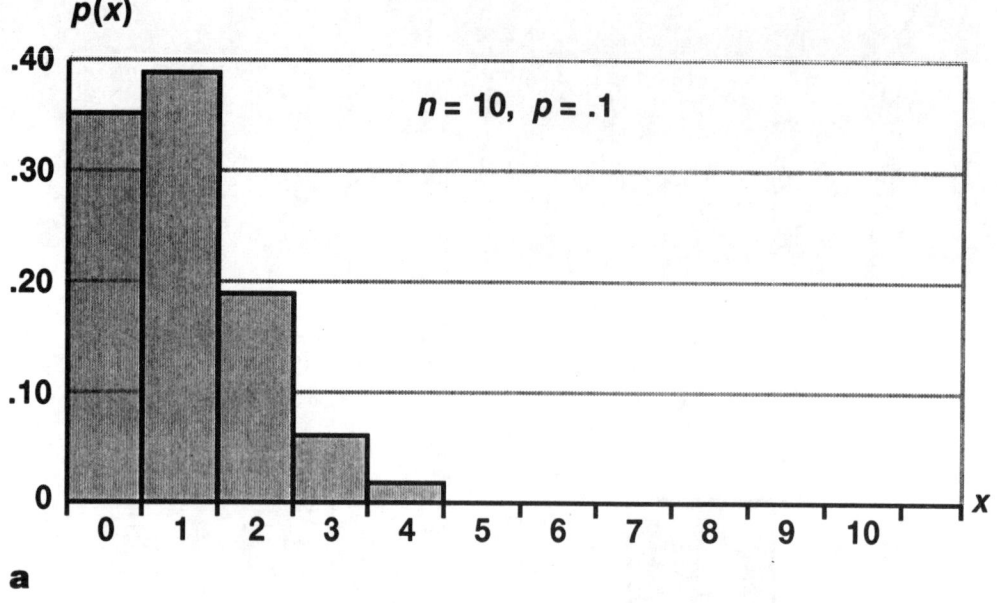
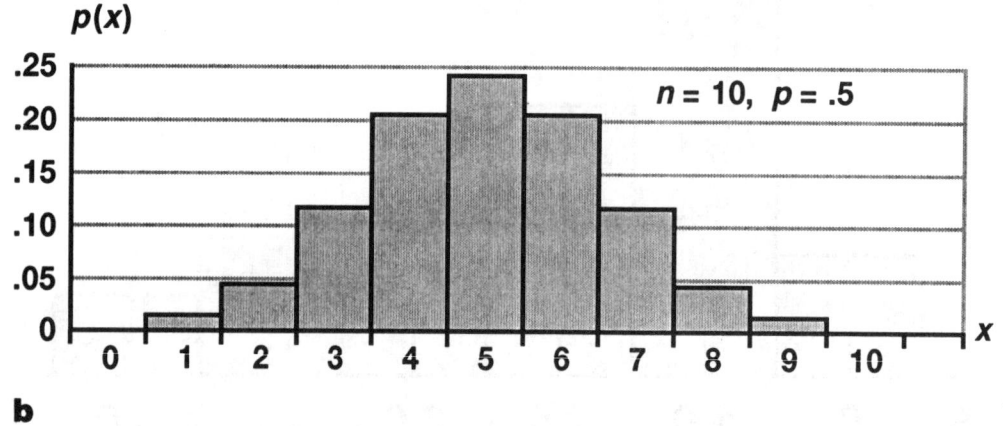
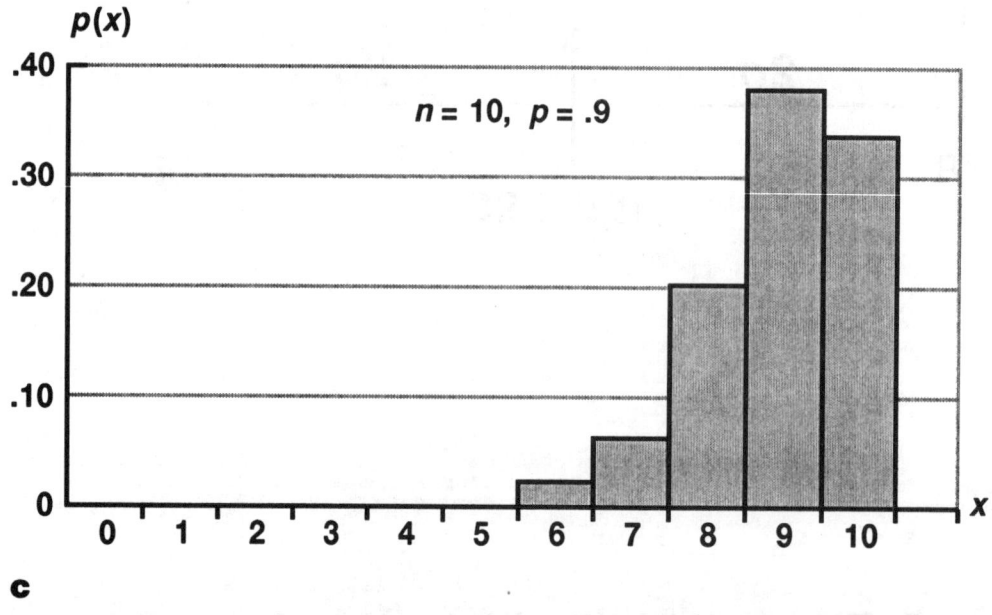

Figure 4.1 Binomial probability distributions.

$n = 5$

a	0.01	0.05	0.10	0.20	0.30	0.40	0.50	0.60	0.70	0.80	0.90	0.95	0.99	a
0	.951	.774	.590	.328	.168	.078	.031	.010	.002	.000	.000	.000	.000	0
1	.999	.977	.919	.737	.528	.337	.188	.087	.031	.007	.000	.000	.000	1
2	1.000	.999	.991	.942	.837	.683	.500	.317	.163	.058	.009	.001	.000	2
3	1.000	1.000	1.000	.993	.969	.913	.812	.663	.472	.263	.081	.023	.001	3
4	1.000	1.000	1.000	1.000	.998	.990	.969	.922	.832	.672	.410	.226	.049	4
5	1.000	1.000	1.000	1.000	1.000	1.000	1.000	1.000	1.000	1.000	1.000	1.000	1.000	5

Appendix II, Table 1 Cumulative binomial probabilities.

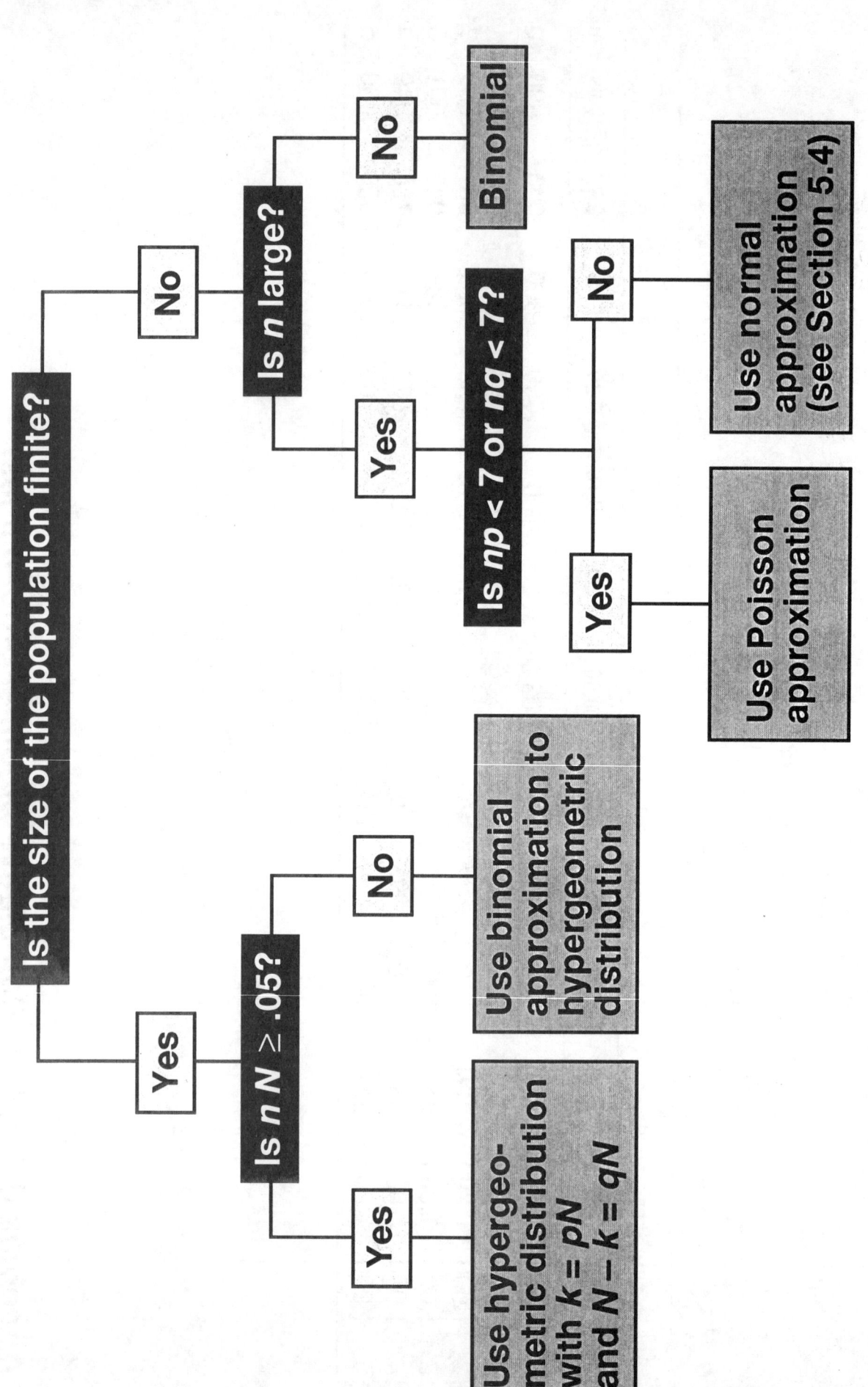

Figure 4.3 Decision tree.

Tabulated values are $P(x \leq a) = \sum_{x=0}^{a} p(x)$. (Computations are rounded at the third decimal place.)

	Mean									
a	0.250	0.500	0.750	1.000	1.250	1.500	1.750	2.000	2.250	2.500
0	0.779	0.607	0.472	0.368	0.287	0.223	0.174	0.135	0.105	0.082
1	0.974	0.910	0.827	0.736	0.645	0.558	0.478	0.406	0.343	0.287
2	0.998	0.986	0.959	0.920	0.868	0.809	0.744	0.677	0.609	0.544
3	1.000	0.998	0.993	0.981	0.962	0.934	0.899	0.857	0.809	0.758
4	1.000	1.000	0.999	0.996	0.991	0.981	0.967	0.947	0.922	0.891
5	1.000	1.000	1.000	0.999	0.998	0.996	0.991	0.983	0.973	0.958
6	1.000	1.000	1.000	1.000	1.000	0.999	0.998	0.995	0.992	0.986
7	1.000	1.000	1.000	1.000	1.000	1.000	1.000	0.999	0.998	0.996
8	1.000	1.000	1.000	1.000	1.000	1.000	1.000	1.000	0.999	0.999
9	1.000	1.000	1.000	1.000	1.000	1.000	1.000	1.000	1.000	1.000
10	1.000	1.000	1.000	1.000	1.000	1.000	1.000	1.000	1.000	1.000
11	1.000	1.000	1.000	1.000	1.000	1.000	1.000	1.000	1.000	1.000
12	1.000	1.000	1.000	1.000	1.000	1.000	1.000	1.000	1.000	1.000
13	1.000	1.000	1.000	1.000	1.000	1.000	1.000	1.000	1.000	1.000
14	1.000	1.000	1.000	1.000	1.000	1.000	1.000	1.000	1.000	1.000

	Mean									
a	2.750	3.000	3.250	3.500	3.750	4.000	4.250	4.500	4.750	5.000
0	0.064	0.050	0.039	0.030	0.024	0.018	0.014	0.011	0.009	0.007
1	0.240	0.199	0.165	0.136	0.112	0.092	0.075	0.061	0.050	0.040
2	0.481	0.423	0.370	0.321	0.277	0.238	0.204	0.174	0.147	0.125
3	0.703	0.647	0.591	0.537	0.484	0.433	0.386	0.342	0.302	0.265
4	0.855	0.815	0.772	0.725	0.678	0.629	0.580	0.532	0.485	0.440
5	0.939	0.916	0.889	0.858	0.823	0.785	0.745	0.703	0.660	0.616
6	0.978	0.966	0.952	0.935	0.914	0.889	0.862	0.831	0.798	0.762
7	0.993	0.988	0.982	0.973	0.962	0.949	0.933	0.913	0.891	0.867
8	0.998	0.996	0.994	0.990	0.985	0.979	0.970	0.960	0.947	0.932
9	0.999	0.999	0.998	0.997	0.995	0.992	0.988	0.983	0.976	0.968
10	1.000	1.000	0.999	0.999	0.998	0.997	0.996	0.993	0.990	0.986
11	1.000	1.000	1.000	1.000	0.999	0.999	0.998	0.998	0.996	0.995
12	1.000	1.000	1.000	1.000	1.000	1.000	1.000	0.999	0.999	0.998
13	1.000	1.000	1.000	1.000	1.000	1.000	1.000	1.000	1.000	0.999
14	1.000	1.000	1.000	1.000	1.000	1.000	1.000	1.000	1.000	1.000

Appendix II, Table 2 Cumulative probabilities of the Poisson distribution.

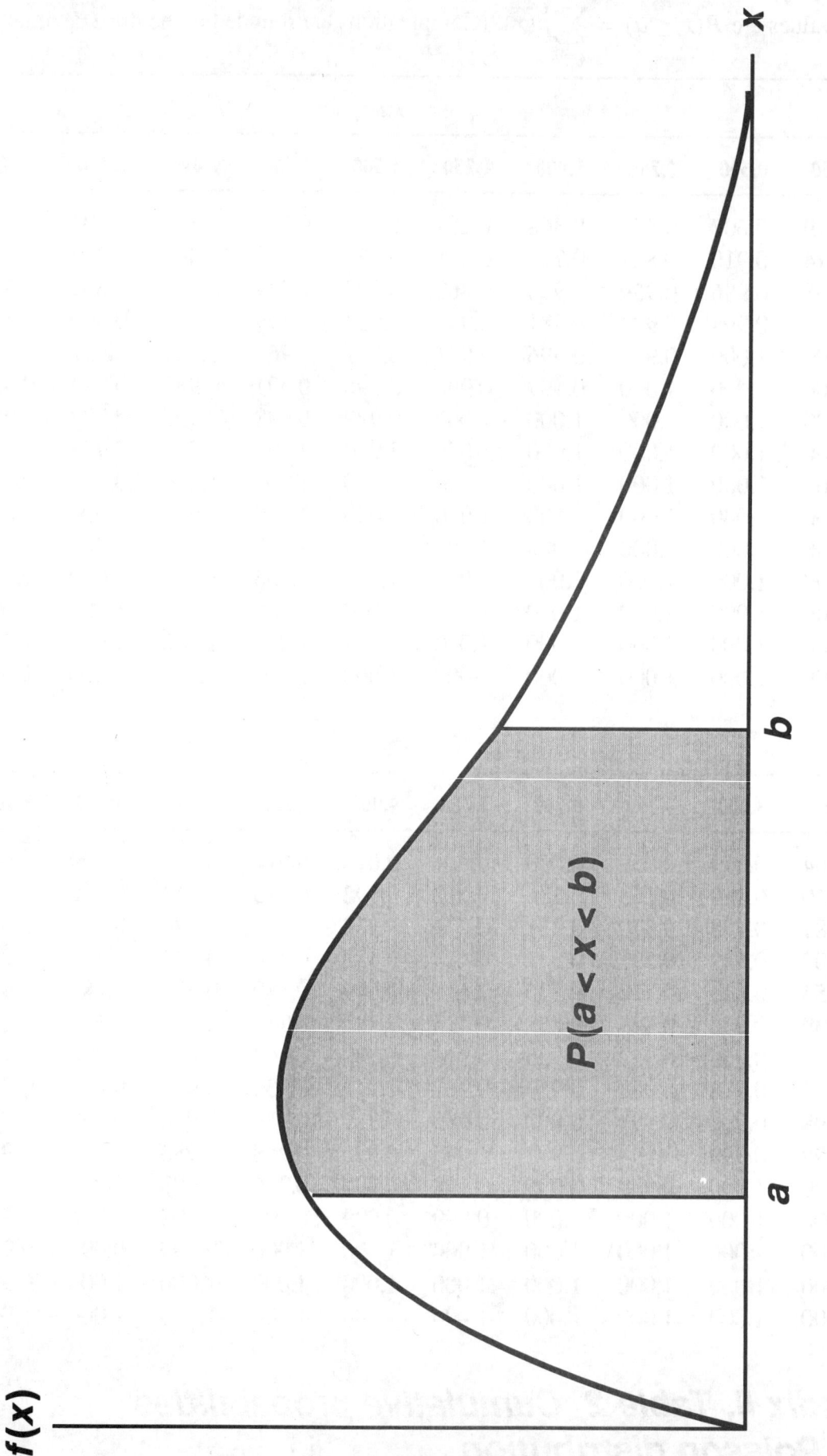

Figure 5.2 The probability distribution for a continuous random variable.

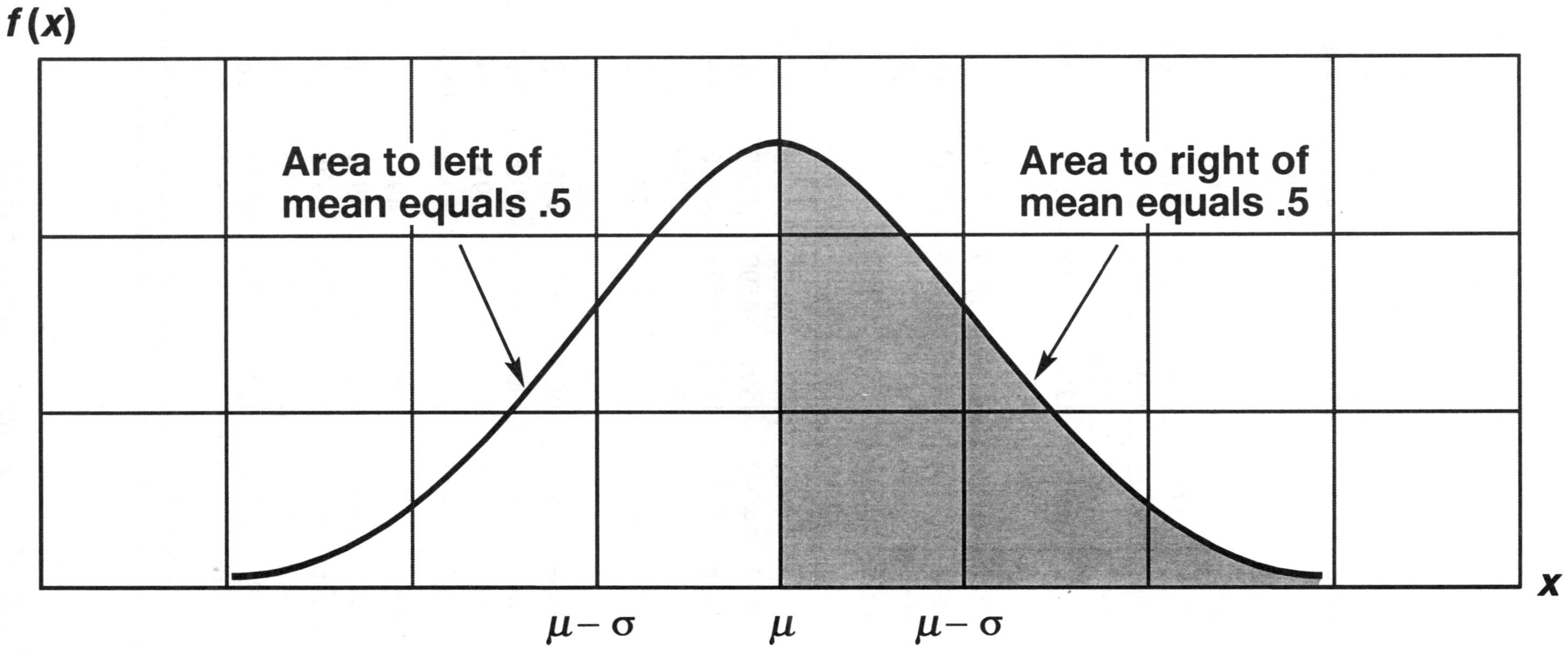

Figure 5.3 *Normal probability distribution.*

z	.00	.01	.02	.03	.04	.05	.06	.07	.08	.09
0.0	.0000	.0040	.0080	.0120	.0160	.0199	.0239	.0279	.0319	.0359
0.1	.0398	.0438	.0478	.0517	.0557	.0596	.0636	.0675	.0714	.0753
0.2	.0793	.0832	.0871	.0910	.0948	.0987	.1026	.1064	.1103	.1141
0.3	.1179	.1217	.1255	.1293	.1331	.1368	.1406	.1443	.1480	.1517
0.4	.1554	.1591	.1628	.1664	.1700	.1736	.1772	.1808	.1844	.1879
0.5	.1915	.1950	.1985	.2019	.2054	.2088	.2123	.2157	.2190	.2224
0.6	.2257	.2291	.2324	.2357	.2389	.2422	.2454	.2486	.2517	.2549
0.7	.2580	.2611	.2642	.2673	.2704	.2734	.2764	.2794	.2823	.2852
0.8	.2881	.2910	.2939	.2967	.2995	.3023	.3051	.3078	.3106	.3133
0.9	.3159	.3186	.3212	.3238	.3264	.3289	.3315	.3340	.3365	.3389
1.0	.3413	.3438	.3461	.3485	.3508	.3531	.3554	.3577	.3599	.3621
1.1	.3643	.3665	.3686	.3708	.3729	.3749	.3770	.3790	.3810	.3830
1.2	.3849	.3869	.3888	.3907	.3925	.3944	.3962	.3980	.3997	.4015
1.3	.4032	.4049	.4066	.4082	.4099	.4115	.4131	.4147	.4162	.4177
1.4	.4192	.4207	.4222	.4236	.4251	.4265	.4279	.4292	.4306	.4319
1.5	.4332	.4345	.4357	.4370	.4382	.4394	.4406	.4418	.4429	.4441
1.6	.4452	.4463	.4474	.4484	.4495	.4505	.4515	.4525	.4535	.4545
1.7	.4554	.4564	.4573	.4582	.4591	.4599	.4608	.4616	.4625	.4633
1.8	.4641	.4649	.4656	.4664	.4671	.4678	.4686	.4693	.4699	.4706
1.9	.4713	.4719	.4726	.4732	.4738	.4744	.4750	.4756	.4761	.4767
2.0	.4772	.4778	.4783	.4788	.4793	.4798	.4803	.4808	.4812	.4817
2.1	.4821	.4826	.4830	.4834	.4838	.4842	.4846	.4850	.4854	.4857
2.2	.4861	.4864	.4868	.4871	.4875	.4878	.4881	.4884	.4887	.4890
2.3	.4893	.4896	.4898	.4901	.4904	.4906	.4909	.4911	.4913	.4916
2.4	.4918	.4920	.4922	.4925	.4927	.4929	.4931	.4932	.4934	.4936
2.5	.4938	.4940	.4941	.4943	.4945	.4946	.4948	.4949	.4951	.4952
2.6	.4953	.4955	.4956	.4957	.4959	.4960	.4961	.4962	.4963	.4964
2.7	.4965	.4966	.4967	.4968	.4969	.4970	.4971	.4972	.4973	.4974
2.8	.4974	.4975	.4976	.4977	.4977	.4978	.4979	.4979	.4980	.4981
2.9	.4981	.4982	.4982	.4983	.4984	.4984	.4985	.4985	.4986	.4986
3.0	.4987	.4987	.4987	.4988	.4988	.4989	.4989	.4989	.4990	.4990

Source: This table is abridged from Table 1 of *Statistical Tables and Formulas,* by A. Hald (New York: Wiley, 1952). Reproduced by permission of A. Hald and the publisher, John Wiley & Sons, Inc.

Appendix II, Table 3 Normal curve areas.

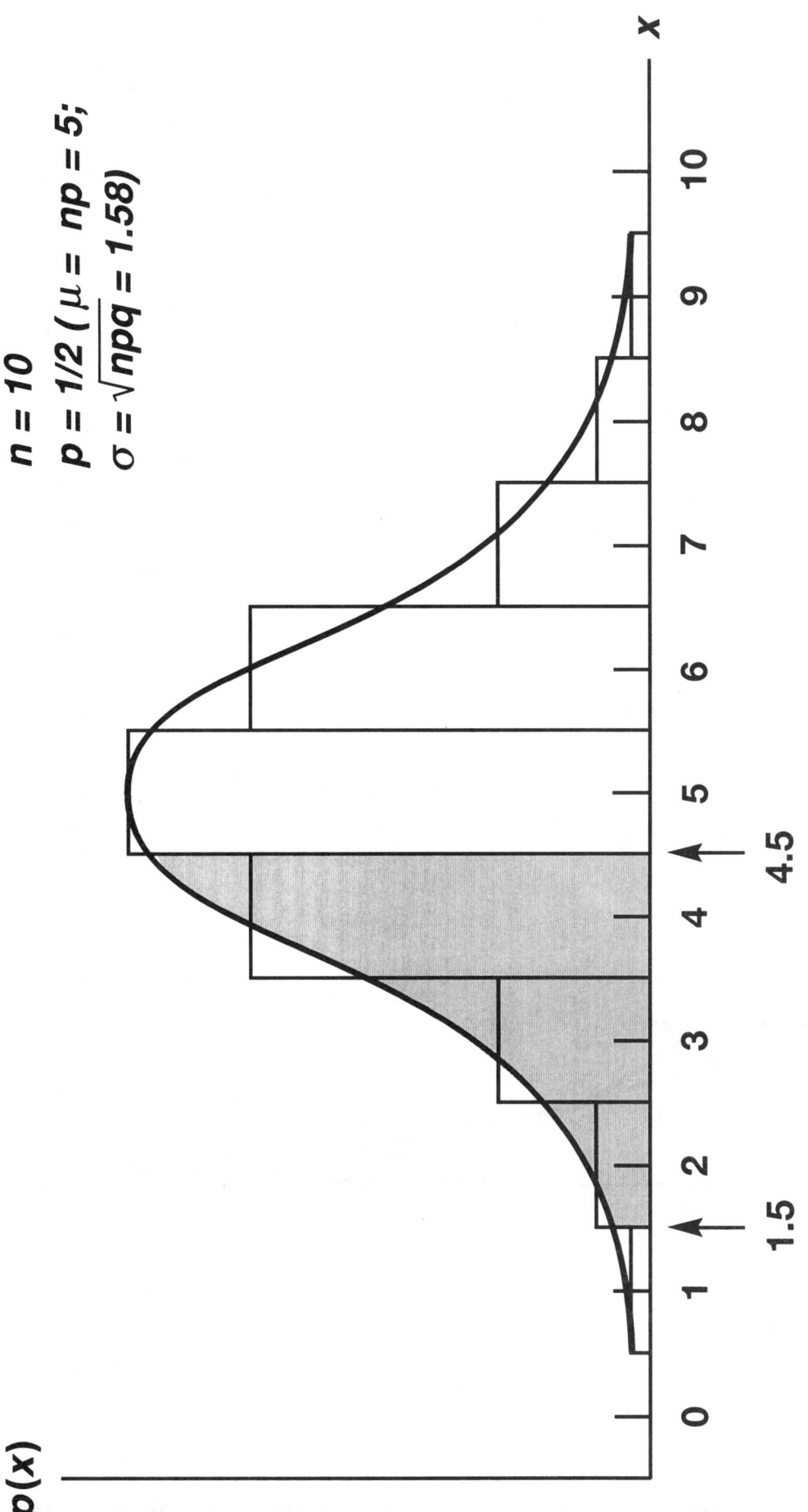

Figure 5.12 Comparison of a binomial probability distribution and the approximating normal distribution.

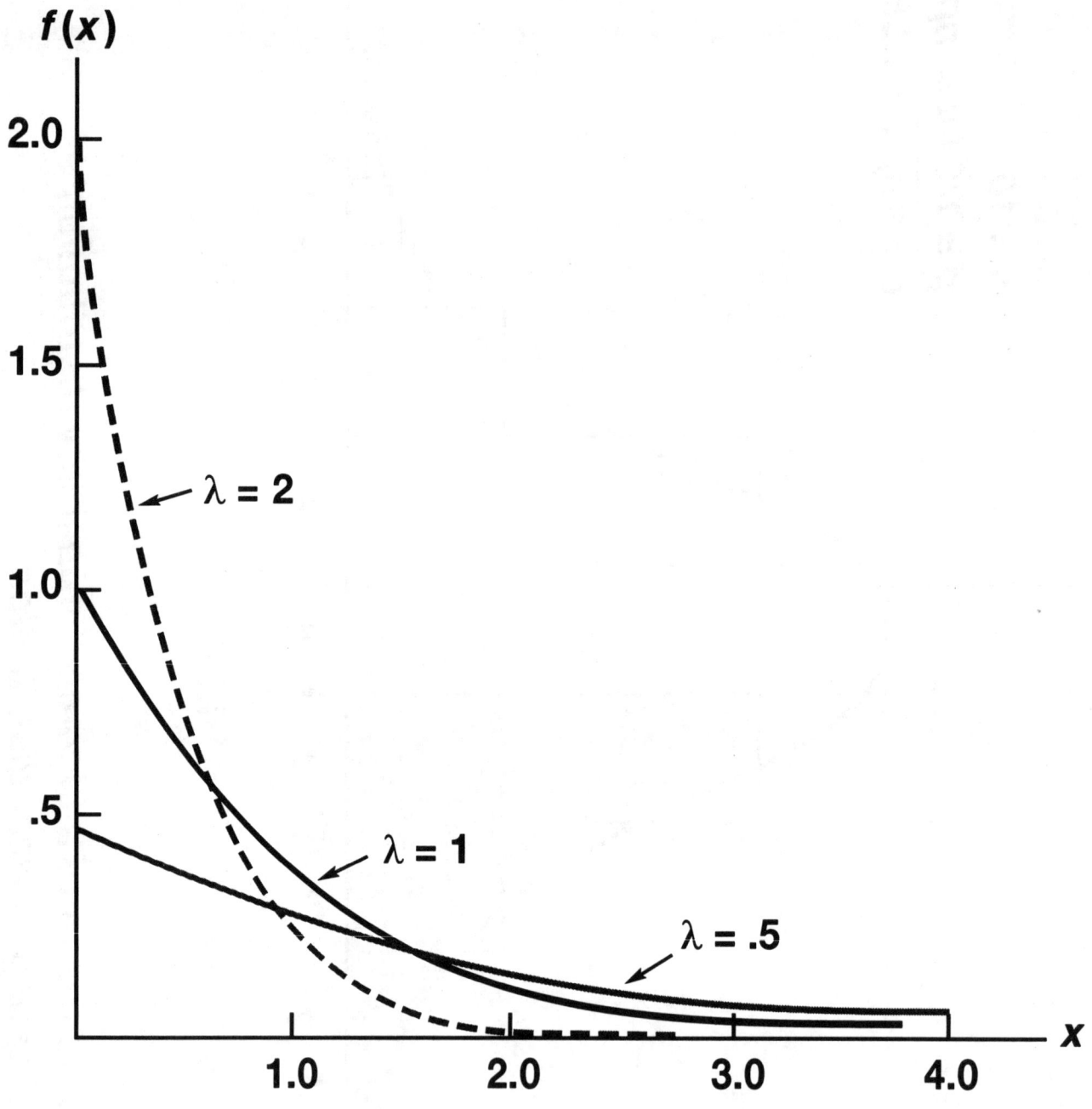

Figure 5.17 The exponential probability density function.

a

Sample	Observations in Sample	$\bar{x}$
1	4,4	4
2	4,2	3
3	4,5	4.5
4	4,1	2.5
5	2,4	3
6	2,2	2
7	2,5	3.5
8	2,1	1.5

Sample	Observations in Sample	$\bar{x}$
9	5,4	4.5
10	5,2	3.5
11	5,5	5
12	5,1	3
13	1,4	2.5
14	1,2	1.5
15	1,5	3
16	1,1	1

b

$\bar{x}$	$p(\bar{x})$
1	1/16
1.5	2/16
2	1/16
2.5	2/16
3	4/16
3.5	2/16
4	1/16
4.5	2/16
5	1/16

(a) Table 6.1 *Calculation of $\bar{x}$ for 16 possible samples of size n = 2.*
(b) Table 6.2 *Sampling distribution for $\bar{x}$.*

Transparency 18 © 1996 Wadsworth Publishing Company/ITP

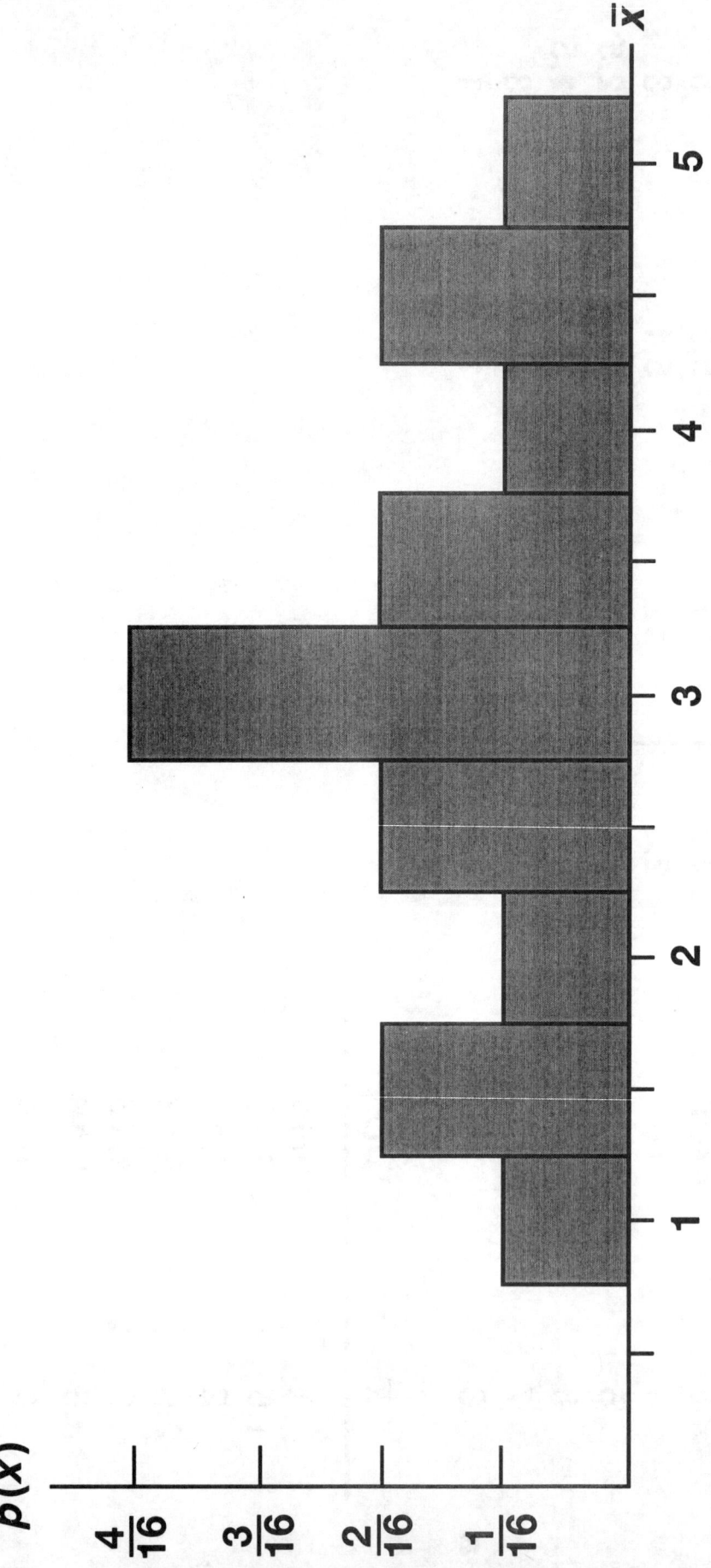

Figure 6.1 Sampling distribution for $\bar{x}$.

Sample	Sample Measurements				Sample Mean $\bar{x}$
1	.992	1.007	1.016	.991	1.00150
2	1.015	.984	.976	1.000	.99375
3	.988	.993	1.011	.981	.99325
4	.996	1.020	1.004	.999	1.00475
5	1.015	1.006	1.002	1.001	1.00600
6	1.000	.982	1.005	.989	.99400
7	.989	1.009	1.019	.994	1.00275
8	.994	1.010	1.009	.990	1.00075
9	1.018	1.016	.990	1.011	1.00875
10	.997	1.005	.989	1.001	.99800
11	1.020	.986	1.002	.989	.99925
12	1.007	.986	.981	.995	.99225
13	1.016	1.002	1.010	.999	1.00675
14	.982	.995	1.011	.987	.99375
15	1.001	1.000	.983	1.002	.99650
16	.992	1.008	1.001	.996	.99925
17	1.020	.988	1.015	.986	1.00225
18	.993	.987	1.006	1.001	.99675
19	.978	1.006	1.002	.982	.99200
20	.984	1.009	.983	.986	.99050
21	.990	1.012	1.010	1.007	1.00475
22	1.015	.983	1.003	.989	.99750
23	.983	.990	.977	1.002	.99300
24	1.011	1.012	.912	1.008	1.00550
25	.987	.987	1.007	.995	.99400

Table 6.4 25 hourly samples of bearing diameters.

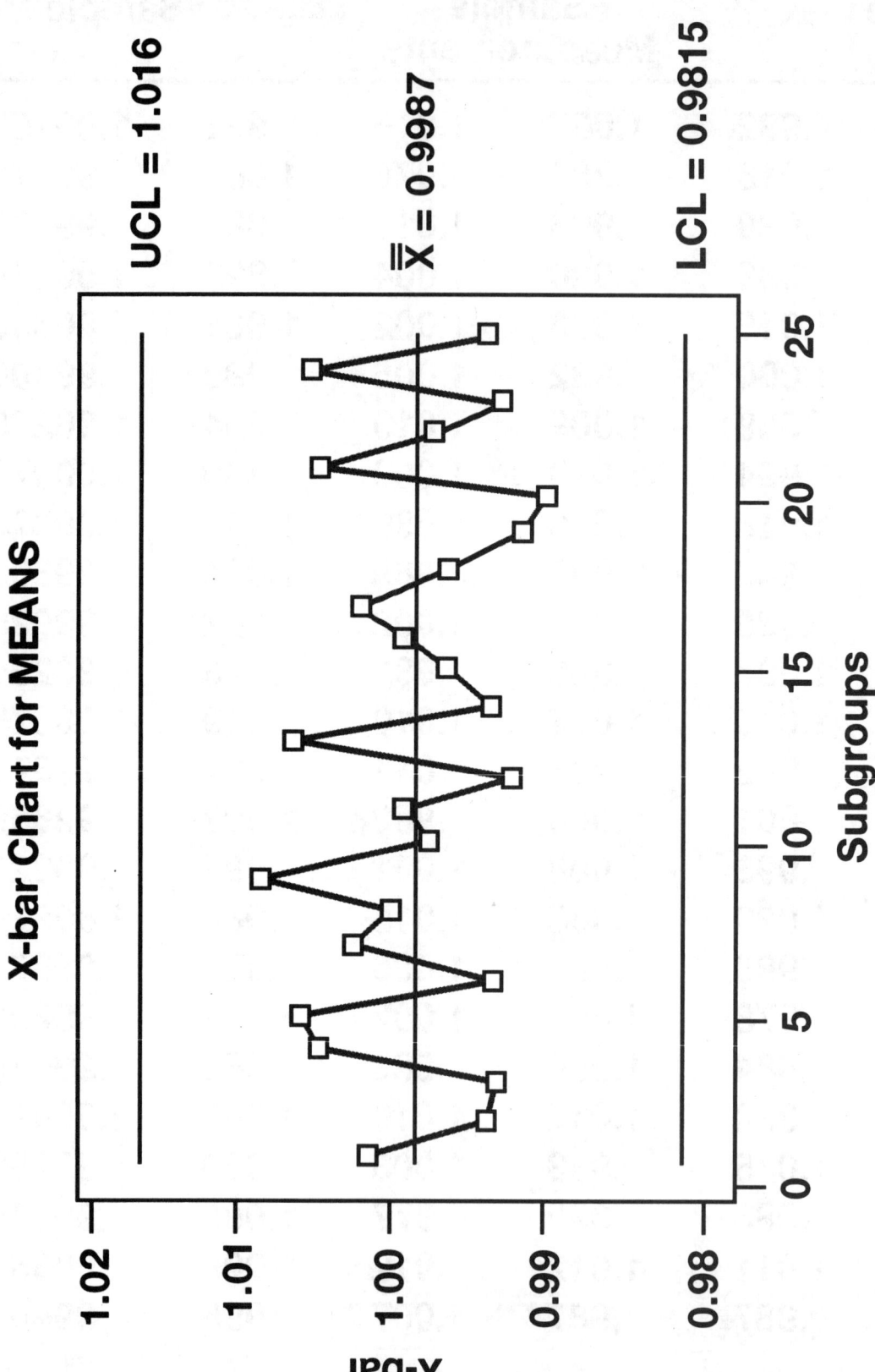

Figure 6.10 Minitab $\bar{x}$ chart for Example 6.5.

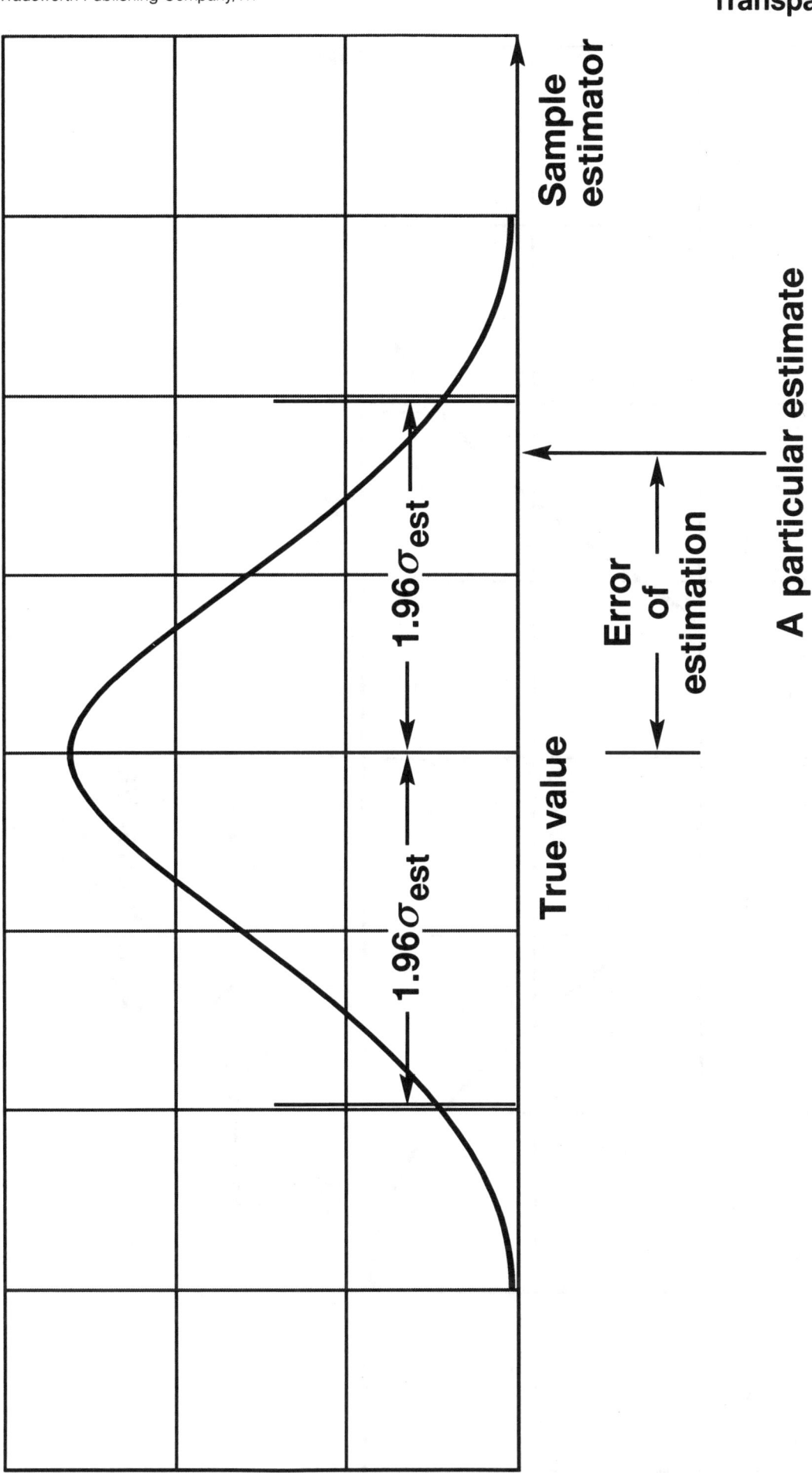

Figure 7.3 *Sampling distribution of an unbiased estimator.*

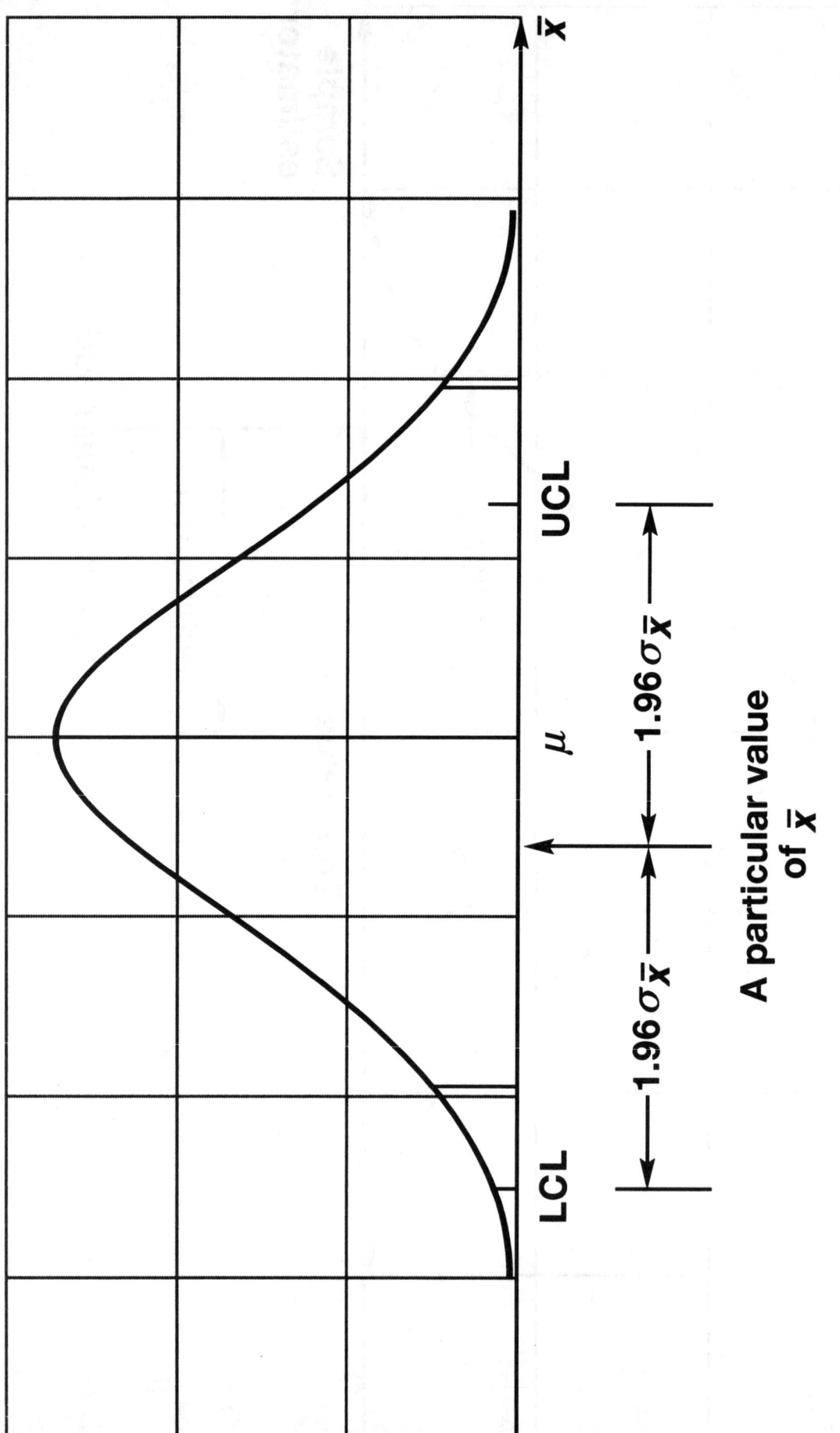

Figure 7.4 95% confidence limits for a population mean.

d.f.	$t_{.100}$	$t_{.050}$	$t_{.025}$	$t_{.010}$	$t_{.005}$	d.f.
1	3.078	6.314	12.706	31.821	63.657	1
2	1.886	2.920	4.303	6.965	9.925	2
3	1.638	2.353	3.182	4.541	5.841	3
4	1.533	2.132	2.776	3.747	4.604	4
5	1.476	2.015	2.571	3.365	4.032	5
6	1.440	1.943	2.447	3.143	3.707	6
7	1.415	1.895	2.365	2.998	3.499	7
8	1.397	1.860	2.306	2.896	3.355	8
9	1.383	1.833	2.262	2.821	3.250	9
10	1.372	1.812	2.228	2.764	3.169	10
11	1.363	1.796	2.201	2.718	3.106	11
12	1.356	1.782	2.179	2.681	3.055	12
13	1.350	1.771	2.160	2.650	3.012	13
14	1.345	1.761	2.145	2.624	2.977	14
15	1.341	1.753	2.131	2.602	2.947	15
16	1.337	1.746	2.120	2.583	2.921	16
17	1.333	1.740	2.110	2.567	2.898	17
18	1.330	1.734	2.101	2.552	2.878	18
19	1.328	1.729	2.093	2.539	2.861	19
20	1.325	1.725	2.086	2.528	2.845	20
21	1.323	1.721	2.080	2.518	2.831	21
22	1.321	1.717	2.074	2.508	2.819	22
23	1.319	1.714	2.069	2.500	2.807	23
24	1.318	1.711	2.064	2.492	2.797	24
25	1.316	1.708	2.060	2.485	2.787	25
26	1.315	1.706	2.056	2.479	2.779	26
27	1.314	1.703	2.052	2.473	2.771	27
28	1.313	1.701	2.048	2.467	2.763	28
29	1.311	1.699	2.045	2.462	2.756	29
inf.	1.282	1.645	1.960	2.326	2.576	inf.

Source: From "Table of Percentage Points of the *t*-Distribution," *Biometrika* 32 (1941) 300. Reproduced by permission of the *Biometrika* Trustees.

Appendix II, Table 4 Critical values of t.

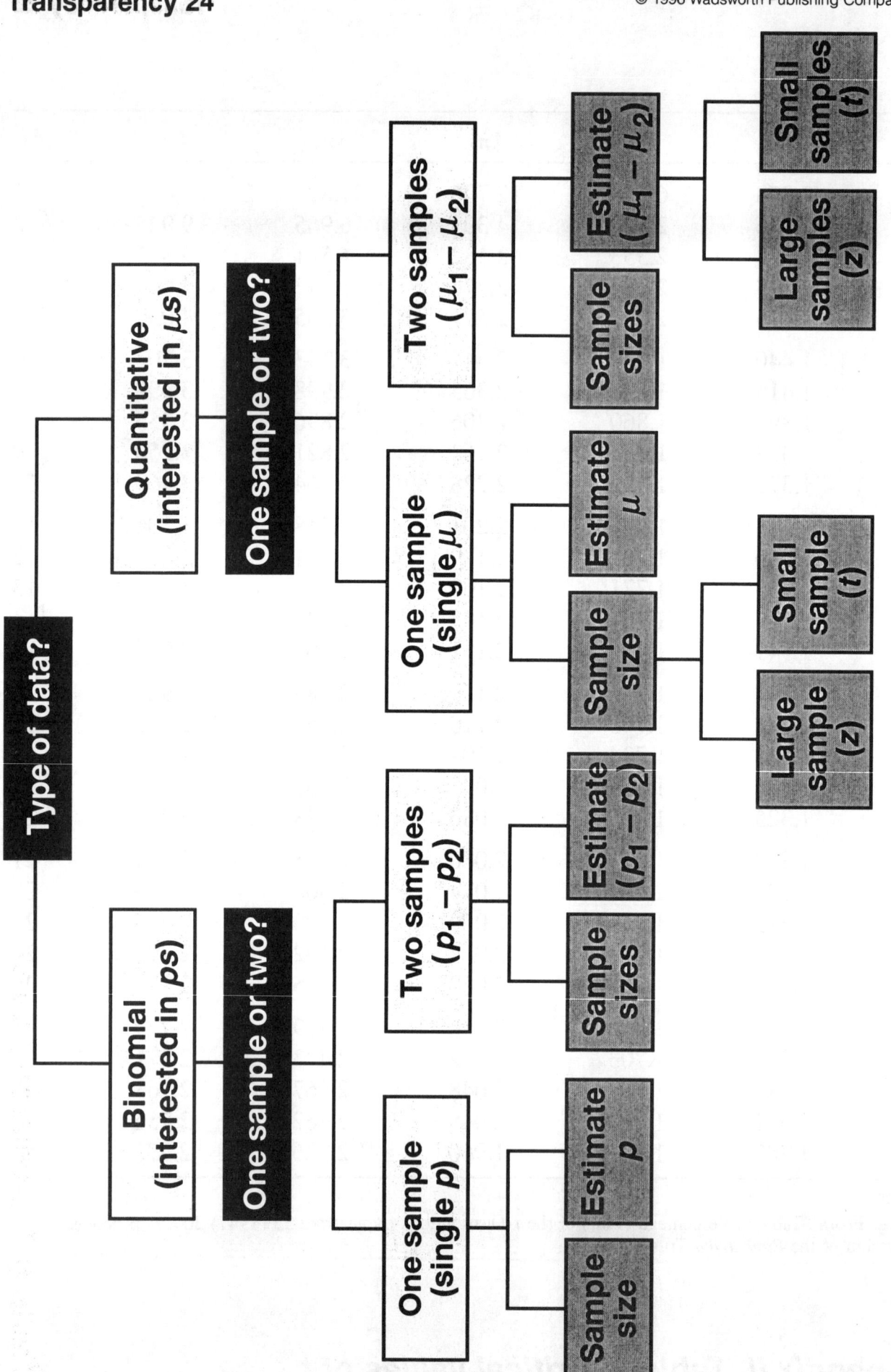

Figure 7.10 Decision tree.

	Null Hypothesis	
Decision	**True**	**False**
Reject H_0	Type I error	Correct decision
Accept H_0	Correct decision	Type II error

Table 8.2 Decision table.

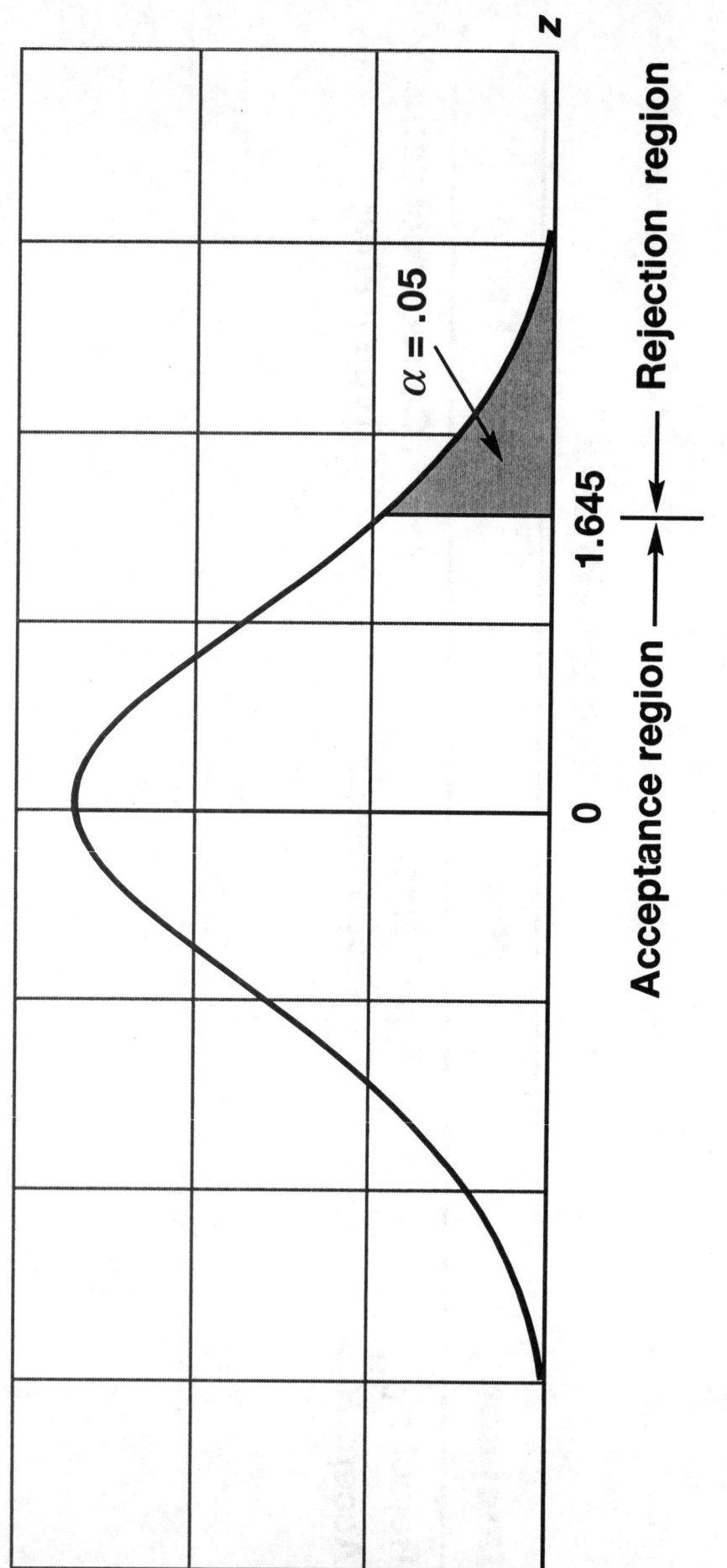

Figure 8.2 Distribution of $z = \dfrac{\bar{x} - \mu_0}{\sigma/n}$ when H_0 is true.

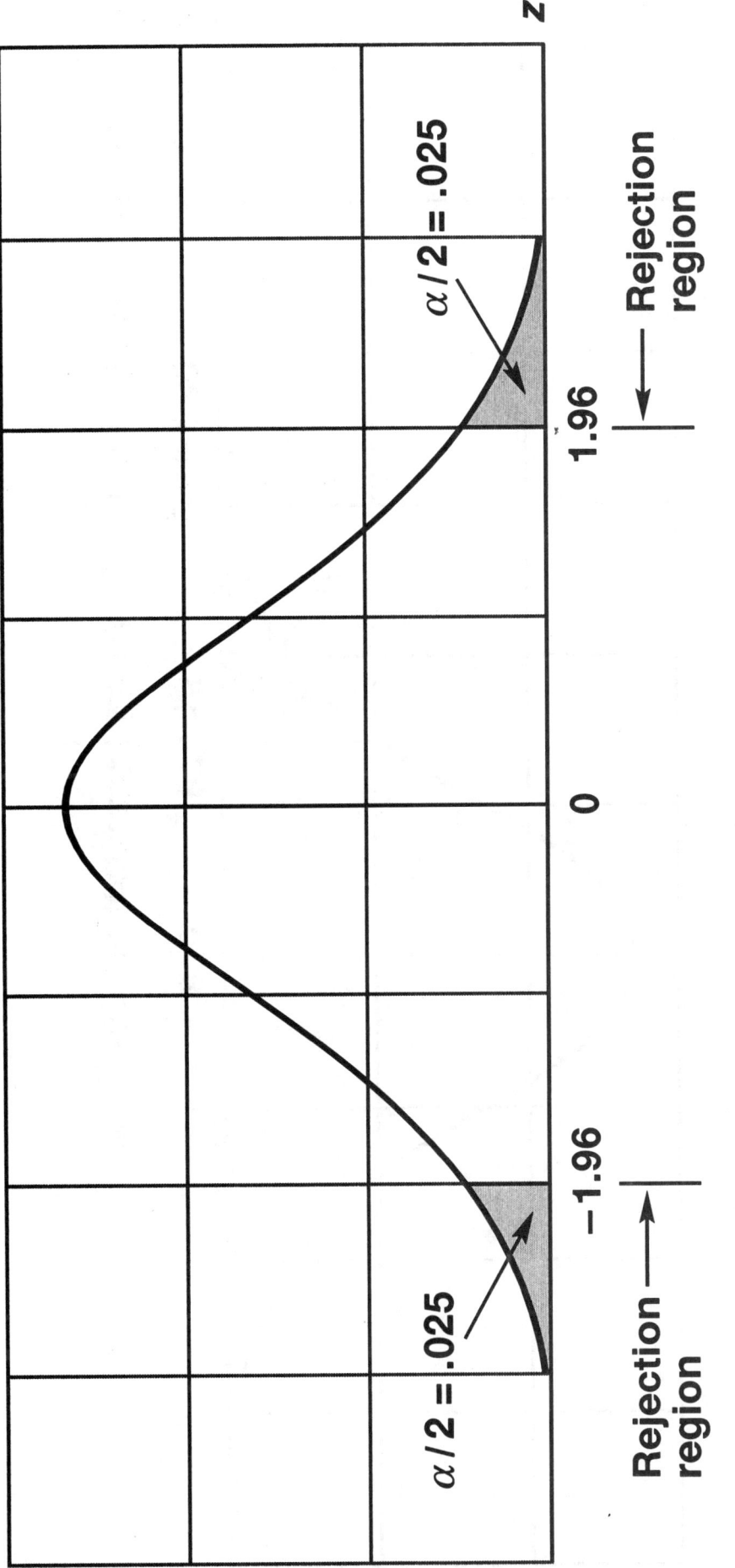

Figure 8.3 The rejection region for a two-tailed test with $\alpha = .05$.

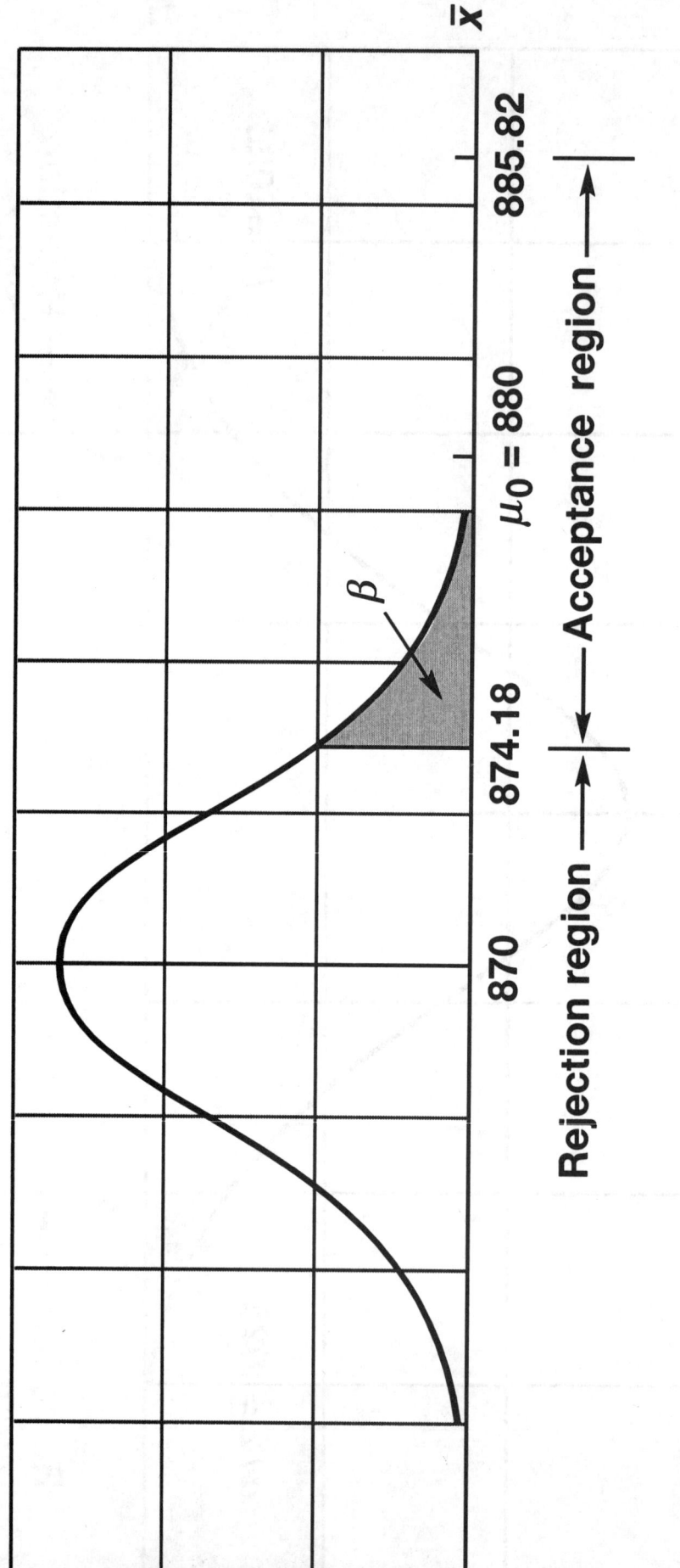

Figure 8.4 *Calculating β in Example 8.5.*

d.f.	$\chi^2_{0.995}$	$\chi^2_{0.990}$	$\chi^2_{0.975}$	$\chi^2_{0.950}$	$\chi^2_{0.900}$
1	0.0000393	0.0001571	0.0009821	0.0039321	0.0157908
2	0.0100251	0.0201007	0.0506356	0.102587	0.210720
3	0.0717212	0.114832	0.215795	0.351846	0.584375
4	0.206990	0.297110	0.484419	0.710721	1.063623
5	0.411740	0.554300	0.831211	1.145476	1.61031
6	0.675727	0.872085	1.237347	0.63539	2.20413
7	0.989265	1.239043	1.68987	2.16735	2.83311
8	1.344419	1.646482	2.17973	2.73264	3.48954
9	1.734926	2.087912	2.70039	3.32511	4.16816
10	2.15585	2.55821	3.24697	3.94030	4.86518
11	2.60321	3.05347	3.81575	4.57481	5.57779
12	3.07382	3.57056	4.40379	5.22603	6.30380
13	3.56503	4.10691	5.00874	5.89186	7.04150
14	4.07468	4.66043	5.62872	6.57063	7.78953
15	4.60094	5.22935	6.26214	7.26094	8.54675
16	5.14224	5.81221	6.90766	7.96164	9.31223
17	5.69724	6.40776	7.56418	8.67176	10.0852
18	6.26481	7.01491	8.23075	9.39046	10.8649
19	6.84398	7.63273	8.90655	10.1170	11.6509
20	7.43386	8.26040	9.59083	10.8508	12.4426
21	8.03366	8.89720	10.28293	11.5913	13.2396
22	8.64272	9.54249	10.9823	12.3380	14.0415
23	9.26042	10.19567	11.6885	13.0905	14.8479
24	9.88623	10.8564	12.4011	13.8484	15.6587
25	10.5197	11.5240	13.1197	14.6114	16.4734
26	11.1603	12.1981	13.8439	15.3791	17.2919
27	11.8076	12.8786	14.5733	16.1513	18.1138
28	12.4613	13.5648	15.3079	16.9279	18.9392
29	13.1211	14.2565	16.0471	17.7083	19.7677
30	13.7867	14.9535	16.7908	18.4926	20.5992
40	20.7065	22.1643	24.4331	26.5093	29.0505
50	27.9907	29.7067	32.3574	34.7642	37.6886
60	35.5346	37.4848	40.4817	43.1879	46.4589
70	43.2752	45.4418	48.7576	51.7393	55.3290
80	51.1720	53.5400	57.1532	60.3915	64.2778
90	59.1963	61.7541	65.6466	69.1260	73.2912
100	67.3276	70.0648	74.2219	77.9295	82.3581

Source: From "Tables of the Percentage Points of the χ^2-Distribution," *Biometrika Tables for Statisticians* 1, 3d ed. (1966). Reproduced by permission of the *Biometrika* Trustees.

Appendix II, Table 5 Critical vlues of chi-square.

$\chi^2_{1.000}$	$\chi^2_{0.050}$	$\chi^2_{0.025}$	$\chi^2_{0.010}$	$\chi^2_{0.005}$	d.f.
2.70554	3.84146	5.02389	6.63490	7.87944	1
4.60517	5.99147	7.37776	9.21034	10.5966	2
6.25139	7.81473	9.34840	11.3449	12.8381	3
7.77944	9.48773	11.1433	13.2767	14.8602	4
9.23635	11.0705	12.8325	15.0863	16.7496	5
10.6446	12.5916	14.4494	16.8119	18.5476	6
12.0170	14.0671	16.0128	18.4753	20.2777	7
13.3616	15.5073	17.5346	20.0902	21.9550	8
14.6837	16.9190	19.0228	21.6660	23.5893	9
15.9871	18.3070	20.4831	23.2093	25.1882	10
17.2750	19.6751	21.9200	24.7250	26.7569	11
18.5494	21.0261	23.3367	26.2170	28.2995	12
19.8119	22.3621	24.7356	27.6883	29.8194	13
21.0642	23.6848	26.1190	29.1413	31.3193	14
22.3072	24.9958	27.4884	30.5779	32.8013	15
23.5418	26.2962	28.8485	31.9999	34.2672	16
24.7690	27.8571	30.1910	33.4087	35.7185	17
25.9894	28.8693	31.5264	34.8053	37.1564	18
27.2036	30.1435	32.8523	36.1908	38.5822	19
28.4120	31.4104	34.1696	37.5662	39.9968	20
29.6151	32.6705	35.4789	38.9321	41.4010	21
30.8133	33.9244	36.7807	40.2894	42.7956	22
32.0069	35.1725	38.0757	41.6384	44.1813	23
33.1963	36.4151	39.3641	42.9798	45.5585	24
34.3816	37.6525	40.6465	44.3141	46.9278	25
35.5631	38.8852	41.9232	45.6417	48.2899	26
36.7412	40.1133	43.1944	46.9630	49.6449	27
37.9159	41.3372	44.4607	48.2782	50.9933	28
39.0875	42.5569	45.7222	49.5879	52.3356	29
40.2560	43.7729	46.9792	50.8922	53.6720	30
51.8050	55.7585	59.3417	63.6907	66.7659	40
63.1671	67.5048	71.4202	76.1539	79.4900	50
74.3970	79.0819	83.2976	88.3794	91.9517	60
85.5271	90.5312	95.0231	100.425	104.215	70
96.5782	101.879	106.629	112.329	116.321	80
107.565	113.145	118.136	124.116	128.299	90
118.498	124.342	129.561	135.807	140.169	100

Appendix II, Table 5 Critical values of chi-square (continued).

		v_1								
v_2	a	1	2	3	4	5	6	7	8	9
1	.100	39.86	49.50	53.59	55.83	57.24	58.20	58.91	59.44	59.86
	.050	161.4	199.5	215.7	224.6	230.2	234.0	236.8	238.9	240.5
	.025	647.8	799.5	864.2	899.6	921.8	937.1	948.2	956.7	963.3
	.010	4052	4999.5	5403	5625	5764	5859	5928	5982	6022
	.005	16211	20000	21615	22500	23056	23437	23715	23925	24091
2	.100	8.53	9.00	9.16	9.24	9.29	9.33	9.35	9.37	9.38
	.050	18.51	19.00	19.16	19.25	19.30	19.33	19.35	19.37	19.38
	.025	38.51	39.00	39.17	39.25	39.30	39.33	39.36	39.37	39.39
	.010	98.50	99.00	99.17	99.25	99.30	99.33	99.36	99.37	99.39
	.005	198.5	199.0	199.2	199.2	199.3	199.3	199.4	199.4	199.4
3	.100	5.54	5.46	5.39	5.34	5.31	5.28	5.27	5.25	5.24
	.050	10.13	9.55	9.28	9.12	9.01	8.94	8.89	8.85	8.81
	.025	17.44	16.04	15.44	15.10	14.88	14.73	14.62	14.54	14.47
	.010	34.12	30.82	29.46	28.71	28.24	27.91	27.67	27.49	27.35
	.005	55.55	49.80	47.47	46.19	45.39	44.84	44.43	44.13	43.88
9	.100	3.36	3.01	2.81	2.69	2.61	2.55	2.51	2.47	2.44
	.050	5.12	4.26	3.86	3.63	3.48	**3.37**	3.29	3.23	3.18
	.025	7.21	5.71	5.08	4.72	4.48	4.32	4.20	4.10	4.03
	.010	10.56	8.02	6.99	6.42	6.06	**5.80**	5.61	5.47	5.35
	.005	13.61	10.11	8.72	7.96	7.47	7.13	6.88	6.69	6.54
10	.100	3.29	2.92	2.73	2.61	2.52	2.46	2.41	2.38	2.35
	.050	4.96	4.10	3.71	3.48	3.33	3.22	3.14	3.07	3.02
	.025	6.94	5.46	4.83	4.47	4.24	4.07	3.95	3.85	3.78
	.010	10.04	7.56	6.55	5.99	5.64	5.39	5.20	5.06	4.94
	.005	12.83	9.43	8.08	7.34	6.87	6.54	6.30	6.12	5.97
11	.100	3.23	2.86	2.66	2.54	2.45	2.39	2.34	2.30	2.27
	.050	4.84	3.98	3.59	3.36	3.20	3.09	3.01	**2.95**	2.90
	.025	6.72	5.26	4.63	4.28	4.04	3.88	3.76	3.66	3.59
	.010	9.65	7.21	6.22	5.67	5.32	5.07	4.89	4.74	4.63
	.005	12.23	8.91	7.60	6.88	6.42	6.10	5.86	5.68	5.54
12	.100	3.18	2.81	2.61	2.48	2.39	2.33	2.28	2.24	2.21
	.050	4.75	3.89	3.49	3.26	3.11	3.00	2.91	2.85	2.80
	.025	6.55	5.10	4.47	4.12	3.89	3.73	3.61	3.51	3.44
	.010	9.33	6.93	5.95	5.41	5.06	4.82	4.64	4.50	4.39
	.005	11.75	8.51	7.23	6.52	6.07	5.76	5.52	5.35	5.20

Table 8.8 Format of the F table from Table 6 in Appendix II.

Transparency 31 © 1996 Wadsworth Publishing Company/ITP

ANALYSIS OF VARIANCE PROCEDURE

DEPENDENT VARIABLE: COST

SOURCE	DF	SUM OF SQUARES	MEAN SQUARE	F VALUE	PR > F	R-SQUARE	C.V.
MODEL	4	6742554.48000000	1685638.62000000	0.98	0.4281	0.080122	72.0381
ERROR	45	77411264.40000000	1720250.32000000		ROOT MSE		COST MEAN
CORRECTED TOTAL	49	84153818.88000000			1311.58313499		1820.68000000

SOURCE	DF	ANOVA SS	F VALUE	PR > F
GROUP	4	6742554.48000000	0.98	0.4281

a

```
MTB > AOVONEWAY C1-C5

ANALYSIS OF VARIANCE
SOURCE   DF      SS        MS        F       P
FACTOR    4  6742554   1685638     0.98   0.428
ERROR    45 77411264   1720250
TOTAL    49 84153816
```

LEVEL	N	MEAN	STDEV
C1	10	1960	1659
C2	10	1996	1384
C3	10	1194	839
C4	10	1675	1066
C5	10	2279	1447

POOLED STDEV = 1312

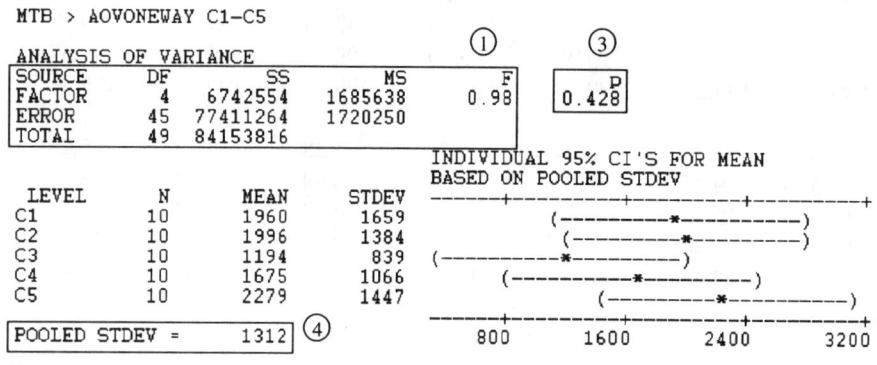

b

	A	B	C	D	E	F	G	H
1			Analysis of Many Samples for CBS9-1.Claims by CBS9-1.Group					
2								
3	Class	Value	Sample Size		Mean		Standard Deviation	
4	1	1	10		1959.8		1658.8	
5	2	2	10		1996.1		1384.0	
6	3	3	10		1193.9		838.9	
7	4	4	10		1674.7		1066.3	
8	5	5	10		2278.9		1446.9	
9								
10			Oneway ANOVA for CBS9-1.Claims					
11	Source of Variation		Sum of Squares		D.F.	Mean Square	F-Ratio	P Value
12	Group		6742550		4	1685640	0.9800	0.4281
13	Error		77411300		45	1720250		
14	Total (corr.)		84153850		49			
15								
16			95% Confidence Intervals (pooled)					
17	Class	Value	Sample Size		Mean	+/-	Interval	
18	1	1	10		1959.8		835.4	
19	2	2	10		1996.1		835.4	
20	3	3	10		1193.9		835.4	
21	4	4	10		1674.7		835.4	
22	5	5	10		2278.9		835.4	

c

(a) Table 9.3 SAS computer printout for an analysis of variance. **(b) Table 9.4** Minitab computer printout of an analysis of variance. **(c) Table 9.5** Excel computer printout for an analysis of variance.

ANALYSIS OF VARIANCE PROCEDURE

DEPENDENT VARIABLE: PRICE

SOURCE	DF	SUM OF SQUARES	MEAN SQUARE	F VALUE	PR > F	R-SQUARE	C.V.
MODEL	6	5.02352000	0.83725333	99.32	0.0001	0.986753	2.1382
ERROR	8	0.06744000	0.00843000		ROOT MSE		PRICE TIME
CORRECTED TOTAL	14	5.09096000			0.09181503		4.29400000

SOURCE	DF	ANOVA SS	F VALUE	PR > F
ESTIMATOR	2	0.13456000	7.98	0.0124
PROJECT	4	4.88896000	144.99	0.0001

a

```
MTB > ANOVA C1 = C3 C2

Factor    Type  Levels  Values
ESTIMATR  fixed    3     1   2   3
PROJECT   fixed    5     1   2   3   4   5

Analysis of Variance for C1

Source    DF      SS        MS        F       P
ESTIMATR   2   0.13456   0.06728    7.98   0.012
PROJECT    4   4.88896   1.22224  144.99   0.000
Error      8   0.06744   0.00843
Total     14   5.09096
```

b

Analysis of Variance for PRICE - Type III Sums of Squares					
Source	Sum of Squares	Df	Mean Square	F-Ratio	P-Value
MAIN EFFECTS					
A:ESTIMATOR	0.13456	2	0.06728	7.98	0.0124
B:PROJECT	4.88896	4	1.22224	144.99	0.0000
RESIDUAL	0.06744	8	0.00843		
TOTAL (Corrected)	5.09096	14			

Table of Least Squares Means for PRICE with 95.0% Confidence Intervals					
Level	Count	Mean	Stnd Error	Lower Limit	Upper Limit
GRAND MEAN	15	4.294			
ESTIMATOR					
1	5	4.294	0.041	4.19931	4.38869
2	5	4.178	0.041	4.08331	4.27269
3	5	4.410	0.041	4.31531	4.50469
PROJECT					
1	3	3.517	0.053	3.39443	3.63891
2	3	4.807	0.053	4.68443	4.92891
3	3	3.967	0.053	3.84443	4.08891
4	3	5.080	0.053	4.95776	5.20224
5	3	4.100	0.053	3.97776	4.22224

c

(a) Table 9.7 SAS ANOVA printout. (b) Table 9.8 Minitab ANOVA printout. (c) Table 9.9 Excel ANOVA printout.

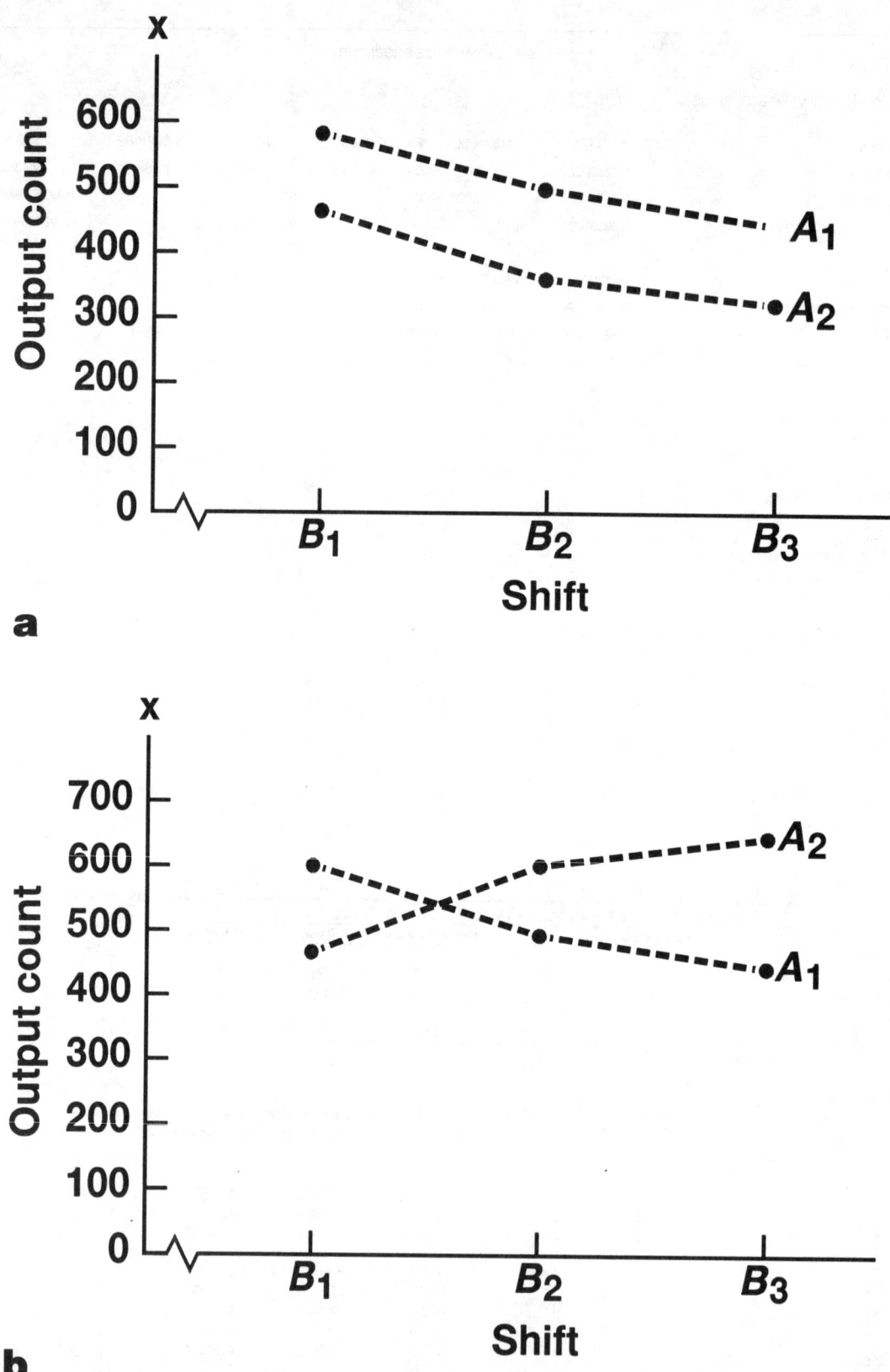

(a) Figure 9.5 Plots indicating that the factors affect output independently of each other.
(b) Figure 9.6 Output counts indicating factor interaction.

ANALYSIS OF VARIANCE PROCEDURE

DEPENDENT VARIABLE: OUTPUT

SOURCE	DF	SUM OF SQUARES	MEAN SQUARE	F VALUE	PR > F	R-SQUARE	C.V.
MODEL	5	100179.16666667	20035.83333333	27.85	0.0001	0.920659	4.8842
ERROR	12	8633.33333333	719.44444444		ROOT MSE		OUTPUT MEAN
CORRECTED TOTAL	17	108812.50000000			26.82246157		549.16666667

SOURCE	DF	ANOVA SS	F VALUE	PR > F
FOREMAN	1	19012.50000000	26.43	0.0002
SHIFT	2	258.33333333	0.18	0.8379
SHIFT*FOREMAN	2	80908.33333333	56.23	0.0001

a

```
MTB > ANOVA C1 = C2 C3 C2*C3

Factor    Type  Levels  Values
FOREMAN   fixed    2      1   2
SHIFT     fixed    3      1   2   3

Analysis of Variance for C1

Source          DF      SS        MS       F       P
FOREMAN          1   19012     19012   26.43   0.000
SHIFT            2     258       129    0.18   0.838
FOREMAN*SHIFT    2   80908     40454   56.23   0.000
Error           12    8633       719
Total           17  108813
```

b

	A	B	C	D	E	F	G	H	I	J
1	Analysis of Variance for OUTPUT - Type III Sums of Squares									
2	Source			Sum of Squares	Df		Mean Square	F-Ratio		P-Value
3	MAIN EFFECTS									
4	A:FOREMAN			19012.5	1		19012.5	26.43		0.0002
5	B:SHIFT			258.3	2		129.2	0.18		0.8379
6										
7	INTERACTIONS									
8	AB			80908.3	2		40454.2	56.23		
9										
10	RESIDUAL			8633.3	12		719.4			
11	TOTAL (Corrected)			108813	17					
12										
13	Table of Least Squares Means for OUTPUT with 95.0% Confidence Intervals									
14	Level			Count	Mean		Stnd Error	Lower Limit		Upper Limit
15	GRAND MEAN			18	549.167					
16	FOREMAN									
17	1			9	516.667		8.941	497.186		536.147
18	2			9	581.667		8.941	562.186		601.147
19	SHIFT									
20	1			6	544.167		10.950	520.308		568.025
21	2			6	550.000		10.950	526.141		573.859
22	3			6	553.333		10.950	529.475		577.192
23	FOREMAN by SHIFT									
24	1	1		3	601.667		15.486	567.926		635.408
25	1	2		3	498.333		15.486	464.592		532.074
26	1	3		3	450.000		15.486	416.259		483.741
27	2	1		3	486.667		15.486	452.926		520.408
28	2	2		3	601.667		15.486	567.926		635.408
29	2	3		3	656.667		15.486	622.926		690.408
30										

c

(a) Table 9.11 SAS ANOVA printout. (b) Table 9.12 Minitab ANOVA printout. (c) Table 9.13 Excel printout.

	t									
v	2	3	4	5	6	7	8	9	10	11
1	17.97	26.98	32.82	37.08	40.41	43.12	45.40	47.36	49.07	50.59
2	6.08	8.33	9.80	10.88	11.74	12.44	13.03	13.54	13.99	14.39
3	4.50	5.91	6.82	7.50	8.04	8.48	8.85	9.18	9.46	9.72
4	3.93	5.04	5.76	6.29	6.71	7.05	7.35	7.60	7.83	8.03
5	3.64	4.60	5.22	5.67	6.03	6.33	6.58	6.80	6.99	7.17
6	3.46	4.34	4.90	5.30	5.63	5.90	6.12	6.32	6.49	6.65
7	3.34	4.16	4.68	5.06	5.36	5.61	5.82	6.00	6.16	6.30
8	3.26	4.04	4.53	4.89	5.17	5.40	5.60	5.77	5.92	6.05
9	3.20	3.95	4.41	4.76	5.02	5.24	5.43	5.59	5.74	5.87
10	3.15	3.88	4.33	4.65	4.91	5.12	5.30	5.46	5.60	5.72
11	3.11	3.82	4.26	4.57	4.82	5.03	5.20	5.35	5.49	5.61
12	3.08	3.77	4.20	4.51	4.75	4.95	5.12	5.27	5.39	5.51
13	3.06	3.73	4.15	4.45	4.69	4.88	5.05	5.19	5.32	5.43
14	3.03	3.70	4.11	4.41	4.64	4.83	4.99	5.13	5.25	5.36
15	3.01	3.67	4.08	4.37	4.60	4.78	4.94	5.08	5.20	5.31
16	3.00	3.65	4.05	4.33	4.56	4.74	4.90	5.03	5.15	5.26
17	2.98	3.63	4.02	4.30	4.52	4.70	4.86	4.99	5.11	5.21
18	2.97	3.61	4.00	4.28	4.49	4.67	4.82	4.96	5.07	5.17
19	2.96	3.59	3.98	4.25	4.47	4.65	4.79	4.92	5.04	5.14
20	2.95	3.58	3.96	4.23	4.45	4.62	4.77	4.90	5.01	5.11
24	2.92	3.53	3.90	4.17	4.37	4.54	4.68	4.81	4.92	5.01
30	2.89	3.49	3.85	4.10	4.30	4.46	4.60	4.72	4.82	4.92
40	2.86	3.44	3.79	4.04	4.23	4.39	4.52	4.63	4.73	4.82
60	2.83	3.40	3.74	3.98	4.16	4.31	4.44	4.55	4.65	4.73
120	2.80	3.36	3.68	3.92	4.10	4.24	4.36	4.47	4.56	4.64
∞	2.77	3.31	3.63	3.86	4.03	4.17	4.29	4.39	4.47	4.55

Appendix II, Table 7 Percentage points of the studentized range, $q(t, v)$; upper 5% points.

				t					
12	13	14	15	16	17	18	19	20	ν
51.96	53.20	54.33	55.36	56.32	57.22	58.04	58.83	59.56	1
14.75	15.08	15.38	15.65	15.91	16.14	16.37	16.57	16.77	2
9.95	10.15	10.35	10.52	10.69	10.84	10.98	11.11	11.24	3
8.21	8.37	8.52	8.66	8.79	8.91	9.03	9.13	9.23	4
7.32	7.47	7.60	7.72	7.83	7.93	8.03	8.12	8.21	5
6.79	6.92	7.03	7.14	7.24	7.34	7.43	7.51	7.59	6
6.43	6.55	6.66	6.76	6.85	6.94	7.02	7.10	7.17	7
6.18	6.29	6.39	6.48	6.57	6.65	6.73	6.80	6.87	8
5.98	6.09	6.19	6.28	6.36	6.44	6.51	6.58	6.64	9
5.83	5.93	6.03	6.11	6.19	6.27	6.34	6.40	6.47	10
5.71	5.81	5.90	5.98	6.06	6.13	6.20	6.27	6.33	11
5.61	5.71	5.80	5.88	5.95	6.02	6.09	6.15	6.21	12
5.53	5.63	5.71	5.79	5.86	5.93	5.99	6.05	6.11	13
5.46	5.55	5.64	5.71	5.79	5.85	5.91	5.97	6.03	14
5.40	5.49	5.57	5.65	5.72	5.78	5.85	5.90	5.96	15
5.35	5.44	5.52	5.59	5.66	5.73	5.79	5.84	5.90	16
5.31	5.39	5.47	5.54	5.61	5.67	5.73	5.79	5.84	17
5.27	5.35	5.43	5.50	5.57	5.63	5.69	5.74	5.79	18
5.23	5.31	5.39	5.46	5.53	5.59	5.65	5.70	5.75	19
5.20	5.28	5.36	5.43	5.49	5.55	5.61	5.66	5.71	20
5.10	5.18	5.25	5.32	5.38	5.44	5.49	5.55	5.59	24
5.00	5.08	5.15	5.21	5.27	5.33	5.38	5.43	5.47	30
4.90	4.98	5.04	5.11	5.16	5.22	5.27	5.31	5.36	40
4.81	4.88	4.94	5.00	5.06	5.11	5.15	5.20	5.24	60
4.71	4.78	4.84	4.90	4.95	5.00	5.04	5.09	5.13	120
4.62	4.68	4.74	4.80	4.85	4.89	4.93	4.97	5.01	∞

Source: From *Biometrika Tables for Statisticians*, Vol. 1, 3rd ed., edited by E. S. Pearson and H. O. Hartley (Cambridge University Press, 1966). Reproduced by permission of the Biometrika Trustees.

Appendix II, Table 7 Percentage points of the studentized range, $q(t, v)$; upper 5% points (continued).

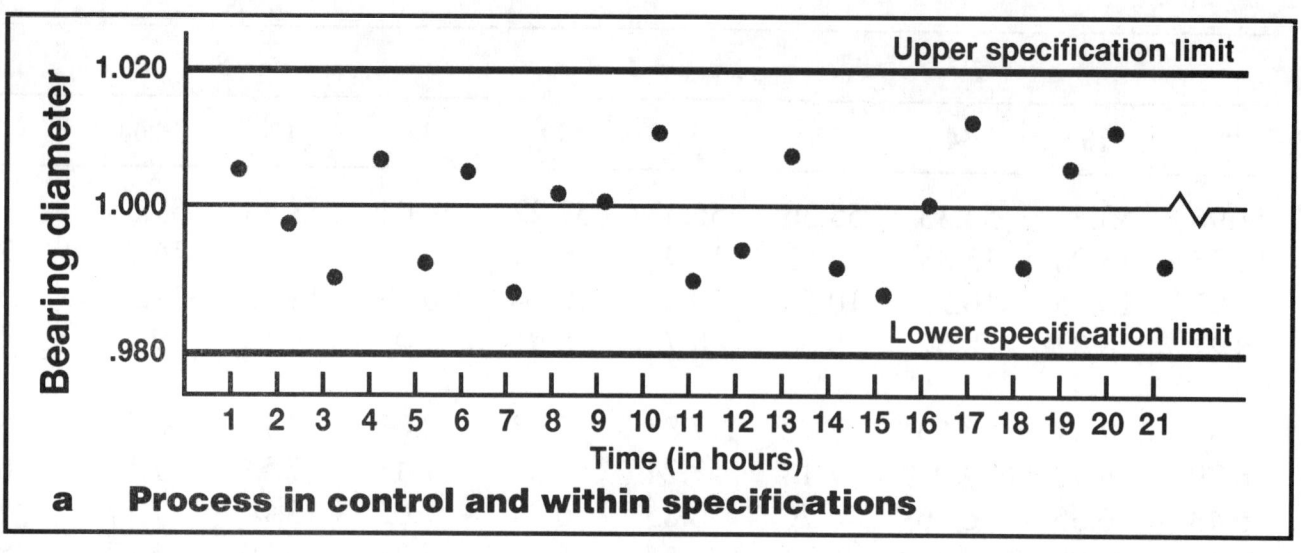
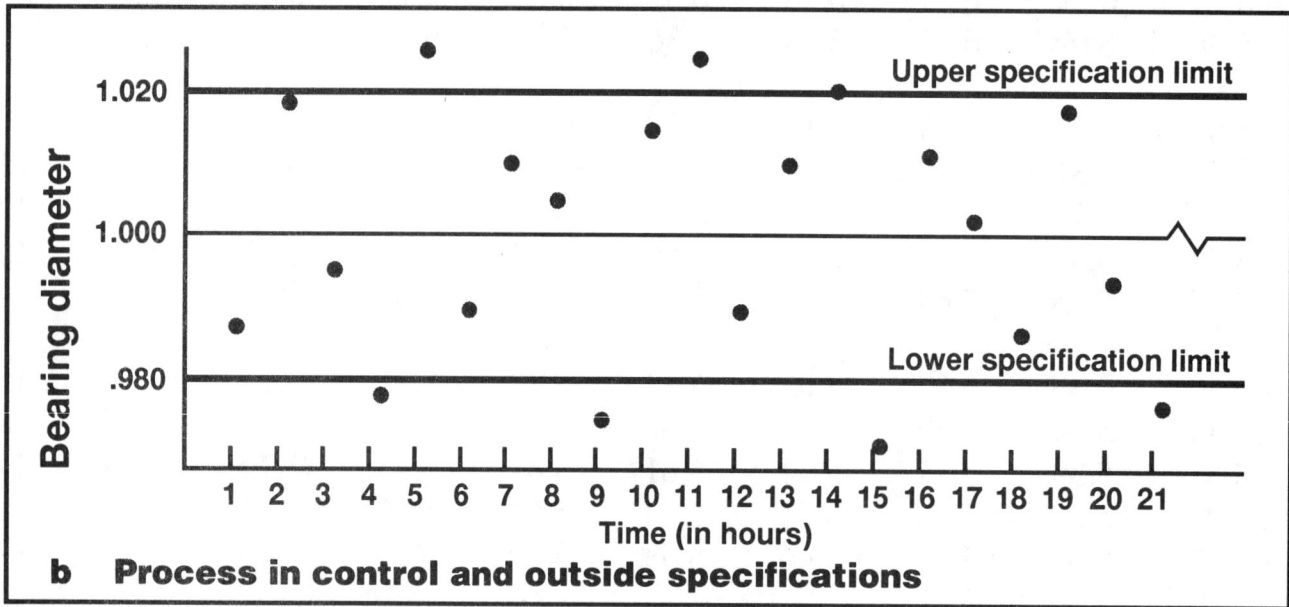
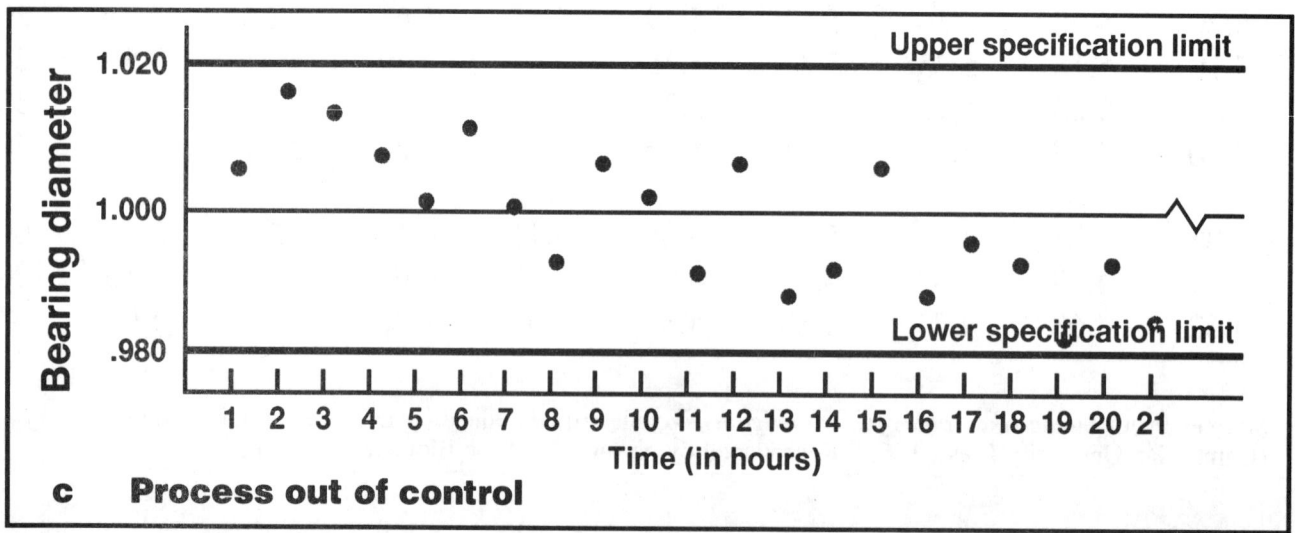

Figure 10.1 Plots of bearing diameters over time.

	Chart for Averages			Chart for Standard Deviations						Chart for Ranges						
		Factors for Control Limits		Factors for Central Line		Factors for Control Limits				Factors for Central Line		Factors for Control Limits				
Number of Observations in Sample, n	A	A_1	A_2	c_2	$1/c_2$	B_1	B_2	B_3	B_4	d_2	$1/d_2$	d_3	D_1	D_2	D_3	D_4
2	2.121	3.760	1.880	.5642	1.7725	0	1.843	0	3.267	1.128	.8865	.853	0	3.686	0	3.276
3	1.732	2.394	1.023	.7236	1.3820	0	1.858	0	2.568	1.693	.5907	.888	0	4.358	0	2.575
4	1.501	1.880	.729	.7979	1.2533	0	1.808	0	2.266	2.059	.4857	.880	0	4.698	0	2.282
5	1.342	1.596	.577	.8407	1.1894	0	1.756	0	2.089	2.326	.4299	.864	0	4.918	0	2.115
6	1.225	1.410	.483	.8686	1.1512	.026	1.711	.030	1.970	2.534	.3946	.848	0	5.078	0	2.004
7	1.134	1.277	.419	.8882	1.1259	.105	1.672	.118	1.882	2.704	.3698	.833	.205	5.203	.076	1.924
8	1.061	1.175	.373	.9027	1.1078	.167	1.638	.185	1.815	2.847	.3512	.820	.387	5.307	.136	1.864
9	1.000	1.094	.337	.9139	1.0942	.219	1.609	.239	1.761	2.970	.3367	.808	.546	5.394	.184	1.816
10	.949	1.028	.308	.9227	1.0837	.262	1.584	.284	1.716	3.078	.3249	.797	.687	5.469	.223	1.777
11	.905	.973	.285	.9300	1.0753	.299	1.561	.321	1.679	3.173	.3152	.787	.812	5.534	.256	1.744
12	.866	.925	.266	.9359	1.0684	.331	1.541	.354	1.646	3.258	.3069	.778	.924	5.592	.284	1.719
13	.832	.884	.249	.9410	1.0627	.359	1.523	.382	1.618	3.336	.2998	.770	1.026	5.646	.308	1.692
14	.802	.848	.235	.9453	1.0579	.384	1.507	.406	1.594	3.407	.2935	.762	1.121	5.693	.329	1.671
15	.775	.816	.223	.9490	1.0537	.406	1.492	.428	1.572	3.472	.2880	.755	1.207	5.737	.348	1.652
16	.750	.788	.212	.9523	1.0501	.427	1.478	.448	1.552	3.532	.2831	.749	1.285	5.779	.364	1.636
17	.728	.762	.203	.9551	1.0470	.445	1.465	.466	1.534	3.588	.2787	.743	1.359	5.817	.379	1.621
18	.707	.738	.194	.9576	1.0442	.461	1.454	.482	1.518	3.640	.2747	.738	1.426	5.854	.392	1.608
19	.688	.717	.187	.9599	1.0418	.477	1.443	.497	1.503	3.689	.2711	.733	1.490	5.888	.404	1.596
20	.671	.697	.180	.9619	1.0396	.491	1.433	.510	1.490	3.735	.2677	.729	1.548	5.922	.414	1.586
21	.655	.679	.173	.9638	1.0376	.504	1.424	.523	1.477	3.778	.2647	.724	1.606	5.950	.425	1.575
22	.640	.662	.167	.9655	1.0358	.516	1.415	.534	1.466	3.819	.2618	.720	1.659	5.979	.434	1.566
23	.626	.647	.162	.9670	1.0342	.527	1.407	.545	1.455	3.858	.2592	.716	1.710	6.006	.443	1.557
24	.612	.632	.157	.9684	1.0327	.538	1.399	.555	1.445	3.895	.2567	.712	1.759	6.031	.452	1.548
25	.600	.619	.153	.9696	1.0313	.548	1.392	.565	1.435	3.931	.2544	.709	1.804	6.058	.459	1.541
Over 25	$\frac{3}{\sqrt{n}}$	$\frac{3}{\sqrt{n}}$	—	—	—	*	†	*	†	—	—	—	—	—	—	—

*$1 - 3/\sqrt{2n}$ †$1 + 3/\sqrt{2n}$

Source: Reproduced by permission from *ASTM Manual on Quality Control of Materials*, American Society for Testing Materials, Philadelphia, PA, 1951.

Appendix II, Table 9 Factors used when constructing control charts.

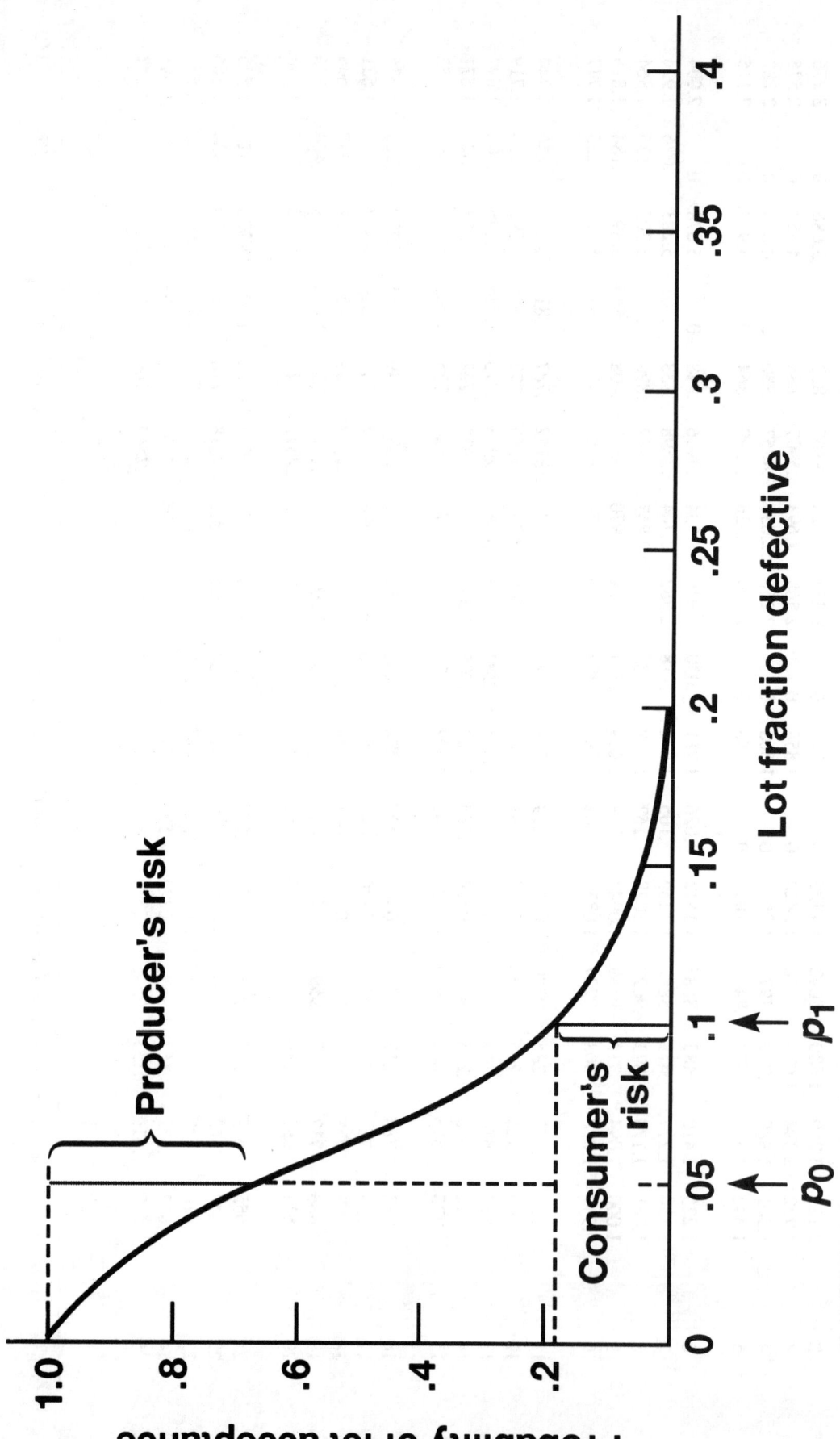

Figure 10.16 Operating characteristic curve showing the producer's risk for AQL = p_0 = .05 and the consumer's risk for p_1 = .10.

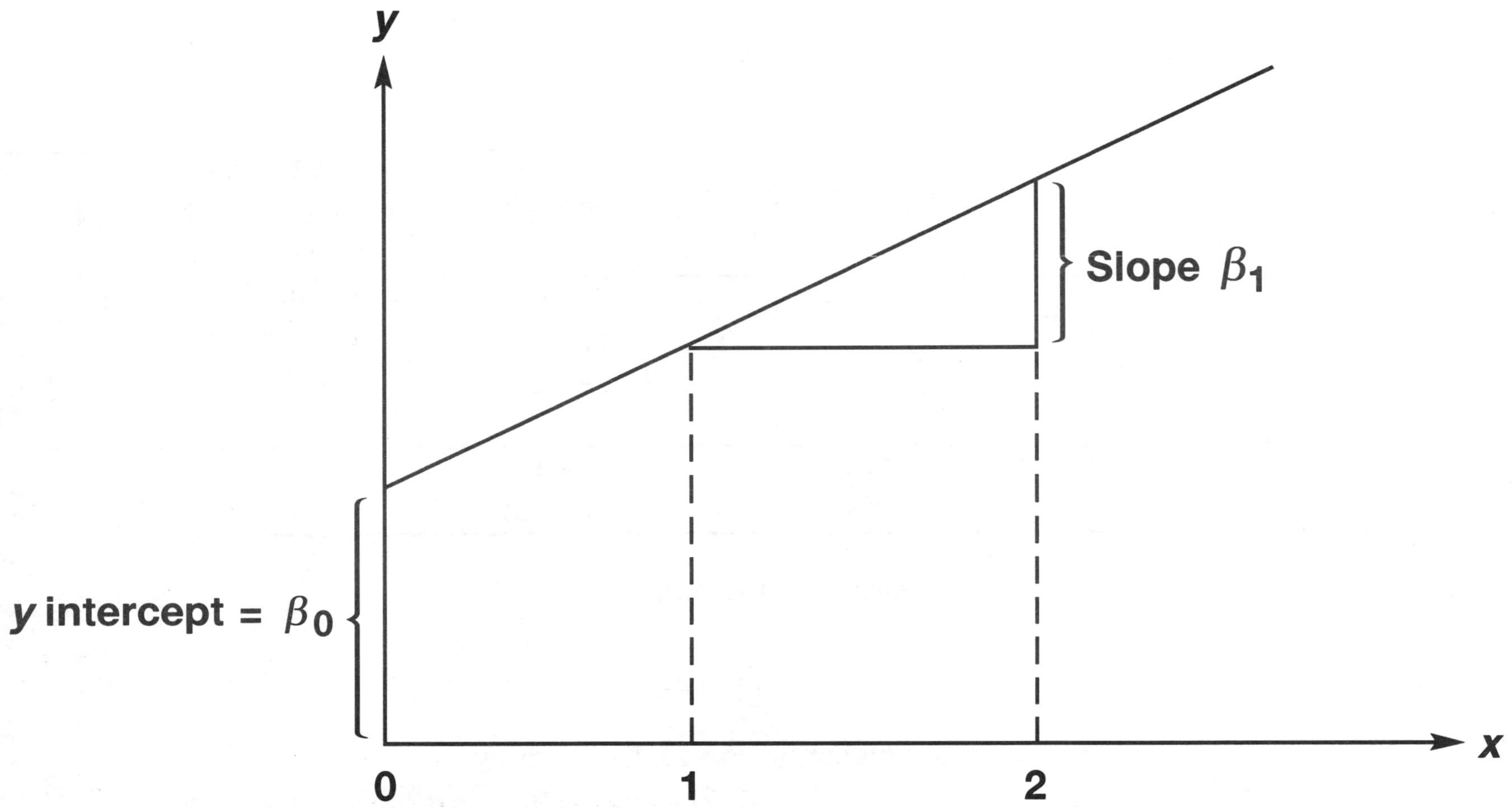

Figure 11.3 *The y intercept and slope for a line.*

	A	B	C	D	E	F	G	H
1			Simple Regression Analysis for CBS11-1					
2								
3		Linear Model: RATING = 40.7842 + 0.765562 * SCORE						
4								
5				Table of Estimates				
6					Estimate	Standard Error	t Value	P Value
7		Intercept			40.7842	8.50686	4.79	0.0014
8		Slope			0.765562	0.17499	4.38	0.0024
9								
10		R-Squared =			70.52 %			
11		Correlation Coef. =			0.84			
12		Std Error of Estimation =			8.70363			
13		Durbin-Watson Statistic =			1.17368			
14		Mean Absolute Error =			6.97801			
15		Sample Size (n) =			10			

Table 11.5 Excel printout for Example 11.2.

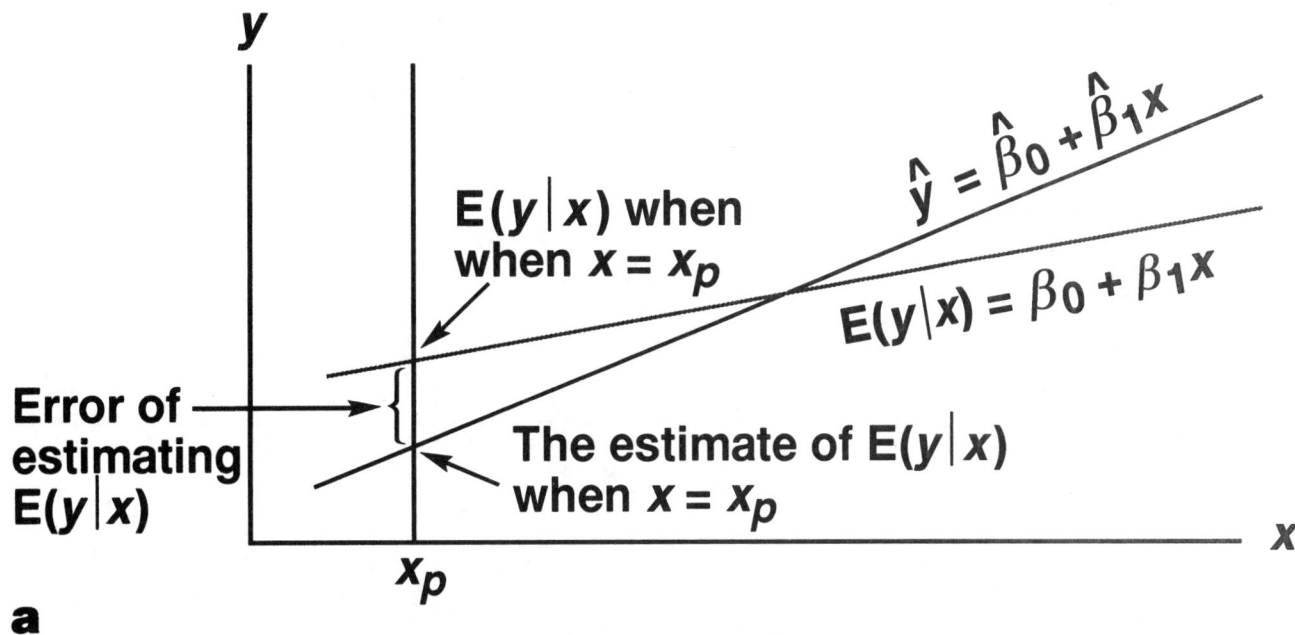

a

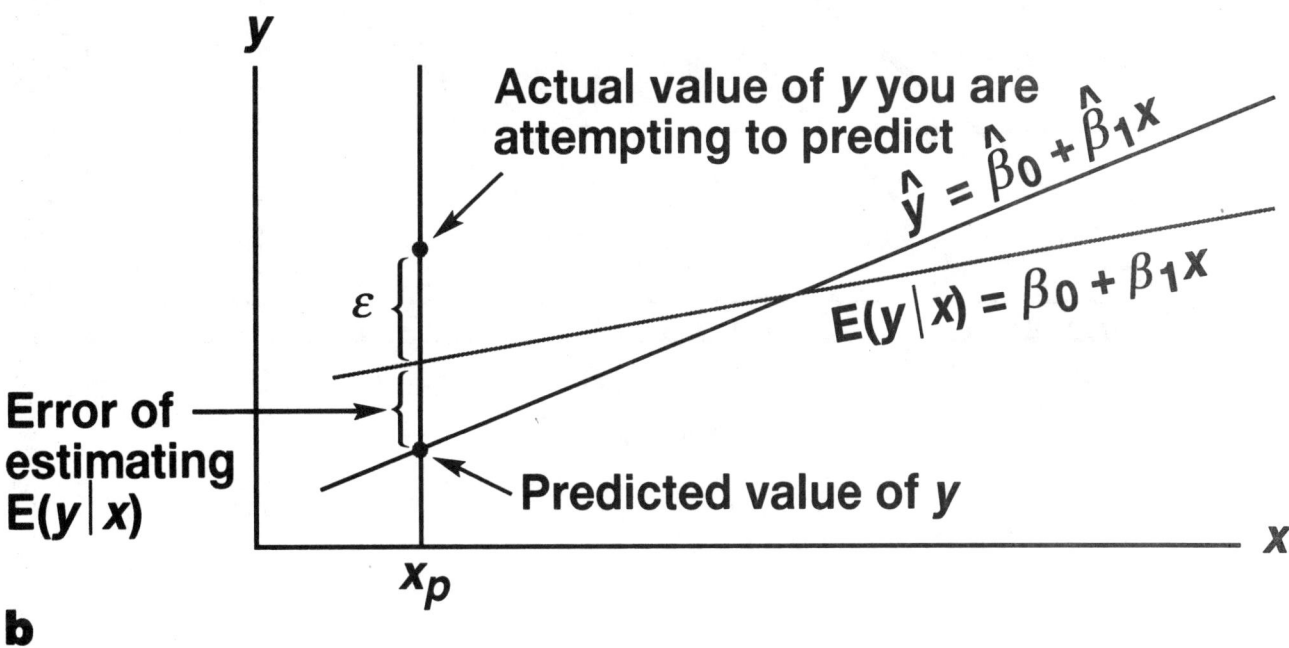

b

(a) Figure 11.7 Estimating E (y/x) when x = x_p.
(b) Figure 11.8 Error in predicting a particular value of y.

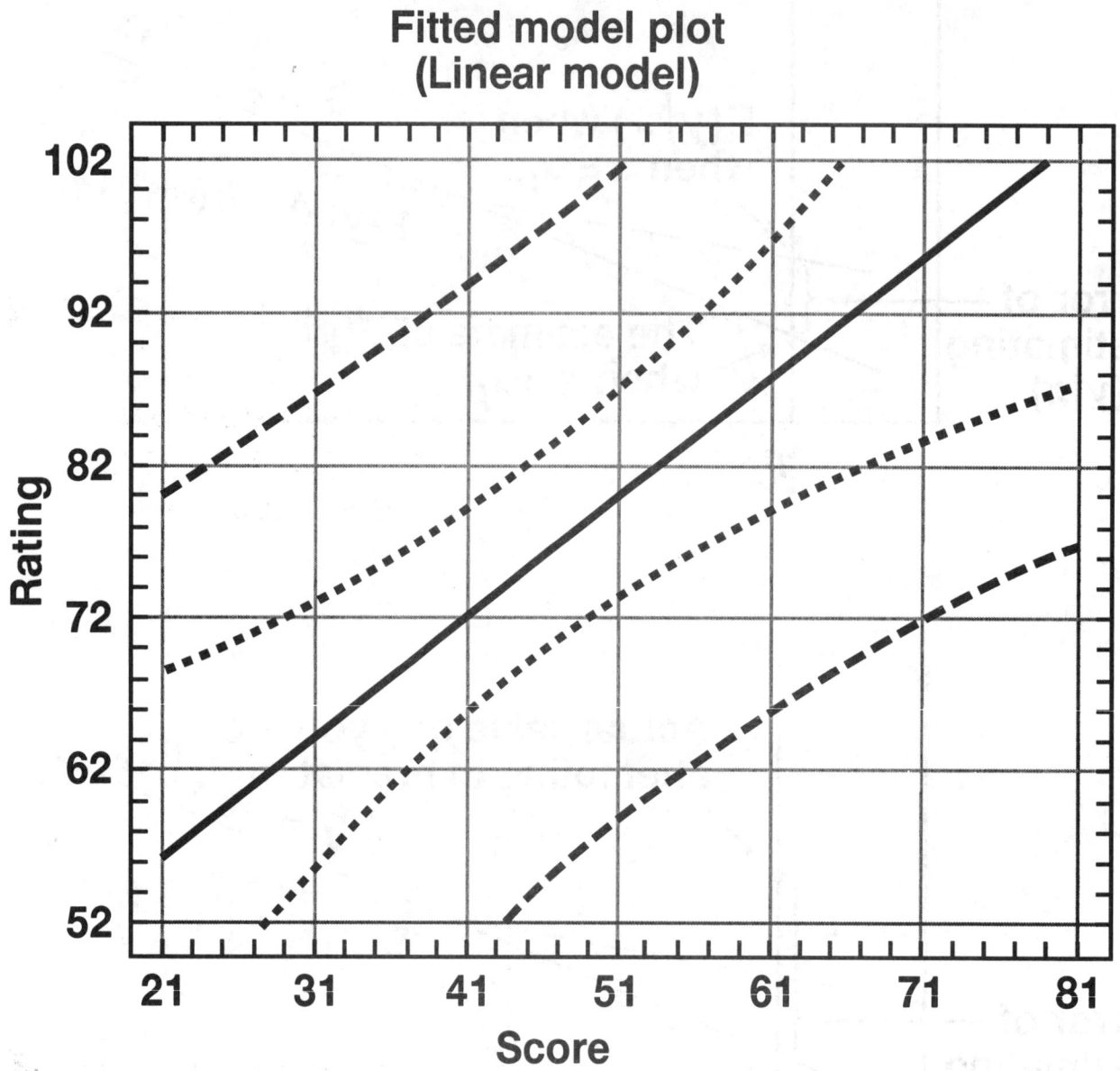

Figure 11.9 Confidence intervals for E(y/x) and prediction intervals for y based on data in Table 11.2.

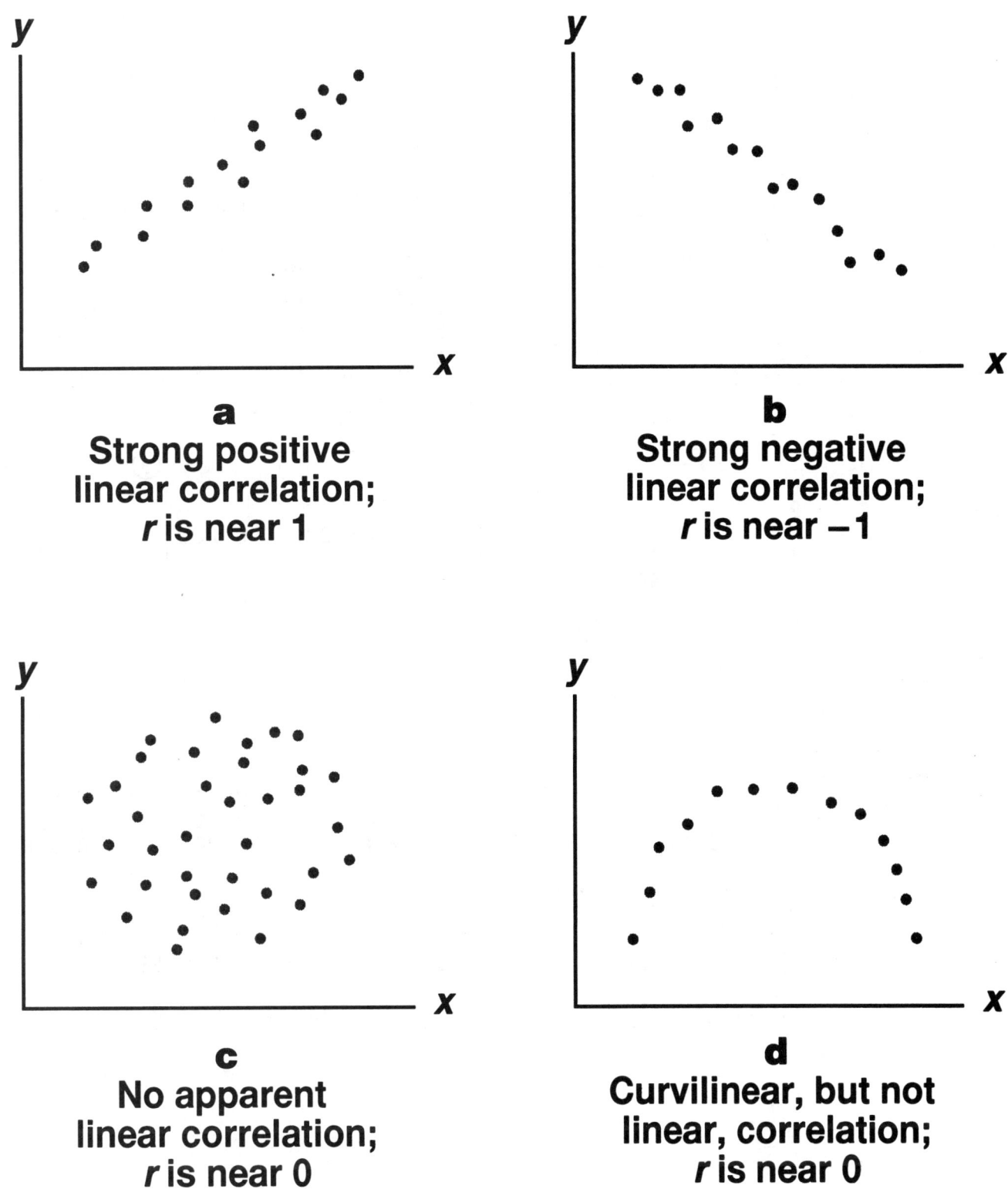

Figure 11.10 Some typical scatterplots with approximate values of r.

```
MTB > REGRESS C8 4 C1-C4;
SUBC>   PREDICT 10 1 3 2;
SUBC>   PREDICT 14 2 3 2.5.
```
①

The regression equation is
LPRICE = - 16.6 + 7.84 SQFT - 34.4 NUMFLRS - 7.99 BDRMS + 54.9 BATHS

Predictor	Coef ②	Stdev ③	t-ratio ④	p ⑤
Constant	-16.58	18.88	-0.88	0.389
SQFT	7.839	1.234	6.35	0.000
NUMFLRS	-34.39	11.15	-3.09	0.005
BDRMS	-7.990	8.249	-0.97	0.342
BATHS	54.93	13.52	4.06	0.000

s = 16.58 ⑥   R-sq = 88.2% ⑦   R-sq(adj) = 86.2% ⑧

Analysis of Variance ⑨

SOURCE	DF	SS	MS	F	p
Regression	4	49359	12340	44.88	0.000
Error	24	6599	275		
Total	28	55958			

SOURCE	DF	SEQ SS ⑩
SQFT	1	44444
NUMFLRS	1	59
BDRMS	1	321
BATHS	1	4536

Unusual Observations

Obs.	SQFT	LPRICE	Fit	Stdev.Fit	Residual	St.Resid
1	6.0	69.00	35.02	9.45	33.98	2.49R
2	8.0	11.50	50.70	9.45	-39.20	-2.88R

R denotes an obs. with a large st. resid.

Fit	Stdev.Fit	95% C.I.	95% P.I.	⑪
113.32	5.80	(101.34, 125.30)	(77.05, 149.59)	
137.75	5.48	(126.44, 149.07)	(101.70, 173.81)	

Table 12.3 Minitab regression analysis for the data in Example 12.1.

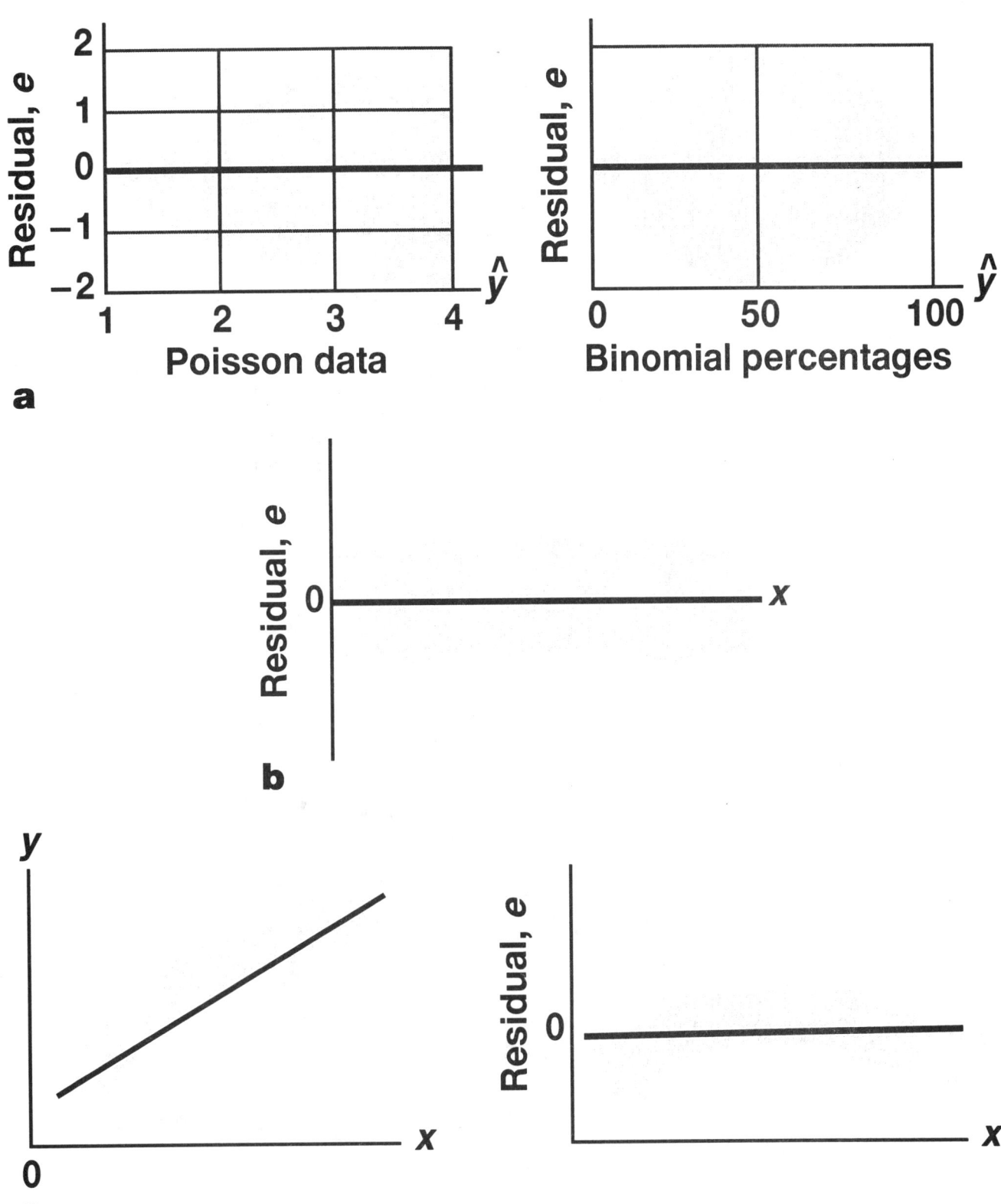

(a) **Figure 12.7** Plots of residuals against $\hat{y}$. *(b) Residual plot when the model provides a good approximation to reality. (c) Data and residual plots for a model that does not agree with reality.*

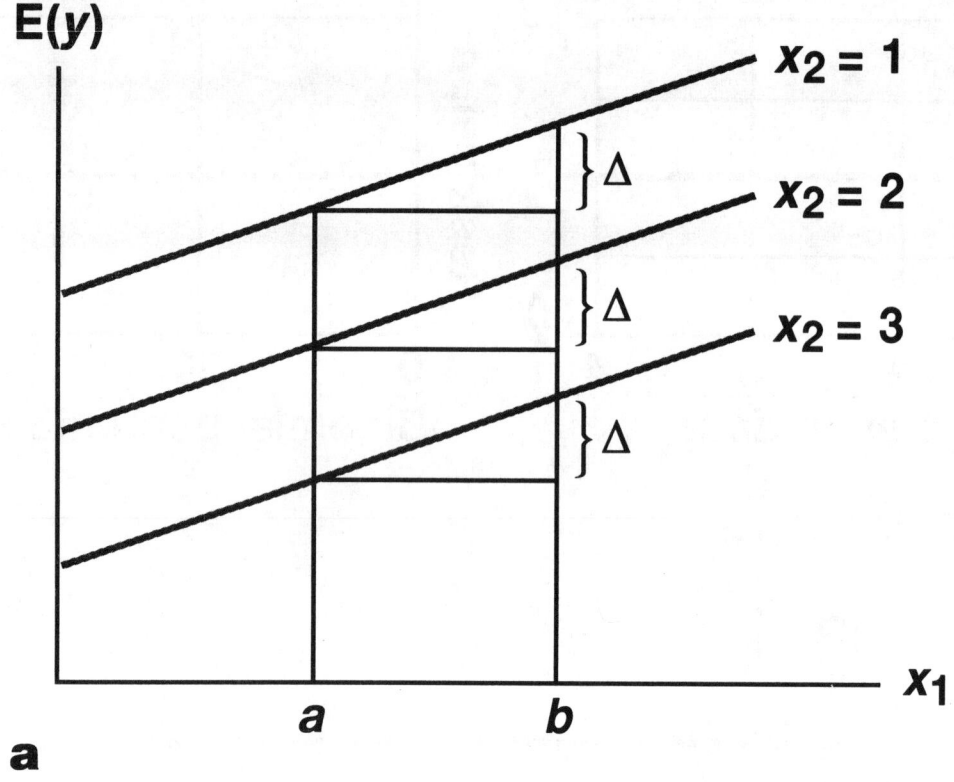

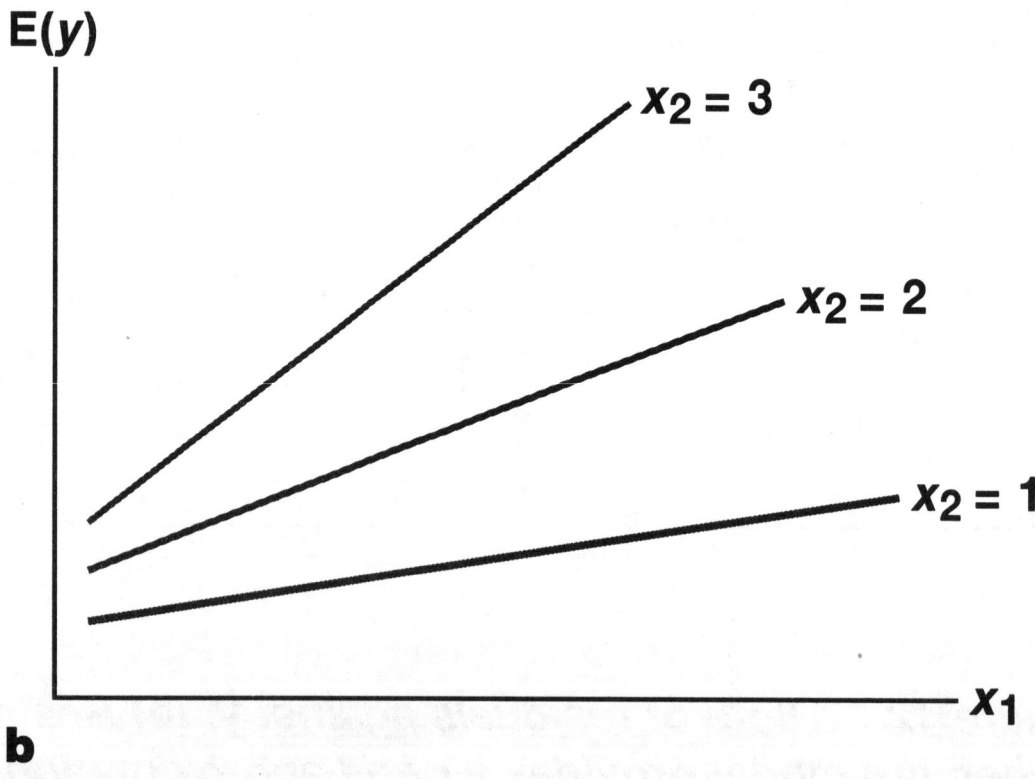

**(a) Figure 12.13 No interaction between x_1 and x_2.
(b) Interaction between x_1 and x_2.**

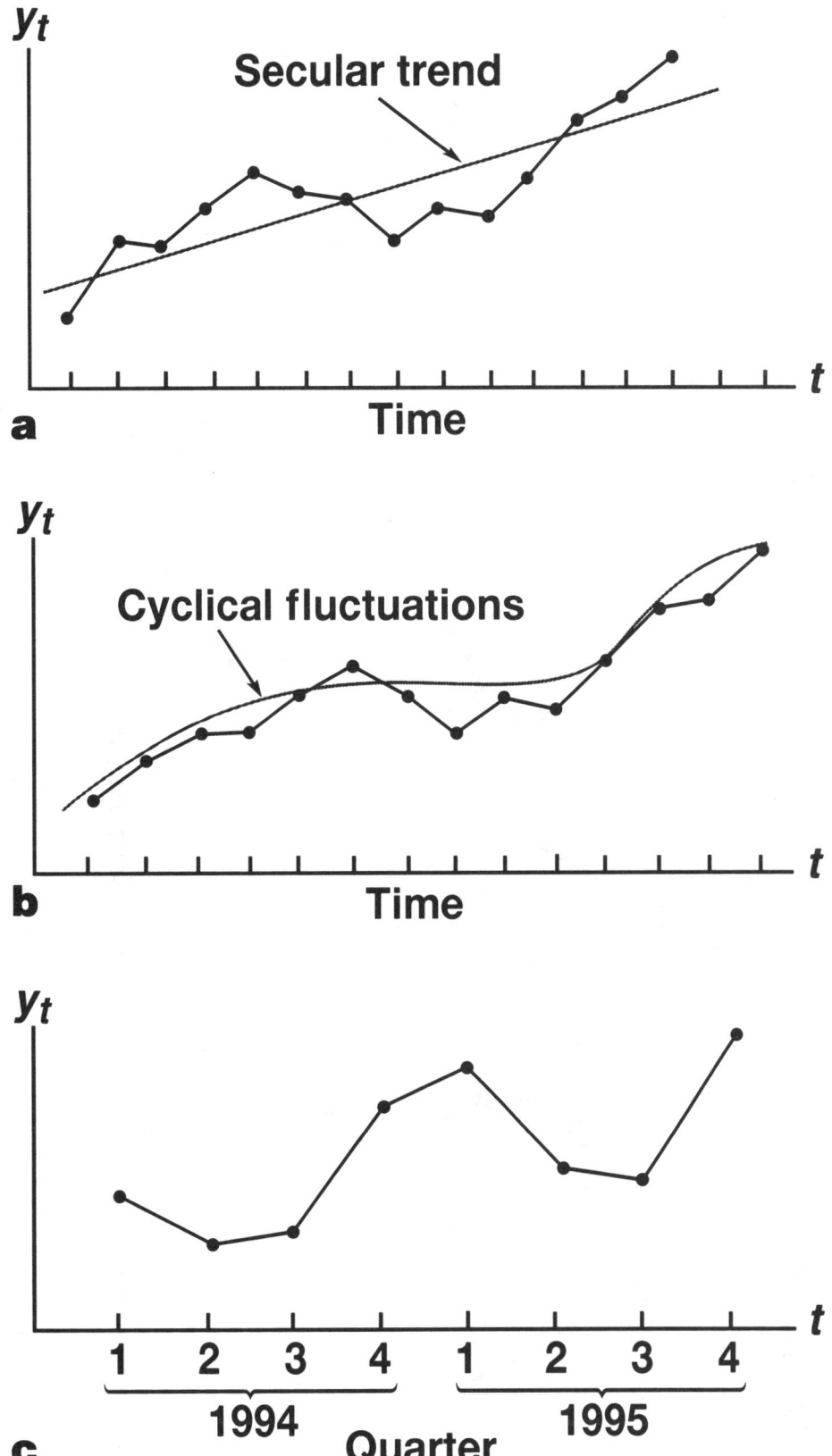

(a) Figure 13.6 The secular or long-term trend in a time series. (b) Figure 13.7 The cyclical component of a time series. (c) The seasonal component of a time series.

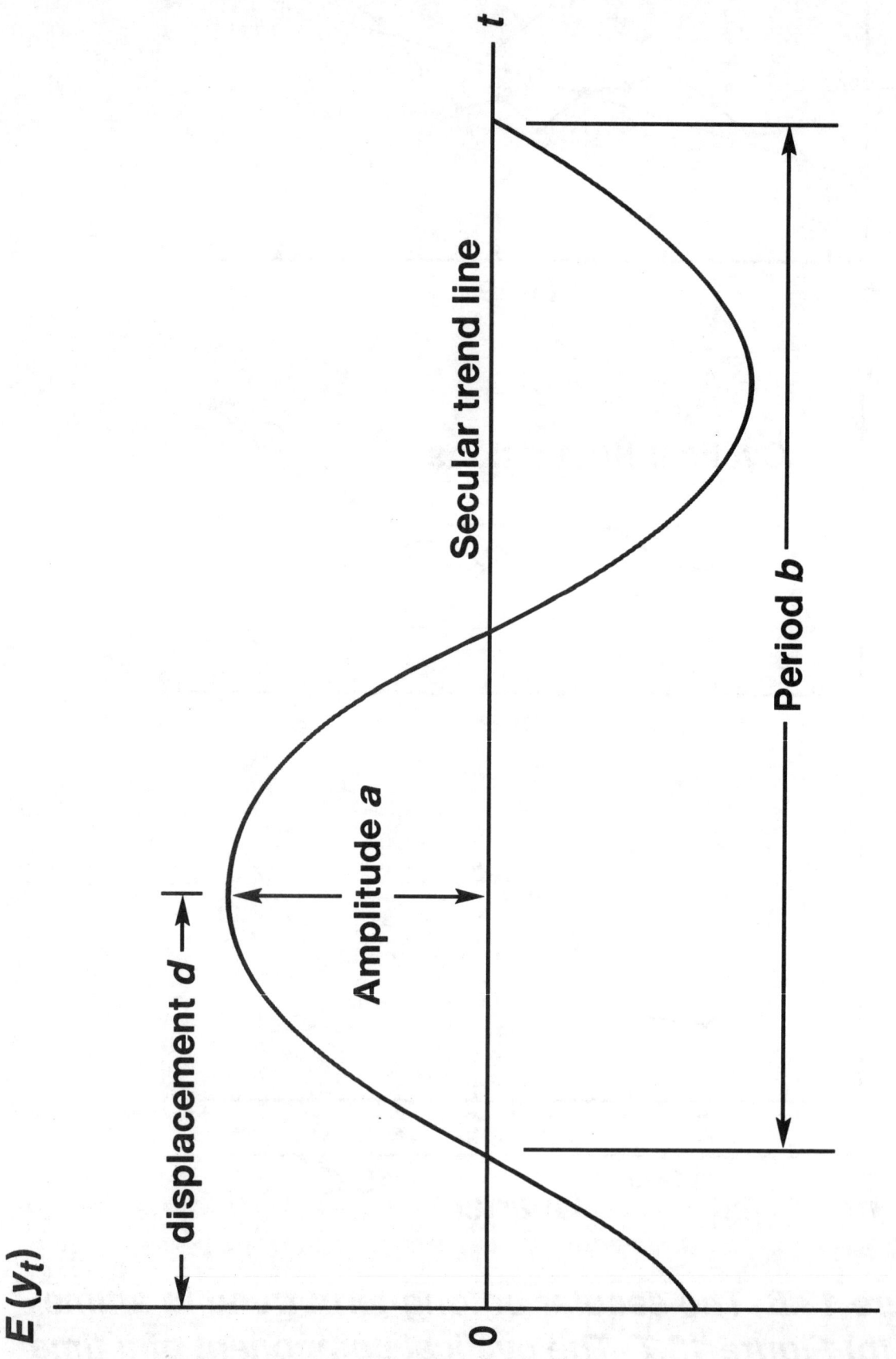

Figure 13.16 Contribution of cyclical terms, $\beta_2 \cos(2\pi t/b) + \beta_3 \sin(2\pi t/b)$.